STATA SURVIVAL ANALYSIS AND EPIDEMIOLOGICAL TABLES REFERENCE MANUAL
RELEASE 11

A Stata Press Publication
StataCorp LP
College Station, Texas

Copyright © 1985–2009 by StataCorp LP
All rights reserved
Version 11

Published by Stata Press, 4905 Lakeway Drive, College Station, Texas 77845
Typeset in TEX
Printed in the United States of America
10 9 8 7 6 5 4 3 2 1

ISBN-10: 1-59718-061-0
ISBN-13: 978-1-59718-061-0

This manual is protected by copyright. All rights are reserved. No part of this manual may be reproduced, stored in a retrieval system, or transcribed, in any form or by any means—electronic, mechanical, photocopy, recording, or otherwise—without the prior written permission of StataCorp LP unless permitted by the license granted to you by StataCorp LP to use the software and documentation. No license, express or implied, by estoppel or otherwise, to any intellectual property rights is granted by this document.

StataCorp provides this manual "as is" without warranty of any kind, either expressed or implied, including, but not limited to, the implied warranties of merchantability and fitness for a particular purpose. StataCorp may make improvements and/or changes in the product(s) and the program(s) described in this manual at any time and without notice.

The software described in this manual is furnished under a license agreement or nondisclosure agreement. The software may be copied only in accordance with the terms of the agreement. It is against the law to copy the software onto DVD, CD, disk, diskette, tape, or any other medium for any purpose other than backup or archival purposes.

The automobile dataset appearing on the accompanying media is Copyright © 1979 by Consumers Union of U.S., Inc., Yonkers, NY 10703-1057 and is reproduced by permission from CONSUMER REPORTS, April 1979.

Stata and Mata are registered trademarks and NetCourse is a trademark of StataCorp LP.

Other brand and product names are registered trademarks or trademarks of their respective companies.

For copyright information about the software, type `help copyright` within Stata.

The suggested citation for this software is

StataCorp. 2009. *Stata: Release 11*. Statistical Software. College Station, TX: StataCorp LP.

Table of contents

intro	Introduction to survival analysis manual	1
survival analysis	Introduction to survival analysis & epidemiological tables commands	4
ct	Count-time data	11
ctset	Declare data to be count-time data	12
cttost	Convert count-time data to survival-time data	18
discrete	Discrete-time survival analysis	21
epitab	Tables for epidemiologists	23
ltable	Life tables for survival data	78
snapspan	Convert snapshot data to time-span data	91
st	Survival-time data	94
st_is	Survival analysis subroutines for programmers	96
stbase	Form baseline dataset	101
stci	Confidence intervals for means and percentiles of survival time	111
stcox	Cox proportional hazards model	121
stcox PH-assumption tests	Tests of proportional-hazards assumption	149
stcox postestimation	Postestimation tools for stcox	164
stcrreg	Competing-risks regression	195
stcrreg postestimation	Postestimation tools for stcrreg	221
stcurve	Plot survivor, hazard, cumulative hazard, or cumulative incidence function	231
stdescribe	Describe survival-time data	241
stfill	Fill in by carrying forward values of covariates	245
stgen	Generate variables reflecting entire histories	249
stir	Report incidence-rate comparison	255
stpower	Sample-size, power, and effect-size determination for survival analysis	258
stpower cox	Sample size, power, and effect size for the Cox proportional hazards model	270
stpower exponential	Sample size and power for the exponential test	286
stpower logrank	Sample size, power, and effect size for the log-rank test	315
stptime	Calculate person-time, incidence rates, and SMR	335
strate	Tabulate failure rates and rate ratios	342
streg	Parametric survival models	353

streg postestimation	Postestimation tools for streg	386
sts	Generate, graph, list, and test the survivor and cumulative hazard functions	396
sts generate	Create variables containing survivor and related functions	412
sts graph	Graph the survivor and cumulative hazard functions	415
sts list	List the survivor or cumulative hazard function	434
sts test	Test equality of survivor functions	439
stset	Declare data to be survival-time data	454
stsplit	Split and join time-span records	496
stsum	Summarize survival-time data	514
sttocc	Convert survival-time data to case–control data	520
sttoct	Convert survival-time data to count-time data	525
stvary	Report whether variables vary over time	527
Glossary		530
Subject and author index		541

Cross-referencing the documentation

When reading this manual, you will find references to other Stata manuals. For example,

[U] **26 Overview of Stata estimation commands**
[R] **regress**
[D] **reshape**

The first example is a reference to chapter 26, *Overview of Stata estimation commands*, in the *User's Guide*; the second is a reference to the `regress` entry in the *Base Reference Manual*; and the third is a reference to the `reshape` entry in the *Data-Management Reference Manual*.

All the manuals in the Stata Documentation have a shorthand notation:

[GSM]	*Getting Started with Stata for Mac*
[GSU]	*Getting Started with Stata for Unix*
[GSW]	*Getting Started with Stata for Windows*
[U]	*Stata User's Guide*
[R]	*Stata Base Reference Manual*
[D]	*Stata Data-Management Reference Manual*
[G]	*Stata Graphics Reference Manual*
[XT]	*Stata Longitudinal-Data/Panel-Data Reference Manual*
[MI]	*Stata Multiple-Imputation Reference Manual*
[MV]	*Stata Multivariate Statistics Reference Manual*
[P]	*Stata Programming Reference Manual*
[SVY]	*Stata Survey Data Reference Manual*
[ST]	*Stata Survival Analysis and Epidemiological Tables Reference Manual*
[TS]	*Stata Time-Series Reference Manual*
[I]	*Stata Quick Reference and Index*
[M]	*Mata Reference Manual*

Detailed information about each of these manuals may be found online at

http://www.stata-press.com/manuals/

Title

intro — Introduction to survival analysis manual

Description

This entry describes this manual and what has changed since Stata 10. See the next entry, [ST] **survival analysis**, for an introduction to Stata's survival analysis capabilities.

Remarks

This manual documents commands for survival analysis and epidemiological tables and is referred to as [ST] in cross-references. Following this entry, [ST] **survival analysis** provides an overview of the commands.

This manual is arranged alphabetically. If you are new to Stata's survival analysis and epidemiological tables commands, we recommend that you read the following sections first:

[ST] **survival analysis**   Introduction to survival analysis & epidemiological tables commands
[ST] **st**   Survival-time data
[ST] **stset**   Set variables for survival data

Stata is continually being updated, and Stata users are always writing new commands. To find out about the latest survival analysis features, type `search survival` after installing the latest official updates; see [R] **update**. To find out about the latest epidemiological features, type `search epi`.

What's new

This section is intended for previous Stata users. If you are new to Stata, you may as well skip it.

1. New command `stcrreg` fits competing-risks regression models for survival data, according to the method of Fine and Gray (1999). In a competing risks model, subjects are at risk of failure because of two or more separate and possibly correlated causes. See [ST] **stcrreg**. Existing command `stcurve` will now graph cumulative incidence functions after `stcrreg`; see [ST] **stcurve**.

2. Stata's new multiple-imputation commands for dealing with missing values may be used with `stcox`, `streg`, and `stcrreg`; see [MI] **intro**. Either `stset` your data before using `mi set`, or use mi's `mi stset` to `stset` your data afterward.

3. Factor variables may now be used with `stcox`, `streg`, and `stcrreg`. See [U] **11.4.3 Factor variables**.

4. New reporting options `baselevels` and `allbaselevels` control how base levels of factor variables are displayed in output tables. New reporting option `noemptycells` controls whether missing cells in interactions are displayed.

   These new options are supported by estimation commands `stcox`, `streg`, and `stcrreg`, and by existing postestimation commands `estat summarize` and `estat vce`. See [R] **estimation options**.

5. New reporting option `noomitted` controls whether covariates that are dropped because of collinearity are reported in output tables. By default, Stata now includes a line in estimation and related output tables for collinear covariates and marks those covariates as "(omitted)". `noomitted` suppresses those lines.

noomitted is supported by estimation commands stcox, streg, and stcrreg, and by existing postestimation commands estat summarize and estat vce. See [R] **estimation options**.

6. New option vsquish eliminates blank lines in estimation and related tables. Many output tables now set off factor variables and time-series–operated variables with a blank line. vsquish removes these lines.

   vsquish is supported by estimation commands stcox, streg, and stcrreg, and by existing postestimation command estat summarize. See [R] **estimation options**.

7. Estimation commands stcox, streg, and stcrreg support new option coeflegend to display the coefficients' legend rather than the coefficient table. The legend shows how you would type a coefficient in an expression, in a test command, or in a constraint definition. See [R] **estimation options**.

8. Estimation commands streg and stcrreg support new option nocnsreport to suppress reporting constraints; see [R] **estimation options**.

9. Concerning predict:

   a. predict after stcox offers three new diagnostic measures of influence: DFBETAs, likelihood displacement values, and LMAX statistics. See [ST] **stcox postestimation**.

   b. predict after stcox can now calculate diagnostic statistics basesurv(), basechazard(), basehc(), mgale(), effects(), esr(), schoenfeld(), and scaledsch(). Previously, you had to request these statistics when you fit the model by specifying the option with the stcox command. Now you obtain them by using predict after estimation. The options continue to work with stcox directly but are no longer documented. See [ST] **stcox postestimation**.

   c. predict after stcox and streg now produces subject-level residuals by default. Previously, record-level or partial results were produced, although there was an inconsistency. This affects multiple-record data only because there is no difference between subject-level and partial residuals in single-record data. This change affects predict's options mgale, csnell, deviance, and scores after stcox (and new options ldisplace, lmax, and dfbeta, of course); and it affects mgale and deviance after streg. predict, deviance was the inconsistency; it always produced subject-level results.

   For instance, in previous Stata versions you typed

   . predict cs, csnell

   to obtain partial Cox–Snell residuals. One statistic per record was produced. To obtain subject-level residuals, for which there is one per subject and which predict stored on each subject's last record, you typed

   . predict ccs, ccsnell

   In Stata 11, when you type

   . predict cs, csnell

   you obtain the subject-level residual. To obtain the partial, you use the new partial option:

   . predict cs, csnell partial

   The same applies to all the other residuals. Concerning the inconsistency, partial deviances are now available.

   Not affected is predict, scores after streg. Log-likelihood scores in parametric models are mathematically defined at the record level and are meaningful only if evaluated at that level.

Prior behavior is restored under version control. See [ST] **stcox postestimation**, [ST] **streg postestimation**, and [ST] **stcrreg postestimation**.

10. `stcox` now allows up to 100 time-varying covariates as specified in option `tvc()`. The previous limit was 10. See [ST] **stcox**.

11. Existing commands `stcurve` and `estat phtest` no longer require that you specify the appropriate options to `stcox` before using them. The commands automatically generate the statistics they require. See [ST] **stcurve** and [ST] **stcox PH-assumption tests**.

12. Existing epitab commands `ir`, `cs`, `cc`, and `mhodds` now treat missing categories of variables in `by()` consistently. By default, missing categories are now excluded from the computation. This may be overridden by specifying `by()`'s new option `missing`. See [ST] **epitab**.

13. Existing command `sts list` has new option `saving()` that creates a dataset containing the results. See [ST] **sts list**.

For a complete list of all the new features in Stata 11, see [U] **1.3 What's new**.

Reference

Fine, J. P., and R. J. Gray. 1999. A proportional hazards model for the subdistribution of a competing risk. *Journal of the American Statistical Association* 94: 496–509.

Also see

[U] **1.3 What's new**

[R] **intro** — Introduction to base reference manual

Title

survival analysis — Introduction to survival analysis & epidemiological tables commands

Description

Stata's survival analysis routines are used to compute sample size, power, and effect size and to declare, convert, manipulate, summarize, and analyze survival data. Survival data is time-to-event data, and survival analysis is full of jargon: truncation, censoring, hazard rates, etc. See the glossary in this manual. For a good Stata-specific introduction to survival analysis, see Cleves et al. (2008).

Stata also has several commands for analyzing contingency tables resulting from various forms of observational studies, such as cohort or matched case–control studies.

This manual documents the following commands, which are described in detail in their respective manual entries.

Declaring and converting count data

ctset	[ST] **ctset**	Declare data to be count-time data
cttost	[ST] **cttost**	Convert count-time data to survival-time data

Converting snapshot data

snapspan	[ST] **snapspan**	Convert snapshot data to time-span data

Declaring and summarizing survival-time data

stset	[ST] **stset**	Declare data to be survival-time data
stdescribe	[ST] **stdescribe**	Describe survival-time data
stsum	[ST] **stsum**	Summarize survival-time data

Manipulating survival-time data

stvary	[ST] **stvary**	Report whether variables vary over time
stfill	[ST] **stfill**	Fill in by carrying forward values of covariates
stgen	[ST] **stgen**	Generate variables reflecting entire histories
stsplit	[ST] **stsplit**	Split time-span records
stjoin	[ST] **stsplit**	Join time-span records
stbase	[ST] **stbase**	Form baseline dataset

Obtaining summary statistics, confidence intervals, tables, etc.

sts	[ST] **sts**	Generate, graph, list, and test the survivor and cumulative hazard functions
stir	[ST] **stir**	Report incidence-rate comparison
stci	[ST] **stci**	Confidence intervals for means and percentiles of survival time
strate	[ST] **strate**	Tabulate failure rate
stptime	[ST] **stptime**	Calculate person-time, incidence rates, and SMR
stmh	[ST] **strate**	Calculate rate ratios with the Mantel–Haenszel method
stmc	[ST] **strate**	Calculate rate ratios with the Mantel–Cox method
ltable	[ST] **ltable**	Display and graph life tables

Fitting regression models

stcox	[ST] **stcox**	Cox proportional hazards model
estat concordance	[ST] **stcox postestimation**	Calculate Harrell's C
estat phtest	[ST] **stcox PH-assumption tests**	Test Cox proportional-hazards assumption
stphplot	[ST] **stcox PH-assumption tests**	Graphically assess the Cox proportional-hazards assumption
stcoxkm	[ST] **stcox PH-assumption tests**	Graphically assess the Cox proportional-hazards assumption
streg	[ST] **streg**	Parametric survival models
stcurve	[ST] **stcurve**	Plot survivor, hazard, cumulative hazard, or cumulative incidence function
stcrreg	[ST] **stcrreg**	Competing-risks regression

Sample-size and power determination for survival analysis

stpower cox	[ST] **stpower cox**	Sample size, power, and effect size for the Cox proportional hazards model
stpower exponential	[ST] **stpower exponential**	Sample size and power for the exponential test
stpower logrank	[ST] **stpower logrank**	Sample size, power, and effect size for the log-rank test

Converting survival-time data

sttocc	[ST] **sttocc**	Convert survival-time data to case–control data
sttoct	[ST] **sttoct**	Convert survival-time data to count-time data

Programmer's utilities

st_*	[ST] **st_is**	Survival analysis subroutines for programmers

Epidemiological tables

ir	[ST] **epitab**	Incidence rates for cohort studies
iri	[ST] **epitab**	Immediate form of ir
cs	[ST] **epitab**	Risk differences, risk ratios, and odds ratios for cohort studies
csi	[ST] **epitab**	Immediate form of cs
cc	[ST] **epitab**	Odds ratios for case–control data
cci	[ST] **epitab**	Immediate form of cc
tabodds	[ST] **epitab**	Tests of log odds for case–control data
mhodds	[ST] **epitab**	Odds ratios controlled for confounding
mcc	[ST] **epitab**	Analysis of matched case–control data
mcci	[ST] **epitab**	Immediate form of mcc

Remarks

Remarks are presented under the following headings:

Introduction
Declaring and converting count data
Converting snapshot data
Declaring and summarizing survival-time data
Manipulating survival-time data
Obtaining summary statistics, confidence intervals, tables, etc.
Fitting regression models
Sample size and power determination for survival analysis
Converting survival-time data
Programmer's utilities
Epidemiological tables

Introduction

All but one entry in this manual deals with the analysis of survival data, which is used to measure the time to an event of interest such as death or failure. Survival data can be organized in two ways. The first way is as *count data*, which refers to observations on populations, whether people or generators, with observations recording the number of units at a given time that failed or were lost because of censoring. The second way is as *survival-time*, or *time-span*, data. In survival-time data, the observations represent periods and contain three variables that record the start time of the period, the end time, and an indicator of whether failure or right-censoring occurred at the end of the period. The representation of the response of these three variables makes survival data unique in terms of implementing the statistical methods in the software.

Survival data may also be organized as *snapshot data* (a small variation of the survival-time format), in which observations depict an instance in time rather than an interval. When you have snapshot data, you simply use the `snapspan` command to convert it to survival-time data before proceeding.

Stata commands that begin with ct are used to convert count data to survival-time data. Survival-time data are analyzed using Stata commands that begin with st, known in our terminology as st commands. You can express all the information contained in count data in an equivalent survival-time dataset, but the converse is not true. Thus Stata commands are made to work with survival-time data because it is the more general representation.

The one remaining entry is [ST] **epitab**, which describes epidemiological tables. [ST] **epitab** covers many commands dealing with analyzing contingency tables arising from various observational studies, such as case–control or cohort studies. [ST] **epitab** is included in this manual because the concepts presented there are related to concepts of survival analysis, and both topics use the same terminology and are of equal interest to many researchers.

Declaring and converting count data

Count data must first be converted to survival-time data before Stata's st commands can be used. Count data can be thought of as aggregated survival-time data. Rather than having observations that are specific to a subject and a period, you have data that, at each recorded time, record the number lost because of failure and, optionally, the number lost because of right-censoring.

`ctset` is used to tell Stata the names of the variables in your count data that record the time, the number failed, and the number censored. You `ctset` your data before typing `cttost` to convert it to survival-time data. Because you `ctset` your data, you can type `cttost` without any arguments to perform the conversion. Stata remembers how the data are `ctset`.

Converting snapshot data

Snapshot data are data in which each observation records the status of a given subject at a certain point in time. Usually you have multiple observations on each subject that chart the subject's progress through the study.

Before using Stata's survival analysis commands with snapshot data, you must first convert the data to survival-time data; that is, the observations in the data should represent intervals. When you convert snapshot data, the existing time variable in your data is used to record the end of a time span, and a new variable is created to record the beginning. Time spans are created using the recorded snapshot times as breakpoints at which new intervals are to be created. Before converting snapshot data to time-span data, you must understand the distinction between *enduring variables* and *instantaneous variables*. Enduring variables record characteristics of the subject that endure throughout the time span, such as sex or smoking status. Instantaneous variables describe events that occur at the end of a time span, such as failure or censoring. When you convert snapshots to intervals, enduring variables obtain their values from the previous recorded snapshot or are set to missing for the first interval. Instantaneous variables obtain their values from the current recorded snapshot because the existing time variable now records the end of the span.

Stata's `snapspan` makes this whole process easy. You specify an ID variable identifying your subjects, the snapshot time variable, the name of the new variable to hold the beginning times of the spans, and any variables that you want to treat as instantaneous variables. Stata does the rest for you.

Declaring and summarizing survival-time data

Stata does not automatically recognize survival-time data, so you must declare your survival-time data to Stata by using `stset`. Every st command relies on the information that is provided when you `stset` your data. Survival-time data come in different forms. For example, your time variables may be dates, time measured from a fixed date, or time measured from some other point unique to each subject, such as enrollment in the study. You can also consider the following questions. What is the onset of risk for the subjects in your data? Is it time zero? Is it enrollment in the study or some other event, such as a heart transplant? Do you have censoring, and if so, which variable records it? What values does this variable record for censoring/failure? Do you have delayed entry? That is, were some subjects at risk of failure before you actually observed them? Do you have simple data and wish to treat everyone as entering and at risk at time zero?

Whatever the form of your data, you must first `stset` it before analyzing it, and so if you are new to Stata's st commands, we highly recommend that you take the time to learn about `stset`. It is really easy once you get the hang of it, and [ST] **stset** has many examples to help. For more discussion of `stset`, see chapter 6 of Cleves et al. (2008).

Once you `stset` the data, you can use `stdescribe` to describe the aspects of your survival data. For example, you will see the number of subjects you were successful in declaring, the total number of records associated with these subjects, the total time at risk for these subjects, time gaps for any of these subjects, any delayed entry, etc. You can use `stsum` to summarize your survival data, for example, to obtain the total time at risk and the quartiles of time-to-failure in analysis-time units.

Manipulating survival-time data

Once your data have been `stset`, you may want to clean them up a bit before beginning your analysis. Suppose that you had an enduring variable and `snapspan` recorded it as missing for the interval leading up to the first recorded snapshot time. You can use `stfill` to fill in missing values of covariates, either by carrying forward the values from previous periods or by making the covariate equal to its earliest recorded (nonmissing) value for all time spans. You can use `stvary` to check for time-varying covariates or to confirm that certain variables, such as sex, are not time varying. You can use `stgen` to generate new covariates based on functions of the time spans for each given subject. For example, you can create a new variable called `eversmoked` that equals one for all a subject's observations, if the variable `smoke` in your data is equal to one for *any* of the subject's time spans. Think of `stgen` as just a convenient way to do things that could be done using by *subject_id*: with survival-time data.

`stsplit` is useful for creating data that have multiple records per subject from data that have one record per subject. Suppose that you have already `stset` your data and wish to introduce a time-varying covariate. You would first need to `stsplit` your data so that separate time spans could be created for each subject, allowing the new covariate to assume different values over time within a subject. `stjoin` is the opposite of `stsplit`. Suppose that you have data with multiple records per subject but then realize that the data could be collapsed into single-subject records with no loss of information. Using `stjoin` would speed up any subsequent analysis using the st commands without changing the results.

`stbase` can be used to set every variable in your multiple-record st data to the value at baseline, defined as the earliest time at which each subject was observed. It can also be used to convert st data to cross-sectional data.

Obtaining summary statistics, confidence intervals, tables, etc.

Stata provides several commands for nonparametric analysis of survival data that can produce a wide array of summary statistics, inference, tables, and graphs. `sts` is a truly powerful command, used to obtain nonparametric estimates, inference, tests, and graphs of the survivor function, the cumulative hazard function, and the hazard function. You can compare estimates across groups, such as smoking versus nonsmoking, and you can adjust these estimates for the effects of other covariates in your data. `sts` can present these estimates as tables and graphs. `sts` can also be used to test the equality of survivor functions across groups.

`stir` is used to estimate incidence rates and to compare incidence rates across groups. `stci` is the survival-time data analog of `ci` and is used to obtain confidence intervals for means and percentiles of time to failure. `strate` is used to tabulate failure rates. `stptime` is used to calculate person-time and standardized mortality/morbidity ratios (SMRs). `stmh` calculates rate ratios by using the Mantel–Haenszel method, and `stmc` calculates rate ratios by using the Mantel–Cox method.

`ltable` displays and graphs life tables for individual-level or aggregate data.

Fitting regression models

Stata has commands for fitting both semiparametric and parametric regression models to survival data. `stcox` fits the Cox proportional hazards model and `predict` after `stcox` can be used to retrieve estimates of the baseline survivor function, the baseline cumulative hazard function, and the baseline hazard contributions. `predict` after `stcox` can also calculate a myriad of Cox regression diagnostic quantities, such as martingale residuals, efficient score residuals, and Schoenfeld residuals. `stcox` has four options for handling tied failures. `stcox` can be used to fit stratified Cox models, where the baseline hazard is allowed to differ over the strata, and it can be used to model multivariate survival data by using a *shared-frailty* model, which can be thought of as a Cox model with random effects. After `stcox`, you can use `estat phtest` to test the proportional-hazards assumption or `estat concordance` to calculate Harrell's C. With `stphplot` and `stcoxkm`, you can graphically assess the proportional-hazards assumption.

Stata offers six parametric regression models for survival data: exponential, Weibull, lognormal, loglogistic, Gompertz, and gamma. All six models are fit using streg, and you can specify the model you want with the distribution() option. All these models, except for the exponential, have ancillary parameters that are estimated (along with the linear predictor) from the data. By default, these ancillary parameters are treated as constant, but you may optionally model the ancillary parameters as functions of a linear predictor. Stratified models may also be fit using streg. You can also fit frailty models with streg and specify whether you want the frailties to be treated as spell-specific or shared across groups of observations.

stcrreg fits a semiparametric regression model for survival data in the presence of competing risks. Competing risks impede the failure event under study from occurring. An analysis of such competing-risks data focuses on the *cumulative incidence function*, the probability of failure in the presence of competing events that prevent that failure. stcrreg provides an analogue to stcox for such data. The baseline *subhazard function*—that which generates failures under competing risks—is left unspecified, and covariates act multiplicatively on the baseline subhazard.

stcurve is for use after stcox, streg, and stcrreg and will plot the estimated survivor, hazard, cumulative hazard, and cumulative incidence function for the fitted model. Covariates, by default, are held fixed at their mean values, but you can specify other values if you wish. stcurve is useful for comparing these functions across different levels of covariates.

Sample size and power determination for survival analysis

Stata has commands for computing sample size, power, and effect size for survival analysis using the log-rank test, the Cox proportional hazards model, and the exponential test comparing exponential hazard rates.

stpower logrank estimates required sample size, power, and effect size for survival analysis comparing survivor functions in two groups using the log-rank test. It provides options to account for unequal allocation of subjects between the two groups, possible withdrawal of subjects from the study (loss to follow-up), and uniform accrual of subjects into the study.

stpower cox estimates required sample size, power, and effect size for survival analysis using Cox proportional hazards (PH) models with possibly multiple covariates. It provides options to account for possible correlation between the covariate of interest and other predictors and for withdrawal of subjects from the study.

stpower exponential estimates required sample size and power for survival analysis comparing two exponential survivor functions using the exponential test (in particular, the Wald test of the difference between hazards or, optionally, of the difference between log hazards). It accommodates unequal allocation between the two groups, flexible accrual of subjects into the study (uniform and truncated exponential), and group-specific losses to follow-up.

The stpower commands allow automated production of customizable tables and have options to assist with the creation of graphs of power curves and more.

Converting survival-time data

Stata has commands for converting survival-time data to case–control and count data. These commands are rarely used, because most of the analyses are performed using data in the survival-time format. sttocc is useful for converting survival data to case–control data suitable for estimation with clogit. sttoct is the opposite of cttost and will convert survival-time data to count data.

Programmer's utilities

Stata also provides routines for programmers interested in writing their own st commands. These are basically utilities for setting, accessing, and verifying the information saved by stset. For example, st_is verifies that the data have in fact been stset and gives the appropriate error if not. st_show is used to preface the output of a program with key information on the st variables used in the analysis. Programmers interested in writing st code should see [ST] **st_is**.

Epidemiological tables

See the *Description* section of [ST] **epitab** for an overview of Stata's commands for calculating statistics and performing lists that are useful for epidemiologists.

Reference

Cleves, M. A., W. W. Gould, R. G. Gutierrez, and Y. Marchenko. 2008. *An Introduction to Survival Analysis Using Stata.* 2nd ed. College Station, TX: Stata Press.

Also see

[ST] **stset** — Declare data to be survival-time data

[ST] **intro** — Introduction to survival analysis manual

Title

> **ct** — Count-time data

Description

The term *ct* refers to count-time data and the commands—all of which begin with the letters "ct"—for analyzing them. If you have data on populations, whether people or generators, with observations recording the number of units under test at time t (subjects alive) and the number of subjects that failed or were lost because of censoring, you have what we call count-time data.

If, on the other hand, you have data on individual subjects with observations recording that this subject came under observation at time t_0 and that later, at t_1, a failure or censoring was observed, you have what we call survival-time data. If you have survival-time data, see [ST] **st**.

Do not confuse count-time data with counting-process data, which can be analyzed using the st commands; see [ST] **st**.

There are two ct commands:

    ctset   [ST] **ctset**   Declare data to be count-time data
    cttost   [ST] **cttost**   Convert count-time data to survival-time data

The key is the cttost command. Once you have converted your count-time data to survival-time data, you can use the st commands to analyze the data. The entire process is as follows:

1. ctset your data so that Stata knows that they are count-time data; see [ST] **ctset**.
2. Type cttost to convert your data to survival-time data; see [ST] **cttost**.
3. Use the st commands; see [ST] **st**.

Also see

[ST] **ctset** — Declare data to be count-time data

[ST] **cttost** — Convert count-time data to survival-time data

[ST] **st** — Survival-time data

[ST] **survival analysis** — Introduction to survival analysis & epidemiological tables commands

Title

ctset — Declare data to be count-time data

Syntax

Declare data in memory to be count-time data and run checks on data

    ctset *timevar nfailvar* [*ncensvar* [*nentvar*]] [, by(*varlist*) no<u>s</u>how]

Specify whether to display identities of key ct variables

    ctset, { <u>s</u>how | no<u>s</u>how }

Clear ct setting

    ctset, clear

Display identity of key ct variables and rerun checks on data

    { ctset | ct }

where *timevar* refers to the time of failure, censoring, or entry. It should contain times ≥ 0.
nfailvar records the number failing at time *timevar*.
ncensvar records the number censored at time *timevar*.
nentvar records the number entering at time *timevar*.
Stata sequences events at the same time as

at *timevar*	*nfailvar* failures occurred,
then at *timevar* + 0	*ncensvar* censorings occurred,
finally at *timevar* + 0 + 0	*nentvar* subjects entered the data.

Menu

Statistics > Survival analysis > Setup and utilities > Declare data to be count-time data

Description

    ct refers to count-time data and is described here and in [ST] **ct**. Do not confuse count-time data with counting-process data, which can be analyzed using the st commands; see [ST] **st**.

    In the first syntax, ctset declares the data in memory to be ct data, informing Stata of the key variables. When you ctset your data, ctset also checks that what you have declared makes sense.

    In the second syntax, ctset changes the value of show/noshow. In show mode—the default—the other ct commands display the identities of the key ct variables before their normal output. If you type ctset, noshow, they will not do this. If you type ctset, noshow and then wish to restore the default behavior, type ctset, show.

In the third syntax, ctset, clear causes Stata to no longer consider the data to be ct data. The dataset itself remains unchanged. It is not necessary to type ctset, clear before doing another ctset. ctset, clear is used mostly by programmers.

In the fourth syntax, ctset—which can be abbreviated ct here—displays the identities of the key ct variables and reruns the checks on your data. Thus ct can remind you of what you have ctset (especially if you have ctset, noshow) and reverify your data if you make changes to the data.

Options

by(*varlist*) indicates that counts are provided by group. For instance, consider data containing records such as

t	fail	cens	sex	agecat
5	10	2	0	1
5	6	1	1	1
5	12	0	0	2

These data indicate that, in the category sex = 0 and agecat = 1, 10 failed and 2 were censored at time 5; for sex = 1, 1 was censored and 6 failed; and so on.

The above data would be declared

. ctset t fail cens, by(sex agecat)

The order of the records is not important, nor is it important that there be a record at every time for every group or that there be only one record for a time and group. However, the data must contain the full table of events.

noshow and show specify whether the identities of the key ct variables are to be displayed at the start of every ct command. Some users find the report reassuring; others find it repetitive. In any case, you can set and unset show, and you can always type ct to see the summary.

clear makes Stata no longer consider the data to be ct data.

Remarks

Remarks are presented under the following headings:

> *Examples*
> *Data errors flagged by ctset*

Examples

About all you can do with ct data in Stata is convert it to survival-time (st) data so that you can use the survival analysis commands. To analyze count-time data with Stata,

. ctset ...
. cttost
. (*now use any of the st commands*)

(Continued on next page)

▷ Example 1: Simple ct data

We have data on generators that are run until they fail:

```
. use http://www.stata-press.com/data/r11/ctset1
. list, sep(0)
```

	failtime	fail
1.	22	1
2.	30	1
3.	40	2
4.	52	1
5.	54	4
6.	55	2
7.	85	7
8.	97	1
9.	100	3
10.	122	2
11.	140	1

For instance, at time 54, four generators failed. To ctset these data, we could type

```
. ctset failtime fail
    dataset name:  http://www.stata-press.com/data/r11/ctset1.dta
           time:  failtime
        no. fail:  fail
        no. lost:  --                  (meaning 0 lost)
       no. enter:  --                  (meaning all enter at time 0)
```

It is not important that there be only 1 observation per failure time. For instance, according to our data, at time 85 there were seven failures. We could remove that observation and substitute two in its place—one stating that at time 85 there were five failures and another that at time 85 there were two more failures. ctset would interpret that data just as it did the previous data.

In more realistic examples, the generators might differ from one another. For instance, the following data show the number failing with old-style (bearings = 0) and new-style (bearings = 1) bearings:

```
. use http://www.stata-press.com/data/r11/ctset2
. list, sepby(bearings)
```

	bearings	failtime	fail
1.	0	22	1
2.	0	40	2
3.	0	54	1
4.	0	84	2
5.	0	97	2
6.	0	100	1
7.	1	30	1
8.	1	52	1
9.	1	55	1
10.	1	100	3
11.	1	122	2
12.	1	140	1

That the data are sorted on bearings is not important. The ctset command for these data is

```
. ctset failtime fail, by(bearings)
  dataset name:  http://www.stata-press.com/data/r11/ctset2.dta
          time:  failtime
      no. fail:  fail
      no. lost:  --            (meaning 0 lost)
     no. enter:  --            (meaning all enter at time 0)
            by:  bearings
```
◁

▷ Example 2: ct data with censoring

In real data, not all units fail in the time allotted. Say that the generator experiment was stopped after 150 days. The data might be

```
. use http://www.stata-press.com/data/r11/ctset3
. list
```

	bearings	failtime	fail	censored
1.	0	22	1	0
2.	0	40	2	0
3.	0	54	1	0
4.	0	84	2	0
5.	1	97	2	0
6.	0	100	1	0
7.	0	150	0	2
8.	1	30	1	0
9.	1	52	1	0
10.	1	55	1	0
11.	1	122	2	0
12.	1	140	1	0
13.	1	150	0	3

The ctset command for these data is

```
. ctset failtime fail censored, by(bearings)
  dataset name:  http://www.stata-press.com/data/r11/ctset3.dta
          time:  failtime
      no. fail:  fail
      no. lost:  censored
     no. enter:  --            (meaning all enter at time 0)
            by:  bearings
```

In some other data, observations might also be censored along the way; that is, the value of censored would not be 0 before time 150. For instance, a record might read

```
        bearings    failtime        fail    censored
               0          84           2           1
```

This would mean that at time 84, two failed and one was lost because of censoring. The failure and censoring occurred at the same time, and when we analyze these data, Stata will assume that the censored observation could have failed, that is, that the censoring occurred after the two failures.

◁

▷ Example 3: ct data with delayed entry

Data on survival time of patients with a particular kind of cancer are collected. Time is measured as time since diagnosis. After data collection started, the sample was enriched with some patients from hospital records who had been previously diagnosed. Some of the data are

time	die	cens	ent	other variables
0	0	0	50	
1	0	0	5	...
⋮				
30	0	0	3	...
31	0	1	2	...
32	1	0	1	...
⋮				
100	1	1	0	...
⋮				

Fifty patients entered at time 0 (time of diagnosis); five patients entered 1 day after diagnosis; and three, two, and one patients entered 30, 31, and 32 days after diagnosis, respectively. On the 32nd day, one of the previously entered patients died.

If the other variables are named sex and agecat, the ctset command for these data is

```
. ctset time die cens ent, by(sex agecat)
       time:  time
   no. fail:  die
   no. lost:  cens
  no. enter:  ent
         by:  sex agecat
```

◁

The count-time format is an inferior way to record data like these—data in which every subject does not enter at time 0—because some information is already lost. When did the patient who died on the 32nd day enter? There is no way of telling.

For traditional survival analysis calculations, it does not matter. More modern methods of estimating standard errors, however, seek to identify each patient, and these data do not support using such methods.

This issue concerns the robust estimates of variance and the vce(robust) options on some of the st analysis commands. After converting the data, you must not use the vce(robust) option, even if an st command allows it, because the identities of the subjects—tying together when a subject starts and ceases to be at risk—are assigned randomly by cttost when you convert your ct to st data. When did the patient who died on the 32nd day enter? For conventional calculations, it does not matter, and cttost chooses a time randomly from the available entry times.

Data errors flagged by ctset

ctset requires only two things of your data: that the counts all be positive or zero and, if you specify an entry variable, that the entering and exiting subjects (failure + censored) balance.

If all subjects enter at time 0, we recommend that you do not specify a number-that-enter variable. ctset can determine for itself the number who enter at time 0 by summing the failures and censorings.

Methods and formulas

ctset is implemented as an ado-file.

Also see

[ST] **ct** — Count-time data

[ST] **cttost** — Convert count-time data to survival-time data

Title

cttost — Convert count-time data to survival-time data

Syntax

cttost [, *options*]

options	description
t0(*t0var*)	name of entry-time variable
wvar(*wvar*)	name of frequency-weighted variable
clear	overwrite current data in memory
† nopreserve	do not save the original data; programmer's command

† nopreserve is not shown in the dialog box.
You must ctset your data before using cttost; see [ST] **ctset**.

Menu

Statistics > Survival analysis > Setup and utilities > Convert count-time data to survival-time data

Description

cttost converts count-time data to their survival-time format so that they can be analyzed with Stata. Do not confuse count-time data with counting-process data, which can also be analyzed with the st commands; see [ST] **ctset** for a definition and examples of count data.

Options

t0(*t0var*) specifies the name of the new variable to create that records entry time. (For most ct data, no entry-time variable is necessary because everyone enters at time 0.)

Even if an entry-time variable is necessary, you need not specify this option. cttost will, by default, choose t0, time0, or etime according to which name does not already exist in the data.

wvar(*wvar*) specifies the name of the new variable to be created that records the frequency weights for the new pseudo-observations. Count-time data are actually converted to frequency-weighted st data, and a variable is needed to record the weights. This sounds more complicated than it is. Understand that cttost needs a new variable name, which will become a permanent part of the st data.

If you do not specify wvar(), cttost will, by default, choose w, pop, weight, or wgt according to which name does not already exist in the data.

clear specifies that it is okay to proceed with the conversion, even though the current dataset has not been saved on disk.

The following option is available with cttost but is not shown in the dialog box:

nopreserve speeds the conversion by not saving the original data that can be restored should things go wrong or should you press *Break*. nopreserve is intended for use by programmers who use cttost as a subroutine. Programmers can specify this option if they have already preserved the original data. nopreserve does not affect the conversion.

Remarks

Converting ct to st data is easy. We have some count-time data,

```
. use http://www.stata-press.com/data/r11/cttost
. ct
       dataset name:  http://www.stata-press.com/data/r11/cttost.dta
               time:  time
           no. fail:  ndead
           no. lost:  ncens
          no. enter:  --                   (meaning all enter at time 0)
                 by:  agecat treat
. list in 1/5
```

	agecat	treat	time	ndead	ncens
1.	2	1	464	4	0
2.	3	0	268	3	1
3.	2	0	638	2	0
4.	1	0	803	1	4
5.	1	0	431	2	0

and to convert it, we type cttost:

```
. cttost
       failure event:  ndead != 0 & ndead < .
  obs. time interval:  (0, time]
   exit on or before:  failure
              weight:  [fweight=w]

         33  total obs.
          0  exclusions

         33  physical obs. remaining, equal to
         82  weighted obs., representing
         39  failures in single record/single failure data
      48726  total analysis time at risk, at risk from t =         0
                                    earliest observed entry t =    0
                                         last observed exit t = 1227
```

(Continued on next page)

Now that it is converted, we can use any of the st commands:

```
. sts test treat, logrank
        failure _d:  ndead
   analysis time _t:  time
            weight:  [fweight=w]
```

Log-rank test for equality of survivor functions

treat	Events observed	Events expected
0	22	17.05
1	17	21.95
Total	39	39.00

            chi2(1) =   2.73
            Pr>chi2 =   0.0986

Methods and formulas

cttost is implemented as an ado-file.

Also see

[ST] **ct** — Count-time data

[ST] **ctset** — Declare data to be count-time data

Title

discrete — Discrete-time survival analysis

As of the date that this manual was printed, Stata does not have a suite of built-in commands for discrete-time survival models matching the `st` suite for continuous-time models, but a good case could be made that it should. Instead, these models can be fit easily using other existing estimation commands and data manipulation tools.

Discrete-time survival analysis concerns analysis of time-to-event data whenever survival times are either a) intrinsically discrete (e.g., numbers of machine cycles) or b) grouped into discrete intervals of time ("interval censoring"). If intervals are of equal length, the same methods can be applied to both a) and b); survival times will be positive integers.

You can fit discrete-time survival models with the maximum likelihood method. Data may contain completed or right-censored spells, and late entry (left truncation) can also be handled, as well as unobserved heterogeneity (also termed "frailty"). Estimation makes use of the property that the sample likelihood can be rewritten in a form identical to the likelihood for a binary dependent variable multiple regression model and applied to a specially organized dataset (Allison 1984, Jenkins 1995). For models without frailty, you can use, for example, `logistic` (or `logit`) to fit the discrete-time logistic hazard model or `cloglog` to fit the discrete-time proportional hazards model (Prentice and Gloeckler 1978). Models incorporating normal frailty may be fit using `xtlogit` and `xtcloglog`. A model with gamma frailty (Meyer 1990) may be fit using `pgmhaz` (Jenkins 1997).

Estimation consists of three steps:

1. *Data organization*: The dataset must be organized so that there is 1 observation for each period when a subject is at risk of experiencing the transition event. For example, if the original dataset contains one row for each subject, i, with information about their spell length, T_i, the new dataset requires T_i rows for each subject, one row for each period at risk. This may be accomplished using `expand` or `stsplit`. (This step is episode splitting at each and every interval.) The result is data of the same form as a discrete panel (`xt`) dataset with repeated observations on each panel (subject).

2. *Variable creation*: You must create at least three types of variables. First, you will need an interval identification variable, which is a sequence of positive integers $t = 1, \ldots, T_i$. For example,

    . sort subject_id
    . by subject_id: generate t = _n

    Second, you need a period-specific censoring indicator, d_i. If $d_i = 1$ if subject i's spell is complete and $d_i = 0$ if the spell is right-censored, the new indicator $d_{it}^* = 1$ if $d_i = 1$ and $t = T_i$, and $d_{it}^* = 0$ otherwise.

    Third, you must define variables (as functions of t) to summarize the pattern of duration dependence. These variables are entered as covariates in the regression. For example, for a duration dependence pattern analogous to that in the continuous-time Weibull model, you could define a new variable $x_1 = \log t$. For a quadratic specification, you define variables $x_1 = t$ and $x_2 = t^2$. We can achieve a piecewise constant specification by defining a set of dummy variables, with each group of periods sharing the same hazard rate, or a semiparametric model (analogous to the Cox regression model for continuous survival-time data) using separate dummy variables for each and every duration interval. No duration variable need be defined if you want to fit a model with a constant hazard rate.

    In addition to these three essentials, you may define other time-varying covariates.

3. *Estimation*: You fit a binary dependent variable multiple regression model, with d_{it}^* as the dependent variable and covariates, including the duration variables and any other covariates.

For estimation using spell data with late entry, the stages are the same as those outlined above, with one modification and one warning. To fit models without frailty, you must drop all intervals prior to each subject's entry to the study. For example, if entry is in period e_i, you drop it if $t < e_i$. If you want to fit frailty models on the basis of discrete-time data with late entry, then be aware that the estimation procedure outlined does not lead to correct estimates. (The sample likelihood in the reorganized data does not account for conditioning for late entry here. You will need to write your own likelihood function by using ml; see [R] **maximize**.)

To derive predicted hazard rates, use the predict command. For example, after logistic or cloglog, use predict, pr. After xtlogit or xtcloglog, use predict, pu0 (which predicts the hazard assuming the individual effect is equal to the mean value). Estimates of the survivor function, S_{it}, can then be derived from the predicted hazard rates, p_{it}, because $S_{it} = (1-p_{i1})(1-p_{i2})(\cdots)(1-p_{it})$.

Acknowledgment

We thank Stephen Jenkins, University of Essex, UK, for drafting this initial entry.

References

Allison, P. D. 1984. *Event History Analysis: Regression for Longitudinal Event Data*. Newbury Park, CA: Sage.

Jenkins, S. P. 1995. Easy estimation methods for discrete-time duration models. *Oxford Bulletin of Economics and Statistics* 57: 129–138.

———. 1997. sbe17: Discrete time proportional hazards regression. *Stata Technical Bulletin* 39: 22–32. Reprinted in *Stata Technical Bulletin Reprints*, vol. 7, pp. 109–121. College Station, TX: Stata Press.

Meyer, B. D. 1990. Unemployment insurance and unemployment spells. *Econometrica* 58: 757–782.

Prentice, R. L., and L. Gloeckler. 1978. Regression analysis of grouped survival data with application to breast cancer data. *Biometrics* 34: 57–67.

Also see

[D] **expand** — Duplicate observations

[ST] **stcox** — Cox proportional hazards model

[ST] **stcrreg** — Competing-risks regression

[ST] **streg** — Parametric survival models

[R] **cloglog** — Complementary log-log regression

[R] **logistic** — Logistic regression, reporting odds ratios

[XT] **xtcloglog** — Random-effects and population-averaged cloglog models

[XT] **xtlogit** — Fixed-effects, random-effects, and population-averaged logit models

Title

epitab — Tables for epidemiologists

Syntax

Cohort studies

    ir var_{case} $var_{exposed}$ var_{time} [*if*] [*in*] [*weight*] [, *ir_options*]

    iri $\#_a$ $\#_b$ $\#_{N_1}$ $\#_{N_2}$ [, tb <u>l</u>evel(*#*)]

    cs var_{case} $var_{exposed}$ [*if*] [*in*] [*weight*] [, *cs_options*]

    csi $\#_a$ $\#_b$ $\#_c$ $\#_d$ [, *csi_options*]

Case–control studies

    cc var_{case} $var_{exposed}$ [*if*] [*in*] [*weight*] [, *cc_options*]

    cci $\#_a$ $\#_b$ $\#_c$ $\#_d$ [, *cci_options*]

    tabodds var_{case} [*expvar*] [*if*] [*in*] [*weight*] [, *tabodds_options*]

    mhodds var_{case} *expvar* [*vars*$_{adjust}$] [*if*] [*in*] [*weight*] [, *mhodds_options*]

Matched case–control studies

    mcc $var_{exposed\_case}$ $var_{exposed\_control}$ [*if*] [*in*] [*weight*] [, tb <u>l</u>evel(*#*)]

    mcci $\#_a$ $\#_b$ $\#_c$ $\#_d$ [, tb <u>l</u>evel(*#*)]

ir_options	description
Options	
by(*varname*[, <u>miss</u>ing])	stratify on *varname*
<u>es</u>tandard	combine external weights with within-stratum statistics
<u>is</u>tandard	combine internal weights with within-stratum statistics
<u>s</u>tandard(*varname*)	combine user-specified weights with within-stratum statistics
<u>po</u>ol	display pooled estimate
<u>noc</u>rude	do not display crude estimate
<u>noh</u>om	do not display homogeneity test
ird	calculate standard incidence-rate difference
tb	calculate test-based confidence intervals
<u>l</u>evel(*#*)	set confidence level; default is level(95)

cs_options	description
Options	
by(*varlist*[, missing])	stratify on *varlist*
estandard	combine external weights with within-stratum statistics
istandard	combine internal weights with within-stratum statistics
standard(*varname*)	combine user-specified weights with within-stratum statistics
pool	display pooled estimate
nocrude	do not display crude estimate
nohom	do not display homogeneity test
rd	calculate standardized risk difference
binomial(*varname*)	number of subjects variable
or	report odds ratio
woolf	use Woolf approximation to calculate SE and CI of the odds ratio
tb	calculate test-based confidence intervals
exact	calculate Fisher's exact p
level(#)	set confidence level; default is level(95)

csi_options	description
or	report odds ratio
woolf	use Woolf approximation to calculate SE and CI of the odds ratio
tb	calculate test-based confidence intervals
exact	calculate Fisher's exact p
level(#)	set confidence level; default is level(95)

cc_options	description
Options	
by(*varname*[, missing])	stratify on *varname*
estandard	combine external weights with within-stratum statistics
istandard	combine internal weights with within-stratum statistics
standard(*varname*)	combine user-specified weights with within-stratum statistics
pool	display pooled estimate
nocrude	do not display crude estimate
nohom	do not display homogeneity test
bd	perform Breslow–Day homogeneity test
tarone	perform Tarone's homogeneity test
binomial(*varname*)	number of subjects variable
cornfield	use Cornfield approximation to calculate CI of the odds ratio
woolf	use Woolf approximation to calculate SE and CI of the odds ratio
tb	calculate test-based confidence intervals
exact	calculate Fisher's exact p
level(#)	set confidence level; default is level(95)

cci_options	description
<u>c</u>ornfield	use Cornfield approximation to calculate CI of the odds ratio
<u>w</u>oolf	use Woolf approximation to calculate SE and CI of the odds ratio
tb	calculate test-based confidence intervals
<u>e</u>xact	calculate Fisher's exact p
<u>l</u>evel(#)	set confidence level; default is level(95)

tabodds_options	description
Main	
<u>b</u>inomial(*varname*)	number of subjects variable
<u>l</u>evel(#)	set confidence level; default is level(95)
or	report odds ratio
<u>ad</u>just(*varlist*)	report odds ratios adjusted for the variables in *varlist*
base(#)	reference group of control variable for odds ratio
<u>c</u>ornfield	use Cornfield approximation to calculate CI of the odds ratio
<u>w</u>oolf	use Woolf approximation to calculate SE and CI of the odds ratio
tb	calculate test-based confidence intervals
<u>g</u>raph	graph odds against categories
<u>ci</u>plot	same as graph option, except include confidence intervals
CI plot	
<u>ci</u>opts(*rcap_options*)	affect rendition of the confidence bands
Plot	
marker_options	change look of markers (color, size, etc.)
marker_label_options	add marker labels; change look or position
cline_options	affect rendition of the plotted points
Add plots	
addplot(*plot*)	add other plots to the generated graph
Y axis, X axis, Titles, Legend, Overall	
twoway_options	any options other than by() documented in [G] *twoway_options*

mhodds_options	description
Options	
by(*varlist*[, <u>missing</u>])	stratify on *varlist*
<u>b</u>inomial(*varname*)	number of subjects variable
<u>c</u>ompare(v_1, v_2)	override categories of the control variable
<u>l</u>evel(#)	set confidence level; default is level(95)

fweights are allowed; see [U] **11.1.6 weight**.

Menu

ir

Statistics > Epidemiology and related > Tables for epidemiologists > Incidence-rate ratio

iri

Statistics > Epidemiology and related > Tables for epidemiologists > Incidence-rate ratio calculator

cs

Statistics > Epidemiology and related > Tables for epidemiologists > Cohort study risk-ratio etc.

csi

Statistics > Epidemiology and related > Tables for epidemiologists > Cohort study risk-ratio etc. calculator

cc

Statistics > Epidemiology and related > Tables for epidemiologists > Case-control odds ratio

cci

Statistics > Epidemiology and related > Tables for epidemiologists > Case-control odds-ratio calculator

tabodds

Statistics > Epidemiology and related > Tables for epidemiologists > Tabulate odds of failure by category

mhodds

Statistics > Epidemiology and related > Tables for epidemiologists > Ratio of odds of failure for two categories

mcc

Statistics > Epidemiology and related > Tables for epidemiologists > Matched case-control studies

mcci

Statistics > Epidemiology and related > Tables for epidemiologists > Matched case-control calculator

Description

ir is used with incidence-rate (incidence-density or person-time) data. It calculates point estimates and confidence intervals for the incidence-rate ratio and difference, along with attributable or prevented fractions for the exposed and total population. iri is the immediate form of ir; see [U] **19 Immediate commands**. Also see [R] **poisson** and [ST] **stcox** for related commands.

cs is used with cohort study data with equal follow-up time per subject and sometimes with cross-sectional data. Risk is then the proportion of subjects who become cases. It calculates point estimates and confidence intervals for the risk difference, risk ratio, and (optionally) the odds ratio, along with attributable or prevented fractions for the exposed and total population. csi is the immediate form of cs; see [U] **19 Immediate commands**. Also see [R] **logistic** and [R] **glogit** for related commands.

cc is used with case–control and cross-sectional data. It calculates point estimates and confidence intervals for the odds ratio, along with attributable or prevented fractions for the exposed and total population. cci is the immediate form of cc; see [U] **19 Immediate commands**. Also see [R] **logistic** and [R] **glogit** for related commands.

tabodds is used with case–control and cross-sectional data. It tabulates the odds of failure against a categorical explanatory variable *expvar*. If *expvar* is specified, tabodds performs an approximate χ^2 test of homogeneity of odds and a test for linear trend of the log odds against the numerical code used for the categories of *expvar*. Both tests are based on the score statistic and its variance; see *Methods and formulas*. When *expvar* is absent, the overall odds are reported. The variable var_{case} is coded 0/1 for individual and simple frequency records and equals the number of cases for binomial frequency records.

Optionally, tabodds tabulates adjusted or unadjusted odds ratios, using either the lowest levels of *expvar* or a user-defined level as the reference group. If adjust(*varlist*) is specified, it produces odds ratios adjusted for the variables in *varlist* along with a (score) test for trend.

mhodds is used with case–control and cross-sectional data. It estimates the ratio of the odds of failure for two categories of *expvar*, controlled for specified confounding variables, $vars_{adjust}$, and tests whether this odds ratio is equal to one. When *expvar* has more than two categories but none are specified with the compare() option, mhodds assumes that *expvar* is a quantitative variable and calculates a 1-degree-of-freedom test for trend. It also calculates an approximate estimate of the log odds-ratio for a one-unit increase in *expvar*. This is a one-step Newton–Raphson approximation to the maximum likelihood estimate calculated as the ratio of the score statistic, U, to its variance, V (Clayton and Hills 1993, 103).

mcc is used with matched case–control data. It calculates McNemar's chi-squared; point estimates and confidence intervals for the difference, ratio, and relative difference of the proportion with the factor; and the odds ratio and its confidence interval. mcci is the immediate form of mcc; see [U] **19 Immediate commands**. Also see [R] **clogit** and [R] **symmetry** for related commands.

Options

Options are listed in the order that they appear in the syntax tables above. The commands for which the option is valid are indicated in parentheses immediately after the option name.

Options (ir, cs, cc, and mhodds) / Main (tabodds)

by(*varname*[, missing]) (ir, cs, cc, and mhodds) specifies that the tables be stratified on *varname*. Missing categories in *varname* are omitted from the stratified analysis, unless option missing is specified within by(). Within-stratum statistics are shown and then combined with Mantel–Haenszel weights. If estandard, istandard, or standard() is also specified (see below), the weights specified are used in place of Mantel–Haenszel weights.

estandard, istandard, and standard(*varname*) (ir, cs, and cc) request that within-stratum statistics be combined with external, internal, or user-specified weights to produce a standardized estimate. These options are mutually exclusive and can be used only when by() is also specified. (When by() is specified without one of these options, Mantel–Haenszel weights are used.)

estandard external weights are the person-time for the unexposed (ir), the total number of unexposed (cs), or the number of unexposed controls (cc).

istandard internal weights are the person-time for the exposed (ir), the total number of exposed (cs), or the number of exposed controls (cc). istandard can be used to produce, among other things, standardized mortality ratios (SMRs).

standard(*varname*) allows user-specified weights. *varname* must contain a constant within stratum and be nonnegative. The scale of *varname* is irrelevant.

pool (ir, cs, and cc) specifies that, in a stratified analysis, the directly pooled estimate also
be displayed. The pooled estimate is a weighted average of the stratum-specific estimates using
inverse-variance weights, which are the inverse of the variance of the stratum-specific estimate.
pool is relevant only if by() is also specified.

nocrude (ir, cs, and cc) specifies that in a stratified analysis the crude estimate—an estimate
obtained without regard to strata—not be displayed. nocrude is relevant only if by() is also
specified.

nohom (ir, cs, and cc) specifies that a χ^2 test of homogeneity not be included in the output of
a stratified analysis. This tests whether the exposure effect is the same across strata and can be
performed for any pooled estimate—directly pooled or Mantel–Haenszel. nohom is relevant only
if by() is also specified.

ird (ir) may be used only with estandard, istandard, or standard(). It requests that ir
calculate the standardized incidence-rate difference rather than the default incidence-rate ratio.

rd (cs) may be used only with estandard, istandard, or standard(). It requests that cs calculate
the standardized risk difference rather than the default risk ratio.

bd (cc) specifies that Breslow and Day's χ^2 test of homogeneity be included in the output of a
stratified analysis. This tests whether the exposure effect is the same across strata. bd is relevant
only if by() is also specified.

tarone (cc) specifies that Tarone's χ^2 test of homogeneity, which is a correction to the Breslow–Day
test, be included in the output of a stratified analysis. This tests whether the exposure effect is the
same across strata. tarone is relevant only if by() is also specified.

binomial(*varname*) (cs, cc, tabodds, and mhodds) supplies the number of subjects (cases plus
controls) for binomial frequency records. For individual and simple frequency records, this option
is not used.

or (cs, csi, and tabodds), for cs and csi, reports the calculation of the odds ratio in addition to
the risk ratio if by() is not specified. With by(), or specifies that a Mantel–Haenszel estimate
of the combined odds ratio be made rather than the Mantel–Haenszel estimate of the risk ratio.
In either case, this is the same calculation that would be made by cc and cci. Typically, cc, cci,
or tabodds is preferred for calculating odds ratios. For tabodds, or specifies that odds ratios be
produced; see base() for details about selecting a reference category. By default, tabodds will
calculate odds.

adjust(*varlist*) (tabodds) specifies that odds ratios adjusted for the variables in *varlist* be calculated.

base(#) (tabodds) specifies that the #th category of *expvar* be used as the reference group for
calculating odds ratios. If base() is not specified, the first category, corresponding to the minimum
value of *expvar*, is used as the reference group.

cornfield (cc, cci, and tabodds) requests that the Cornfield (1956) approximation be used to
calculate the confidence interval of the odds ratio. By default, cc and cci report an exact interval
and tabodds reports a standard-error–based interval, with the standard error coming from the
square root of the variance of the score statistic.

woolf (cs, csi, cc, cci, and tabodds) requests that the Woolf (1955) approximation, also known
as the Taylor expansion, be used for calculating the standard error and confidence interval for the
odds ratio. By default, cs and csi with the or option report the Cornfield (1956) interval; cc
and cci report an exact interval; and tabodds reports a standard-error–based interval, with the
standard error coming from the square root of the variance of the score statistic.

tb (ir, iri, cs, csi, cc, cci, tabodds, mcc, and mcci) requests that test-based confidence intervals (Miettinen 1976) be calculated wherever appropriate in place of confidence intervals based on other approximations or exact confidence intervals. We recommend that test-based confidence intervals be used only for pedagogical purposes and never for research work.

exact (cs, csi, cc, and cci) requests that Fisher's exact p be calculated rather than the χ^2 and its significance level. We recommend specifying exact whenever samples are small. When the least-frequent cell contains 1,000 cases or more, there will be no appreciable difference between the exact significance level and the significance level based on the χ^2, but the exact significance level will take considerably longer to calculate. exact does not affect whether exact confidence intervals are calculated. Commands always calculate exact confidence intervals where they can, unless cornfield, woolf, or tb is specified.

compare(v_1, v_2) (mhodds) indicates the categories of *expvar* to be compared; v_1 defines the numerator and v_2, the denominator. When compare() is not specified and there are only two categories, the second is compared to the first; when there are more than two categories, an approximate estimate of the odds ratio for a unit increase in *expvar*, controlled for specified confounding variables, is given.

level(#) (ir, iri, cs, csi, cc, cci, tabodds, mhodds, mcc, and mcci) specifies the confidence level, as a percentage, for confidence intervals. The default is level(95) or as set by set level; see [R] **level**.

The following options are for use only with tabodds.

Main

graph (tabodds) produces a graph of the odds against the numerical code used for the categories of *expvar*. All graph options except connect() are allowed. This option is not allowed with the or option or the adjust() option.

ciplot (tabodds) produces the same plot as the graph option, except that it also includes the confidence intervals. This option may not be used with either the or option or the adjust() option.

CI plot

ciopts(*rcap_options*) (tabodds) is allowed only with the ciplot option. It affects the rendition of the confidence bands; see [G] ***rcap_options***.

Plot

marker_options (tabodds) affect the rendition of markers drawn at the plotted points, including their shape, size, color, and outline; see [G] ***marker_options***.

marker_label_options (tabodds) specify if and how the markers are to be labeled; see [G] ***marker_label_options***.

cline_options (tabodds) affect whether lines connect the plotted points and the rendition of those lines; see [G] ***cline_options***.

Add plots

addplot(*plot*) (tabodds) provides a way to add other plots to the generated graph; see [G] ***addplot_option***.

Y axis, X axis, Titles, Legend, Overall

twoway_options (tabodds) are any of the options documented in [G] ***twoway_options***, excluding by(). These include options for titling the graph (see [G] ***title_options***) and options for saving the graph to disk (see [G] ***saving_option***).

Remarks

Remarks are presented under the following headings:

> Incidence-rate data
> Stratified incidence-rate data
> Standardized estimates with stratified incidence-rate data
> Cumulative incidence data
> Stratified cumulative incidence data
> Standardized estimates with stratified cumulative incidence data
> Case–control data
> Stratified case–control data
> Case–control data with multiple levels of exposure
> Case–control data with confounders and possibly multiple levels of exposure
> Standardized estimates with stratified case–control data
> Matched case–control data

To calculate appropriate statistics and suppress inappropriate statistics, the ir, cs, cc, tabodds, mhodds, and mcc commands, along with their immediate counterparts, are organized in the way epidemiologists conceptualize data. ir processes incidence-rate data from prospective studies; cs, cohort study data with equal follow-up time (cumulative incidence); cc, tabodds, and mhodds, case–control or cross-sectional (prevalence) data; and mcc, matched case–control data. With the exception of mcc, these commands work with both simple and stratified tables.

Epidemiological data are often summarized in a contingency table from which various statistics are calculated. The rows of the table reflect cases and noncases or cases and person-time, and the columns reflect exposure to a *risk factor*. To an epidemiologist, *cases* and *noncases* refer to the outcomes of the process being studied. For instance, a case might be a person with cancer and a noncase might be a person without cancer.

A *factor* is something that might affect the chances of being ultimately designated a case or a noncase. Thus a case might be a cancer patient and the factor, smoking behavior. A person is said to be *exposed* or *unexposed* to the factor. Exposure can be classified as a dichotomy, smokes or does not smoke, or as multiple levels, such as number of cigarettes smoked per week.

For an introduction to epidemiological methods, see Walker (1991). For an intermediate treatment, see Clayton and Hills (1993) and Lilienfeld and Stolley (1994). For other advanced discussions, see van Belle et al. (2004); Kleinbaum, Kupper, and Morgenstern (1982); and Rothman, Greenland, and Lash (2008). For an anthology of writings on epidemiology since World War II, see Greenland (1987). See Jewell (2004) for a text aimed at graduate students in the medical professions that uses Stata for much of the analysis. See Dohoo, Martin, and Stryhn (2003) for a graduate-level text on the principles and methods of veterinary epidemiologic research; Stata datasets and do-files are available.

Incidence-rate data

In incidence-rate data from a prospective study, you observe the transformation of noncases into cases. Starting with a group of noncase subjects, you monitor them to determine whether they become cases (e.g., stricken with cancer). You monitor two populations: those exposed and those unexposed to the factor (e.g., multiple X-rays). A summary of the data is

	Exposed	Unexposed	Total
Cases	a	b	$a+b$
Person-time	N_1	N_0	$N_1 + N_0$

▷ Example 1: iri

It will be easiest to understand these commands if we start with the immediate forms. Remember, in the immediate form, we specify the data on the command line rather than specifying names of variables containing the data; see [U] **19 Immediate commands**. We have data (Boice Jr. and Monson [1977]; reported in Rothman, Greenland, and Lash [2008, 244]) on breast cancer *cases* and person-years of observation for women with tuberculosis repeatedly *exposed* to multiple X-ray fluoroscopies, and those not so exposed:

	X-ray fluoroscopy	
	Exposed	Unexposed
Breast cancer cases	41	15
Person-years	28,010	19,017

Using `iri`, the immediate form of `ir`, we specify the values in the table following the command:

```
. iri 41 15 28010 19017
```

	Exposed	Unexposed	Total
Cases	41	15	56
Person-time	28010	19017	47027
Incidence rate	.0014638	.0007888	.0011908

	Point estimate	[95% Conf. Interval]	
Inc. rate diff.	.000675	.0000749	.0012751
Inc. rate ratio	1.855759	1.005684	3.6093 (exact)
Attr. frac. ex.	.4611368	.0056519	.722938 (exact)
Attr. frac. pop	.337618		

(midp) Pr(k>=41) =		0.0177 (exact)
(midp) 2*Pr(k>=41) =		0.0355 (exact)

`iri` shows the table, reports the incidence rates for the exposed and unexposed populations, and then shows the point estimates of the difference and ratio of the two incidence rates along with their confidence intervals. The incidence rate is simply the frequency with which noncases are transformed into cases.

Next `iri` reports the attributable fraction among the exposed population, an estimate of the proportion of exposed cases attributable to exposure. We estimate that 46.1% of the 41 breast cancer cases among the exposed were due to exposure. (Had the incidence-rate ratio been less than 1, `iri` would have reported the prevented fraction in the exposed population, an estimate of the net proportion of all potential cases in the exposed population that was prevented by exposure; see the following technical note.)

After that, the table shows the attributable fraction in the total population, which is the net proportion of all cases attributable to exposure. This number, of course, depends on the proportion of cases that are exposed in the base population, which here is taken to be 41/56 and may not be relevant in all situations. We estimate that 33.8% of the 56 cases were due to exposure. We estimate that 18.9 cases were caused by exposure; that is, $0.338 \times 56 = 0.461 \times 41 = 18.9$.

At the bottom of the table, `iri` reports both one- and two-sided exact significance tests. For the one-sided test, the probability that the number of exposed cases is 41 or greater is 0.0177. This is a "midp" calculation; see *Methods and formulas* below. The two-sided test is $2 \times 0.0177 = 0.0354$.

◁

❑ Technical note

When the incidence-rate ratio is less than 1, `iri` (and `ir`, `cs`, `csi`, `cc`, and `cci`) substitutes the prevented fraction for the attributable fraction. Let's reverse the roles of exposure in the above data, treating as exposed a person who did not receive the X-ray fluoroscopy. You can think of this as a new treatment for preventing breast cancer—the suggested treatment being not to use fluoroscopy.

```
. iri 15 41 19017 28010
```

	Exposed	Unexposed		Total		
Cases	15	41		56		
Person-time	19017	28010		47027		
Incidence rate	.0007888	.0014638		.0011908		
	Point estimate			[95% Conf. Interval]		
Inc. rate diff.	−.000675			−.0012751	−.0000749	
Inc. rate ratio	.5388632			.277062	.9943481	(exact)
Prev. frac. ex.	.4611368			.0056519	.722938	(exact)
Prev. frac. pop	.1864767					

(midp)	Pr(k<=15) =	0.0177 (exact)
(midp)	2*Pr(k<=15) =	0.0355 (exact)

The prevented fraction among the exposed is the net proportion of all potential cases in the exposed population that were prevented by exposure. We estimate that 46.1% of potential cases among the women receiving the new "treatment" were prevented by the treatment. (Previously, we estimated that the same percentage of actual cases among women receiving the X-rays was caused by the X-rays.)

The prevented fraction for the population, which is the net proportion of all potential cases in the total population that was prevented by exposure, as with the attributable fraction, depends on the proportion of cases that are exposed in the base population—here taken as 15/56—so it may not be relevant in all situations. We estimate that 18.6% of the potential cases were prevented by exposure.

See Greenland and Robins (1988) for a discussion of how to interpret attributable and prevented fractions.

❑

▷ Example 2: ir

ir works like iri, except that it obtains the entries in the tables by summing data. You specify three variables: the first represents the number of cases represented by this observation, the second indicates whether the observation is for subjects exposed to the factor, and the third records the total time the subjects in this observation were observed. An observation may reflect one subject or a group of subjects.

For instance, here is a 2-observation dataset for the table in the previous example:

```
. use http://www.stata-press.com/data/r11/irxmpl
. list
```

	cases	exposed	time
1.	41	0	28010
2.	15	1	19017

If we typed `ir cases exposed time`, we would obtain the same output that we obtained above. Another way the data might be recorded is

```
. use http://www.stata-press.com/data/r11/irxmpl2
. list
```

	cases	exposed	time
1.	20	0	14000
2.	21	0	14010
3.	15	1	19017

Here the first 2 observations will be automatically summed by ir because both are exposed. Finally, the data might be individual-level data:

```
. use http://www.stata-press.com/data/r11/irxmpl3
. list in 1/5
```

	cases	exposed	time
1.	1	1	10
2.	0	1	8
3.	0	0	9
4.	1	0	2
5.	0	1	1

The first observation represents a woman who got cancer, was exposed, and was observed for 10 years. The second is a woman who did not get cancer, was exposed, and was observed for 8 years, and so on.

◁

❑ Technical note

ir (and all the other commands) assumes that a subject was exposed if the exposed variable is nonzero and not missing, assumes the subject was not exposed if the variable is zero, and ignores the observation if the variable is missing. For ir, the case variable and the time variable are restricted to nonnegative integers and are summed within the exposed and unexposed groups to obtain the entries in the table.

❑

Stratified incidence-rate data

▷ Example 3: ir with stratified data

ir can work with stratified tables, as well as with single tables. For instance, Rothman (1986, 185) discusses data from Rothman and Monson (1973) on mortality by sex and age for patients with trigeminal neuralgia:

	Age through 64		Age 65+	
	Males	Females	Males	Females
Deaths	14	10	76	121
Person-years	1516	1701	949	2245

Entering the data into Stata, we have the following dataset:

```
. use http://www.stata-press.com/data/r11/rm
(Rothman and Monson 1973 data)

. list
```

	age	male	deaths	pyears
1.	<65	1	14	1516
2.	<65	0	10	1701
3.	65+	1	76	949
4.	65+	0	121	2245

The stratified analysis of the incidence-rate ratio is

```
. ir deaths male pyears, by(age)
```

Age category	IRR	[95% Conf. Interval]		M-H Weight	
<65	1.570844	.6489373	3.952809	4.712465	(exact)
65+	1.485862	1.100305	1.99584	35.95147	(exact)
Crude	1.099794	.831437	1.449306		(exact)
M-H combined	1.49571	1.141183	1.960377		

Test of homogeneity (M-H)   chi2(1) =   0.02   Pr>chi2 = 0.8992

The row labeled M-H combined reflects the combined Mantel–Haenszel estimates.

As with the previous example, it is not important that each entry in the table correspond to 1 observation in the data—ir sums the time (pyears) and case (deaths) variables within the exposure (male) category.

The difference between the unadjusted crude estimate and the Mantel–Haenszel estimate suggests confounding by age: women in the study are older, and older patients are more likely to die. But we should not use the Mantel–Haenszel estimate without checking its homogeneity assumption. The chi-squared test of homogeneity gives a p-value of 0.8992, so we have no evidence that the exposure effect (the effect of being male) differs across age categories. We are justified in using the Mantel–Haenszel estimate.

◁

❑ Technical note

Stratification is one way to deal with confounding; that is, perhaps sex affects the incidence of trigeminal neuralgia and so does age, so the table was stratified by age in an attempt to uncover the sex effect. (We are concerned that age may confound the true association between sex and the incidence of trigeminal neuralgia because the age distributions are so different for males and females. If age affects incidence, the difference in the age distributions would induce different incidences for males and females and thus confound the true effect of sex.)

We do not, however, have to use tables to uncover effects; the estimation alternative when we have aggregate data is Poisson regression, and we can use the same data on which we ran `ir` with `poisson`. Poisson regression also works with individual-level data.

(Although `age` in the previous example appears to be a string, it is actually a numeric variable taking on values 0 and 1. We attached a value label to produce the labels <65 and 65+ to make `ir`'s output look better; see [U] **12.6.3 Value labels**. Stata's estimation commands will ignore this labeling.)

```
. poisson deaths male age, exposure(pyears) irr
Iteration 0:   log likelihood = -10.836732
Iteration 1:   log likelihood = -10.734087
Iteration 2:   log likelihood = -10.733944
Iteration 3:   log likelihood = -10.733944

Poisson regression                                Number of obs   =          4
                                                  LR chi2(2)      =     164.01
                                                  Prob > chi2     =     0.0000
Log likelihood = -10.733944                       Pseudo R2       =     0.8843

------------------------------------------------------------------------------
      deaths |        IRR   Std. Err.      z    P>|z|     [95% Conf. Interval]
-------------+----------------------------------------------------------------
        male |   1.495096   .2060997     2.92   0.004     1.141118     1.95888
         age |   8.888775   1.934943    10.04   0.000     5.801616    13.61867
      pyears |  (exposure)
------------------------------------------------------------------------------
```

Compare these results with the Mantel–Haenszel estimates produced by `ir`:

Source	IR Ratio	95% Conf. Int.	
Mantel–Haenszel (ir)	1.50	1.14	1.96
poisson	1.50	1.14	1.96

The results from `poisson` agree with the Mantel–Haenszel estimates to two decimal places. But `poisson` also estimates an incidence-rate ratio for age. Here the estimate is not of much interest, because the outcome variable is total mortality and we already knew that older people have a higher mortality rate. In other contexts, however, the estimate might be of greater interest.

See [R] **poisson** for an explanation of the `poisson` command.

❑

❑ Technical note

Both the model fit above and the preceding table asserted that exposure effects are the same across age categories and, if they are not, then both of the previous results are equally inappropriate. The table presented a test of homogeneity, reassuring us that the exposure effects do indeed appear to be constant. The Poisson-regression alternative can be used to reproduce that test by including interactions between the age groups and exposure:

```
. poisson deaths male age male#c.age, exposure(pyears) irr
Iteration 0:   log likelihood = -10.898799
Iteration 1:   log likelihood = -10.726225
Iteration 2:   log likelihood = -10.725904
Iteration 3:   log likelihood = -10.725904

Poisson regression                              Number of obs   =          4
                                                LR chi2(3)      =     164.03
                                                Prob > chi2     =     0.0000
Log likelihood = -10.725904                     Pseudo R2       =     0.8843
```

deaths	IRR	Std. Err.	z	P>|z|	[95% Conf. Interval]	
male	1.660688	1.396496	0.60	0.546	.3195218	8.631283
age	9.167973	3.01659	6.73	0.000	4.810583	17.47226
male#c.age						
1	.9459	.41539	-0.13	0.899	.3999832	2.236911
pyears	(exposure)					

The significance level of the male#c.age effect is 0.899, the same as previously reported by ir.

Here forming the male-times-age interaction was easy because there were only two age groups. Had there been more groups, the test would have been slightly more difficult—see the following technical note.

❑

❑ Technical note

A word of caution is in order when applying poisson (or any estimation technique) to more than two age categories. Say that in our data, we had three age categories, which we will call categories 0, 1, and 2, and that they are stored in the variable agecat. We might think of the categories as corresponding to age less than 35, 35–64, and 65 and above.

With such data, we might type ir deaths male pyears, by(agecat), but we would *not* type poisson deaths male agecat, exposure(pyears) to obtain the equivalent Poisson-regression estimated results. Such a model might be reasonable, but it is not equivalent because we would be constraining the age effect in category 2 to be (multiplicatively) twice the effect in category 1.

To poisson (and all of Stata's estimation commands other than anova), agecat is simply one variable, and only one estimated coefficient is associated with it. Thus the model is

$$\text{Poisson index} = P = \beta_0 + \beta_1 \text{male} + \beta_2 \text{agecat}$$

The expected number of deaths is then e^P, and the incidence-rate ratio associated with a variable is e^β; see [R] **poisson**. Thus the value of the Poisson index when male==0 and agecat==1 is $\beta_0 + \beta_2$, and the possibilities are

	male==0	male==1
agecat==0	β_0	$\beta_0 + \beta_1$
agecat==1	$\beta_0 + \beta_2$	$\beta_0 + \beta_2 + \beta_1$
agecat==2	$\beta_0 + 2\beta_2$	$\beta_0 + 2\beta_2 + \beta_1$

The age effect for agecat==2 is constrained to be twice the age effect for agecat==1—the only difference between lines 3 and 2 of the table is that β_2 is replaced with $2\beta_2$. Under certain circumstances, such a constraint might be reasonable, but it does not correspond to the assumptions made in generating the Mantel–Haenszel combined results.

To obtain results equivalent to the Mantel–Haenszel result, we must estimate a separate effect for each age group, meaning that we must replace $2\beta_2$, the constrained effect, with β_3, a new coefficient that is free to take on any value. We can achieve this by creating two new variables and using them in place of `agecat`. `agecat1` will take on the value 1 when `agecat` is 1 and 0 otherwise; `agecat2` will take on the value 1 when `agecat` is 2 and 0 otherwise:

```
. generate agecat1 = (agecat==1)
. generate agecat2 = (agecat==2)
. poisson deaths male agecat1 agecat2 [freq=pop], exposure(pyears) irr
```

In Stata, we do not have to generate these variables for ourselves. We could use factor variables:

```
. poisson deaths male i.agecat [freq=pop], exposure(pyears) irr
```

See [U] **11.4.3 Factor variables**.

To reproduce the homogeneity test with multiple age categories, we could type

```
. poisson deaths agecat##male [freq=pop], exp(pyears) irr
. testparm agecat#male
```

Poisson regression combined with factor variables generalizes to multiway tables. Suppose that there are three exposure categories. Assume exposure variable `burn` takes on the values 1, 2, and 3 for first-, second-, and third-degree burns. The table itself is estimated by typing

```
. poisson deaths i.burn i.agecat [freq=pop], exp(pyears) irr
```

and the test of homogeneity is estimated by typing

```
. poisson deaths burn##agecat [freq=pop], exp(pyears) irr
. testparm burn#agecat
```

❑

Standardized estimates with stratified incidence-rate data

The `by()` option specifies that the data are stratified and, by default, will produce a Mantel–Haenszel combined estimate of the incidence-rate ratio. With the `estandard`, `istandard`, or `standard(`*varname*`)` options, you can specify your own weights and obtain standardized estimates of the incidence-rate ratio or difference.

▷ Example 4: ir with stratified data, using standardized estimates

Rothman, Greenland, and Lash (2008, 264) report results from Doll and Hill (1966) on age-specific coronary disease deaths among British male doctors from cigarette smoking:

Age	Smokers		Nonsmokers	
	Deaths	Person-years	Deaths	Person-years
35–44	32	52,407	2	18,790
45–54	104	43,248	12	10,673
55–64	206	28,612	28	5,710
65–74	186	12,663	28	2,585
75–84	102	5,317	31	1,462

We have entered these data into Stata:

```
. use http://www.stata-press.com/data/r11/dollhill2
. list

     | age    smokes  deaths  pyears |
  1. | 35-44       1      32   52407 |
  2. | 35-44       0       2   18790 |
  3. | 45-54       1     104   43248 |
  4. | 45-54       0      12   10673 |
  5. | 55-64       1     206   28612 |
     |------------------------------|
  6. | 55-64       0      28    5710 |
  7. | 65-74       1     186   12663 |
  8. | 65-74       0      28    2585 |
  9. | 75-84       1     102    5317 |
 10. | 75-84       0      31    1462 |
```

We can obtain the Mantel–Haenszel combined estimate along with the crude estimate for ignoring stratification of the incidence-rate ratio and 90% confidence intervals by typing

```
. ir deaths smokes pyears, by(age) level(90)
         age |      IRR     [90% Conf. Interval]    M-H Weight
       35-44 | 5.736638    1.704271   33.61646    1.472169   (exact)
       45-54 | 2.138812    1.274552   3.813282    9.624747   (exact)
       55-64 | 1.46824     1.044915   2.110422    23.34176   (exact)
       65-74 | 1.35606     .9626026   1.953505    23.25315   (exact)
       75-84 | .9047304    .6375194   1.305412    24.31435   (exact)
             |
       Crude | 1.719823    1.437544   2.0688                 (exact)
 M-H combined| 1.424682    1.194375   1.699399

     Test of homogeneity (M-H)   chi2(4) =    10.41  Pr>chi2 = 0.0340
```

Note the presence of heterogeneity revealed by the test; the effect of smoking is not the same across age categories. Moreover, the listed stratum-specific estimates show an effect that appears to be declining with age. (Even if the test of homogeneity is not significant, you should always examine estimates carefully when stratum-specific effects occur on both sides of 1 for ratios and 0 for differences.)

Rothman, Greenland, and Lash (2008, 269) obtain the standardized incidence-rate ratio and 90% confidence intervals, weighting each age category by the population of the exposed group, thus producing the standardized mortality ratio (SMR). This calculation can be reproduced by specifying by(age) to indicate that the table is stratified and istandard to specify that we want the internally standardized rate. We may also specify that we would like to see the pooled estimate (weighted average where the weights are based on the variance of the strata calculations):

```
. ir deaths smokes pyears, by(age) level(90) istandard pool
         age |      IRR     [90% Conf. Interval]    Weight
       35-44 | 5.736638    1.704271   33.61646    52407   (exact)
       45-54 | 2.138812    1.274552   3.813282    43248   (exact)
       55-64 | 1.46824     1.044915   2.110422    28612   (exact)
       65-74 | 1.35606     .9626026   1.953505    12663   (exact)
       75-84 | .9047304    .6375194   1.305412     5317   (exact)
             |
       Crude | 1.719823    1.437544   2.0688              (exact)
Pooled (direct)| 1.355343  1.134356   1.619382
I. Standardized| 1.417609  1.186541   1.693676

     Test of homogeneity (direct) chi2(4) =    10.20  Pr>chi2 = 0.0372
```

We obtained the simple pooled results because we specified the `pool` option. Note the significance of the homogeneity test; it provides the motivation for standardizing the rate ratios.

If we wanted the externally standardized ratio (weights proportional to the population of the unexposed group), we would substitute `estandard` for `istandard` in the above command.

We are not limited to incidence-rate ratios; `ir` can also estimate incidence-rate differences. Differences may be standardized internally or externally. We will obtain the internally weighted difference (Rothman, Greenland, and Lash 2008, 266–267):

```
. ir deaths smokes pyears, by(age) level(90) istandard ird
```

age	IRD	[90% Conf. Interval]		Weight
35-44	.0005042	.0002877	.0007206	52407
45-54	.0012804	.0006205	.0019403	43248
55-64	.0022961	.0005628	.0040294	28612
65-74	.0038567	.0000521	.0076614	12663
75-84	-.0020201	-.0090201	.00498	5317
Crude	.0018537	.001342	.0023654	
I. Standardized	.0013047	.000712	.0018974	

▷ Example 5: ir with user-specified weights

In addition to calculating results by using internal or external weights, `ir` (and `cs` and `cc`) can calculate results for arbitrary weights. If we wanted to obtain the incidence-rate ratio weighting each age category equally, we would type

```
. generate conswgt=1
. ir deaths smokes pyears, by(age) standard(conswgt)
```

age	IRR	[95% Conf. Interval]		Weight	
35-44	5.736638	1.463557	49.40468	1	(exact)
45-54	2.138812	1.173714	4.272545	1	(exact)
55-64	1.46824	.9863624	2.264107	1	(exact)
65-74	1.35606	.9081925	2.096412	1	(exact)
75-84	.9047304	.6000757	1.399687	1	(exact)
Crude	1.719823	1.391992	2.14353		(exact)
Standardized	1.155026	.9006199	1.481295		

❑ Technical note

`estandard` and `istandard` are convenience features; they do nothing different from what you could accomplish by creating the appropriate weights and using the `standard()` option. For instance, we could duplicate the previously shown results of `istandard` (example before last) by typing

(*Continued on next page*)

```
. sort age smokes
. by age: generate wgt=pyears[_N]
. list in 1/4

     age    smokes   deaths   pyears   conswgt    wgt
1.  35-44      0        2      18790      1      52407
2.  35-44      1       32      52407      1      52407
3.  45-54      0       12      10673      1      43248
4.  45-54      1      104      43248      1      43248

. ir deaths smokes pyears, by(age) level(90) standard(wgt) ird
  (output omitted)
```

sort age smokes made the exposed group (smokes = 1) the last observation within each age category. by age: gen wgt=pyears[_N] created wgt equal to the last observation in each age category.

❑

Cumulative incidence data

Cumulative incidence data are "follow-up data with denominators consisting of persons rather than person-time" (Rothman 1986, 172). A group of noncases is monitored for some time, during which some become cases. Each subject is also known to be exposed or unexposed. A summary of the data is

	Exposed	Unexposed	Total
Cases	a	b	$a+b$
Noncases	c	d	$c+d$
Total	$a+c$	$b+d$	$a+b+c+d$

Data of this type are generally summarized using the risk ratio, $\{a/(a+c)\}/\{b/(b+d)\}$. A ratio of 2 means that an exposed subject is twice as likely to become a case than is an unexposed subject, a ratio of one-half means half as likely, and so on. The "null" value—the number corresponding to no effect—is a ratio of 1. If cross-sectional data are analyzed in this format, the risk ratio becomes a prevalence ratio.

▷ Example 6: csi

We have data on diarrhea during a 10-day follow-up period among 30 breast-fed infants colonized with *Vibrio cholerae* 01 according to antilipopolysaccharide antibody titers in the mother's breast milk (Glass et al. [1983]; reported in Rothman, Greenland, and Lash [2008, 248]):

| | Antibody level ||
	High	Low
Diarrhea	7	12
No diarrhea	9	2

The csi command works much like the iri command. Our sample is small, so we will specify the exact option.

```
. csi 7 12 9 2, exact
                 |   Exposed    Unexposed  |     Total
-----------------+--------------------------+------------
          Cases  |        7           12   |        19
       Noncases  |        9            2   |        11
-----------------+--------------------------+------------
          Total  |       16           14   |        30
                 |
           Risk  |    .4375      .8571429  |  .6333333
                 |
                 |   Point estimate        | [95% Conf. Interval]
-----------------+--------------------------+----------------------
Risk difference  |      -.4196429          | -.7240828   -.1152029
     Risk ratio  |       .5104167          |  .2814332    .9257086
  Prev. frac. ex.|       .4895833          |  .0742914    .7185668
  Prev. frac. pop|       .2611111          |
                 +-----------------------------------------------
                    1-sided Fisher's exact P = 0.0212
                    2-sided Fisher's exact P = 0.0259
```

We find that high antibody levels reduce the risk of diarrhea (the risk falls from 0.86 to 0.44). The difference is just significant at the 2.59% two-sided level. (Had we not specified the exact option, a χ^2 value and its significance level would have been reported in place of Fisher's exact p. The calculated χ^2 two-sided significance level would have been 0.0173, but this calculation is inferior for small samples.)

◁

❑ Technical note

By default, cs and csi do not report the odds ratio, but they will if you specify the or option. If you want odds ratios, however, use the cc or cci commands—the commands appropriate for case–control data—because cs and csi calculate the attributable (prevented) fraction with the risk ratio, even if you specify or:

```
. csi 7 12 9 2, or exact
                 |   Exposed    Unexposed  |     Total
-----------------+--------------------------+------------
          Cases  |        7           12   |        19
       Noncases  |        9            2   |        11
-----------------+--------------------------+------------
          Total  |       16           14   |        30
                 |
           Risk  |    .4375      .8571429  |  .6333333
                 |
                 |   Point estimate        | [95% Conf. Interval]
-----------------+--------------------------+----------------------
Risk difference  |      -.4196429          | -.7240828   -.1152029
     Risk ratio  |       .5104167          |  .2814332    .9257086
  Prev. frac. ex.|       .4895833          |  .0742914    .7185668
  Prev. frac. pop|       .2611111          |
     Odds ratio  |       .1296296          |  .0246233    .7180882   (Cornfield)
                 +-----------------------------------------------
                    1-sided Fisher's exact P = 0.0212
                    2-sided Fisher's exact P = 0.0259
```

❑

❑ Technical note

As with `iri` and `ir`, `csi` and `cs` report either the attributable or the prevented fraction for the exposed and total populations; see the discussion under *Incidence-rate data* above. In example 6, we estimated that 49% of potential cases in the exposed population were prevented by exposure. We also estimated that exposure accounted for a 26% reduction in cases over the entire population, but that is based on the exposure distribution of the (small) population (16/30) and probably is of little interest.

Fleiss, Levin, and Paik (2003, 128) report infant mortality by birthweight for 72,730 live white births in 1974 in New York City:

```
. csi 618 422 4597 67093
```

	Exposed	Unexposed	Total
Cases	618	422	1040
Noncases	4597	67093	71690
Total	5215	67515	72730
Risk	.1185043	.0062505	.0142995

	Point estimate	[95% Conf. Interval]	
Risk difference	.1122539	.1034617	.121046
Risk ratio	18.95929	16.80661	21.38769
Attr. frac. ex.	.9472554	.9404996	.9532441
Attr. frac. pop	.5628883		

```
                    chi2(1) =  4327.92  Pr>chi2 = 0.0000
```

In these data, exposed means a premature baby (birthweight ≤2,500 g), and a case is a baby who is dead at the end of one year. We find that being premature accounts for 94.7% of deaths among the premature population. We also estimate, paraphrasing from Fleiss, Levin, and Paik (2003, 128), that 56.3% of all white infant deaths in New York City in 1974 could have been prevented if prematurity had been eliminated. (Moreover, Fleiss, Levin, and Paik put a standard error on the attributable fraction for the population. The formula is given in *Methods and formulas* but is appropriate only for the population on which the estimates are based because other populations may have different probabilities of exposure.)

❑

▷ Example 7: cs

`cs` works like `csi`, except that it obtains its information from the data. The data equivalent to typing `csi 7 12 9 2` are

```
. use http://www.stata-press.com/data/r11/csxmpl, clear
. list
```

	case	exp	pop
1.	1	1	7
2.	1	0	12
3.	0	1	9
4.	0	0	2

We could then type `cs case exp [freq=pop]`. If we had individual-level data, so that each observation reflected a patient and we had 30 observations, we would type `cs case exp`.

◁

Stratified cumulative incidence data

▷ Example 8: cs with stratified data

Rothman, Greenland, and Lash (2008, 260) reprint the following age-specific information for deaths from all causes for tolbutamide and placebo treatment groups (University Group Diabetes Program 1970):

	Age through 54		Age 55 and above	
	Tolbutamide	Placebo	Tolbutamide	Placebo
Dead	8	5	22	16
Surviving	98	115	76	69

The data corresponding to these results are

```
. use http://www.stata-press.com/data/r11/ugdp
. list
```

	age	case	exposed	pop
1.	<55	0	0	115
2.	<55	0	1	98
3.	<55	1	0	5
4.	<55	1	1	8
5.	55+	0	0	69
6.	55+	0	1	76
7.	55+	1	0	16
8.	55+	1	1	22

The order of the observations is unimportant. If we were now to type `cs case exposed [freq=pop]`, we would obtain a summary for all the data, ignoring the stratification by age. To incorporate the stratification, we type

```
. cs case exposed [freq=pop], by(age)
```

Age category	RR	[95% Conf. Interval]		M-H Weight
<55	1.811321	.6112044	5.367898	2.345133
55+	1.192602	.6712664	2.11883	8.568306
Crude	1.435574	.8510221	2.421645	
M-H combined	1.325555	.797907	2.202132	

Test of homogeneity (M-H)   chi2(1) =   0.447   Pr>chi2 = 0.5037

Mantel–Haenszel weights are appropriate when the risks may differ according to the strata but the risk ratio is believed to be the same (homogeneous across strata). Under these assumptions, Mantel–Haenszel weights are designed to use the information efficiently. They are not intended to measure a composite risk ratio when the within-stratum risk ratios differ. Then we want a standardized ratio (see below).

The risk ratios above appear to differ markedly, but the confidence intervals are also broad because of the small sample sizes. The test of homogeneity shows that the differences can be attributed to chance; the use of the Mantel–Haenszel combined test is sensible.

◁

❑ **Technical note**

Stratified cumulative incidence tables are not the only way to control for confounding. Another way is logistic regression. However, logistic regression measures effects with odds ratios, not with risk ratios. So before we fit a logistic model, let's use `cs` to estimate the Mantel–Haenszel odds ratio:

```
. cs case exposed [freq=pop], by(age) or
```

Age category	OR	[95% Conf. Interval]	M-H Weight	
<55	1.877551	.6238165 5.637046	2.168142	(Cornfield)
55+	1.248355	.6112772 2.547411	6.644809	(Cornfield)
Crude	1.510673	.8381198 2.722012		
M-H combined	1.403149	.7625152 2.582015		

```
Test of homogeneity (M-H)       chi2(1) =    0.347  Pr>chi2 = 0.5556
                 Test that combined OR = 1:
                            Mantel-Haenszel chi2(1) =     1.19
                                           Pr>chi2 =   0.2750
```

The Mantel–Haenszel odds ratio is 1.40. It measures the association between death and treatment while adjusting for age. A more general way to adjust for age is logistic regression; the outcome variable is `case`, and it is explained by `age` and `exposed`. (As in the incidence-rate example, `age` may appear to be a string variable in our data—we listed the data in the previous example—but it is actually a numeric variable taking on values 0 and 1 with value labels disguising that fact; see [U] **12.6.3 Value labels**.)

```
. logistic case exposed age [freq=pop]
Logistic regression                             Number of obs   =        409
                                                LR chi2(2)      =      22.47
                                                Prob > chi2     =     0.0000
Log likelihood = -142.6212                      Pseudo R2       =     0.0730
```

case	Odds Ratio	Std. Err.	z	P>|z|	[95% Conf. Interval]
exposed	1.404674	.4374454	1.09	0.275	.7629451 2.586175
age	4.216299	1.431519	4.24	0.000	2.167361 8.202223

Compare these results with the Mantel–Haenszel estimates obtained with `cs`:

Source	Odds Ratio	95% Conf. Int.	
Mantel–Haenszel (cs)	1.40	0.76	2.58
logistic	1.40	0.76	2.59

They are virtually identical.

Logistic regression has advantages over the stratified-table approach. First, we obtained an estimate of the age effect: being 55 years or over significantly increases the odds of death. In addition to the point estimate, 4.22, we have a confidence interval for the effect: 2.17 to 8.20.

A discrete effect at age 55 is not a plausible model of aging. It would be more reasonable to assume that a 54-year-old patient has a higher probability of death, due merely to age, than does a 53-year-old patient; a 53-year-old, a higher probability than a 52-year-old patient; and so on. If we had the underlying data, where each patient's age is presumably known, we could include the actual age in the model and so better control for the age effect. This would improve our estimate of the effect of being exposed to tolbutamide.

See [R] **logistic** for an explanation of the `logistic` command. Also see the technical note in *Stratified incidence-rate data* concerning categorical variables, which applies to logistic regression as well as Poisson regression.

❏

Standardized estimates with stratified cumulative incidence data

As with `ir`, `cs` can produce standardized estimates, and the method is basically the same, although the options for which estimates are to be combined or standardized make it confusing. We showed above that `cs` can produce Mantel–Haenszel weighted estimates of the risk ratio (the default) or the odds ratio (obtained by specifying `or`). `cs` can also produce standardized estimates of the risk ratio (the default) or the risk difference (obtained by specifying `rd`).

▷ Example 9: cs with stratified data, using standardized estimates

To produce an estimate of the internally standardized risk ratio by using our age-specific data on deaths from all causes for tolbutamide and placebo treatment groups (example above), we type

```
. cs case exposed [freq=pop], by(age) istandard
```

Age category	RR	[95% Conf. Interval]		Weight
<55	1.811321	.6112044	5.367898	106
55+	1.192602	.6712664	2.11883	98
Crude	1.435574	.8510221	2.421645	
I. Standardized	1.312122	.7889772	2.182147	

We could obtain externally standardized estimates by substituting `estandard` for `istandard`.

To produce an estimate of the risk ratio weighting each age category equally, we could type

```
. generate wgt=1
. cs case exposed [freq=pop], by(age) standard(wgt)
```

Age category	RR	[95% Conf. Interval]		Weight
<55	1.811321	.6112044	5.367898	1
55+	1.192602	.6712664	2.11883	1
Crude	1.435574	.8510221	2.421645	
Standardized	1.304737	.7844994	2.169967	

If we instead wanted the risk difference, we would type

```
. cs case exposed [freq=pop], by(age) standard(wgt) rd
```

Age category	RD	[95% Conf. Interval]		Weight
<55	.033805	-.0278954	.0955055	1
55+	.0362545	-.0809204	.1534294	1
Crude	.0446198	-.0192936	.1085332	
Standardized	.0350298	-.0311837	.1012432	

If we wanted to weight the less-than-55 age group five times as heavily as the 55-and-over group, we would create `wgt` to contain 5 for the first age group and 1 for the second (or 10 for the first group and 2 for the second—the scale of the weights does not matter).

◁

Case–control data

In case–control data, you select a sample on the basis of the outcome under study; i.e., cases and noncases are sampled at different rates. If you were examining the link between coffee consumption and heart attacks, for instance, you could select a sample of subjects with and without the heart problem and then examine their coffee-drinking behavior. A subject who has suffered a heart attack is called a *case* just as with cohort study data. A subject who has never suffered a heart attack, however, is called a *control* rather than merely a noncase, emphasizing that the sampling was performed with respect to the outcome.

In case–control data, all hope of identifying the risk (i.e., incidence) of the outcome (heart attacks) associated with the factor (coffee drinking) vanishes, at least without information on the underlying sampling fractions, but you can examine the proportion of coffee drinkers among the two populations and reason that, if there is a difference, coffee drinking may be associated with the risk of heart attacks. Remarkably, even without the underlying sampling fractions, you can also measure the ratio of the odds of heart attacks if a subject drinks coffee to the odds if a subject does not—the so-called odds ratio.

What is lost is the ability to compare absolute rates, which is not always the same as comparing relative rates; see Fleiss, Levin, and Paik (2003, 123).

▷ Example 10: cci

cci calculates the odds ratio and the attributable risk associated with a 2×2 table. Rothman et al. (1979; reprinted in Rothman [1986, 161], and Rothman, Greenland, and Lash [2008, 251]) present case–control data on the history of chlordiazopoxide use in early pregnancy for mothers of children born with and without congenital heart defects:

	Chlordiazopoxide use	
	Yes	No
Case mothers	4	386
Control mothers	4	1250

```
. cci 4 386 4 1250, level(90)
```

	Exposed	Unexposed	Total	Proportion Exposed
Cases	4	386	390	0.0103
Controls	4	1250	1254	0.0032
Total	8	1636	1644	0.0049
	Point estimate		[90% Conf. Interval]	
Odds ratio	3.238342		.7698467	13.59664 (exact)
Attr. frac. ex.	.6912		-.2989599	.9264524 (exact)
Attr. frac. pop	.0070892			

chi2(1) = 3.07 Pr>chi2 = 0.0799

We obtain a point estimate of the odds ratio as 3.24 and a χ^2 value, which is a test that the odds ratio is 1, significant at the 10% level.

◁

❑ Technical note

The epitab commands can calculate four different confidence intervals for the odds ratio: the exact, Woolf, Cornfield, and test-based intervals. The exact interval, illustrated in example 10, is the default. The interval is "exact" because it uses an exact sampling distribution—a distribution with no unknown parameters under the null hypothesis. An exact interval does not use a normal or chi-squared approximation. "Exact" does not describe the coverage probability; the coverage probability of a 90% exact interval is not exactly 90%. The coverage probability is actually bounded below by 90% (Agresti 2002, 99), so a 90% exact interval will always cover the odds ratio with probability at least 90% (if the model is correct).

The Woolf, Cornfield, and test-based intervals, on the other hand, are approximate. They approximate the exact sampling distribution with a normal model and are not guaranteed to maintain their nominal coverage: the coverage probability of a 90% approximate interval fluctuates above and below 90%. The coverage approaches 90% only in the limit as the sample size increases. Exact intervals are conservative; approximate intervals can be conservative or anticonservative (Agresti 2002, 100).

If you wish to maintain nominal coverage, then you should use the exact interval. But you will pay a price for the coverage: the exact interval will usually be wider than the approximate intervals. Example 10 is no exception:

Method	90% Conf. Int.	Command
exact	0.77 13.60	cci
Woolf	1.01 10.40	cci, woolf
test-based	1.07 9.77	cci, tb
Cornfield	1.07 9.83	cci, cornfield

The exact interval is the widest of the four—so wide that it includes the null value of one—even though the chi-squared p-value of 0.0799 was significant at the 10% level. The exact interval and chi-squared test come from different models, so we should not expect them to always agree on sharp conclusions such as statistical significance.

The odds-ratio intervals are all frequentist methods, so we cannot compare them rigorously with one example. See Brown (1981), Gart and Thomas (1982), and Agresti (1999) for more rigorous comparisons. Agresti (1999) found that the Woolf interval performed well, even for small samples.

❑

❑ Technical note

By default, cc and cci report exact confidence intervals but an approximate significance test. You can replace the approximate test with Fisher's exact test by specifying the exact option. We recommend specifying exact whenever any cell count is less than 1,000.

(Continued on next page)

```
. cci 4 386 4 1250, exact level(90)
                                                        Proportion
                   |   Exposed     Unexposed  |   Total   Exposed
             ------+---------------------------+--------------------
             Cases |         4           386   |     390    0.0103
          Controls |         4          1250   |    1254    0.0032
             ------+---------------------------+--------------------
             Total |         8          1636   |    1644    0.0049

                   |  Point estimate   |  [90% Conf. Interval]
             ------+-------------------+-------------------------
        Odds ratio |        3.238342   |   .7698467    13.59664  (exact)
     Attr. frac. ex.|          .6912   |  -.2989599    .9264524  (exact)
    Attr. frac. pop |       .0070892   |
                                        1-sided Fisher's exact P = 0.0964
                                        2-sided Fisher's exact P = 0.0964
```

In this table, the one- and two-sided significance values are equal. This is not a mistake, but it does not happen often. Exact significance values are calculated by summing the probabilities for tables that have the same marginals (row and column sums) but that are less likely (given an odds ratio of 1) than the observed table. When considering each possible table, we might ask if the table is in the same or opposite tail as the observed table. If it is in the same tail, we would count the table under consideration in the one-sided test and, either way, we would count it in the two-sided test. Here all the tables more extreme than this table are in the same tail, so the one- and two-sided tests are the same.

The *p*-value of 0.0964 is significant at the 10% level, but the exact confidence interval is not (it includes the null odds ratio of one). It was not surprising that the exact interval disagreed with the chi-squared test; after all, they come from different models. Now the exact interval and Fisher's exact test also disagree, even though they come from the same model!

The test and interval disagree because the exact sampling distribution is asymmetric, and the test and interval handle the asymmetry differently. The two-sided test, as we have seen, sums the probabilities of all tables at least as unlikely as the observed table, and in example 10, all the unlikely tables fall in the same tail of the distribution. The other tail does not contribute to the *p*-value. The exact interval, on the other hand, must always use both tails of the distribution, because the interval inverts two one-sided tests, not one two-sided test (Breslow and Day 1980, 128–129).

❑ Technical note

The reported value of the attributable or prevented fraction among the exposed is calculated using the odds ratio as a proxy for the risk ratio. This can be justified only if the outcome is rare in the population. The extrapolation to the attributable or prevented fraction for the population assumes that the control group is a random sample of the corresponding group in the underlying population.

▷ Example 11: cc equivalent to cci

Equivalent to typing `cci 4 386 4 1250` would be typing `cc case exposed [freq=pop]` with the following data:

```
. use http://www.stata-press.com/data/r11/ccxmpl, clear
. list

     case   exposed   pop
1.     1       1        4
2.     1       0      386
3.     0       1        4
4.     0       0     1250
```

◁

Stratified case–control data

▷ Example 12: cc with stratified data

cc can work with stratified tables. Rothman, Greenland, and Lash (2008, 276) reprint and discuss data from a case–control study on infants with congenital heart disease and Down syndrome and healthy controls, according to maternal spermicide use before conception and maternal age at delivery (Rothman 1982):

	Maternal age to 34		Maternal age 35+	
	Spermicide used	not used	Spermicide used	not used
Down syndrome	3	9	1	3
Controls	104	1059	5	86

The data corresponding to these tables are

```
. use http://www.stata-press.com/data/r11/downs
. list

     case   exposed   pop    age
1.     1       1        3    <35
2.     1       0        9    <35
3.     0       1      104    <35
4.     0       0     1059    <35
5.     1       1        1    35+

6.     1       0        3    35+
7.     0       1        5    35+
8.     0       0       86    35+
```

The stratified results for the odds ratio are

```
. cc case exposed [freq=pop], by(age) woolf

     Maternal age |     OR      [95% Conf. Interval]    M-H Weight
     -------------+--------------------------------------------------
              <35 | 3.394231    .9048403   12.73242    .7965957 (Woolf)
              35+ | 5.733333    .5016418   65.52706    .1578947 (Woolf)
     -------------+--------------------------------------------------
            Crude | 3.501529    1.110362   11.04208              (Woolf)
     M-H combined | 3.781172    1.18734    12.04142
     -------------+--------------------------------------------------
     Test of homogeneity (M-H)       chi2(1) =    0.14   Pr>chi2 = 0.7105

                     Test that combined OR = 1:
                                Mantel-Haenszel chi2(1) =     5.81
                                                Pr>chi2 =   0.0159
```

For no particular reason, we also specified the `woolf` option to obtain Woolf approximations to the within-stratum confidence intervals rather than the default. Had we wanted test-based confidence intervals and Tarone's test of homogeneity, we would have used

```
. cc case exposed [freq=pop], by(age) tb tarone
    Maternal age |       OR      [95% Conf. Interval]     M-H Weight
  ---------------+------------------------------------------------------
             <35 |   3.394231     .976611   11.79672      .7965957 (tb)
             35+ |   5.733333     .6402941  51.33752      .1578947 (tb)
  ---------------+------------------------------------------------------
           Crude |   3.501529    1.189946   10.30358                (tb)
    M-H combined |   3.781172    1.282056   11.15183                (tb)
  ---------------+------------------------------------------------------
  Test of homogeneity (M-H)       chi2(1) =     0.14    Pr>chi2 = 0.7105
  Test of homogeneity (Tarone)    chi2(1) =     0.14    Pr>chi2 = 0.7092

               Test that combined OR = 1:
                            Mantel-Haenszel chi2(1) =      5.81
                                            Pr>chi2 =    0.0159
```

We recommend that test-based confidence intervals be used only for pedagogical reasons and never for research work.

Whatever method you choose for calculating confidence intervals, Stata will report a test of homogeneity, which here is $\chi^2(1) = 0.14$ and not significant. That is, the odds of Down syndrome might vary with maternal age, but we cannot reject the hypothesis that the association between Down syndrome and spermicide is the same in the two maternal age strata. This is thus a test to reject the appropriateness of the single, Mantel–Haenszel combined odds ratio—a rejection not justified by these data.

◁

❑ Technical note

The `cc` command includes four tests of homogeneity: Mantel–Haenszel (the default); directly pooled, also known as the Woolf test (available with the `pool` option); Tarone (available with the `tarone` option); and Breslow–Day (available with the `bd` option). The preferred test is Tarone's (Tarone 1985, 94), which corrected an error in the Breslow–Day test; see Breslow (1996, 17–18) for details of the error and Tarone's correction.

The other two homogeneity tests, the Mantel–Haenszel and directly pooled, are less useful: they use the logs of the stratum-specific odds ratios, so they are undefined when any stratum has a zero cell. The epitab commands deal with the problem differently: `cs` omits the offending strata, while `cc` substitutes the Tarone test. The Tarone test does not use the stratum-specific odds ratios, so it can still be calculated when there are zero cells.

None of the tests are appropriate for finely-stratified (many strata with only a few observations each) studies (Rothman, Greenland, and Lash 2008, 280). If you have fine stratification, one alternative is multilevel logistic regression; see [XT] **xtmelogit**.

❑

❑ Technical note

As with cohort study data, an alternative to stratified tables for uncovering effects is logistic regression. From the logistic point of view, case–control data are no different from cohort study data—you must merely ignore the estimated intercept, which is not reported by `logistic` in any case. (`logit`, on the other hand, makes the same estimates as `logistic` but displays the coefficients rather than transforming them to odds ratios, so it does display the estimated intercept. The intercept is meaningless in case–control data because it reflects the baseline prevalence of the outcome, which you controlled by sampling.)

The data we used with cc can be used directly by logistic. (The age variable, which appears to be a string, is really numeric with an associated value label; see [U] **12.6.3 Value labels**. age takes on the value 0 for the age-less-than-35 group and 1 for the 35+ group.)

```
. logistic case exposed age [freq=pop]
Logistic regression                             Number of obs   =       1270
                                                LR chi2(2)      =       8.74
                                                Prob > chi2     =     0.0127
Log likelihood = -81.517532                     Pseudo R2       =     0.0509
```

case	Odds Ratio	Std. Err.	z	P>|z|	[95% Conf. Interval]
exposed	3.787779	2.241922	2.25	0.024	1.187334	12.0836
age	4.582857	2.717351	2.57	0.010	1.433594	14.65029

We compare the results with those presented by cc in the previous example:

Source	Odds ratio	95% CI	
Mantel–Haenszel (cc)	3.78	1.19	12.04
logistic	3.79	1.19	12.08

As with the cohort study data in example 8, results are virtually identical, and all the same comments we made previously apply once again.

To demonstrate an advantage of logistic regression, let's now ask a question that would be difficult to answer on the basis of a stratified table analysis. We now know that spermicide use appears to increase the risk of having a baby with Down syndrome, and we know that the mother's age also increases the risk. Is the effect of spermicide use statistically different for mothers in the two age groups?

```
. logistic case exposed age c.age#exposed [freq=pop]
Logistic regression                             Number of obs   =       1270
                                                LR chi2(3)      =       8.87
                                                Prob > chi2     =     0.0311
Log likelihood = -81.451332                     Pseudo R2       =     0.0516
```

case	Odds Ratio	Std. Err.	z	P>|z|	[95% Conf. Interval]
exposed	3.394231	2.289544	1.81	0.070	.9048403	12.73242
age	4.104651	2.774868	2.09	0.037	1.091034	15.44237
exposed# c.age 1	1.689141	2.388785	0.37	0.711	.1056563	27.0045

The answer is no. The odds ratio and confidence interval reported for exposed now measure the spermicide effect for an age==0 (age < 35) mother. The odds ratio and confidence interval reported for c.age#exposed are the (multiplicative) difference in the spermicide odds ratio for an age==1 (age 35+) mother relative to a young mother. The point estimate is that the effect is larger for older mothers, suggesting grounds for future research, but the difference is not significant.

See [R] **logistic** for an explanation of the logistic command. Also see the technical note under *Incidence-rate data* above concerning Poisson regression, which applies equally to logistic regression.

❑

Case–control data with multiple levels of exposure

In a case–control study, subjects with the disease of interest (cases) are compared to disease-free individuals (controls) to assess the relationship between exposure to one or more risk factors and disease incidence. Often exposure is measured qualitatively at several discrete levels or measured on a continuous scale and then grouped into three or more levels. The data can be summarized as

	\multicolumn{5}{c}{Exposure level}
	1	2	...	k	Total
Cases	a_1	a_2	...	a_k	M_1
Controls	c_1	c_2	...	c_k	M_0
Total	N_1	N_2	...	N_k	T

An advantage afforded by having multiple levels of exposure is the ability to examine dose–response relationships. If the association between a risk factor and a disease or outcome is real, we expect the strength of that association to increase with the level and duration of exposure. A dose–response relationship provides strong support for a direct or even causal relationship between the risk factor and the outcome. On the other hand, the lack of a dose–response is usually seen as an argument against causality.

We can use the `tabodds` command to tabulate the odds of failure or odds ratios against a categorical exposure variable. The test for trend calculated by `tabodds` can serve as a test for dose–response if the exposure variable is at least ordinal. If the exposure variable has no natural ordering, the trend test is meaningless and should be ignored. See the technical note at the end of this section for more information regarding the test for trend.

Before looking at an example, consider three possible data arrangements for case–control and prevalence studies. The most common data arrangement is individual records, where each subject in the study has his or her own record. Closely related are frequency records where identical individual records are included only once, but with a variable giving the frequency with which the record occurs. The `fweight` *weight* option is used for these data to specify the frequency variable. Data can also be arranged as binomial frequency records where each record contains a variable, D, the number of cases; another variable, N, the number of total subjects (cases plus controls); and other variables. An advantage of binomial frequency records is that large datasets can be entered succinctly into a Stata database.

▷ Example 13: tabodds

Consider the following data from the Ille-et-Vilaine study of esophageal cancer, discussed in Breslow and Day (1980, chap. 4 and app. I), corresponding to subjects age 55–64 who use from 0 to 9 g of tobacco per day:

	\multicolumn{5}{c}{Alcohol consumption (g/day)}
	0–39	40–79	80–119	120+	Total
Cases	2	9	9	5	25
Controls	47	31	9	5	92
Total	49	40	18	10	117

The study included 24 such tables, each representing one of four levels of tobacco use and one of six age categories. We can create a binomial frequency-record dataset by typing

```
. input alcohol D N agegrp tobacco
       alcohol        D         N       agegrp     tobacco
  1.         1        2        49            4           1
  2.         2        9        40            4           1
  3.         3        9        18            4           1
  4.         4        5        10            4           1
  5. end
```

where D is the number of esophageal cancer cases and N is the number of total subjects (cases plus controls) for each combination of six age groups (agegrp), four levels of alcohol consumption in g/day (alcohol), and four levels of tobacco use in g/day (tobacco).

Both the tabodds and mhodds commands can correctly handle all three data arrangements. Binomial frequency records require that the number of total subjects (cases plus controls) represented by each record N be specified with the binomial() option.

We could also enter the data as frequency-weighted data:

```
. input alcohol case freq agegrp tobacco
       alcohol      case       freq      agegrp     tobacco
  1.         1         1          2           4           1
  2.         1         0         47           4           1
  3.         2         1          9           4           1
  4.         2         0         31           4           1
  5.         3         1          9           4           1
  6.         3         0          9           4           1
  7.         4         1          5           4           1
  8.         4         0          5           4           1
  9. end
```

If you are planning on using any of the other estimation commands, such as poisson or logistic, we recommend that you enter your data either as individual records or as frequency-weighted records and not as binomial frequency records, because the estimation commands currently do not recognize the binomial() option.

We have entered all the esophageal cancer data into Stata as a frequency-weighted record dataset as previously described. In our data, case indicates the esophageal cancer cases and controls, and freq is the number of subjects represented by each record (the weight).

We added value labels to the agegrp, alcohol, and tobacco variables in our dataset to ease interpretation in outputs, but these variables are numeric.

We are interested in the association between alcohol consumption and esophageal cancer. We first use tabodds to tabulate the odds of esophageal cancer against alcohol consumption:

(*Continued on next page*)

```
. use http://www.stata-press.com/data/r11/bdesop, clear
. tabodds case alcohol [fweight=freq]
```

alcohol	cases	controls	odds	[95% Conf. Interval]	
0-39	29	386	0.07513	0.05151	0.10957
40-79	75	280	0.26786	0.20760	0.34560
80-119	51	87	0.58621	0.41489	0.82826
120+	45	22	2.04545	1.22843	3.40587

```
Test of homogeneity (equal odds):   chi2(3)  =    158.79
                                    Pr>chi2  =    0.0000
Score test for trend of odds:       chi2(1)  =    152.97
                                    Pr>chi2  =    0.0000
```

The test of homogeneity clearly indicates that the odds of esophageal cancer differ by level of alcohol consumption, and the test for trend indicates a significant increase in odds with increasing alcohol use. This suggests a strong dose–response relation. The graph option can be used to study the shape of the relationship of the odds with alcohol consumption. Most of the heterogeneity in these data can be "explained" by the linear increase in risk of esophageal cancer with increased dosage (alcohol consumption).

We also could have requested that the odds ratios at each level of alcohol consumption be calculated by specifying the or option. For example, tabodds case alcohol [fweight=freq], or would produce odds ratios using the minimum value of alcohol—i.e., alcohol = 1 (0–39)—as the reference group, and the command tabodds case alcohol [fweight=freq], or base(2) would use alcohol = 2 (40–79) as the reference group.

Although our results appear to provide strong evidence supporting an association between alcohol consumption and esophageal cancer, we need to be concerned with the possible existence of confounders, specifically age and tobacco use, in our data. We can again use tabodds to tabulate the odds of esophageal cancer against age and against tobacco use, independently:

```
. tabodds case agegrp [fweight=freq]
```

agegrp	cases	controls	odds	[95% Conf. Interval]	
25-34	1	115	0.00870	0.00121	0.06226
35-44	9	190	0.04737	0.02427	0.09244
45-54	46	167	0.27545	0.19875	0.38175
55-64	76	166	0.45783	0.34899	0.60061
65-74	55	106	0.51887	0.37463	0.71864
75+	13	31	0.41935	0.21944	0.80138

```
Test of homogeneity (equal odds):   chi2(5)  =     96.94
                                    Pr>chi2  =    0.0000
Score test for trend of odds:       chi2(1)  =     83.37
                                    Pr>chi2  =    0.0000
```

```
. tabodds case tobacco [fweight=freq]
```

tobacco	cases	controls	odds	[95% Conf. Interval]
0-9	78	447	0.17450	0.13719 0.22194
10-19	58	178	0.32584	0.24228 0.43823
20-29	33	99	0.33333	0.22479 0.49428
30+	31	51	0.60784	0.38899 0.94983

```
Test of homogeneity (equal odds):  chi2(3) =    29.33
                                   Pr>chi2 =   0.0000
Score test for trend of odds:      chi2(1) =    26.93
                                   Pr>chi2 =   0.0000
```

We can see that there is evidence to support our concern that both age and tobacco use are potentially important confounders. Clearly, before we can make any statements regarding the association between esophageal cancer and alcohol use, we must examine and, if necessary, adjust for the effect of any confounder. We will return to this example in the following section.

◁

❑ Technical note

The score test for trend performs a test for linear trend of the log odds against the numerical code used for the exposure variable. The test depends not only on the relationship between dose level and the outcome but also on the numeric values assigned to each level or, to be more accurate, to the distance between the numeric values assigned. For example, the trend test on a dataset with four exposure levels coded 1, 2, 3, and 4 gives the same results as coding the levels 10, 20, 30, and 40 because the distance between the levels in each case is constant. In the first case, the distance is one unit, and in the second case, it is 10 units. However, if we code the exposure levels as 1, 10, 100, and 1,000, we would obtain different results because the distance between exposure levels is not constant. Thus be careful when assigning values to exposure levels. You must determine whether equally spaced numbers make sense for your data or if other more meaningful values should be used.

Remember that we are testing whether a log-linear relationship exists between the odds and the exposure variable. For your particular problem, this relationship may not be correct or even make sense, so you must be careful in interpreting the output of this trend test.

❑

Case–control data with confounders and possibly multiple levels of exposure

In the esophageal cancer data example introduced earlier, we determined that the apparent association between alcohol consumption and esophageal cancer could be confounded by age and tobacco use. You can adjust for the effect of possible confounding factors by stratifying on these factors. This is the method used by both tabodds and mhodds to adjust for other variables in the dataset. We will compare and contrast these two commands in the following example.

▷ Example 14: tabodds, adjusting for confounding factors

We begin by using tabodds to tabulate unadjusted odds ratios.

```
. tabodds case alcohol [fweight=freq], or
```

alcohol	Odds Ratio	chi2	P>chi2	[95% Conf. Interval]	
0-39	1.000000	.	.	.	.
40-79	3.565271	32.70	0.0000	2.237981	5.679744
80-119	7.802616	75.03	0.0000	4.497054	13.537932
120+	27.225705	160.41	0.0000	12.507808	59.262107

```
Test of homogeneity (equal odds):  chi2(3) =  158.79
                                   Pr>chi2 =  0.0000
Score test for trend of odds:      chi2(1) =  152.97
                                   Pr>chi2 =  0.0000
```

The alcohol = 1 group (0–39) was used by tabodds as the reference category for calculating the odds ratios. We could have selected a different group by specifying the base() option; however, because the lowest dosage level is most often the appropriate reference group, as it is in these data, the base() option is seldom used.

We use tabodds with the adjust() option to tabulate Mantel–Haenszel age-adjusted odds ratios:

```
. tabodds case alcohol [fweight=freq], adjust(age)
Mantel-Haenszel odds ratios adjusted for age
```

alcohol	Odds Ratio	chi2	P>chi2	[95% Conf. Interval]	
0-39	1.000000	.	.	.	.
40-79	4.268155	37.36	0.0000	2.570025	7.088314
80-119	8.018305	59.30	0.0000	4.266893	15.067922
120+	28.570426	139.70	0.0000	12.146409	67.202514

```
Score test for trend of odds:  chi2(1) =  135.09
                               Pr>chi2 =  0.0000
```

We observe that the age-adjusted odds ratios are just slightly higher than the unadjusted ones, so it appears that age is not as strong a confounder as it first appeared. Even after adjusting for age, the dose–response relationship, as measured by the trend test, remains strong.

We now perform the same analysis but this time adjust for tobacco use instead of age.

```
. tabodds case alcohol [fweight=freq], adjust(tobacco)
Mantel-Haenszel odds ratios adjusted for tobacco
```

alcohol	Odds Ratio	chi2	P>chi2	[95% Conf. Interval]	
0-39	1.000000	.	.	.	.
40-79	3.261178	28.53	0.0000	2.059764	5.163349
80-119	6.771638	62.54	0.0000	3.908113	11.733306
120+	19.919526	123.93	0.0000	9.443830	42.015528

```
Score test for trend of odds:  chi2(1) =  135.04
                               Pr>chi2 =  0.0000
```

Again we observe a significant dose–response relationship and not much difference between the adjusted and unadjusted odds ratios. We could also adjust for the joint effect of both age and tobacco use by specifying adjust(tobacco age), but we will not bother here.

◁

A different approach to analyzing these data is to use the mhodds command. This command estimates the ratio of the odds of failure for two categories of an exposure variable, controlling for any specified confounding variables, and it tests whether this odds ratio is equal to one. For multiple exposures, if two exposure levels are not specified with compare(), then mhodds assumes that exposure is quantitative and calculates a 1-degree-of-freedom test for trend. This test for trend is the same one that tabodds reports.

▷ Example 15: mhodds, controlling for confounding factors

We first use mhodds to estimate the effect of alcohol controlled for age:

```
. mhodds case alcohol agegrp [fweight=freq]
Score test for trend of odds with alcohol
controlling for agegrp
(The Odds Ratio estimate is an approximation to the odds ratio
for a one unit increase in alcohol)
```

Odds Ratio	chi2(1)	P>chi2	[95% Conf. Interval]	
2.845895	135.09	0.0000	2.385749	3.394792

Because alcohol has more than two levels, mhodds estimates and reports an approximate age-adjusted odds ratio for a one-unit increase in alcohol consumption. The χ^2 value reported is identical to that reported by tabodds for the score test for trend on the previous page.

We now use mhodds to estimate the effect of alcohol controlled for age, and while we are at it, we do this by levels of tobacco consumption:

```
. mhodds case alcohol agegrp [fweight=freq], by(tobacco)
Score test for trend of odds with alcohol
controlling for agegrp
by tobacco
note: only 19 of the 24 strata formed in this analysis contribute
      information about the effect of the explanatory variable
(The Odds Ratio estimate is an approximation to the odds ratio
for a one unit increase in alcohol)
```

tobacco	Odds Ratio	chi2(1)	P>chi2	[95% Conf. Interval]	
0-9	3.579667	75.95	0.0000	2.68710	4.76871
10-19	2.303580	25.77	0.0000	1.66913	3.17920
20-29	2.364135	13.27	0.0003	1.48810	3.75589
30+	2.217946	8.84	0.0029	1.31184	3.74992

Mantel-Haenszel estimate controlling for agegrp and tobacco

Odds Ratio	chi2(1)	P>chi2	[95% Conf. Interval]	
2.751236	118.37	0.0000	2.292705	3.301471

Test of homogeneity of ORs (approx): chi2(3) =   5.46
                                     Pr>chi2 = 0.1409

Again, because alcohol has more than two levels, mhodds estimates and reports an approximate Mantel–Haenszel age and tobacco-use adjusted odds ratio for a one-unit increase in alcohol consumption. The χ^2 test for trend reported with the Mantel–Haenszel estimate is again the same one that tabodds produces if adjust(agegrp tobacco) is specified.

The results from this analysis also show an effect of alcohol, controlled for age, of about ×2.7, which is consistent across different levels of tobacco consumption. Similarly,

```
. mhodds case tobacco agegrp [fweight=freq], by(alcohol)
Score test for trend of odds with tobacco
controlling for agegrp
by alcohol
note: only 18 of the 24 strata formed in this analysis contribute
      information about the effect of the explanatory variable
(The Odds Ratio estimate is an approximation to the odds ratio
for a one unit increase in tobacco)
```

alcohol	Odds Ratio	chi2(1)	P>chi2	[95% Conf. Interval]	
0-39	2.420650	15.61	0.0001	1.56121	3.75320
40-79	1.427713	5.75	0.0165	1.06717	1.91007
80-119	1.472218	3.38	0.0659	0.97483	2.22339
120+	1.214815	0.59	0.4432	0.73876	1.99763

Mantel-Haenszel estimate controlling for agegrp and alcohol

Odds Ratio	chi2(1)	P>chi2	[95% Conf. Interval]	
1.553437	20.07	0.0000	1.281160	1.883580

```
Test of homogeneity of ORs (approx): chi2(3)  =    5.26
                                     Pr>chi2  =  0.1540
```

shows an effect of tobacco, controlled for age, of about ×1.5, which is consistent across different levels of alcohol consumption.

Comparisons between particular levels of alcohol and tobacco consumption can be made by generating a new variable with levels corresponding to all combinations of alcohol and tobacco, as in

```
. egen alctob = group(alcohol tobacco)
. mhodds case alctob [fweight=freq], compare(16,1)
Maximum likelihood estimate of the odds ratio
Comparing alctob==16 vs. alctob==1
```

Odds Ratio	chi2(1)	P>chi2	[95% Conf. Interval]	
93.333333	103.21	0.0000	14.766136	589.938431

which yields an odds ratio of 93 between subjects with the highest levels of alcohol and tobacco and those with the lowest levels. Similar results can be obtained simultaneously for all levels of alctob using alctob = 1 as the comparison group by specifying tabodds D alctob, bin(N) or.

◁

Standardized estimates with stratified case–control data

▷ Example 16: cc with stratified data, using standardized estimates

You obtain standardized estimates (here for the odds ratio) by using cc just as you obtain standardized estimates by using ir or cs. Along with the by() option, you specify one of estandard, istandard, or standard(*varname*).

Case–control studies can provide standardized rate-ratio estimates when density sampling is used, or when the disease is rare (Rothman, Greenland, and Lash 2008, 269). Rothman, Greenland, and Lash (2008, 276) report the SMR for the case–control study on infants with congenital heart disease and Down syndrome. We can reproduce their estimates along with the pooled estimates by typing

```
. use http://www.stata-press.com/data/r11/downs, clear
. cc case exposed [freq=pop], by(age) istandard pool
```

Maternal age	OR	[95% Conf. Interval]		Weight	
<35	3.394231	.5812415	13.87412	104	(exact)
35+	5.733333	.0911619	85.89602	5	(exact)
Crude	3.501529	.8080857	11.78958		(exact)
Pooled (direct)	3.824166	1.196437	12.22316		
I. Standardized	3.779749	1.180566	12.10141		

Test of homogeneity (direct)   chi2(1) =   0.14   Pr>chi2 = 0.7109

Using the distribution of the nonexposed subjects in the source population as the standard, we can obtain an estimate of the standardized rate ratio (SRR):

```
. cc case exposed [freq=pop], by(age) estandard
```

Maternal age	OR	[95% Conf. Interval]		Weight	
<35	3.394231	.5812415	13.87412	1059	(exact)
35+	5.733333	.0911619	85.89602	86	(exact)
Crude	3.501529	.8080857	11.78958		(exact)
E. Standardized	3.979006	1.176096	13.46191		

Finally, if we wanted to weight the two age groups equally, we could type

```
. generate wgt=1
. cc case exposed [freq=pop], by(age) standard(wgt)
```

Maternal age	OR	[95% Conf. Interval]		Weight	
<35	3.394231	.5812415	13.87412	1	(exact)
35+	5.733333	.0911619	85.89602	1	(exact)
Crude	3.501529	.8080857	11.78958		(exact)
Standardized	5.275104	.6233794	44.6385		

◁

Matched case–control data

Matched case–control studies are performed to gain sample-size efficiency and to control for important confounding factors. In a matched case–control design, each case is matched with a control on the basis of demographic characteristics, clinical characteristics, etc. Thus their difference with respect to the outcome must be due to something other than the matching variables. If the only difference between them was exposure to the factor, we could attribute any difference in outcome to the factor.

A summary of the data is

	Controls		
Cases	Exposed	Unexposed	Total
Exposed	a	b	M_1
Unexposed	c	d	M_0
Total	N_1	N_0	$T = a + b + c + d$

Each entry in the table represents the number of case–control pairs. For instance, in a of the pairs, both members were exposed; in b of the pairs, the case was exposed but the control was not; and so on. In total, T pairs were observed.

▷ Example 17: mcci

Rothman (1986, 257) discusses data from Jick et al. (1973) on a matched case–control study of myocardial infarction and drinking six or more cups of coffee per day (persons drinking from one to five cups per day were excluded):

	Controls	
Cases	6+ cups	0 cups
6+ cups	8	8
0 cups	3	8

mcci analyzes matched case–control data:

```
. mcci 8 8 3 8
                 |     Controls
          Cases  |  Exposed   Unexposed  |   Total
         -------+------------------------+--------
        Exposed |        8           8   |     16
      Unexposed |        3           8   |     11
         -------+------------------------+--------
          Total |       11          16   |     27

McNemar's chi2(1) =      2.27    Prob > chi2 = 0.1317
Exact McNemar significance probability      = 0.2266

Proportion with factor
        Cases        .5925926
        Controls     .4074074         [95% Conf. Interval]
                                   ---------------------
        difference   .1851852       -.0822542    .4526246
        ratio       1.454545         .891101    2.374257
        rel. diff.   .3125          -.0243688    .6493688

        odds ratio  2.666667         .6400364   15.6064    (exact)
```

The point estimate states that the odds of drinking 6 or more cups of coffee per day is 2.67 times greater among the myocardial infarction patients. The confidence interval is wide, however, and the p-value of 0.1317 from McNemar's test is not statistically significant.

◁

mcc works like the other nonimmediate commands but does not handle stratified data. If you have stratified matched case–control data, you can use conditional logistic regression to estimate odds ratios; see [R] **clogit**.

Matched case–control studies can also be analyzed using mhodds by controlling on the variable used to identify the matched sets. For example, if the variable set is used to identify the matched set for each subject,

. mhodds fail xvar set

will do the job. Any attempt to control for further variables will restrict the analysis to the comparison of cases and matched controls that share the same values of these variables. In general, this would lead to the omission of many records from the analysis. Similar considerations usually apply when investigating effect modification by using the by() option. An important exception to this rule is that a variable used in matching cases to controls may appear in the by() option without loss of data.

▷ Example 18: mhodds with matched case–control data

Let's use mhodds to analyze matched case–control studies using the study of endometrial cancer and exposure to estrogen described in Breslow and Day (1980, chap. 5). In this study, there are four controls matched to each case. Cases and controls are matched on age, marital status, and time living in the community. The data collected include information on the daily dose of conjugated estrogen therapy. Breslow and Day created four levels of the dose variable and began by analyzing the 1:1 study formed by using the first control in each set. We examine the effect of exposure to estrogen:

```
. use http://www.stata-press.com/data/r11/bdendo11, clear
. describe

Contains data from http://www.stata-press.com/data/r11/bdendo11.dta
  obs:           126
 vars:            13                          3 Mar 2009 23:29
 size:         2,898 (99.9% of memory free)

              storage  display     value
variable name   type   format      label       variable label

set             int    %8.0g                   Set number
fail            byte   %8.0g                   Case=1/Control=0
gall            byte   %8.0g                   Gallbladder dis
hyp             byte   %8.0g                   Hypertension
ob              byte   %8.0g                   Obesity
est             byte   %8.0g                   Estrogen
dos             byte   %8.0g                   Ordinal dose
dur             byte   %8.0g                   Ordinal duration
non             byte   %8.0g                   Non-estrogen drug
duration        int    %8.0g                   months
age             int    %8.0g                   years
cest            byte   %8.0g                   Conjugated est dose
agegrp          float  %9.0g                   age group of set

Sorted by:  set
```

```
. mhodds fail est set
Mantel-Haenszel estimate of the odds ratio
Comparing est==1 vs. est==0, controlling for set
note: only 32 of the 63 strata formed in this analysis contribute
      information about the effect of the explanatory variable
```

Odds Ratio	chi2(1)	P>chi2	[95% Conf. Interval]	
9.666667	21.12	0.0000	2.944702	31.733072

For the 1:1 matched study, the Mantel–Haenszel methods are equivalent to conditional likelihood methods. The maximum conditional likelihood estimate of the odds ratio is given by the ratio of the off-diagonal frequencies in the two-way (case–control) table below. The data must be in the 1-observation-per-group format; that is, the matched case and control must appear in 1 observation (the same format as required by the mcc command; see also [R] **clogit**).

```
. keep fail est set
. reshape wide est, i(set) j(fail)
(note: j = 0 1)
```

Data	long	->	wide
Number of obs.	126	->	63
Number of variables	3	->	3
j variable (2 values)	fail	->	(dropped)
xij variables:			
	est	->	est0 est1

```
. rename est1 case
. rename est0 control
. label variable case case
. label variable control control
. tabulate case control
```

	control		
case	0	1	Total
0	4	3	7
1	29	27	56
Total	33	30	63

The odds ratio is $29/3 = 9.67$, which agrees with the value obtained from mhodds. In the more general 1:m matched study, however, the Mantel–Haenszel methods are no longer equivalent to maximum conditional likelihood, although they are usually close.

To illustrate the use of the by() option in matched case–control studies, we look at the effect of exposure to estrogen, stratified by age3, which codes the sets into three age groups (55–64, 65–74, and 75+) as follows:

```
. use http://www.stata-press.com/data/r11/bdendo11, clear
. generate age3 = agegrp
. recode age3 1/2=1 3/4=2 5/6=3
(age3: 124 changes made)
```

```
. mhodds fail est set, by(age3)
Mantel-Haenszel estimate of the odds ratio
Comparing est==1 vs. est==0, controlling for set
by age3

note: only 32 of the 63 strata formed in this analysis contribute
      information about the effect of the explanatory variable
```

age3	Odds Ratio	chi2(1)	P>chi2	[95% Conf. Interval]	
1	6.000000	3.57	0.0588	0.72235	49.83724
2	15.000000	12.25	0.0005	1.98141	113.55557
3	8.000000	5.44	0.0196	1.00059	63.96252

```
Mantel-Haenszel estimate controlling for set and age3
```

Odds Ratio	chi2(1)	P>chi2	[95% Conf. Interval]	
9.666667	21.12	0.0000	2.944702	31.733072

```
Test of homogeneity of ORs (approx): chi2(2) =   0.41
                                     Pr>chi2 = 0.8128
```

There is no further loss of information when we stratify by age3 because age was one of the matching variables.

The full set of matched controls can be used in the same way. For example, the effect of exposure to estrogen is obtained (using the full dataset) with

```
. use http://www.stata-press.com/data/r11/bdendo, clear
. mhodds fail est set
Mantel-Haenszel estimate of the odds ratio
Comparing est==1 vs. est==0, controlling for set

note: only 58 of the 63 strata formed in this analysis contribute
      information about the effect of the explanatory variable
```

Odds Ratio	chi2(1)	P>chi2	[95% Conf. Interval]	
8.461538	31.16	0.0000	3.437773	20.826746

The effect of exposure to estrogen, stratified by age3, is obtained with

```
. generate age3 = agegrp
. recode age3 1/2=1 3/4=2 5/6=3
(age3: 310 changes made)
. mhodds fail est set, by(age3)
Mantel-Haenszel estimate of the odds ratio
Comparing est==1 vs. est==0, controlling for set
by age3

note: only 58 of the 63 strata formed in this analysis contribute
      information about the effect of the explanatory variable
```

age3	Odds Ratio	chi2(1)	P>chi2	[95% Conf. Interval]	
1	3.800000	3.38	0.0660	0.82165	17.57438
2	10.666667	18.69	0.0000	2.78773	40.81376
3	13.500000	9.77	0.0018	1.59832	114.02620

```
Mantel-Haenszel estimate controlling for set and age3
─────────────────────────────────────────────────────────
 Odds Ratio    chi2(1)      P>chi2     [95% Conf. Interval]
─────────────────────────────────────────────────────────
   8.461538     31.16       0.0000     3.437773   20.826746
─────────────────────────────────────────────────────────

  Test of homogeneity of ORs (approx): chi2(2) =   1.41
                                       Pr>chi2 = 0.4943
```

◁

Saved results

ir and iri save the following in r():

Scalars

r(p)	one-sided p-value	r(afe)	attributable (prev.) fraction among exposed
r(ird)	incidence-rate difference	r(lb_afe)	lower bound of CI for afe
r(lb_ird)	lower bound of CI for ird	r(ub_afe)	upper bound of CI for afe
r(ub_ird)	upper bound of CI for ird	r(afp)	attributable fraction for the population
r(irr)	incidence-rate ratio	r(chi2_mh)	Mantel–Haenszel heterogeneity χ^2 (ir only)
r(lb_irr)	lower bound of CI for irr	r(chi2_p)	pooled heterogeneity χ^2 (pool only)
r(ub_irr)	upper bound of CI for irr	r(df)	degrees of freedom (ir only)

cs and csi save the following in r():

Scalars

r(p)	two-sided p-value	r(ub_or)	upper bound of CI for or
r(rd)	risk difference	r(afe)	attributable (prev.) fraction among exposed
r(lb_rd)	lower bound of CI for rd	r(lb_afe)	lower bound of CI for afe
r(ub_rd)	upper bound of CI for rd	r(ub_afe)	upper bound of CI for afe
r(rr)	risk ratio	r(afp)	attributable fraction for the population
r(lb_rr)	lower bound of CI for rr	r(chi2_mh)	Mantel–Haenszel heterogeneity χ^2 (cs only)
r(ub_rr)	upper bound of CI for rr	r(chi2_p)	pooled heterogeneity χ^2 (pool only)
r(or)	odds ratio	r(df)	degrees of freedom
r(lb_or)	lower bound of CI for or	r(chi2)	χ^2

cc and cci save the following in r():

Scalars

r(p)	two-sided p-value	r(ub_afe)	upper bound of CI for afe
r(p1_exact)	χ^2 or one-sided exact significance	r(afp)	attributable fraction for the population
r(p_exact)	two-sided significance	r(chi2_p)	pooled heterogeneity χ^2
r(or)	odds ratio	r(chi2_bd)	Breslow–Day χ^2
r(lb_or)	lower bound of CI for or	r(df_bd)	degrees of freedom for Breslow–Day χ^2
r(ub_or)	upper bound of CI for or	r(chi2_t)	Tarone χ^2
r(afe)	attributable (prev.) fraction among exposed	r(df_t)	degrees of freedom for Tarone χ^2
		r(df)	degrees of freedom
r(lb_afe)	lower bound of CI for afe	r(chi2)	χ^2

tabodds saves the following in r():

Scalars

r(odds)	odds	r(p_hom)	p-value for test of homogeneity
r(lb_odds)	lower bound for odds	r(df_hom)	degrees of freedom for χ^2 test of homogeneity
r(ub_odds)	upper bound for odds	r(chi2_tr)	χ^2 for score test for trend
r(chi2_hom)	χ^2 test of homogeneity	r(p_trend)	p-value for score test for trend

mhodds saves the following in r():

Scalars

r(p)	two-sided p-value	r(chi2_hom)	χ^2 test of homogeneity
r(or)	odds ratio	r(df_hom)	degrees of freedom for χ^2 test of homogeneity
r(lb_or)	lower bound of CI for or	r(chi2)	χ^2
r(ub_or)	upper bound of CI for or		

mcc and mcci save the following in r():

Scalars

r(p_exact)	two-sided significance	r(R_f)	ratio of proportion with factor
r(or)	odds ratio	r(lb_R_f)	lower bound of CI for R_f
r(lb_or)	lower bound of CI for or	r(ub_R_f)	upper bound of CI for R_f
r(ub_or)	upper bound of CI for or	r(RD_f)	relative difference in proportion with factor
r(D_f)	difference in proportion with factor	r(lb_RD_f)	lower bound of CI for RD_f
		r(ub_RD_f)	upper bound of CI for RD_f
r(lb_D_f)	lower bound of CI for D_f	r(chi2)	χ^2
r(ub_D_f)	upper bound of CI for D_f		

Methods and formulas

All the epitab commands are implemented as ado-files.

The notation for incidence-rate data is

	Exposed	Unexposed	Total
Cases	a	b	M_1
Person-time	N_1	N_0	T

The notation for 2×2 tables is

	Exposed	Unexposed	Total
Cases	a	b	M_1
Controls	c	d	M_0
Total	N_1	N_0	T

The notation for $2 \times k$ tables is

	Exposure level				
	1	2	...	k	Total
Cases	a_1	a_2	...	a_k	M_1
Controls	c_1	c_2	...	c_k	M_0
Total	N_1	N_2	...	N_k	T

If the tables are stratified, all quantities are indexed by i, the stratum number.

We will refer to Fleiss, Levin, and Paik (2003); Kleinbaum, Kupper, and Morgenstern (1982); and Rothman (1986) so often that we will adopt the notation F-23 to mean Fleiss, Levin, and Paik (2003) page 23; KKM-52 to mean Kleinbaum, Kupper, and Morgenstern (1982) page 52; and R-164 to mean Rothman (1986) page 164.

We usually avoid making the continuity corrections to χ^2 statistics, following the advice of KKM-292: "... the use of a continuity correction has been the subject of considerable debate in the statistical literature .... On the basis of our evaluation of this debate and other evidence, we do *not* recommend the use of the continuity correction." Breslow and Day (1980, 133), on the other hand, argue for inclusion of the correction, but not strongly. Their summary is that for small datasets, one should use exact statistics. In practice, we believe that the adjustment makes little difference for reasonably sized datasets.

Methods and formulas are presented under the following headings:

Unstratified incidence-rate data (ir and iri)
Unstratified cumulative incidence data (cs and csi)
Unstratified case–control data (cc and cci)
Unstratified matched case–control data (mcc and mcci)
Stratified incidence-rate data (ir with the by() option)
Stratified cumulative incidence data (cs with the by() option)
Stratified case–control data (cc with by() option, mhodds, tabodds)

Unstratified incidence-rate data (ir and iri)

The incidence-rate difference is defined as $I_d = a/N_1 - b/N_0$ (R-164). The standard error of the difference is $s_{I_d} \approx \sqrt{a/N_1^2 + b/N_0^2}$ (R-170), from which confidence intervals are calculated. For test-based confidence intervals (obtained with the `tb` option), define

$$\chi = \frac{a - N_1 M_1/T}{\sqrt{M_1 N_1 N_0/T^2}}$$

(R-155). Test-based confidence intervals are $I_d(1 \pm z/\chi)$ (R-171), where z is obtained from the normal distribution.

The incidence-rate ratio is defined as $I_r = (a/N_1)/(b/N_0)$ (R-164). Let p_l and p_u be the exact confidence interval of the binomial probability for observing a successes in M_1 trials (obtained from `cii`; see [R] **ci**). The exact confidence interval for the incidence ratio is then $(p_l N_0)/\{(1-p_l)N_1\}$ to $(p_u N_0)/\{(1-p_u)N_1\}$ (R-166). Test-based confidence intervals are $I_r^{1 \pm z/\chi}$ (R-172).

The attributable fraction among the exposed is defined as AFE $= (I_r - 1)/I_r$ for $I_r \geq 1$ (KKM-164; R-38); the confidence interval is obtained by similarly transforming the interval values of I_r. The attributable fraction for the population is AF $=$ AFE $\cdot a/M_1$ (KKM-161); no confidence interval is reported. For $I_r < 1$, the prevented fraction among the exposed is defined as PFE $= 1 - I_r$ (KKM-166; R-39); the confidence interval is obtained by similarly transforming the interval values of I_r. The prevented fraction for the population is PF $=$ PFE $\cdot N_1/T$ (KKM-165); no confidence interval is reported.

The "midp" one-sided exact significance (R-155) is calculated as the binomial probability (with $n = M_1$ and $p = N_1/T$) $\Pr(k = a)/2 + \Pr(k > a)$ if $I_r \geq 1$ and $\Pr(k = a)/2 + \Pr(k < a)$ otherwise. The two-sided significance is twice the one-sided significance (R-155). If preferred, you can obtain nonmidp exact probabilities (and, to some ways of thinking, a more reasonable definition of two-sided significance) using `bitest`; see [R] **bitest**.

Unstratified cumulative incidence data (cs and csi)

The risk difference is defined as $R_d = a/N_1 - b/N_0$ (R-164). Its standard error is

$$s_{R_d} \approx \left\{ \frac{ac}{N_1^3} + \frac{bd}{N_0^3} \right\}^{1/2}$$

(R-172), from which confidence intervals are calculated. For test-based confidence intervals (obtained with the tb option), define

$$\chi = \frac{a - N_1 M_1/T}{\sqrt{(M_1 M_0 N_1 N_0)/\{T^2(T-1)\}}}$$

(R-163). Test-based confidence intervals are $R_d(1 \pm z/\chi)$ (R-172).

The risk ratio is defined as $R_r = (a/N_1)/(b/N_0)$ (R-165). The standard error of $\ln R_r$ is

$$s_{\ln R_r} \approx \left(\frac{c}{aN_1} + \frac{d}{bN_0} \right)^{1/2}$$

(R-173), from which confidence intervals are calculated. Test-based confidence intervals are $R_r^{1 \pm z/\chi}$ (R-173).

For $R_r \geq 1$, the attributable fraction among the exposed is calculated as AFE $= (R_r - 1)/R_r$ (KKM-164; R-38); the confidence interval is obtained by similarly transforming the interval values for R_r. The attributable fraction for the population is calculated as AF $=$ AFE $\cdot a/M_1$ (KKM-161); no confidence interval is reported, but F-128 provides

$$\left\{ \frac{c + (a+d)\text{AF}}{bT} \right\}^{1/2}$$

as the approximate standard error of $\ln(1 - \text{AF})$.

For $R_r < 1$, the prevented fraction among the exposed is calculated as PFE $= 1 - R_r$ (KKM-166; R-39); the confidence interval is obtained by similarly transforming the interval values for R_r. The prevented fraction for the population is calculated as PF $=$ PFE $\cdot N_1/T$; no confidence interval is reported.

The odds ratio, available with the or option, is defined as $\psi = (ad)/(bc)$ (R-165). Several confidence intervals are available. The default interval for cs and csi is the Cornfield (1956) approximate interval. If we let z_α be the index from a normal distribution for an α significance level, the Cornfield interval (ψ_l, ψ_u) is calculated from

$$\psi_l = a_l(M_0 - N_1 + a_l) \Big/ \{(N_1 - a_l)(M_1 - a_l)\}$$

$$\psi_u = a_u(M_0 - N_1 + a_u) \Big/ \{(N_1 - a_u)(M_1 - a_u)\}$$

where a_u and a_l are determined iteratively from

$$a_{i+1} = a \pm z_\alpha \left(\frac{1}{a_i} + \frac{1}{N_1 - a_i} + \frac{1}{M_1 - a_i} + \frac{1}{M_0 - N_1 + a_i} \right)^{-1/2}$$

(Newman 2001, sec. 4.4). a_{i+1} converges to a_u using the plus sign and a_l using the minus sign. a_0 is taken as a. With small numbers, the iterative technique may fail. It is then restarted by decrementing (a_l) or incrementing (a_u) a_0. If that fails, a_0 is again decremented or incremented and iterations restarted, and so on, until a terminal condition is met ($a_0 < 0$ or $a_0 > M_1$), at which point the value is not calculated.

Two other odds-ratio confidence intervals are available with cs and csi: the Woolf and test-based intervals. The Woolf method (Woolf 1955; R-173; Schlesselman 1982, 176), available with the woolf option, estimates the standard error of $\ln\psi$ by

$$s_{\ln\psi} = \left(\frac{1}{a} + \frac{1}{b} + \frac{1}{c} + \frac{1}{d}\right)^{1/2}$$

from which confidence intervals are calculated. The Woolf interval cannot be calculated when there exists a zero cell. Sometimes the Woolf interval is called the "logit interval" (Breslow and Day 1980, 134).

Test-based intervals are available with the tb option; the formula used is $\psi^{1\pm z/\chi}$ (R-174).

The χ^2 statistic, reported by default, can be calculated as

$$\chi^2 = \frac{(ad - bc)^2 T}{M_1 M_0 N_1 N_0}$$

(Schlesselman 1982, 179).

Fisher's exact test, available with the exact option, is calculated as described in [R] **tabulate twoway**.

Unstratified case–control data (cc and cci)

cc and cci report by default the same odds ratio, ψ, that is available with the or option in cs and csi. But cc and cci calculate the confidence interval differently: they default to the exact odds-ratio interval, not the Cornfield interval, but you can request the Cornfield interval with the cornfield option. The $1 - \alpha$ exact interval $(\underline{R}, \overline{R})$ is calculated from

$$\alpha/2 = \frac{\sum_{k=a}^{\min(N_1, M_1)} \binom{N_1}{k}\binom{N_0}{M_1-k}\underline{R}^k}{\sum_{k=\max(0, M_1-N_0)}^{\min(N_1, M_1)} \binom{N_1}{k}\binom{N_0}{M_1-k}\underline{R}^k}$$

and

$$1 - \alpha/2 = \frac{\sum_{k=a+1}^{\min(N_1, M_1)} \binom{N_1}{k}\binom{N_0}{M_1-k}\overline{R}^k}{\sum_{k=\max(0, M_1-N_0)}^{\min(N_1, M_1)} \binom{N_1}{k}\binom{N_0}{M_1-k}\overline{R}^k}$$

(R-169). The equations invert two one-sided Fisher exact tests.

cc and cci also report the same tests of significance as cs and csi: the χ^2 statistic is the default, and Fisher's exact test is obtained with the exact option. The odds ratio, ψ, is used as an estimate of the risk ratio in calculating attributable or prevented fractions. For $\psi \geq 1$, the attributable fraction among the exposed is calculated as AFE $= (\psi - 1)/\psi$ (KKM-164); the confidence interval is obtained by similarly transforming the interval values for ψ. The attributable fraction for the population is calculated as AF $=$ AFE $\cdot a/M_1$ (KKM-161). No confidence interval is reported; however, F-152 provides

$$\left(\frac{a}{M_1 b} + \frac{c}{M_0 d}\right)^{1/2}$$

as the standard error of $\ln(1 - \text{AF})$.

For $\psi < 1$, the prevented fraction among the exposed is calculated as PFE $= 1 - \psi$ (KKM-166); the confidence interval is obtained by similarly transforming the interval values for ψ. The prevented fraction for the population is calculated as PF $= \{(a/M_1)\text{PFE}\}/\{(a/M_1)\text{PFE} + \psi\}$ (KKM-165); no confidence interval is reported.

Unstratified matched case–control data (mcc and mcci)

Referring to the table at the beginning of *Methods and formulas*, the columns of the 2×2 table indicate controls; the rows are cases. Each entry in the table reflects a pair of a matched case and control.

McNemar's (1947) χ^2 is defined as

$$\chi^2 = \frac{(b-c)^2}{b+c}$$

(KKM-389).

The proportion of controls with the factor is $p_1 = N_1/T$, and the proportion of cases with the factor is $p_2 = M_1/T$.

The difference in the proportions is $P_d = p_2 - p_1$. An estimate of its standard error when the two underlying proportions are *not* hypothesized to be equal is

$$s_{P_d} \approx \frac{\{(a+d)(b+c) + 4bc\}^{1/2}}{T^{3/2}}$$

(F-378), from which confidence intervals are calculated. The confidence interval uses a continuity correction (F-378, eq. 13.15).

The ratio of the proportions is $P_r = p_2/p_1$ (R-276, R-278). The standard error of $\ln P_r$ is

$$s_{\ln P_r} \approx \left(\frac{b+c}{M_1 N_1}\right)^{1/2}$$

(R-276), from which confidence intervals are calculated.

The relative difference in the proportions is $P_e = (b-c)/(b+d)$ (F-379). Its standard error is

$$s_{P_e} \approx (b+d)^{-2} \{(b+c+d)(bc+bd+cd) - bcd\}^{1/2}$$

(F-379), from which confidence intervals are calculated.

The odds ratio is $\psi = b/c$ (F-376), and the exact Fisher confidence interval is obtained by transforming into odds ratios the exact binomial confidence interval for the binomial parameter from observing b successes in $b+c$ trials (R-264). Binomial confidence limits are obtained from `cii` (see [R] **ci**) and are transformed by $p/(1-p)$. Test-based confidence intervals are $\psi^{1 \pm z/\chi}$ (R-267), where $\chi = (b-c)/\sqrt{b+c}$ is the square root of McNemar's χ^2.

The exact McNemar significance probability is a two-tailed exact test of $H_0: \psi = 1$. The *p*-value, calculated from the binomial distribution, is

$$\min\left\{1, 2\sum_{k=0}^{\min(b,c)} \binom{b+c}{k}\left(\frac{1}{2}\right)^{b+c}\right\}$$

(Agresti 2002, 412).

Stratified incidence-rate data (ir with the by() option)

Statistics presented for each stratum are calculated independently according to the formulas in *Unstratified incidence-rate data (ir and iri)* above. Within strata, the Mantel–Haenszel style weight is $W_i = b_i N_{1i}/T_i$, and the Mantel–Haenszel combined incidence-rate ratio (Rothman and Boice Jr. 1982) is

$$I_{\mathrm{mh}} = \frac{\sum_i a_i N_{0i}/T_i}{\sum_i W_i}$$

(R-196). The standard error for the log of the incidence-rate ratio was derived by Greenland and Robins (1985, 63) and appears in R-213:

$$s_{\ln I_{\mathrm{mh}}} \approx \left\{\frac{\sum_i M_{1i} N_{1i} N_{0i}/T_i^2}{\left(\sum_i a_i N_{0i}/T_i\right)\left(\sum_i b_i N_{1i}/T_i\right)}\right\}^{1/2}$$

The confidence interval is calculated first on the log scale and then is transformed.

For standardized rates, let w_i be the user-specified weight within stratum i. The standardized rate difference (the ird option) and rate ratio are defined as

$$\mathrm{SRD} = \frac{\sum_i w_i(R_{1i} - R_{0i})}{\sum_i w_i}$$

$$\mathrm{SRR} = \frac{\sum_i w_i R_{1i}}{\sum_i w_i R_{0i}}$$

(R-229). The standard error of SRD is

$$s_{\mathrm{SRD}} \approx \left\{\frac{1}{(\sum_i w_i)^2}\sum_i w_i^2\left(\frac{a_i}{N_{1i}^2} + \frac{b_i}{N_{0i}^2}\right)\right\}^{1/2}$$

(R-231), from which confidence intervals are calculated. The standard error of ln(SRR) is

$$s_{\ln(\mathrm{SRR})} \approx \left\{\frac{\sum_i w_i^2 a_i/N_{1i}^2}{(\sum_i w_i R_{1i})^2} + \frac{\sum_i w_i^2 b_i/N_{0i}^2}{(\sum_i w_i R_{0i})^2}\right\}^{1/2}$$

(R-231), from which confidence intervals are calculated.

Internally and externally standardized measures are calculated using $w_i = N_{1i}$ and $w_i = N_{0i}$, respectively, and are obtained with the istandard and estandard options, respectively.

Directly pooled estimates are available with the pool option. The directly pooled estimate is a weighted average of stratum-specific estimates; each weight, w_i, is inversely proportional to the variance of the estimate for stratum i. The variances for rate differences come from the formulas in *Unstratified incidence-rate data (ir and iri)*, while the variances of log rate-ratios are estimated by $(1/a_i + 1/b_i)$ (R-184). Ratios are averaged in the log scale before being exponentiated. The standard error of the directly pooled estimate is calculated as $1/\sqrt{\sum w_i}$, from which confidence intervals are calculated (R-183–185); the calculation for ratios again uses the log scale.

For rate differences, the χ^2 test of homogeneity is calculated as $\sum(R_{di} - \widehat{R}_d)^2/\text{var}(R_{di})$, where R_{di} are the stratum-specific rate differences and $\widehat{R}_d$ is the directly pooled estimate. The number of degrees of freedom is one less than the number of strata (R-222).

For rate ratios, the same calculation is made, except that it is made on a logarithmic scale using $\ln(R_{ri})$ (R-222), and $\ln(\widehat{R}_d)$ may be the log of either the directly pooled estimate or the Mantel–Haenszel estimate.

Stratified cumulative incidence data (cs with the by() option)

Statistics presented for each stratum are calculated independently according to the formulas in *Unstratified cumulative incidence data (cs and csi)* above. The Mantel–Haenszel χ^2 test (Mantel and Haenszel 1959) is

$$\chi^2_{\text{mh}} = \frac{\{\sum_i (a_i - N_{1i}M_{1i}/T_i)\}^2}{\sum_i (N_{1i}N_{0i}M_{1i}M_{0i})/\{T_i^2(T_i - 1)\}}$$

(R-206).

For the odds ratio (available with the or option), the Mantel–Haenszel weight is $W_i = b_i c_i/T_i$, and the combined odds ratio (Mantel and Haenszel 1959) is

$$\psi_{\text{mh}} = \frac{\sum_i a_i d_i/T_i}{\sum_i W_i}$$

(R-195). The standard error (Robins, Breslow, and Greenland 1986) is

$$s_{\ln\psi_{\text{mh}}} \approx \left\{ \frac{\sum_i P_i R_i}{2(\sum_i R_i)^2} + \frac{\sum_i P_i S_i + Q_i R_i}{2 \sum_i R_i \sum_i S_i} + \frac{\sum_i Q_i S_i}{2(\sum_i S_i)^2} \right\}^{1/2}$$

where

$$P_i = (a_i + d_i)/T_i$$
$$Q_i = (b_i + c_i)/T_i$$
$$R_i = a_i d_i/T_i$$
$$S_i = b_i c_i/T_i$$

(R-220). Alternatively, test-based confidence intervals are calculated as $\psi_{\text{mh}}^{1\pm z/\chi}$ (R-220).

For the risk ratio (the default), the Mantel–Haenszel-style weight is $W_i = b_i N_{1i}/T_i$, and the combined risk ratio (Rothman and Boice Jr. 1982) is

$$R_{\mathrm{mh}} = \frac{\sum_i a_i N_{0i}/T_i}{\sum_i W_i}$$

(R-196). The standard error (Greenland and Robins 1985) is

$$s_{\ln R_{\mathrm{mh}}} \approx \left\{ \frac{\sum_i (M_{1i} N_{1i} N_{0i} - a_i b_i T_i)/T_i^2}{\left(\sum_i a_i N_{0i}/T_i\right)\left(\sum_i b_i N_{1i}/T_i\right)} \right\}^{1/2}$$

(R-216), from which confidence intervals are calculated.

For standardized rates, let w_i be the user-specified weight within stratum i. The standardized rate difference (SRD, the rd option) and rate ratios (SRR, the default) are defined as in *Stratified incidence-rate data (ir with the by() option)*, where the individual risks are defined $R_{1i} = a_i/N_{1i}$ and $R_{0i} = b_i/N_{0i}$. The standard error of SRD is

$$s_{\mathrm{SRD}} \approx \left[\frac{1}{(\sum_i w_i)^2} \sum_i w_i^2 \left\{ \frac{a_i(N_{1i} - a_i)}{N_{1i}^3} + \frac{b_i(N_{0i} - b_i)}{N_{0i}^3} \right\} \right]^{1/2}$$

(R-231), from which confidence intervals are calculated. The standard error of ln(SRR) is

$$s_{\ln(\mathrm{SRR})} \approx \left\{ \frac{\sum_i w_i^2 a_i (N_{1i} - a_i)/N_{1i}^3}{(\sum_i w_i R_{1i})^2} + \frac{\sum_i w_i^2 b_i (N_{0i} - b_i)/N_{0i}^3}{(\sum_i w_i R_{0i})^2} \right\}^{1/2}$$

(R-231), from which confidence intervals are calculated.

Internally and externally standardized measures are calculated using $w_i = N_{1i}$ and $w_i = N_{0i}$, respectively, and are obtained with the istandard and estandard options, respectively.

Directly pooled estimates of the odds ratio are available when you specify both the pool and or options. The directly pooled estimate is a weighted average of stratum-specific log odds-ratios; each weight, w_i, is inversely proportional to the variance of the log odds-ratio for stratum i. The variances of the log odds-ratios are estimated by Woolf's method, described under *Unstratified cumulative incidence data (cs and csi)*. The standard error of the directly pooled log odds-ratio is calculated as $1/\sqrt{\sum w_i}$, from which confidence intervals are calculated and then exponentiated (Kahn and Sempos 1989, 113–115).

Direct pooling is also available for risk ratios and risk differences; the variance formulas may be found in *Unstratified cumulative incidence data (cs and csi)*. The directly pooled risk ratio is provided when the pool option is specified. The directly pooled risk difference is provided only when you specify the pool and rd options, and one of the estandard, istandard, and standard() options.

For risk differences, the χ^2 test of homogeneity is calculated as $\sum (R_{di} - \widehat{R}_d)^2 / \mathrm{var}(R_{di})$, where R_{di} are the stratum-specific risk differences and $\widehat{R}_d$ is the directly pooled estimate. The number of degrees of freedom is one less than the number of strata (R-222).

For risk and odds ratios, the same calculation is made, except that it is made in the log scale using $\ln(R_{ri})$ or $\ln(\psi_i)$ (R-222), and $\ln(\widehat{R}_d)$ may be the log of either the directly pooled estimate or the Mantel–Haenszel estimate.

Stratified case–control data (cc with by() option, mhodds, tabodds)

Statistics presented for each stratum are calculated independently according to the formulas in *Unstratified cumulative incidence data (cs and csi)* above. The combined odds ratio, ψ_{mh}, and the test that $\psi_{\mathrm{mh}} = 1$ (χ^2_{mh}) are calculated as described in *Stratified cumulative incidence data (cs with the by() option)* above.

For standardized weights, let w_i be the user-specified weight within stratum i. The standardized odds ratio (the `standard()` option) is calculated as

$$\mathrm{SOR} = \frac{\sum_i w_i a_i / c_i}{\sum_i w_i b_i / d_i}$$

(Greenland 1986, 473). The standard error of ln(SOR) is

$$s_{\ln(\mathrm{SOR})} = \left\{ \frac{\sum_i (w_i a_i / c_i)^2 \left(\frac{1}{a_i} + \frac{1}{b_i} + \frac{1}{c_i} + \frac{1}{d_i} \right)}{\left(\sum_i w_i a_i / c_i \right)^2} \right\}^{1/2}$$

from which confidence intervals are calculated. The internally and externally standardized odds ratios are calculated using $w_i = c_i$ and $w_i = d_i$, respectively.

The directly pooled estimate of the odds ratio (the `pool` option) is calculated as described in *Stratified cumulative incidence data (cs with the by() option)* above.

The directly pooled and Mantel–Haenszel χ^2 tests of homogeneity are calculated as $\sum \{\ln(R_{ri}) - \ln(\widehat{R}_r)\}^2 / \mathrm{var}\{\ln(R_{ri})\}$, where R_{ri} are the stratum-specific odds ratios and $\widehat{R}_r$ is the pooled estimate (Mantel–Haenszel or directly pooled). The number of degrees of freedom is one less than the number of strata (R-222).

The Breslow–Day χ^2 test of homogeneity is available with the `bd` option. Let $\widehat{\psi}$ be the Mantel–Haenszel estimate of the common odds ratio, and let $A_i(\widehat{\psi})$ be the fitted count for cell a; $A_i(\widehat{\psi})$ is found by solving the quadratic equation

$$A(M_0 - N_1 + A) = (\widehat{\psi})(M_1 - A)(N_1 - A)$$

and choosing the root that makes all cells in stratum i positive. Let $\mathrm{Var}(a_i; \widehat{\psi})$ be the estimated variance of a_i conditioned on the margins and on an odds ratio of $\widehat{\psi}$:

$$\mathrm{Var}(a_i; \widehat{\psi}) = \left\{ \frac{1}{A_i(\widehat{\psi})} + \frac{1}{M_{1i} - A_i(\widehat{\psi})} + \frac{1}{N_{1i} - A_i(\widehat{\psi})} + \frac{1}{M_{0i} - N_{1i} + A_i(\widehat{\psi})} \right\}^{-1}$$

The Breslow–Day χ^2 statistic is then

$$\sum_i \frac{\{a_i - A_i(\widehat{\psi})\}^2}{\mathrm{Var}(a_i; \widehat{\psi})}$$

The Tarone χ^2 test of homogeneity (the `tarone` option) is calculated as

$$\sum_i \frac{\{a_i - A_i(\widehat{\psi})\}^2}{\mathrm{Var}(a_i; \widehat{\psi})} - \frac{\{\sum_i a_i - \sum_i A_i(\widehat{\psi})\}^2}{\sum_i \mathrm{Var}(a_i; \widehat{\psi})}$$

Tarone (1985) provides this correction to the Breslow–Day statistic to ensure that its distribution is asymptotically chi-squared. Without the correction, the Breslow–Day statistic does not necessarily follow a chi-squared distribution because it is based on the Mantel–Haenszel estimate, $\widehat{\psi}$, which is an inefficient estimator of the common odds ratio.

When the exposure variable has multiple levels, mhodds calculates an approximate estimate of the log odds-ratio for a one-unit increase in exposure as the ratio of the score statistic, U, to its variance, V (Clayton and Hills 1993, 103), which are defined below. This is a one-step Newton–Raphson approximation to the maximum likelihood estimate. Within-stratum estimates are combined with Mantel–Haenszel weights.

By default, both tabodds and mhodds produce test statistics and confidence intervals based on score statistics (Clayton and Hills 1993). tabodds reports confidence intervals for the odds of the ith exposure level, unless the adjust() or or option is specified. The confidence interval for odds$_i$, $i = 1, \ldots, k$, is given by

$$\text{odds}_i \cdot \exp\left(\pm z \sqrt{1/a_i + 1/c_i}\right)$$

The score χ^2 test of homogeneity of odds is calculated as

$$\chi^2_{k-1} = \frac{T(T-1)}{M_1 M_0} \sum_{i=1}^{k} \frac{(a_i - E_i)^2}{N_i}$$

where $E_i = (M_1 N_i)/T$.

Let l_i denote the value of the exposure at the ith level. The score χ^2 test for trend of odds is calculated as

$$\chi^2_1 = \frac{U^2}{V}$$

where

$$U = \frac{M_1 M_0}{T} \left(\sum_{i=1}^{k} \frac{a_i l_i}{M_1} - \sum_{i=1}^{k} \frac{c_i l_i}{M_0} \right)$$

and

$$V = \frac{M_1 M_0}{T} \left\{ \frac{\sum_{i=1}^{k} N_i l_i^2 - (\sum_{i=1}^{k} N_i l_i)^2 / T}{T-1} \right\}$$

Acknowledgments

We thank Hal Morgenstern, Department of Epidemiology, University of Michigan; Ardythe Morrow, Center for Pediatric Research, Norfolk, Virginia; and the late Stewart West, Baylor College of Medicine, for their assistance in designing these commands. We also thank Jonathan Freeman, Department of Epidemiology, Harvard School of Public Health, for encouraging us to extend these commands to include tests for homogeneity, for helpful comments on the default behavior of the commands, and for his comments on an early draft of this section. We also thank David Clayton, Cambridge Institute for Medical Research, and Michael Hills (retired), London School of Hygiene and Tropical Medicine; they wrote the original versions of mhodds and tabodds. Finally, we thank William Dupont and Dale Plummer for their contribution to the implementation of exact confidence intervals for the odds ratios for cc and cci.

> Jerome Cornfield (1912–1979) was born in New York City. He majored in history at New York University and took courses in statistics at the U.S. Department of Agriculture Graduate School but otherwise had little formal training. Cornfield held positions at the Bureau of Labor Statistics, the National Cancer Institute, the National Institutes of Health, Johns Hopkins University, the University of Pittsburgh, and George Washington University. He worked on many problems in biomedical statistics, including the analysis of clinical trials, epidemiology (especially case–control studies), and Bayesian approaches.
>
> Quinn McNemar (1900–1986) was born in West Virginia and attended college there and in Pennsylvania. After a brief spell of high school teaching, he began graduate study of psychology at Stanford and then joined the faculty. McNemar's text *Psychological Statistics*, first published in 1949, was widely influential, and he made many substantive and methodological contributions to the application of statistics in psychology.
>
> Barnet Woolf (1902–1983) was born in London. His parents were immigrants from Lithuania. Woolf was educated at Cambridge, where he studied physiology and biochemistry, and proposed methods for linearizing plots in enzyme chemistry that were later rediscovered by others (see Haldane [1957]). His later career in London, Birmingham, Rothamsted, and Edinburgh included lasting contributions to nutrition, epidemiology, public health, genetics, and statistics. He was also active in left-wing causes and wrote witty poems, songs, and revues.

References

Abramson, J. H., and Z. H. Abramson. 2001. *Making Sense of Data: A Self-Instruction Manual on the Interpretation of Epidemiological Data*. 3rd ed. New York: Oxford University Press.

Agresti, A. 1999. On logit confidence intervals for the odds ratio with small samples. *Biometrics* 55: 597–602.

———. 2002. *Categorical Data Analysis*. 2nd ed. Hoboken, NJ: Wiley.

Boice Jr., J. D., and R. R. Monson. 1977. Breast cancer in women after repeated fluoroscopic examinations of the chest. *Journal of the National Cancer Institute* 59: 823–832.

Breslow, N. E. 1996. Statistics in epidemiology: The case–control study. *Journal of the American Statistical Association* 91: 14–28.

Breslow, N. E., and N. E. Day. 1980. *Statistical Methods in Cancer Research: Vol. 1—The Analysis of Case–Control Studies*. Lyon: IARC.

Brown, C. C. 1981. The validity of approximation methods for interval estimation of the odds ratio. *American Journal of Epidemiology* 113: 474–480.

Carlin, J., and S. Vidmar. 2000. sbe35: Menus for epidemiological statistics. *Stata Technical Bulletin* 56: 15–16. Reprinted in *Stata Technical Bulletin Reprints*, vol. 10, pp. 86–87. College Station, TX: Stata Press.

Clayton, D. G., and M. Hills. 1993. *Statistical Models in Epidemiology*. Oxford: Oxford University Press.

———. 1995. ssa8: Analysis of case-control and prevalence studies. *Stata Technical Bulletin* 27: 26–31. Reprinted in *Stata Technical Bulletin Reprints*, vol. 5, pp. 227–233. College Station, TX: Stata Press.

Cornfield, J. 1956. A statistical problem arising from retrospective studies. In Vol. 4 of *Proceedings of the Third Berkeley Symposium*, ed. J. Neyman, 135–148. Berkeley, CA: University of California Press.

Dohoo, I., W. Martin, and H. Stryhn. 2003. *Veterinary Epidemiologic Research*. Charlottetown, Prince Edward Island: AVC.

Doll, R., and A. B. Hill. 1966. Mortality of British doctors in relation to smoking: Observations on coronary thrombosis. *Journal of the National Cancer Institute, Monographs* 19: 205–268.

Dupont, W. D. 2009. *Statistical Modeling for Biomedical Researchers: A Simple Introduction to the Analysis of Complex Data*. 2nd ed. Cambridge: Cambridge University Press.

Dupont, W. D., and D. Plummer. 1999. sbe31: Exact confidence intervals for odds ratios from case–control studies. *Stata Technical Bulletin* 52: 12–16. Reprinted in *Stata Technical Bulletin Reprints*, vol. 9, pp. 150–154. College Station, TX: Stata Press.

Fleiss, J. L., B. Levin, and M. C. Paik. 2003. *Statistical Methods for Rates and Proportions*. 3rd ed. New York: Wiley.

Gart, J. J., and D. G. Thomas. 1982. The performance of three approximate confidence limit methods for the odds ratio. *American Journal of Epidemiology* 115: 453–470.

Gini, R., and J. Pasquini. 2006. Automatic generation of documents. *Stata Journal* 6: 22–39.

Glass, R. I., A. M. Svennerholm, B. J. Stoll, M. R. Khan, K. M. Hossain, M. I. Huq, and J. Holmgren. 1983. Protection against cholera in breast-fed children by antibodies in breast milk. *New England Journal of Medicine* 308: 1389–1392.

Gleason, J. R. 1999. sbe30: Improved confidence intervals for odds ratios. *Stata Technical Bulletin* 51: 24–27. Reprinted in *Stata Technical Bulletin Reprints*, vol. 9, pp. 146–150. College Station, TX: Stata Press.

Greenhouse, S. W., and J. B. Greenhouse. 1998. Cornfield, Jerome. In Vol. 1 of *Encyclopedia of Biostatistics*, ed. P. Armitage and T. Colton, 955–959. Chichester, UK: Wiley.

Greenland, S. 1986. Estimating variances of standardized estimators in case–control studies and sparse data. *Journal of Chronic Diseases* 39: 473–477.

———. 1987. *Evolution of Epidemiologic Ideas: Annotated Readings on Concepts and Methods*. Newton Lower Falls, MA: Epidemiology Resources.

Greenland, S., and J. M. Robins. 1985. Estimation of a common effect parameter from sparse follow-up data. *Biometrics* 41: 55–68.

———. 1988. Conceptual problems in the definition and interpretation of attributable fractions. *American Journal of Epidemiology* 128: 1185–1197.

Haldane, J. B. S. 1957. Graphical methods in enzyme chemistry. *Nature* 179: 832.

Hastorf, A. H., E. R. Hilgard, and R. R. Sears. 1988. Quinn McNemar (1900–1986). *American Psychologist* 43: 196–197.

Hill, W. G. 1984. Barnet Woolf. *Year Book, Royal Society of Edinburgh 1984* 214–219.

Jewell, N. P. 2004. *Statistics for Epidemiology*. Boca Raton, FL: Chapman & Hall/CRC.

Jick, H., O. S. Miettinen, R. K. Neff, S. Shapiro, O. P. Heinonen, and D. Slone. 1973. Coffee and myocardial infarction. *New England Journal of Medicine* 289: 63–67.

Kahn, H. A., and C. T. Sempos. 1989. *Statistical Methods in Epidemiology*. New York: Oxford University Press.

Kleinbaum, D. G., L. L. Kupper, and H. Morgenstern. 1982. *Epidemiologic Research: Principles and Quantitative Methods (Industrial Health and Safety)*. New York: Wiley.

Lilienfeld, D. E., and P. D. Stolley. 1994. *Foundations of Epidemiology*. 3rd ed. New York: Oxford University Press.

López-Vizcaíno, M. E., M. I. Pérez-Santiago, and L. Abraira-García. 2000a. sbe32.1: Automated outbreak detection from public health surveillance data: Errata. *Stata Technical Bulletin Reprints* 55: 2.

———. 2000b. sbe32: Automated outbreak detection from public health surveillance data. *Stata Technical Bulletin* 54: 23–25. Reprinted in *Stata Technical Bulletin Reprints*, vol. 9, pp. 154–157. College Station, TX: Stata Press.

MacMahon, B., S. Yen, D. Trichopoulos, K. Warren, and G. Nardi. 1981. Coffee and cancer of the pancreas. *New England Journal of Medicine* 304: 630–633.

Mantel, N., and W. Haenszel. 1959. Statistical aspects of the analysis of data from retrospective studies of disease. *Journal of the National Cancer Institute* 22: 719–748. Reprinted in *Evolution of Epidemiologic Ideas*, ed. S. Greenland, pp. 112–141. Newton Lower Falls, MA: Epidemiology Resources.

McNemar, Q. 1947. Note on the sampling error of the difference between correlated proportions or percentages. *Psychometrika* 12: 153–157.

Miettinen, O. S. 1976. Estimability and estimation in case-referent studies. *American Journal of Epidemiology* 103: 226–235. Reprinted in *Evolution of Epidemiologic Ideas*, ed. S. Greenland, pp. 181–190. Newton Lower Falls, MA: Epidemiology Resources.

Newman, S. C. 2001. *Biostatistical Methods in Epidemiology*. New York: Wiley.

Orsini, N., R. Bellocco, M. Bottai, A. Wolk, and S. Greenland. 2008. A tool for deterministic and probabilistic sensitivity analysis of epidemiologic studies. *Stata Journal* 8: 29–48.

Pearce, M. S., and R. Feltbower. 2000. sg149: Tests for seasonal data via the Edwards and Walter & Elwood tests. *Stata Technical Bulletin* 56: 47–49. Reprinted in *Stata Technical Bulletin Reprints*, vol. 10, pp. 214–217. College Station, TX: Stata Press.

Reilly, M., and A. Salim. 2000. sxd2: Computing optimal sampling designs for two-stage studies. *Stata Technical Bulletin* 58: 37–41. Reprinted in *Stata Technical Bulletin Reprints*, vol. 10, pp. 376–382. College Station, TX: Stata Press.

Robins, J. M., N. E. Breslow, and S. Greenland. 1986. Estimators of the Mantel-Haenszel variance consistent in both sparse data and large-strata limiting models. *Biometrics* 42: 311–323.

Rothman, K. J. 1982. Spermicide use and Down's syndrome. *American Journal of Public Health* 72: 399–401.

———. 1986. *Modern Epidemiology*. Boston: Little, Brown.

———. 2002. *Epidemiology: An Introduction*. New York: Oxford University Press.

Rothman, K. J., and J. D. Boice Jr. 1982. *Epidemiologic Analysis with a Programmable Calculator*. Brookline, MA: Epidemiology Resources.

Rothman, K. J., D. C. Fyler, A. Goldblatt, and M. B. Kreidberg. 1979. Exogenous hormones and other drug exposures of children with congenital heart disease. *American Journal of Epidemiology* 109: 433–439.

Rothman, K. J., S. Greenland, and T. L. Lash. 2008. *Modern Epidemiology*. 3rd ed. Philadelphia: Lippincott Williams & Wilkins.

Rothman, K. J., and R. R. Monson. 1973. Survival in trigeminal neuralgia. *Journal of Chronic Diseases* 26: 303–309.

Royston, J. P., and A. Babiker. 2002. A menu-driven facility for complex sample size calculation in randomized controlled trials with a survival or a binary outcome. *Stata Journal* 2: 151–163.

Schlesselman, J. J. 1982. *Case–Control Studies: Design, Conduct, Analysis*. New York: Oxford University Press.

Tarone, R. E. 1985. On heterogeneity tests based on efficient scores. *Biometrika* 72: 91–95.

University Group Diabetes Program. 1970. A study of the effects of hypoglycemic agents on vascular complications in patients with adult-onset diabetes, II: Mortality results. *Diabetes* 19, supplement 2: 789–830.

van Belle, G., L. D. Fisher, P. J. Heagerty, and T. S. Lumley. 2004. *Biostatistics: A Methodology for the Health Sciences*. 2nd ed. New York: Wiley.

Walker, A. M. 1991. *Observation and Inference: An Introduction to the Methods of Epidemiology*. Newton Lower Falls, MA: Epidemiology Resources.

Wang, Z. 1999. sbe27: Assessing confounding effects in epidemiological studies. *Stata Technical Bulletin* 49: 12–15. Reprinted in *Stata Technical Bulletin Reprints*, vol. 9, pp. 134–138. College Station, TX: Stata Press.

———. 2007. Two postestimation commands for assessing confounding effects in epidemiological studies. *Stata Journal* 7: 183–196.

Woolf, B. 1955. On estimating the relation between blood group disease. *Annals of Human Genetics* 19: 251–253. Reprinted in *Evolution of Epidemiologic Ideas: Annotated Readings on Concepts and Methods*, ed. S Greenland, pp. 108–110. Newton Lower Falls, MA: Epidemiology Resources.

Also see

[ST] **stcox** — Cox proportional hazards model

[R] **bitest** — Binomial probability test

[R] **ci** — Confidence intervals for means, proportions, and counts

[R] **clogit** — Conditional (fixed-effects) logistic regression

[R] **dstdize** — Direct and indirect standardization

[R] **glogit** — Logit and probit regression for grouped data

[R] **logistic** — Logistic regression, reporting odds ratios

[R] **poisson** — Poisson regression

[R] **symmetry** — Symmetry and marginal homogeneity tests

[R] **tabulate twoway** — Two-way tables of frequencies

[ST] **Glossary**

Title

ltable — Life tables for survival data

Syntax

> ltable *timevar* [*deadvar*] [*if*] [*in*] [*weight*] [, *options*]

options	description
Main	
<u>nota</u>ble	display graph only; suppress display of table
<u>g</u>raph	present the table graphically, as well as in tabular form
by(*groupvar*)	produce separate tables (or graphs) for each value of *groupvar*
<u>t</u>est	report χ^2 measure of differences between groups (2 tests)
overlay	overlay plots on the same graph
<u>s</u>urvival	display survival table; the default
<u>f</u>ailure	display cumulative failure table
<u>h</u>azard	display hazard table
ci	graph confidence interval
<u>l</u>evel(#)	set confidence level; default is level(95)
<u>noa</u>djust	suppress actuarial adjustment for deaths and censored observations
<u>tv</u>id(*varname*)	subject ID variable to use with time-varying parameters
<u>interv</u>als(w | *numlist*)	time intervals in which data are to be aggregated for tables
<u>sav</u>ing(*filename*[, replace])	save the life-table data to *filename*; use replace to overwrite existing *filename*
Plot	
<u>p</u>lotopts(*plot_options*)	affect rendition of the plotted line and plotted points
<u>p</u>lot#opts(*plot_options*)	affect rendition of the #th plotted line and plotted points; available only with overlay
CI plot	
<u>ci</u>opts(*rspike_options*)	affect rendition of the confidence intervals
<u>ci</u>#opts(*rspike_options*)	affect rendition of the #th confidence interval; available only with overlay
Add plots	
addplot(*plot*)	add other plots to the generated graph
Y axis, X axis, Titles, Legend, Overall	
twoway_options	any options other than by() documented in [G] ***twoway_options***
<u>byopts</u>(*byopts*)	how subgraphs are combined, labeled, etc.

plot_options	description
connect_options	change look of lines or connecting method
marker_options	change look of markers (color, size, etc.)

fweights are allowed; see [U] **11.1.6 weight**.

Menu

Statistics > Survival analysis > Summary statistics, tests, and tables > Life tables for survival data

Description

ltable displays and graphs life tables for individual-level or aggregate data and optionally presents the likelihood-ratio and log-rank tests for equivalence of groups. ltable also allows you to examine the empirical hazard function through aggregation. See also [ST] **sts** for alternative commands.

timevar specifies the time of failure or censoring. If *deadvar* is not specified, all values of *timevar* are interpreted as failure times; otherwise, *timevar* is interpreted as a failure time, where $deadvar \neq 0$, and as a censoring time otherwise. Observations with *timevar* or *deadvar* equal to missing are ignored.

deadvar does *not* specify the *number* of failures. An observation with *deadvar* equal to 1 or 50 has the same interpretation—the observation records one failure. Specify frequency weights for aggregated data (e.g., ltable time [freq=number]).

Options

      ⌐Main⌐

notable suppresses displaying the table. This option is often used with graph.

graph requests that the table be presented graphically, as well as in tabular form; when notable is also specified, only the graph is presented. When you specify graph, only one table can be calculated and graphed at a time; see survival, failure, and hazard below.

    graph may not be specified with hazard. Use sts graph to graph estimates of the hazard function.

by(*groupvar*) creates separate tables (or graphs within the same image) for each value of *groupvar*. *groupvar* may be string or numeric.

test presents two χ^2 measures of the differences between groups, the likelihood-ratio test of homogeneity and the log-rank test for equality of survivor functions. The two groups are identified by the by() option, so by() must also be specified.

overlay causes the plot from each group identified in the by() option to be overlaid on the same graph. The default is to generate a separate graph (within the same image) for each group. This option requires the by() option.

survival, failure, and hazard indicate the table to be displayed. If none is specified, the default is the survival table. Specifying failure displays the cumulative failure table. Specifying survival failure would display both the survival and the cumulative failure table. If graph is specified, multiple tables may not be requested.

ci graphs the confidence intervals around survival, failure, or hazard.

level(#) specifies the confidence level, as a percentage, for confidence intervals. The default is level(95) or as set by set level; see [R] **level**.

noadjust suppresses the actuarial adjustment for deaths and censored observations. The default is to consider the adjusted number at risk at the start of the interval as the total at the start minus (the number dead or censored)/2. If noadjust is specified, the number at risk is simply the total at the start, corresponding to the standard Kaplan–Meier assumption. noadjust should be specified when using ltable to list results corresponding to those produced by sts list; see [ST] **sts list**.

tvid(*varname*) is for use with longitudinal data with time-varying parameters. Each subject appears in the data more than once, and equal values of *varname* identify observations referring to the same subject. When tvid() is specified, only the last observation on each subject is used in making the table. The order of the data does not matter, and *last* here means the last observation chronologically.

intervals(w | *numlist*) specifies the intervals into which the data are to be aggregated for tabular presentation. A numeric argument is interpreted as the width of the interval. For instance, interval(2) aggregates data into the intervals $0 \leq t < 2$, $2 \leq t < 4$, and so on. Not specifying interval() is equivalent to specifying interval(1). Because in most data, failure times are recorded as integers, this amounts to no aggregation except that implied by the recording of the time variable, and so it produces Kaplan–Meier product-limit estimates of the survival curve (with an actuarial adjustment; see the noadjust option below). Also see [ST] **sts list**. Although it is possible to examine survival and failure without aggregation, some form of aggregation is almost always required to examine the hazard.

When more than one argument is specified, intervals are aggregated as specified. For instance, interval(0,2,8,16) aggregates data into the intervals $0 \leq t < 2$, $2 \leq t < 8$, and $8 \leq t < 16$, and (if necessary) the open-ended interval $t \geq 16$.

interval(w) is equivalent to interval(0,7,15,30,60,90,180,360,540,720), corresponding to 1 week, (roughly) 2 weeks, 1 month, 2 months, 3 months, 6 months, 1 year, 1.5 years, and 2 years when failure times are recorded in days. The w suggests widening intervals.

saving(*filename*[, replace]) creates a Stata data file (.dta file) containing the life table. This option will not save the graph to disk; see [G] **graph save** to save the resulting graph to disk.

> replace indicates that *filename* be overwritten, if it exists. This option is not shown in the dialog box.

Plot

plotopts(*plot_options*) affects the rendition of the plotted line and plotted points; see [G] *connect_options* and [G] *marker_options*.

plot#opts(*plot_options*) affects the rendition of the #th plotted line and plotted points; see [G] *connect_options* and [G] *marker_options*. This option is valid only if overlay is specified.

CI plot

ciopts(*rspike_options*) affects the rendition of the confidence intervals for the graphed survival, failure, or hazard; see [G] *rspike_options*.

ci#opts(*rspike_options*) affects the rendition of the #th confidence interval for the graphed survival, failure, or hazard; see [G] *rspike_options*. This option is valid only if overlay is specified.

Add plots

addplot(*plot*) provides a way to add other plots to the generated graph; see [G] ***addplot_option***.

Y axis, X axis, Titles, Legend, and Overall

twoway_options are any of the options documented in [G] ***twoway_options***, excluding by(). These include options for titling the graph (see [G] ***title_options***) and for saving the graph to disk (see [G] ***saving_option***).

byopts(*byopts*) affects the appearance of the combined graph when by() is specified, including the overall graph title and the organization of subgraphs. See [G] ***by_option***.

Remarks

Life tables describe death rates in a given population over time. Such tables date back to the 17th century; Edmund Halley (1693) is often credited with their development. ltable is for use with "cohort" data, and although one often thinks of such tables as monitoring a population from the "birth" of the first member to the "death" of the last, more generally, such tables can be thought of as a reasonable way to list any kind of survival data. For an introductory discussion of life tables, see Pagano and Gauvreau (2000, 489–495); for an intermediate discussion, see Selvin (2004, 335–377); and for a more complete discussion, see Chiang (1984).

▷ Example 1

In Pike (1966), two groups of rats were exposed to a carcinogen, and the number of days to death from vaginal cancer was recorded (reprinted in Kalbfleisch and Prentice 2002, 2):

Group 1	143	164	188	188	190	192	206	209	213	216
	220	227	230	234	246	265	304	216*	244*	
Group 2	142	156	163	198	205	232	232	233	233	233
	233	239	240	261	280	280	296	296	323	204*
	344*									

The '*' on a few of the entries indicates that the observation was censored—as of the recorded day, the rat had still not died because of vaginal cancer but was withdrawn from the experiment for other reasons.

Having entered these data into Stata, we see that the first few observations are

```
. use http://www.stata-press.com/data/r11/rat
. list in 1/5
```

	group	t	died
1.	1	143	1
2.	1	164	1
3.	1	188	1
4.	1	188	1
5.	1	190	1

For example, the first observation records a rat from group 1 that died on the 143rd day. The died variable records whether that rat died or was withdrawn (censored):

```
. list if died==0
```

	group	t	died
18.	1	216	0
19.	1	244	0
39.	2	204	0
40.	2	344	0

Four rats, two from each group, did not die but were withdrawn.

The life table for group 1 is

```
. ltable t died if group==1
```

Interval		Beg. Total	Deaths	Lost	Survival	Std. Error	[95% Conf. Int.]	
143	144	19	1	0	0.9474	0.0512	0.6812	0.9924
164	165	18	1	0	0.8947	0.0704	0.6408	0.9726
188	189	17	2	0	0.7895	0.0935	0.5319	0.9153
190	191	15	1	0	0.7368	0.1010	0.4789	0.8810
192	193	14	1	0	0.6842	0.1066	0.4279	0.8439
206	207	13	1	0	0.6316	0.1107	0.3790	0.8044
209	210	12	1	0	0.5789	0.1133	0.3321	0.7626
213	214	11	1	0	0.5263	0.1145	0.2872	0.7188
216	217	10	1	1	0.4709	0.1151	0.2410	0.6713
220	221	8	1	0	0.4120	0.1148	0.1937	0.6194
227	228	7	1	0	0.3532	0.1125	0.1502	0.5648
230	231	6	1	0	0.2943	0.1080	0.1105	0.5070
234	235	5	1	0	0.2355	0.1012	0.0751	0.4459
244	245	4	0	1	0.2355	0.1012	0.0751	0.4459
246	247	3	1	0	0.1570	0.0931	0.0312	0.3721
265	266	2	1	0	0.0785	0.0724	0.0056	0.2864
304	305	1	1	0	0.0000	.	.	.

The reported survival rates are the survival rates at the end of the interval. Thus, 94.7% of rats survived 144 days or more.

◁

❑ Technical note

If you compare the table just printed with the corresponding table in Kalbfleisch and Prentice (2002, 16), you will notice that the survival estimates differ beginning with the interval 216–217, which is the first interval containing a censored observation. ltable treats censored observations as if they were withdrawn halfway through the interval. The table printed in Kalbfleisch and Prentice treated censored observations as if they were withdrawn at the end of the interval, even though Kalbfleisch and Prentice (2002, 19) mention how results could be adjusted for censoring.

Here the same results as those printed in Kalbfleisch and Prentice could be obtained by incrementing the time of withdrawal by 1 for the four censored observations. We say "here" because there were no deaths on the incremented dates. For instance, one of the rats was withdrawn on the 216th day, a day on which there was also a real death. There were no deaths on day 217, however, so moving the withdrawal forward 1 day is equivalent to assuming that the withdrawal occurred at the end of the day 216–217 interval. If the adjustments are made and ltable is used to calculate survival in both groups, the results are the same as those printed in Kalbfleisch and Prentice, except that for group 2 in the interval 240–241, they report the survival as 0.345 when they mean 0.354.

In any case, the one-half adjustment for withdrawals is generally accepted, but it is only a crude adjustment that becomes cruder the wider the intervals.

❑

▷ Example 2: ltable with aggregated intervals

When you do not specify the intervals, `ltable` uses unit intervals. The only aggregation performed on the data was aggregation due to deaths or withdrawals occurring on the same "day". If we wanted to see the table aggregated into 30-day intervals, we would type

```
. ltable t died if group==1, interval(30)
```

Interval		Beg. Total	Deaths	Lost	Survival	Std. Error	[95% Conf. Int.]	
120	150	19	1	0	0.9474	0.0512	0.6812	0.9924
150	180	18	1	0	0.8947	0.0704	0.6408	0.9726
180	210	17	6	0	0.5789	0.1133	0.3321	0.7626
210	240	11	6	1	0.2481	0.1009	0.0847	0.4552
240	270	4	2	1	0.1063	0.0786	0.0139	0.3090
300	330	1	1	0	0.0000	.	.	.

The interval displayed as 120 150 indicates the interval including 120 and up to, but not including, 150. The reported survival rate is the survival rate just after the close of the interval.

When you specify more than one number as the argument to `interval()`, you specify the cutoff points, not the widths.

```
. ltable t died if group==1, interval(120,180,210,240,330)
```

Interval		Beg. Total	Deaths	Lost	Survival	Std. Error	[95% Conf. Int.]	
120	180	19	2	0	0.8947	0.0704	0.6408	0.9726
180	210	17	6	0	0.5789	0.1133	0.3321	0.7626
210	240	11	6	1	0.2481	0.1009	0.0847	0.4552
240	330	4	3	1	0.0354	0.0486	0.0006	0.2245

If any of the underlying failure or censoring times are larger than the last cutoff specified, then they are treated as being in the open-ended interval:

```
. ltable t died if group==1, interval(120,180,210,240)
```

Interval		Beg. Total	Deaths	Lost	Survival	Std. Error	[95% Conf. Int.]	
120	180	19	2	0	0.8947	0.0704	0.6408	0.9726
180	210	17	6	0	0.5789	0.1133	0.3321	0.7626
210	240	11	6	1	0.2481	0.1009	0.0847	0.4552
240	.	4	3	1	0.0354	0.0486	0.0006	0.2245

Whether the last interval is treated as open ended or not makes no difference for survival and failure tables, but it does affect hazard tables. If the interval is open ended, the hazard is not calculated for it.

◁

▷ Example 3: ltable with separate tables for each group

The by(*varname*) option specifies that separate tables be presented for each value of *varname*. Remember that our rat dataset contains two groups:

```
. ltable t died, by(group) interval(30)
```

Interval	Beg. Total	Deaths	Lost	Survival	Std. Error	[95% Conf. Int.]	
group = 1							
120 150	19	1	0	0.9474	0.0512	0.6812	0.9924
150 180	18	1	0	0.8947	0.0704	0.6408	0.9726
180 210	17	6	0	0.5789	0.1133	0.3321	0.7626
210 240	11	6	1	0.2481	0.1009	0.0847	0.4552
240 270	4	2	1	0.1063	0.0786	0.0139	0.3090
300 330	1	1	0	0.0000	.	.	.
group = 2							
120 150	21	1	0	0.9524	0.0465	0.7072	0.9932
150 180	20	2	0	0.8571	0.0764	0.6197	0.9516
180 210	18	2	1	0.7592	0.0939	0.5146	0.8920
210 240	15	7	0	0.4049	0.1099	0.1963	0.6053
240 270	8	2	0	0.3037	0.1031	0.1245	0.5057
270 300	6	4	0	0.1012	0.0678	0.0172	0.2749
300 330	2	1	0	0.0506	0.0493	0.0035	0.2073
330 360	1	0	1	0.0506	0.0493	0.0035	0.2073

◁

▷ Example 4: ltable for failure tables

A failure table is simply a different way of looking at a survival table; failure is 1 − survival:

```
. ltable t died if group==1, interval(30) failure
```

Interval	Beg. Total	Deaths	Lost	Cum. Failure	Std. Error	[95% Conf. Int.]	
120 150	19	1	0	0.0526	0.0512	0.0076	0.3188
150 180	18	1	0	0.1053	0.0704	0.0274	0.3592
180 210	17	6	0	0.4211	0.1133	0.2374	0.6679
210 240	11	6	1	0.7519	0.1009	0.5448	0.9153
240 270	4	2	1	0.8937	0.0786	0.6910	0.9861
300 330	1	1	0	1.0000	.	.	.

◁

▷ Example 5: Survival rate at start of interval versus end of interval

Selvin (2004, 357) presents follow-up data from Cutler and Ederer (1958) on six cohorts of kidney cancer patients. The goal is to estimate the 5-year survival probability.

Year	Interval	Alive	Deaths	Lost	With-drawn	Year	Interval	Alive	Deaths	Lost	With-drawn
1946	0–1	9	4	1		1948	0–1	21	11	0	
	1–2	4	0	0			1–2	10	1	2	
	2–3	4	0	0			2–3	7	0	0	
	3–4	4	0	0			3–4	7	0	0	7
	4–5	4	0	0		1949	0–1	34	12	0	
	5–6	4	0	0	4		1–2	22	3	3	
1947	0–1	18	7	0			2–3	16	1	0	15
	1–2	11	0	0		1950	0–1	19	5	1	
	2–3	11	1	0			1–2	13	1	1	11
	3–4	10	2	2		1951	0–1	25	8	2	15
	4–5	6	0	0	6						

The following is the Stata dataset corresponding to the table:

```
. use http://www.stata-press.com/data/r11/selvin
. list
```

	year	t	died	pop
1.	1946	.5	1	4
2.	1946	.5	0	1
3.	1946	5.5	0	4
4.	1947	.5	1	7
5.	1947	2.5	1	1

(output omitted)

As summary data may often come in the form shown above, it is worth understanding exactly how the data were translated for use with ltable. t records the time of death or censoring (lost to follow-up or withdrawal). died contains 1 if the observation records a death and 0 if it instead records lost or withdrawn patients. pop records the number of patients in the category. The first line of the original table stated that, in the 1946 cohort, there were nine patients at the start of the interval 0–1, and during the interval, four died and one was lost to follow-up. Thus we entered in observation 1 that at t = 0.5, four patients died and in observation 2 that at t = 0.5, one patient was censored. We ignored the information on the total population because ltable will figure that out for itself.

The second line of the table indicated that in the interval 1–2, four patients were still alive at the beginning of the interval, and during the interval, zero died or were lost to follow-up. Because no patients died or were censored, we entered nothing into our data. Similarly, we entered nothing for lines 3, 4, and 5 of the table. The last line for 1946 stated that, in the interval 5–6, four patients were alive at the beginning of the interval and that those four patients were withdrawn. In observation 3, we entered that there were four censorings at t = 5.5.

It does not matter that we chose to record the times of deaths or censoring as midpoints of intervals; we could just as well have recorded the times as 0.8 and 5.8. By default, ltable will form intervals 0–1, 1–2, and so on, and place observations into the intervals to which they belong. We suggest using 0.5 and 5.5 because those numbers correspond to the underlying assumptions made by ltable in making its calculations. Using midpoints reminds you of these assumptions.

(Continued on next page)

To obtain the survival rates, we type

```
. ltable t died [freq=pop]
              Beg.                              Std.
Interval      Total    Deaths    Lost    Survival    Error      [95% Conf. Int.]
    0    1    126       47       19      0.5966     0.0455      0.5017    0.6792
    1    2     60        5       17      0.5386     0.0479      0.4405    0.6269
    2    3     38        2       15      0.5033     0.0508      0.4002    0.5977
    3    4     21        2        9      0.4423     0.0602      0.3225    0.5554
    4    5     10        0        6      0.4423     0.0602      0.3225    0.5554
    5    6      4        0        4      0.4423     0.0602      0.3225    0.5554
```

We estimate the 5-year survival rate as 0.4423 and the 95% confidence interval as 0.3225 to 0.5554.

Selvin (2004, 361), in presenting these results, lists the survival in the interval 0–1 as 1, in 1–2 as 0.597, in 2–3 as 0.539, and so on. That is, relative to us, he shifted the rates down one row and inserted a 1 in the first row. In his table, the survival rate is the survival rate at the *start* of the interval. In our table, the survival rate is the survival rate at the *end* of the interval (or, equivalently, at the start of the next interval). This is, of course, simply a difference in the way the numbers are presented and not in the numbers themselves. ◁

▷ Example 6: ltable for hazard tables

The discrete hazard function is the rate of failure—the number of failures occurring within a time interval divided by the width of the interval (assuming that there are no censored observations). Although the survival and failure tables are meaningful at the "individual" level—with intervals so narrow that each contains only one failure—that is not true for the discrete hazard. If all intervals contained one death and if all intervals were of equal width, the hazard function would be $1/\Delta t$ and so appear to be a constant!

The empirically determined discrete hazard function can be revealed only by aggregation. Gross and Clark (1975, 37) print data on malignant melanoma at the University of Texas M. D. Anderson Tumor Clinic between 1944 and 1960. The interval is the time from initial diagnosis:

Interval (years)	Number lost to follow-up	Number withdrawn alive	Number dying
0–1	19	77	312
1–2	3	71	96
2–3	4	58	45
3–4	3	27	29
4–5	5	35	7
5–6	1	36	9
6–7	0	17	3
7–8	2	10	1
8–9	0	8	3
9+	0	0	32

For our statistical purposes, there is no difference between the number lost to follow-up (patients who disappeared) and the number withdrawn alive (patients dropped by the researchers)—both are censored. We have entered the data into Stata; here are a few of the data:

```
. use http://www.stata-press.com/data/r11/tumor
. list in 1/6, separator(0)
```

	t	d	pop
1.	.5	1	312
2.	.5	0	19
3.	.5	0	77
4.	1.5	1	96
5.	1.5	0	3
6.	1.5	0	71

We entered each group's time of death or censoring as the midpoint of the intervals and entered the numbers of the table, recording d as 1 for deaths and 0 for censoring. The hazard table is

```
. ltable t d [freq=pop], hazard interval(0(1)9)
```

Interval		Beg. Total	Cum. Failure	Std. Error	Hazard	Std. Error	[95% Conf. Int.]	
0	1	913	0.3607	0.0163	0.4401	0.0243	0.3924	0.4877
1	2	505	0.4918	0.0176	0.2286	0.0232	0.1831	0.2740
2	3	335	0.5671	0.0182	0.1599	0.0238	0.1133	0.2064
3	4	228	0.6260	0.0188	0.1461	0.0271	0.0931	0.1991
4	5	169	0.6436	0.0190	0.0481	0.0182	0.0125	0.0837
5	6	122	0.6746	0.0200	0.0909	0.0303	0.0316	0.1502
6	7	76	0.6890	0.0208	0.0455	0.0262	0.0000	0.0969
7	8	56	0.6952	0.0213	0.0202	0.0202	0.0000	0.0598
8	9	43	0.7187	0.0235	0.0800	0.0462	0.0000	0.1705
9	.	32	1.0000	.	.	.	.	.

We specified the interval() option as we did—and not as interval(1) or omitting the option altogether—to force the last interval to be open ended. Had we not, and if we had recorded t as 9.5 for observations in that interval (as we did), ltable would have calculated a hazard rate for the "interval". Here the result of that calculation would have been 2, but no matter the result, it would have been meaningless because we do not know the width of the interval.

When dealing with the survivor or failure function, you are not limited to merely examining a column of numbers. With the graph option, you can see the result graphically:

(Continued on next page)

```
. ltable t d [freq=pop], i(0(1)9) graph notable ci xlab(0(2)10)
```

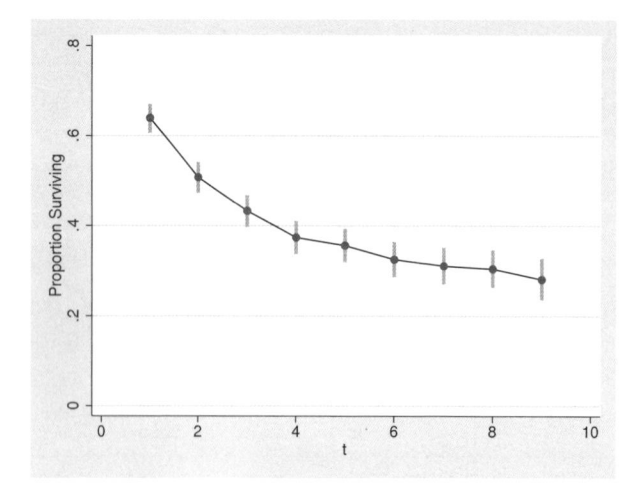

The vertical lines in the graph represent the 95% confidence intervals for the survivor function. Among the options we specified, although it is not required, is notable, which suppressed printing the table, saving us some paper. xlab() was passed through to the graph command (see [G] *twoway_options*) and was unnecessary but made the graph look better.

◁

❑ Technical note

Because many intervals can exist during which no failures occur (in which case the hazard estimate is zero), the estimated hazard is best graphically represented using a kernel smooth. Such an estimate is available in sts graph; see [ST] **sts graph**.

❑

Methods and formulas

ltable is implemented as an ado-file.

Let τ_i be the individual failure or censoring times. The data are aggregated into intervals given by t_j, $j = 1, \ldots, J$, and $t_{J+1} = \infty$ with each interval containing counts for $t_j \leq \tau < t_{j+1}$. Let d_j and m_j be the number of failures and censored observations during the interval and N_j the number alive at the start of the interval. Define $n_j = N_j - m_j/2$ as the adjusted number at risk at the start of the interval. If the noadjust option is specified, $n_j = N_j$.

The product-limit estimate of the survivor function is

$$S_j = \prod_{k=1}^{j} \frac{n_k - d_k}{n_k}$$

(Kalbfleisch and Prentice 2002, 10, 15). Greenwood's formula for the asymptotic standard error of S_j is

$$s_j = S_j \sqrt{\sum_{k=1}^{j} \frac{d_k}{n_k(n_k - d_k)}}$$

(Greenwood 1926; Kalbfleisch and Prentice 2002, 17). s_j is reported as the standard deviation of survival but is not used in generating the confidence intervals because it can produce intervals outside 0 and 1. The "natural" units for the survivor function are $\log(-\log S_j)$, and the asymptotic standard error of that quantity is

$$\widehat{s}_j = \sqrt{\frac{\sum d_k/\{n_k(n_k - d_k)\}}{\left[\sum \log\{(n_k - d_k)/n_k\}\right]^2}}$$

(Kalbfleisch and Prentice 2002, 18). The corresponding confidence intervals are $S_j^{\exp(\pm z_{1-\alpha/2} \widehat{s}_j)}$.

The cumulative failure time is defined as $G_j = 1 - S_j$, and thus the variance is the same as for S_j and the confidence intervals are $1 - S_j^{\exp(\pm z_{1-\alpha/2} \widehat{s}_j)}$.

Both S_j and G_j are graphed against t_{j+1}.

Define the within-interval failure rate as $f_j = d_j/n_j$. The maximum likelihood estimate of the (within-interval) hazard is then

$$\lambda_j = \frac{f_j}{(1 - f_j/2)(t_{j+1} - t_j)}$$

The standard error of λ_j is

$$s_{\lambda_j} = \lambda_j \sqrt{\frac{1 - \{(t_{j+1} - t_j)\lambda_j/2\}^2}{d_j}}$$

from which a confidence interval is calculated.

If the noadjust option is specified, the estimate of the hazard is

$$\lambda_j = \frac{f_j}{t_{j+1} - t_j}$$

and its standard error is

$$s_{\lambda_j} = \frac{\lambda_j}{\sqrt{d_j}}$$

The confidence interval is

$$\left[\frac{\lambda_j}{2d_j}\chi^2_{2d_j,\alpha/2}, \ \frac{\lambda_j}{2d_j}\chi^2_{2d_j,1-\alpha/2}\right]$$

where $\chi^2_{2d_j,q}$ is the qth quantile of the χ^2 distribution with $2d_j$ degrees of freedom (Cox and Oakes 1984, 53–54, 38–40).

For the likelihood-ratio test for homogeneity, let d_g be the total number of deaths in the gth group. Define $T_g = \sum_{i \in g} \tau_i$, where i indexes the individual failure or censoring times. The χ^2 value with $G - 1$ degrees of freedom (where G is the total number of groups) is

$$\chi^2 = 2\left\{ \left(\sum d_g\right) \log\left(\frac{\sum T_g}{\sum d_g}\right) - \sum d_g \log\left(\frac{T_g}{d_g}\right) \right\}$$

(Lawless 2003, 155).

The log-rank test for homogeneity is the test presented by sts test; see [ST] sts.

Acknowledgments

ltable is based on the lftbl command by Henry Krakauer and John Stewart (1991). We also thank Michel Henry-Amar, Centre Regional François Baclesse, Caen, France, for his comments.

References

Chiang, C. L. 1984. *The Life Table and Its Applications*. Malabar, FL: Krieger.

Cox, D. R., and D. Oakes. 1984. *Analysis of Survival Data*. London: Chapman & Hall/CRC.

Cutler, S. J., and F. Ederer. 1958. Maximum utilization of the life table method in analyzing survival. *Journal of Chronic Diseases* 8: 699–712.

Greenwood, M. 1926. The natural duration of cancer. *Reports on Public Health and Medical Subjects* 33: 1–26.

Gross, A. J., and V. A. Clark. 1975. *Survival Distribution: Reliability Applications in the Biomedical Sciences*. New York: Wiley.

Halley, E. 1693. An estimate of the degrees of the mortality of mankind, drawn from curious tables of the births and funerals at the city of Breslaw; with an attempt to ascertain the price of annuities upon lives. *Philosophical Transactions* 17: 596–610.

Kahn, H. A., and C. T. Sempos. 1989. *Statistical Methods in Epidemiology*. New York: Oxford University Press.

Kalbfleisch, J. D., and R. L. Prentice. 2002. *The Statistical Analysis of Failure Time Data*. 2nd ed. New York: Wiley.

Krakauer, H., and J. Stewart. 1991. ssa1: Actuarial or life-table analysis of time-to-event data. *Stata Technical Bulletin* 1: 23–25. Reprinted in *Stata Technical Bulletin Reprints*, vol. 1, pp. 200–202. College Station, TX: Stata Press.

Lawless, J. F. 2003. *Statistical Models and Methods for Lifetime Data*. 2nd ed. New York: Wiley.

Pagano, M., and K. Gauvreau. 2000. *Principles of Biostatistics*. 2nd ed. Belmont, CA: Duxbury.

Pike, M. C. 1966. A method of analysis of a certain class of experiments in carcinogenesis. *Biometrics* 22: 142–161.

Ramalheira, C. 2001. ssa14: Global and multiple causes-of-death life tables. *Stata Technical Bulletin* 59: 29–45. Reprinted in *Stata Technical Bulletin Reprints*, vol. 10, pp. 333–355. College Station, TX: Stata Press.

Selvin, S. 2004. *Statistical Analysis of Epidemiologic Data*. 3rd ed. New York: Oxford University Press.

Also see

[ST] **stcox** — Cox proportional hazards model

Title

snapspan — Convert snapshot data to time-span data

Syntax

snapspan *idvar timevar varlist* [, <u>g</u>enerate(*newt0var*) replace]

Menu

Statistics > Survival analysis > Setup and utilities > Convert snapshot data to time-span data

Description

snapspan converts snapshot data to time-span data. See *Remarks* below for a description of snapshot and time-span data. Time-span data are required for use with survival analysis commands, such as stcox, streg, and stset.

idvar records the subject ID and may be string or numeric.

timevar records the time of the snapshot; it must be numeric and may be recorded on any scale: date, hour, minute, second, etc.

varlist are the "event" variables, meaning that they occur at the instant of *timevar*. *varlist* can also include retrospective variables that are to apply to the time span ending at the time of the current snapshot. The other variables are assumed to be measured at the time of the snapshot and thus apply from the time of the snapshot forward. See *Specifying varlist* below.

Options

generate(*newt0var*) adds *newt0var* to the dataset containing the entry time for each converted time-span record.

replace specifies that it is okay to change the data in memory, even though the dataset has not been saved on disk in its current form.

Remarks

Remarks are presented under the following headings:

> *Snapshot and time-span datasets*
> *Specifying varlist*

Snapshot and time-span datasets

snapspan converts a snapshot dataset to a time-span dataset. A snapshot dataset records a subject *id*, a *time*, and then other variables measured at the *time*:

Snapshot datasets:

idvar	timevar	x1	x2	...
47	12	5	27	...
47	42	5	18	...
47	55	5	19	...

idvar	datevar	x1	x2	...
122	14jul1998	5	27	...
122	12aug1998	5	18	...
122	08sep1998	5	19	...

idvar	year	x1	x2	...
122	1994	5	27	...
122	1995	5	18	...
122	1997	5	19	...

A time-span dataset records a span of time ($time0, time1$):

```
                                          some variables assumed
                                              to occur at time1
                                          |
        |<— other variables assumed constant over span —>|
    ____|_____|_____> time
      time0                                            time1
```

Time-span data are required, for instance, by stset and the st system. The variables assumed to occur at time1 are the failure or event variables. All the other variables are assumed to be constant over the span.

Time-span datasets:

idvar	time0	time1	x1	x2	...	event
47	0	12	5	13	...	0
47	12	42	5	27	...	0
47	42	55	5	18	...	1

idvar	time0	time1	x1	x2	...	event
122	01jan1998	14jul1998	5	13	...	0
122	14jul1998	12aug1998	5	27	...	0
122	12aug1998	08sep1998	5	18	...	1

idvar	time0	time1	x1	x2	...	event
122	1993	1994	5	13	...	0
122	1994	1995	5	27	...	0
122	1995	1997	5	18	...	1

To convert snapshot data to time-span data, you need to distinguish between event and nonevent variables. Event variables happen at an instant.

Say that you have a snapshot dataset containing variable e recording an event (e = 1 might record surgery, death, becoming unemployed, etc.) and the rest of the variables—call them x1, x2, etc.—recording characteristics (such as sex, birth date, blood pressure, or weekly wage). The same data, in snapshot and time-span form, would be

```
In snapshot form:                In time-span form:
id   time    x1   x2    e        id   time0  time   x1   x2    e

 1     5     a1   b1   e1         1     .      5    .    .    e1
 1     7     a2   b2   e2         1     5      7    a1   b1   e2
 1     9     a3   b3   e3         1     7      9    a2   b2   e3
 1    11     a4   b4   e4         1     9     11    a3   b3   e4
```

snapspan converts data from the form on the left to the form on the right:

. snapspan id time e, generate(time0) replace

The form on the right is suitable for use by stcox and stset and the other survival analysis commands.

Specifying varlist

The *varlist*—the third variable on—specifies the "event" variables.

In fact, the *varlist* specifies the variables that apply to the time span ending at the time of the current snapshot. The other variables are assumed to be measured at the time of the snapshot and thus apply from the time of the snapshot forward.

Thus *varlist* should include retrospective variables.

For instance, say that the snapshot recorded bp, blood pressure; smokes, whether the patient smoked in the last 2 weeks; and event, a variable recording examination, surgery, etc. Then *varlist* should include smokes and event. The remaining variables, bp and the rest, would be assumed to apply from the time of the snapshot forward.

Suppose that the snapshot recorded ecs, employment change status (hired, fired, promoted, etc.); wage, the current hourly wage; and ms, current marital status. Then *varlist* should include esc and ms (assuming snapshot records are not generated for reason of ms change). The remaining variables, wage and the rest, would be assumed to apply from the time of the snapshot forward.

Methods and formulas

snapspan is implemented as an ado-file.

Also see

[ST] **stset** — Declare data to be survival-time data

Title

> **st** — Survival-time data

Description

The term *st* refers to survival-time data and the commands—most of which begin with the letters st—for analyzing these data. If you have data on individual subjects with observations recording that this subject came under observation at time t_0 and that later, at t_1, a failure or censoring was observed, you have what we call survival-time data.

If you have subject-specific data, with observations recording not a span of time, but measurements taken on the subject at that point in time, you have what we call a snapshot dataset; see [ST] **snapspan**.

If you have data on populations, with observations recording the number of units under test at time t (subjects alive) and the number of subjects that failed or were lost because of censoring, you have what we call count-time data; see [ST] **ct**.

The st commands are

stset	[ST] **stset**	Declare data to be survival-time data
stdescribe	[ST] **stdescribe**	Describe survival-time data
stsum	[ST] **stsum**	Summarize survival-time data
stvary	[ST] **stvary**	Report whether variables vary over time
stfill	[ST] **stfill**	Fill in by carrying forward values of covariates
stgen	[ST] **stgen**	Generate variables reflecting entire histories
stsplit	[ST] **stsplit**	Split time-span records
stjoin	[ST] **stsplit**	Join time-span records
stbase	[ST] **stbase**	Form baseline dataset
sts	[ST] **sts**	Generate, graph, list, and test the survivor and cumulative hazard functions
stir	[ST] **stir**	Report incidence-rate comparison
stci	[ST] **stci**	Confidence intervals for means and percentiles of survival time
strate	[ST] **strate**	Tabulate failure rate
stptime	[ST] **stptime**	Calculate person-time
stmh	[ST] **strate**	Calculate rate ratios with the Mantel–Haenszel method
stmc	[ST] **strate**	Calculate rate ratios with the Mantel–Cox method
stcox	[ST] **stcox**	Fit Cox proportional hazards model
estat concordance	[ST] **stcox postestimation**	Calculate Harrell's C
estat phtest	[ST] **stcox PH-assumption tests**	Test Cox proportional-hazards assumption
stphplot	[ST] **stcox PH-assumption tests**	Graphically assess the Cox proportional-hazards assumption
stcoxkm	[ST] **stcox PH-assumption tests**	Graphically assess the Cox proportional-hazards assumption

streg	[ST] **streg**	Fit parametric survival models
stcurve	[ST] **stcurve**	Plot survivor, hazard, cumulative hazard, or cumulative incidence function
stcrreg	[ST] **stcrreg**	Fit competing-risks regression models
stpower	[ST] **stpower**	Sample-size, power, and effect-size determination for survival studies
stpower cox	[ST] **stpower cox**	Sample size, power, and effect size for the Cox proportional hazards model
stpower exponential	[ST] **stpower exponential**	Sample size and power for the exponential test
stpower logrank	[ST] **stpower logrank**	Sample size, power, and effect size for the log-rank test
sttocc	[ST] **sttocc**	Convert survival-time data to case–control data
sttoct	[ST] **sttoct**	Convert survival-time data to count-time data
st_*	[ST] **st_is**	Survival analysis subroutines for programmers

The st commands are used for analyzing time-to-absorbing-event (single-failure) data and for analyzing time-to-repeated-event (multiple-failure) data.

You begin an analysis by stsetting your data, which tells Stata the key survival-time variables; see [ST] **stset**. Once you have stset your data, you can use the other st commands. If you save your data after stsetting it, you will not have to stset it again in the future; Stata will remember.

The subsequent st entries are printed in this manual in alphabetical order. You can skip around, but if you want to be an expert on all of Stata's survival analysis capabilities, we suggest the reading order listed above.

Reference

Cleves, M. A. 1999. ssa13: Analysis of multiple failure-time data with Stata. *Stata Technical Bulletin* 49: 30–39. Reprinted in *Stata Technical Bulletin Reprints*, vol. 9, pp. 338–349. College Station, TX: Stata Press.

Also see

[ST] **stset** — Declare data to be survival-time data

[ST] **ct** — Count-time data

[ST] **snapspan** — Convert snapshot data to time-span data

[ST] **survival analysis** — Introduction to survival analysis & epidemiological tables commands

[ST] **Glossary**

Title

> **st_is** — Survival analysis subroutines for programmers

Syntax

Verify that data in memory are survival-time data

> st_is 2 {full | analysis}

Display or do not display summary of survival-time variables

> st_show [noshow]

Risk-group summaries

> st_ct "[byvars]" -> *newtvar newpopvar newfailvar* [*newcensvar* [*newentvar*]]

You must have stset your data before using st_is, st_show, and st_ct; see [ST] **stset**.

Description

These commands are provided for programmers wishing to write new st commands.

st_is verifies that the data in memory are survival-time (st) data. If not, it issues the error message "data not st", r(119).

st is currently "release 2", meaning that this is the second design of the system. Programs written for the previous release continue to work. (The previous release of st corresponds to Stata 5.)

Modern programs code st_is 2 full or st_is 2 analysis. st_is 2 verifies that the dataset in memory is in release 2 format; if it is in the earlier format, it is converted to release 2 format. (Older programs simply code st_is. This verifies that no new features are stset about the data that would cause the old program to break.)

The full and analysis parts indicate whether the dataset may include past, future, or past and future data. Code st_is 2 full if the command is suitable for running on the analysis sample and the past and future data (many data-management commands fall into this category). Code st_is 2 analysis if the command is suitable for use only with the analysis sample (most statistical commands fall into this category). See [ST] **stset** for the definitions of past and future.

st_show displays the summary of the survival-time variables or does nothing, depending on what you specify when stsetting the data. noshow requests that st_show display nothing.

st_ct is a low-level utility that provides risk-group summaries from survival-time data.

Remarks

Remarks are presented under the following headings:

> Definitions of characteristics and st variables
> Outline of an st command
> Using the st_ct utility
> Comparison of st_ct with sttoct
> Verifying data
> Converting data

Definitions of characteristics and st variables

From a programmer's perspective, st is a set of conventions that specify where certain pieces of information are stored and how that information should be interpreted, together with a few subroutines that make it easier to follow the conventions.

At the lowest level, st is nothing more than a set of Stata characteristics that programmers may access:

char _dta[_dta]	st (marks that the data is st)
char _dta[st_ver]	2 (version number)
char _dta[st_id]	*varname* or nothing; id() variable
char _dta[st_bt0]	*varname* or nothing; t0() variable
char _dta[st_bt]	*varname*; t variable from stset t, ...
char _dta[st_bd]	*varname* or nothing; failure() variable
char _dta[st_ev]	list of numbers or nothing; *numlist* from failure(*varname*[==*numlist*])
char _dta[st_enter]	contents of enter() or nothing; *numlist* expanded
char _dta[st_exit]	contents of exit() or nothing; *numlist* expanded
char _dta[st_orig]	contents of origin() or nothing; *numlist* expanded
char _dta[st_bs]	# or 1; scale() value
char _dta[st_o]	_origin or #
char _dta[st_s]	_scale or #
char _dta[st_ifexp]	*exp* or nothing; from stset ... if exp ...
char _dta[st_if]	*exp* or nothing; contents of if()
char _dta[st_ever]	*exp* or nothing; contents of ever()
char _dta[st_never]	*exp* or nothing; contents of never()
char _dta[st_after]	*exp* or nothing; contents of after()
char _dta[st_befor]	*exp* or nothing; contents of before()
char _dta[st_wt]	weight type or nothing; user-specified weight
char _dta[st_wv]	*varname* or nothing; user-specified weighting variable
char _dta[st_w]	[*weighttype*=*weightvar*] or nothing
char _dta[st_show]	noshow or nothing
char _dta[st_t]	_t (for compatibility with release 1)
char _dta[st_t0]	_t0 (for compatibility with release 1)
char _dta[st_d]	_d (for compatibility with release 1)
char _dta[st_n0]	# or nothing; number of st notes
char _dta[st_n1]	text of first note or nothing
char _dta[st_n2]	text of second note or nothing
char _dta[st_set]	text or nothing. If filled in, streset will refuse to execute and present this text as the reason

All st datasets also have the following four variables:

_t0	time of entry (in t units) into risk pool
_t	time of exit (in t units) from risk pool
_d	contains 1 if failure, 0 if censoring
_st	contains 1 if observation is to be used and 0 otherwise

Thus, in a program, you might code

```
display "the failure/censoring base time variable is _t"
display "and its mean in the uncensored subsample is"
summarize _t if _d
```

No matter how simple or complicated the data, these four variables exist and are filled in. For instance, in simple data, _t0 might contain 0 for every observation, and _d might always contain 1.

Some st datasets also contain the variables

>   _origin   evaluated value of origin()
>   _scale   evaluated value of scale()

The _dta[st_o] characteristic contains either the name _origin or a number, often 0. It contains a number when the origin does not vary across observations. _dta[st_s] works the same way with the scale() value. Thus the origin and scale are _dta[st_o] and _dta[st_s]. In fact, these characteristics are seldom used because variables _t and _t0 are already adjusted.

Some st datasets have an id() variable that clusters together records on the same subject. The name of the variable varies, and the name can be obtained from the _dta[st_id] characteristic. If there is no id() variable, the characteristic contains nothing.

Outline of an st command

If you are writing a new st command, place st_is near the top of your code to ensure that your command does not execute on inappropriate data. Also place st_show following the parsing of your command's syntax to display the key st variables. The minimal outline for an st command is

```
program st name
        version 11
        st_is 2 ...
        ... syntax command ...
        ... determined there are no syntax errors ...
        st_show
        ... guts of program ...
end
```

st_is 2 appears even before the input is parsed. This is to avoid irritating users when they type a command, get a syntax error, work hard to eliminate the error, and then learn that "data not st".

A fuller outline for an st command, particularly one that performs analysis on the data, is

```
program st name
        version 11
        st_is 2 ...
        syntax ... [, ... noSHow ... ]
        st_show 'show'
        marksample touse
        quietly replace 'touse' = 0 if _st==0
        ... guts of program ...
end
```

All calculations and actions are to be restricted, at the least, to observations for which $\_st \neq 0$. Observations with $\_st = 0$ are to be ignored.

Using the st_ct utility

st_ct converts the data in memory to observations containing summaries of risk groups. Consider the code

```
st_is 2 analysis
preserve
st_ct "" -> t pop die
```

Typing this would change the data in memory to contain something akin to count-time data. The transformed data would have observations containing

> t   time
> pop population at risk at time t
> die number who fail at time t

There would be one record per time t, and the data would be sorted by t. The original data are discarded, which is why you should code preserve; see [P] **preserve**.

The above three lines of code could be used as the basis for calculating the Kaplan–Meier product-limit survivor-function estimate. The rest of the code is

```
keep if die
gen double hazard = die/pop
gen double km     = 1-hazard         if _n==1
replace    km     = (1-hazard)*km[_n-1] if _n>1
```

st_ct can be used to obtain risk groups separately for subgroups of the population. The code

```
st_is 2 analysis
preserve
st_ct "race sex" -> t pop die
```

would change the data in memory to contain

> race
> sex
> t   time
> pop population at risk at time t
> die number who fail at time t

There would be one observation for each race–sex–t combination, and the data would be sorted by race sex t.

With this dataset, you could calculate the Kaplan–Meier product-limit survivor-function estimate for each race–sex group by coding

```
keep if die
gen double hazard = die/pop
by race sex: gen double km     = 1-hazard         if _n==1
by race sex: replace    km     = (1-hazard)*km[_n-1] if _n>1
```

st_ct is a convenient subroutine. The above code fragment works regardless of the complexity of the underlying survival-time data. It does not matter whether there is one record per subject, no censoring, and one failure per subject, or multiple records per subject, gaps, and recurring failures for the same subject. st_ct forms risk groups that summarize the events recorded by the data.

st_ct can provide the number of censored records and the number who enter the risk group. The code

```
st_ct "" -> t pop die cens ent
```

creates records containing

> t   time
> pop population at risk at time t
> die number who fail at time t
> cens number who are censored at t (after the failures)
> ent number who enter at t (after the censorings)

As before,

```
st_ct "race sex" -> t pop die cens ent
```

would create a similar dataset with records for each race–sex group.

Comparison of st_ct with sttoct

sttoct—see [ST] **sttoct**—is related to st_ct, and in fact, sttoct is implemented in terms of st_ct. The differences between them are that

- sttoct creates ct data, meaning that the dataset is marked as being ct. st_ct merely creates a useful dataset; it does not ctset the data.

- st_ct creates a total population at-risk variable—which is useful in programming—but sttoct creates no such variable.

- sttoct eliminates thrashings—censorings and reentries of the same subject as covariates change—if there are no gaps, strata shifting, etc. st_ct does not do this. Thus, at a particular time, sttoct might show that there are two lost to censoring and none entered, whereas st_ct might show 12 censorings and 10 entries. This makes no difference in calculating the number at risk and the number who fail, which are the major ingredients in survival calculations.

- st_ct is faster.

Verifying data

As long as you code st_is at the top of your program, you need not verify the consistency of the data. That is, you need not verify that subjects do not fail before they enter, etc.

The dataset is verified when you stset it. If you make a substantive change to the data, you must rerun stset (which can be done by typing stset or streset without arguments) to reverify that all is well.

Converting data

If you write a program that converts the data from one form of st data to another, or from st data to something else, be sure to issue the appropriate stset command. For instance, a command we have written, stbase, converts the data from st to a simple cross-section in one instance. In our program, we coded stset, clear so that all other st commands would know that these are no longer st data and that making st calculations on them would be inappropriate.

Even if we had forgotten, other st programs would have found many of the key st variables missing and would have ended with a "[such-and-such] not found" error.

Methods and formulas

st_is is implemented as an ado-file.

Also see

[ST] **stset** — Declare data to be survival-time data

[ST] **sttoct** — Convert survival-time data to count-time data

[ST] **st** — Survival-time data

[ST] **survival analysis** — Introduction to survival analysis & epidemiological tables commands

Title

> **stbase** — Form baseline dataset

Syntax

> stbase [*if*] [*in*] [, *options*]

options	description
Main	
at(#)	convert single/multiple-record st data to cross-sectional dataset at time #
gap(*newvar*)	name of variable containing gap time; default is gap or gaptime
replace	overwrite current data in memory
noshow	do not show st setting information
†nopreserve	programmer's option; see *Options* below

†nopreserve is not shown in the dialog box.
You must stset your data before using stbase; see [ST] **stset**.
fweights, iweights, and pweights may be specified using stset; see [ST] **stset**.

Menu

Statistics > Survival analysis > Setup and utilities > Form baseline dataset

Description

stbase without the at() option converts multiple-record st data to st data with every variable set to its value at baseline, defined as the earliest time at which each subject was observed. stbase without at() does nothing to single-record st data.

stbase, at() converts single- or multiple-record st data to a cross-sectional dataset (not st data), recording the number of failures at the specified time. All variables are given their values at baseline—the earliest time at which each subject was observed. In this form, single-failure data could be analyzed by logistic regression and multiple-failure data by Poisson regression, for instance.

stbase can be used with single- or multiple-record or single- or multiple-failure st data.

Options

☐ Main ☐

at(#) changes what stbase does. Without the at() option, stbase produces another related st dataset. With the at() option, stbase produces a related cross-sectional dataset.

gap(*newvar*) is allowed only with at(); it specifies the name of a new variable to be added to the data containing the amount of time the subject was not at risk after entering and before # as specified in at(). If gap() is not specified, the new variable will be named gap or gaptime, depending on which name does not already exist in the data.

replace specifies that it is okay to change the data in memory, even though the dataset has not been saved to disk in its current form.

noshow prevents stbase from showing the key st variables. This option is rarely used because most people type stset, show or stset, noshow to set once and for all whether they want to see these variables mentioned at the top of the output of every st command; see [ST] **stset**.

The following option is available with stbase but is not shown in the dialog box:

nopreserve is for use by programmers using stbase as a subroutine. It specifies that stbase not preserve the original dataset so that it can be restored should an error be detected or should the user press *Break*. Programmers would specify this option if, in their program, they had already preserved the original data.

Remarks

Remarks are presented under the following headings:

>   *stbase without the at() option*
>   *stbase with the at() option*
>   *Single-failure st data where all subjects enter at time 0*
>   *Single-failure st data where some subjects enter after time 0*
>   *Single-failure st data with gaps and perhaps delayed entry*
>   *Multiple-failure st data*

stbase without the at() option

Once you type stbase, you may not streset your data, even though the data are st. streset will refuse to run because the data have changed, and if the original rules were reapplied, they might produce different, incorrect results. The st commands use four key variables:

>   _t0   the time at which the record came under observation
>   _t   the time at which the record left observation
>   _d   1 if the record left under failure, 0 otherwise
>   _st   whether the observation is to be used (contains 1 or 0)

These variables are adjusted by stbase. The _t0 and _t variables, in particular, are derived from your variables according to options you specified at the time you stset the data, which might include an origin() rule, an entry() rule, and the like. Once intervening observations are eliminated, those rules will not necessarily produce the same results that they did previously.

To illustrate how stbase works, consider multiple-record, time-varying st data, on which you have performed some analysis. You now wish to compare your results with a simpler, non–time-varying analysis. For instance, suppose that variables x1 and x2 measure blood pressure and weight, respectively, and that readings were taken at various times. Perhaps you fit the model

```
. use http://www.stata-press.com/data/r11/mfail
. stset
-> stset t, id(id) failure(d) exit(time .) noshow
                id:  id
     failure event:  d != 0 & d < .
obs. time interval:  (t[_n-1], t]
 exit on or before:  time .
```

```
    1734  total obs.
       0  exclusions

    1734  obs. remaining, representing
     926  subjects
     808  failures in multiple failure-per-subject data
  435855  total analysis time at risk, at risk from t =         0
                                  earliest observed entry t =   0
                                    last observed exit t =    960
```

```
. stcox x1 x2
Iteration 0:   log likelihood = -5034.9569
Iteration 1:   log likelihood = -4978.4198
Iteration 2:   log likelihood = -4978.1915
Iteration 3:   log likelihood = -4978.1914
Refining estimates:
Iteration 0:   log likelihood = -4978.1914

Cox regression -- Breslow method for ties

No. of subjects =          926              Number of obs   =      1734
No. of failures =          808
Time at risk    =       435855
                                             LR chi2(2)      =    113.53
Log likelihood  =   -4978.1914               Prob > chi2     =    0.0000
```

_t	Haz. Ratio	Std. Err.	z	P>|z|	[95% Conf. Interval]
x1	2.273456	.216537	8.62	0.000	1.886311 2.740059
x2	.329011	.0685638	-5.33	0.000	.2186883 .4949888

with these data. You now wish to fit that same model but this time use the values of x1 and x2 at baseline. You do this by typing

```
. stbase, replace
notes:
   1. no gaps
   2. there were multiple failures or reentries after failures
   3. baseline data has multiple records per id(id)
   4. all records have covariate values at baseline
```

(Continued on next page)

```
. stcox x1 x2

Iteration 0:   log likelihood = -7886.9779
Iteration 1:   log likelihood = -7863.9974
Iteration 2:   log likelihood = -7863.9295
Iteration 3:   log likelihood = -7863.9295
Refining estimates:
Iteration 0:   log likelihood = -7863.9295

Cox regression -- Breslow method for ties

No. of subjects =         926              Number of obs   =       1734
No. of failures =        1337
Time at risk    =      435855
                                           LR chi2(2)      =      46.10
Log likelihood  =   -7863.9295             Prob > chi2     =     0.0000
```

_t	Haz. Ratio	Std. Err.	z	P>|z|	[95% Conf. Interval]	
x1	1.413195	.1107945	4.41	0.000	1.211903	1.647921
x2	.4566673	.0765272	-4.68	0.000	.3288196	.6342233

Another way you could perform the analysis is to type

```
. generate x1_0 = x1
. generate x2_0 = x2
. stfill x1_0 x2_0, baseline
. stcox x1 x2
```

See [ST] **stfill**. The method you use makes no difference, but if there were many explanatory variables, stbase would be easier.

stbase changes the data to record the same events but changes the values of all other variables to their values at the earliest time the subject was observed.

stbase also simplifies the st data where possible. Say that one of your subjects has three records in the original data and ends in a failure:

```
                                                                    ——> time
|———| |———| |——X|
```

After running `stbase`, this subject would have one record in the data:

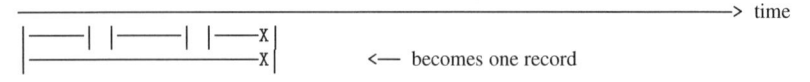

```
                                                                    ——> time
|———| |———| |——X|
|————————————X|          <— becomes one record
```

Here are some other examples of how `stbase` would process records with gaps and multiple failure events:

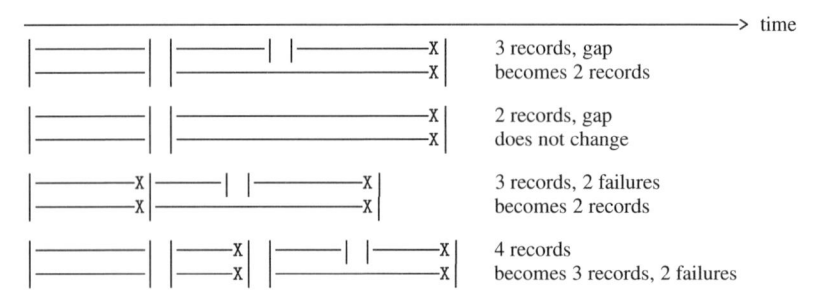

```
                                                                    ——> time
|———|    |——| |————X|      3 records, gap
|——————————————X|           becomes 2 records

|———|       |————————X|     2 records, gap
|———|       |————————X|     does not change

|——X|———| |———————X|        3 records, 2 failures
|——X|———————————X|          becomes 2 records

|———| |—X| |——| |———X|      4 records
|————X|————————X|           becomes 3 records, 2 failures
```

The following example shows numerically what is shown in the diagram above.

```
. use http://www.stata-press.com/data/r11/stbasexmpl, clear
. list id time0 time wgt death, sepby(id)
```

	id	time0	time	wgt	death
1.	1	0	2	114	0
2.	1	3	5	110	0
3.	1	5	11	118	1
4.	2	0	2	120	0
5.	2	3	11	111	1
6.	3	0	2	108	1
7.	3	2	4	105	0
8.	3	4	7	113	1
9.	4	0	2	98	0
10.	4	3	4	101	1
11.	4	5	6	106	0
12.	4	6	11	104	1

```
. stset time, id(id) fail(death) time0(time0) exit(time .)
                id:  id
     failure event:  death != 0 & death < .
obs. time interval:  (time0, time]
 exit on or before:  time .
```

```
        12  total obs.
         0  exclusions
```

```
        12  obs. remaining, representing
         4  subjects
         6  failures in multiple failure-per-subject data
        36  total analysis time at risk, at risk from t =         0
                                   earliest observed entry t =    0
                                        last observed exit t =   11
```

`. list, sepby(id)`

	id	time0	time	wgt	death	_st	_d	_t	_t0
1.	1	0	2	114	0	1	0	2	0
2.	1	3	5	110	0	1	0	5	3
3.	1	5	11	118	1	1	1	11	5
4.	2	0	2	120	0	1	0	2	0
5.	2	3	11	111	1	1	1	11	3
6.	3	0	2	108	1	1	1	2	0
7.	3	2	4	105	0	1	0	4	2
8.	3	4	7	113	1	1	1	7	4
9.	4	0	2	98	0	1	0	2	0
10.	4	3	4	101	1	1	1	4	3
11.	4	5	6	106	0	1	0	6	5
12.	4	6	11	104	1	1	1	11	6

```
. stbase, replace
         failure _d:  death
   analysis time _t:  time
   exit on or before:  time .
                  id:  id
notes:
   1.  there were gaps
   2.  there were multiple failures or reentries after failures
   3.  baseline data has multiple records per id(id)
   4.  all records have covariate values at baseline
. list, sepby(id)
```

	id	time0	time	wgt	death	_st	_d	_t	_t0
1.	1	0	2	114	0	1	0	2	0
2.	1	3	11	114	1	1	1	11	3
3.	2	0	2	120	0	1	0	2	0
4.	2	3	11	120	1	1	1	11	3
5.	3	0	2	108	1	1	1	2	0
6.	3	2	7	108	1	1	1	7	2
7.	4	0	2	98	0	1	0	2	0
8.	4	3	4	98	1	1	1	4	3
9.	4	5	11	98	1	1	1	11	5

stbase with the at() option

stbase, at() produces a cross-sectional dataset recording the status of each subject at the specified time. This new dataset is not st. Four "new" variables are created:

- the first entry time for the subject,
- the time on gap,
- the time at risk, and
- the number of failures during the time at risk.

The names given to those variables depend on how your data are stset. Pretend that your stset command was

```
. stset var1, failure(var2) time0(var3) ...
```

Then

the first entry time	will be named	*var3* or time0 or _t0
the time on gap	will be named	gap() or gap or gaptime
the time at risk	will be named	*var1*
the number of (or whether) failures	will be named	*var2* or failure or _d

The names may vary because, for instance, if you did not specify a *var2* variable when you stset your data, stbase, at() looks around for a name.

You need not memorize this; the names are obvious from the output produced by stbase, at().

Consider the actions of stbase, at() with some particular st datasets. Pretend that the command given is

```
. use http://www.stata-press.com/data/r11/stbasexmpl2, clear
. list id time0 time wgt death, sepby(id)
```

	id	time0	time	wgt	death
1.	1	0	2	114	0
2.	1	2	8	110	0
3.	1	8	11	118	1
4.	2	0	1	120	0
5.	2	1	3	111	0
6.	2	3	8	108	0
7.	2	8	10	98	1

```
. stset time, id(id) fail(death) time0(time0)
                id:  id
     failure event:  death != 0 & death < .
obs. time interval:  (time0, time]
 exit on or before:  failure
```

```
         7  total obs.
         0  exclusions

         7  obs. remaining, representing
         2  subjects
         2  failures in single failure-per-subject data
        21  total analysis time at risk, at risk from t =         0
                                  earliest observed entry t =     0
                                       last observed exit t =    11
. list, sepby(id)
```

	id	time0	time	wgt	death	_st	_d	_t	_t0
1.	1	0	2	114	0	1	0	2	0
2.	1	2	8	110	0	1	0	8	2
3.	1	8	11	118	1	1	1	11	8
4.	2	0	1	120	0	1	0	1	0
5.	2	1	3	111	0	1	0	3	1
6.	2	3	8	108	0	1	0	8	3
7.	2	8	10	98	1	1	1	10	8

(*Continued on next page*)

```
. stbase, at(5) replace
        failure _d:  death
   analysis time _t:  time
                id:  id

         data now cross-section at time 5
```

Variable	description
id	subject identifier
time0	first entry time
gap	time on gap
time	time at risk
death	number of failures during interval time

Variable	Obs	Mean	Std. Dev.	Min	Max
time0	2	0	0	0	0
gap	2	0	0	0	0
time	2	5	0	5	5
death	2	0	0	0	0

```
. list
```

	id	wgt	death	time	time0	gap
1.	1	114	0	5	0	0
2.	2	120	0	5	0	0

thus producing a cross-section at analysis time 5.

Note that the value of time specified with the at() option must correspond to time in the analysis scale, i.e., t. See [ST] **stset** for a definition of analysis time.

Single-failure st data where all subjects enter at time 0

The result of stbase, at(5) would be one record per subject. Any subject who was censored before time 5 would not appear in the data; the rest would. Those that failed after time 5 will be recorded as having been censored at time 5 (*failvar* = 0); those that failed at time 5 or earlier will have *failvar* = 1.

timevar will contain

    for the failures:
        time of failure   if failed on or before time 5 or
        5   if the subject has not failed yet

    for the censored:
        5   if the subject has not failed yet

With such data, you could perform

- logistic regression of *failvar* on any of the characteristics or
- incidence-rate analysis, summing the failures (perhaps within strata) and the time at risk, *timevar*.

With these data, you could examine 5-year survival probabilities.

Single-failure st data where some subjects enter after time 0

The data produced by `stbase, at(5)` would be similar to the above, except

- persons who enter on or after time 5 would not be included in the data (because they have not entered yet) and
- the time at risk, *timevar*, would properly account for the time at which each patient entered.

timevar (the time at risk) will contain

for the failures:	
time of failure or less	if failed on or before time 5 (or less because the subject may not have entered at time 0); or
5 or less	if the subject has not failed yet (or less because the subject may not have entered at time 0)
for the censored:	
5 or less	if the subject has not failed yet (or less because the subject may not have entered at time 0)

Depending on the analysis you are performing, you may have to discard those that enter late. This is easy to do because t0 contains the first time of entry.

With these data, you could perform the following:

- Logistic regression of *failvar* on any of the characteristics, but only if you restricted the sample to `if t0 == 0` because those who entered after time 0 have a lesser risk of failing over the fixed interval.
- Incidence-rate analysis, summing the failures (perhaps within stratum) and the time at risk, *timevar*. Here you would have to do nothing differently from what you did in the previous example. The time-at-risk variable already includes the time of entry for each patient.

Single-failure st data with gaps and perhaps delayed entry

These data will be similar to the delayed-entry, no-gap data, but `gap` will contain 0 only for those observations that have no gap.

If analyzing these data, you could perform

- logistic regression, but the sample must be restricted to `if t0 == 0 & gap == 0`, or
- incidence-rate analysis, and nothing would need to be done differently; the time at risk, *timevar*, accounts for late entry and gaps.

Multiple-failure st data

The multiple-failure case parallels the single-failure case, except that `fail` will not solely contain 0 and 1; it will contain 0, 1, 2, ..., depending on the number of failures observed. Regardless of late entry, gaps, etc., you could perform

- Poisson regression of `fail`, the number of events, but remember to specify `exposure(`*timevar*`)`, and
- incidence-rate analysis.

Methods and formulas

stbase is implemented as an ado-file.

Also see

[ST] **stfill** — Fill in by carrying forward values of covariates

[ST] **stset** — Declare data to be survival-time data

Title

stci — Confidence intervals for means and percentiles of survival time

Syntax

stci [*if*] [*in*] [, *options*]

options	description
Main	
by(*varlist*)	report separate summaries by grouping variables
<u>m</u>edian	calculate median survival times; the default
<u>r</u>mean	calculate mean survival time restricted to longest follow-up time
<u>e</u>mean	calculate the mean survival time by exponentially extending the survival curve to zero
p(*#*)	compute the *#* percentile of survival times
<u>cc</u>orr	calculate the standard error for **rmean** using a continuity correction
<u>nosh</u>ow	do not show st setting information
dd(*#*)	set maximum number of decimal digits to report
<u>l</u>evel(*#*)	set confidence level; default is level(95)
<u>g</u>raph	plot exponentially extended survivor function
<u>t</u>max(*#*)	set maximum analysis time of *#* to be plotted
Plot	
cline_options	affect rendition of the plotted lines
Add plots	
addplot(*plot*)	add other plots to the generated graph
Y axis, X axis, Titles, Legend, Overall	
twoway_options	any options other than by() documented in [G] ***twoway_options***

You must stset your data before using stci; see [ST] **stset**.
by is allowed; see [D] **by**.

Menu

Statistics > Survival analysis > Summary statistics, tests, and tables > CIs for means and percentiles of survival time

Description

stci computes means and percentiles of survival time, standard errors, and confidence intervals. For multiple-event data, survival time is the time until a failure.

stci can be used with single- or multiple-record or single- or multiple-failure st data.

111

Options

☐ Main ☐

by(*varlist*) requests separate summaries for each group, along with an overall total. Observations are in the same group if they have equal values of the variables in *varlist*. *varlist* may contain any number of variables, each of which may be string or numeric.

median specifies median survival times. This is the default.

rmean and emean specify mean survival times. If the longest follow-up time is censored, emean (extended mean) computes the mean survival by exponentially extending the survival curve to zero, and rmean (restricted mean) computes the mean survival time restricted to the longest follow-up time. If the longest follow-up time is a failure, the restricted mean survival time and the extended mean survival time are equal.

p(#) specifies the percentile of survival time to be computed. For example, p(25) will compute the 25th percentile of survival times, and p(75) will compute the 75th percentile of survival times. Specifying p(50) is the same as specifying the median option.

ccorr specifies that the standard error for the restricted mean survival time be computed using a continuity correction. ccorr is valid only with the rmean option.

noshow prevents stci from showing the key st variables. This option is seldom used because most people type stset, show or stset, noshow to set whether they want to see these variables mentioned at the top of the output of every st command; see [ST] **stset**.

dd(#) specifies the maximum number of decimal digits to be reported for standard errors and confidence intervals. This option affects only how values are reported and not how they are calculated.

level(#) specifies the confidence level, as a percentage, for confidence intervals. The default is level(95) or as set by set level; see [U] **20.7 Specifying the width of confidence intervals**.

graph specifies that the exponentially extended survivor function be plotted. This option is valid only when the emean option is also specified and is not valid in conjunction with the by() option.

tmax(#) is for use with the graph option. It specifies the maximum analysis time to be plotted.

☐ Plot ☐

cline_options affect the rendition of the plotted lines; see [G] *cline_options*.

☐ Add plots ☐

addplot(*plot*) provides a way to add other plots to the generated graph; see [G] *addplot_option*.

☐ Y axis, X axis, Titles, Legend, Overall ☐

twoway_options are any of the options documented in [G] *twoway_options*, excluding by(). These include options for titling the graph (see [G] *title_options*) and for saving the graph to disk (see [G] *saving_option*).

Remarks

Remarks are presented under the following headings:

Single-failure data
Multiple-failure data

Single-failure data

Here is an example of stci with single-record survival data:

```
. use http://www.stata-press.com/data/r11/page2
. stset, noshow
. stci
```

	no. of subjects	50%	Std. Err.	[95% Conf. Interval]	
total	40	232	2.562933	213	239

```
. stci, by(group)
```

group	no. of subjects	50%	Std. Err.	[95% Conf. Interval]	
1	19	216	5.171042	190	234
2	21	233	2.179595	232	280
total	40	232	2.562933	213	239

In the example above, we obtained the median survival time, by default.

To obtain the 25th or any other percentile of survival time, specify the p(#) option.

```
. stci, p(25)
```

	no. of subjects	25%	Std. Err.	[95% Conf. Interval]	
total	40	198	10.76878	164	220

```
. stci, p(25) by(group)
```

group	no. of subjects	25%	Std. Err.	[95% Conf. Interval]	
1	19	190	8.411659	143	213
2	21	232	14.88531	142	233
total	40	198	10.76878	164	220

The p-percentile of survival time is the analysis time at which $p\%$ of subjects have failed and $1 - p\%$ have not. In the table above, 25% of subjects in group 1 failed by time 190, whereas 25% of subjects in group 2 failed by time 232, indicating a better survival experience for this group.

(Continued on next page)

We can verify the quantities reported by stci by plotting and examining the Kaplan–Meier survival curves.

```
. sts graph, by(group)
```

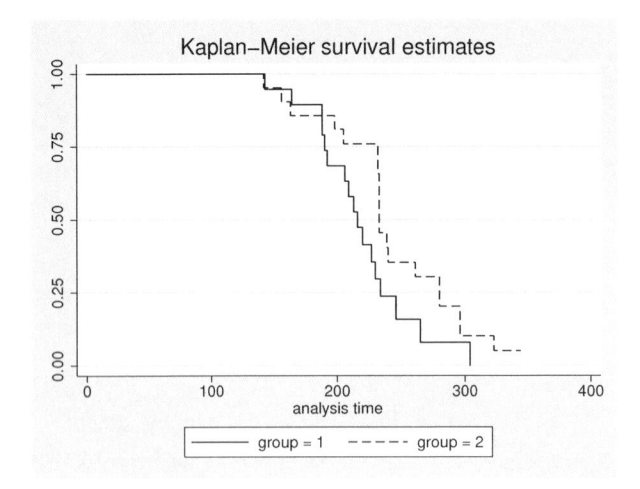

Kaplan–Meier survival estimates

[Graph showing survival probability vs analysis time for group = 1 and group = 2]

The mean survival time reported by rmean is calculated as the area under the Kaplan–Meier survivor function. If the observation with the largest analysis time is censored, the survivor function does not go to zero. Consequently, the area under the curve underestimates the mean survival time.

In the graph above, the survival probability for group = 1 goes to 0 at analysis time 344, but the survivor function for group = 2 never goes to 0. For these data, the mean survival time for group = 1 will be properly estimated, but it will be underestimated for group = 2. When we specify the rmean option, Stata informs us if any of the mean survival times is underestimated.

```
. stci, rmean by(group)
```

group	no. of subjects	restricted mean	Std. Err.	[95% Conf. Interval]
1	19	218.7566	9.122424	200.877 236.636
2	21	241.8571(*)	11.34728	219.617 264.097
total	40	231.3522(*)	7.700819	216.259 246.446

(*) largest observed analysis time is censored, mean is underestimated.

Stata flagged the mean for group = 2 and the overall mean as being underestimated.

If the largest observed analysis time is censored, stci's emean option extends the survivor function from the last observed time to zero by using an exponential function and computes the area under the entire curve.

```
. stci, emean
```

	no. of subjects	extended mean
total	40	234.2557

The resulting area must be evaluated with care because it is an ad hoc approximation that can at times be misleading. We recommend that you plot and examine the extended survivor function. This is facilitated by the use of stci's graph option.

. stci, emean graph

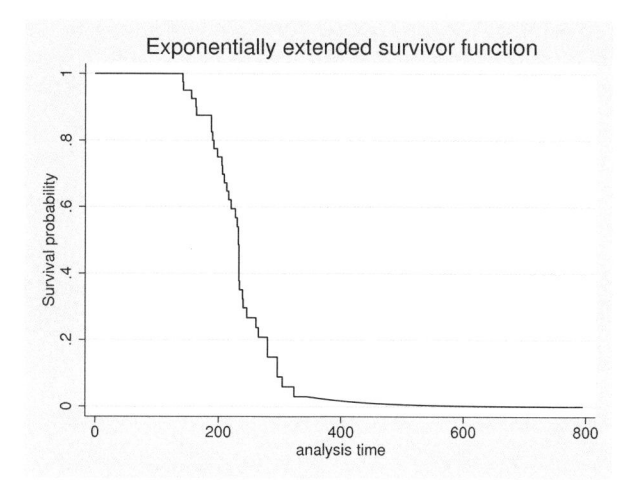

Exponentially extended survivor function

stci also works with multiple-record survival data. Here is a summary of the multiple-record Stanford heart transplant data introduced in [ST] **stset**:

```
. use http://www.stata-press.com/data/r11/stan3
(Heart transplant data)
. stset, noshow
. stci
```

	no. of subjects	50%	Std. Err.	[95% Conf. Interval]	
total	103	100	38.64425	69	219

stci with the **by()** option may produce results with multiple-record data that you might think are in error:

. stci, by(posttran)

posttran	no. of subjects	50%	Std. Err.	[95% Conf. Interval]	
0	103	149	22.16591	69	340
1	69	96	38.01968	45	285
total	103	100	38.64425	69	219

For the number of subjects, $103 + 69 \neq 103$. The **posttran** variable is not constant for the subjects in this dataset:

. stvary posttran

	subjects for whom the variable is				
variable	constant	varying	never missing	always missing	sometimes missing
posttran	34	69	103	0	0

In this dataset, subjects have one or two records. All subjects were eligible for heart transplantation. They have one record if they die or are lost because of censoring before transplantation, and they have two records if the operation was performed. Then the first record records their survival up to transplantation, and the second records their subsequent survival. posttran is 0 in the first record and 1 in the second.

Therefore, all 103 subjects have records with posttran = 0, and when stci reported results for this group, it summarized the pretransplantation survival. The median survival time was 149 days.

The posttran = 1 line of stci's output summarizes the posttransplantation survival: 69 patients underwent transplantation, and the median survival time was 96 days. For these data, this is not 96 more days, but 96 days in total. That is, the clock was not reset on transplantation. Thus, without attributing cause, we can describe the differences between the groups as an increased hazard of death at early times followed by a decreased hazard later.

Multiple-failure data

If you simply type stci with multiple-failure data, the reported survival time is the survival time to the first failure, assuming that the hazard function is not indexed by number of failures.

Here we have some multiple-failure data:

```
. use http://www.stata-press.com/data/r11/mfail2
. st
-> stset t, id(id) failure(d) time0(t0) exit(time .) noshow
              id:  id
   failure event:  d != 0 & d < .
obs. time interval:  (t0, t]
 exit on or before:  time .
. stci
```

	no. of subjects	50%	Std. Err.	[95% Conf. Interval]	
total	926	420	13.42537	394	451

To understand this output, let's also obtain output for each failure separately:

```
. stgen nf = nfailures()
. stci, by(nf)
```

nf	no. of subjects	50%	Std. Err.	[95% Conf. Interval]	
0	926	399	11.50173	381	430
1	529	503	13.68105	425	543
2	221	687	16.83127	549	817
3	58	.	.	.	.
total	926	420	13.42537	394	451

The stgen command added, for each subject, a variable containing the number of previous failures. nf is 0 for a subject, up to and including the first failure. Then nf is 1 up to and including the second failure, and then it is 2, and so on; see [ST] **stgen**.

The first line, corresponding to nf = 0, states that among those who had experienced no failures yet, the median time to first failure is 399.

Similarly, the second line, corresponding to nf = 1, is for those who have already experienced one failure. The median time of second failures is 503.

When we simply typed stci, we obtained the same information shown as the total line of the more detailed output. The total survival time distribution is an estimate of the distribution of the time to first failure, assuming that the hazard function, $h(t)$, is the same across failures—that the second failure is no different from the first failure. This is an odd definition of *same* because the clock, t, is not reset in $h(t)$ upon failure. The hazard of a failure—any failure—at time t is $h(t)$.

Another definition of *same* would have it that the hazard of a failure is given by $h(\tau)$, where τ is the time since the last failure—that the process resets itself. These definitions are different unless $h()$ is a constant function of t.

Let's examine this multiple-failure data, assuming that the process repeats itself. The key variables in this st data are id, t0, t, and d:

```
. st
-> stset t, id(id) failure(d) time0(t0) exit(time .) noshow
              id:  id
   failure event:  d != 0 & d < .
obs. time interval:  (t0, t]
 exit on or before:  time .
```

Our goal, for each subject, is to reset t0 and t to 0 after every failure event. We must trick Stata, or at least trick stset because it will not let us set data where the same subject has multiple records summarizing the overlapping periods. The trick is create a new id variable that is different for every id–nf combination (remember, nf is the variable we previously created that records the number of prior failures). Then each of the "new" subjects can have their clock start at time 0:

```
. egen newid = group(id nf)
. sort newid t
. by newid: replace t = t - t0[1]
(808 real changes made)
. by newid: generate newt0 = t0 - t0[1]
. stset t, failure(d) id(newid) time0(newt0)
              id:  newid
   failure event:  d != 0 & d < .
obs. time interval:  (newt0, t]
 exit on or before:  failure

     1734  total obs.
        0  exclusions

     1734  obs. remaining, representing
     1734  subjects
      808  failures in single failure-per-subject data
   435444  total analysis time at risk, at risk from t =         0
                                 earliest observed entry t =    0
                                      last observed exit t =  797
```

stset no longer thinks that we have multiple-failure data. Whereas with id, subjects had multiple failures, newid gives a unique identity to each id–nf combination. Each "new" subject has at most one failure.

(Continued on next page)

```
. stci, by(nf)

        failure _d:  d
   analysis time _t:  t
                 id:  newid
```

| | no. of | | | | |
nf	subjects	50%	Std. Err.	[95% Conf. Interval]	
0	926	399	11.22457	381	430
1	529	384	9.16775	359	431
2	221	444	7.406977	325	515
3	58	.	.	.	.
total	1734	404	10.29992	386	430

Compare this table with the one we previously obtained. The number of subjects is the same, but the survival times differ because now we measure the times from one failure to the next, whereas previously we measured the time from a fixed point. The time between events in these data appears to be independent of event number.

Similarly, we can obtain the mean survival time for these data restricted to the longest follow-up time:

```
. stci, rmean by(nf)

        failure _d:  d
   analysis time _t:  t
                 id:  newid
```

| | no. of | restricted | | | |
nf	subjects	mean	Std. Err.	[95% Conf. Interval]	
0	926	399.1802	8.872794	381.79	416.571
1	529	397.0077(*)	13.36058	370.821	423.194
2	221	397.8051(*)	25.78559	347.266	448.344
3	58	471(*)	0	471	471
total	1734	404.7006	7.021657	390.938	418.463

(*) largest observed analysis time is censored, mean is underestimated.

Saved results

stci saves the following in r():

Scalars
    r(N_sub)   number of subjects   r(se)   standard error
    r(p#)   #th percentile   r(lb)   lower bound of CI
    r(rmean)   restricted mean   r(ub)   upper bound of CI
    r(emean)   extended mean

Methods and formulas

stci is implemented as an ado-file.

The percentiles of survival times are obtained from $S(t)$, the Kaplan–Meier product-limit estimate of the survivor function. The 25th percentile, for instance, is obtained as the minimum value of t such that $S(t) \leq 0.75$. The restricted mean is obtained as the area under the Kaplan–Meier product-limit survivor curve. The extended mean is obtained by extending the Kaplan–Meier product-limit survivor curve to zero by using an exponentially fitted curve and then computing the area under the entire curve. If the longest follow-up time ends in failure, the Kaplan–Meier product-limit survivor curve goes to zero, and the restricted mean and extended mean are identical.

The large-sample standard error for the pth percentile of the distribution is given by Collett (2003, 35) and Klein and Moeschberger (2003, 122) as

$$\frac{\sqrt{\widehat{g}}}{\widehat{f}(t_p)}$$

where $\widehat{g}$ is the Greenwood pointwise standard-error estimate for $\widehat{S}(t_p)$ and $\widehat{f}(t_p)$ is the estimated density function at the pth percentile.

Confidence intervals, however, are not calculated based on this standard error. For a given confidence level, the upper confidence limit for the pth percentile is defined as the first time at which the upper confidence limit for $S(t)$ (based on a $\ln\{-\ln S(t)\}$ transformation) is less than or equal to $1 - p/100$, and, similarly, the lower confidence limit is defined as the first time at which the lower confidence limit of $S(t)$ is less than or equal to $1 - p/100$.

The restricted mean is obtained as the area under the Kaplan–Meier product-limit survivor curve. The extended mean is obtained by extending the Kaplan–Meier product-limit survivor curve to zero by using an exponentially fitted curve and then computing the area under the entire curve. If the longest follow-up time ends in failure, the Kaplan–Meier product-limit survivor curve goes to zero, and the restricted mean and the extended mean are identical.

The standard error for the estimated restricted mean is computed as given by Klein and Moeschberger (2003, 118) and Collett (2003, 340):

$$\widehat{SE} = \sum_{i=1}^{D} \widehat{A}_i \sqrt{\frac{d_i}{R_i(R_i - d_i)}}$$

where the sum is over all distinct failure times, $\widehat{A}_i$ is the estimated area under the curve from time i to the maximum follow-up time, R_i is the number of subjects at risk at time i, and d_i is the number of failures at time i.

The $100(1 - \alpha)\%$ confidence interval for the estimated restricted mean is computed as

$$\widehat{A}_i \pm Z_{1-\alpha/2} \widehat{SE}$$

References

Collett, D. 2003. *Modelling Survival Data in Medical Research*. 2nd ed. London: Chapman & Hall/CRC.

Klein, J. P., and M. L. Moeschberger. 2003. *Survival Analysis: Techniques for Censored and Truncated Data*. 2nd ed. New York: Springer.

Also see

[ST] **stdescribe** — Describe survival-time data

[ST] **stir** — Report incidence-rate comparison

[ST] **sts** — Generate, graph, list, and test the survivor and cumulative hazard functions

[ST] **stset** — Declare data to be survival-time data

[ST] **stvary** — Report whether variables vary over time

[ST] **stptime** — Calculate person-time, incidence rates, and SMR

Title

> **stcox** — Cox proportional hazards model

Syntax

> stcox [*varlist*] [*if*] [*in*] [, *options*]

options	description
Model	
est̲imate	fit model without covariates
st̲rata(*varnames*)	strata ID variables
sh̲ared(*varname*)	shared-frailty ID variable
off̲set(*varname*)	include *varname* in model with coefficient constrained to 1
br̲eslow	use Breslow method to handle tied failures; the default
ef̲ron	use Efron method to handle tied failures
exactm	use exact marginal-likelihood method to handle tied failures
exactp	use exact partial-likelihood method to handle tied failures
Time varying	
tvc(*varlist*)	time-varying covariates
texp(*exp*)	multiplier for time-varying covariates; default is texp(_t)
SE/Robust	
vce(*vcetype*)	*vcetype* may be oim, r̲obust, c̲luster *clustvar*, bo̲otstrap, or j̲ackknife
no̲adjust	do not use standard degree-of-freedom adjustment
Reporting	
l̲evel(*#*)	set confidence level; default is level(95)
nohr	report coefficients, not hazard ratios
nosh̲ow	do not show st setting information
display_options	control spacing and display of omitted variables and base and empty cells
Maximization	
maximize_options	control the maximization process; seldom used
† coef̲legend	display coefficients' legend instead of coefficient table

121

†coeflegend does not appear in the dialog box.

You must stset your data before using stcox; see [ST] **stset**.

varlist may contain factor variables; see [U] **11.4.3 Factor variables**.

bootstrap, by, fracpoly, jackknife, mfp, mi estimate, nestreg, statsby, stepwise, and svy are allowed; see [U] **11.1.10 Prefix commands**.

vce(bootstrap) and vce(jackknife) are not allowed with the mi estimate prefix.

estimate, shared(), efron, exactm, exactp, tvc(), texp(), vce(), and noadjust are not allowed with the svy prefix.

fweights, iweights, and pweights may be specified using stset; see [ST] **stset**. Weights are not supported with efron and exactp. Also weights may not be specified if you are using the bootstrap prefix with the stcox command.

See [U] **20 Estimation and postestimation commands** for more capabilities of estimation commands.

Menu

Statistics > Survival analysis > Regression models > Cox proportional hazards model

Description

stcox fits, via maximum likelihood, proportional hazards models on st data. stcox can be used with single- or multiple-record or single- or multiple-failure st data.

Options for stcox

⎯⎯⎯⎯⎯⎯⎯[Model]⎯⎯

estimate forces fitting of the null model. All Stata estimation commands redisplay results when the command name is typed without arguments. So does stcox. What if you wish to fit a Cox model on $x_j\beta$, where $x_j\beta$ is defined as 0? Logic says that you would type stcox. There are no explanatory variables, so there is nothing to type after the command. Unfortunately, this looks the same as stcox typed without arguments, which is a request to redisplay results.

To fit the null model, type stcox, estimate.

strata(*varnames*) specifies up to five strata variables. Observations with equal values of the strata variables are assumed to be in the same stratum. Stratified estimates (equal coefficients across strata but with a baseline hazard unique to each stratum) are then obtained.

shared(*varname*) specifies that a Cox model with shared frailty be fit. Observations with equal value of *varname* are assumed to have shared (the same) frailty. Across groups, the frailties are assumed to be gamma-distributed latent random effects that affect the hazard multiplicatively, or, equivalently, the logarithm of the frailty enters the linear predictor as a random offset. Think of a shared-frailty model as a Cox model for panel data. *varname* is a variable in the data that identifies the groups.

Shared-frailty models are discussed more in *Cox regression with shared frailty*.

offset(*varname*); see [R] **estimation options**.

breslow, efron, exactm, and exactp specify the method for handling tied failures in the calculation of the log partial likelihood (and residuals). breslow is the default. Each method is described in *Treatment of tied failure times*. efron and the exact methods require substantially more computer time than the default breslow option. exactm and exactp may not be specified with tvc(), vce(robust), or vce(cluster *clustvar*).

Time varying

tvc(*varlist*) specifies those variables that vary continuously with respect to time, i.e., time-varying covariates. This is a convenience option used to speed up calculations and to avoid having to stsplit the data over many failure times.

texp(*exp*) is used in conjunction with tvc(*varlist*) to specify the function of analysis time that should be multiplied by the time-varying covariates. For example, specifying texp(log(_t)) would cause the time-varying covariates to be multiplied by the logarithm of analysis time. If tvc(*varlist*) is used without texp(*exp*), Stata understands that you mean texp(_t) and thus multiplies the time-varying covariates by the analysis time.

Both tvc(*varlist*) and texp(*exp*) are explained more in the section on *Cox regression with continuous time-varying covariates* below.

SE/Robust

vce(*vcetype*) specifies the type of standard error reported, which includes types that are derived from asymptotic theory, that are robust to some kinds of misspecification, that allow for intragroup correlation, and that use bootstrap or jackknife methods; see [R] **vce_option**.

noadjust is for use with vce(robust) or vce(cluster *clustvar*). noadjust prevents the estimated variance matrix from being multiplied by $N/(N-1)$ or $g/(g-1)$, where g is the number of clusters. The default adjustment is somewhat arbitrary because it is not always clear how to count observations or clusters. In such cases, however, the adjustment is likely to be biased toward 1, so we would still recommend making it.

Reporting

level(*#*); see [R] **estimation options**.

nohr specifies that coefficients be displayed rather than exponentiated coefficients or hazard ratios. This option affects only how results are displayed and not how they are estimated. nohr may be specified at estimation time or when redisplaying previously estimated results (which you do by typing stcox without a variable list).

noshow prevents stcox from showing the key st variables. This option is seldom used because most people type stset, show or stset, noshow to set whether they want to see these variables mentioned at the top of the output of every st command; see [ST] **stset**.

display_options: <u>noomit</u>ted, vsquish, <u>noempty</u>cells, <u>base</u>levels, <u>allbase</u>levels; see [R] **estimation options**.

Maximization

maximize_options: <u>iterate</u>(*#*), [<u>no</u>]<u>log</u>, <u>trace</u>, <u>tol</u>erance(*#*), <u>ltol</u>erance(*#*), <u>nrtol</u>erance(*#*), <u>nonrtol</u>erance; see [R] **maximize**. These options are seldom used.

The following option is available with stcox but is not shown in the dialog box:

coeflegend; see [R] **estimation options**.

(*Continued on next page*)

Remarks

Remarks are presented under the following headings:

> Cox regression with uncensored data
> Cox regression with censored data
> Treatment of tied failure times
> Cox regression with discrete time-varying covariates
> Cox regression with continuous time-varying covariates
> Robust estimate of variance
> Cox regression with multiple-failure data
> Stratified estimation
> Cox regression with shared frailty

What follows is a summary of what can be done with stcox. For a complete tutorial, see Cleves et al. (2008), which devotes three chapters to this topic.

In the Cox proportional hazards model (Cox 1972), the hazard is assumed to be

$$h(t) = h_0(t) \exp(\beta_1 x_1 + \cdots + \beta_k x_k)$$

The Cox model provides estimates of $\beta_1, \ldots, \beta_k$ but provides no direct estimate of $h_0(t)$—the baseline hazard. Formally, the function $h_0(t)$ is not directly estimated, but it is possible to recover an estimate of the cumulative hazard $H_0(t)$ and, from that, an estimate of the baseline survivor function $S_0(t)$.

stcox fits the Cox proportional hazards model; that is, it provides estimates of β and its variance–covariance matrix. Estimates of $H_0(t)$, $S_0(t)$, and other predictions and diagnostics are obtained with predict after stcox; see [ST] **stcox postestimation**.

stcox with the strata() option will produce stratified Cox regression estimates. In the stratified estimator, the hazard at time t for a subject in group i is assumed to be

$$h_i(t) = h_{0i}(t) \exp(\beta_1 x_1 + \cdots + \beta_k x_k)$$

That is, the coefficients are assumed to be the same, regardless of group, but the baseline hazard can be group specific.

Regardless of whether you specify strata(), the default variance estimate is to calculate the conventional, inverse matrix of negative second derivatives. The theoretical justification for this estimator is based on likelihood theory. The vce(robust) option instead switches to the robust measure developed by Lin and Wei (1989). This variance estimator is a variant of the estimator discussed in [U] **20.16 Obtaining robust variance estimates**.

stcox with the shared() option fits a Cox model with shared frailty. A *frailty* is a group-specific latent random effect that multiplies into the hazard function. The distribution of the frailties is gamma with mean 1 and variance to be estimated from the data. Shared-frailty models are used to model within-group correlation. Observations within a group are correlated because they share the same frailty.

We give examples below with uncensored, censored, time-varying, and recurring failure data, but it does not matter in terms of what you type. Once you have stset your data, to fit a model you type stcox followed by the names of the explanatory variables. You do this whether your dataset has single or multiple records, includes censored observations or delayed entry, or even has single or multiple failures. You use stset to describe the properties of the data, and then that information is available to stcox—and all the other st commands—so that you do not have to specify it again.

Cox regression with uncensored data

▷ Example 1

We wish to analyze an experiment testing the ability of emergency generators with a new-style bearing to withstand overloads. For this experiment, the overload protection circuit was disabled, and the generators were run overloaded until they burned up. Here are our data:

```
. use http://www.stata-press.com/data/r11/kva
(Generator experiment)

. list
```

	failtime	load	bearings
1.	100	15	0
2.	140	15	1
3.	97	20	0
4.	122	20	1
5.	84	25	0
6.	100	25	1
7.	54	30	0
8.	52	30	1
9.	40	35	0
10.	55	35	1
11.	22	40	0
12.	30	40	1

Twelve generators, half with the new-style bearings and half with the old, were allocated to this destructive test. The first observation reflects an old-style generator (bearings = 0) under a 15-kVA overload. It stopped functioning after 100 hours. The second generator had new-style bearings (bearings = 1) and, under the same overload condition, lasted 140 hours. Paired experiments were also performed under overloads of 20, 25, 30, 35, and 40 kVA.

We wish to fit a Cox proportional hazards model in which the failure rate depends on the amount of overload and the style of the bearings. That is, we assume that bearings and load do not affect the shape of the overall hazard function, but they do affect the relative risk of failure. To fit this model, we type

```
. stset failtime
  (output omitted)
. stcox load bearings
         failure _d:  1 (meaning all fail)
   analysis time _t:  failtime

Iteration 0:   log likelihood = -20.274897
Iteration 1:   log likelihood = -10.515114
Iteration 2:   log likelihood = -8.8700259
Iteration 3:   log likelihood = -8.5915211
Iteration 4:   log likelihood = -8.5778991
Iteration 5:   log likelihood =  -8.577853
Refining estimates:
Iteration 0:   log likelihood =  -8.577853
```

```
Cox regression -- Breslow method for ties
No. of subjects =           12                Number of obs   =         12
No. of failures =           12
Time at risk    =          896
                                              LR chi2(2)      =      23.39
Log likelihood  =    -8.577853                Prob > chi2     =     0.0000

          _t | Haz. Ratio   Std. Err.      z    P>|z|     [95% Conf. Interval]
        load |   1.52647    .2188172     2.95   0.003     1.152576    2.021653
    bearings |  .0636433    .0746609    -2.35   0.019     .0063855     .6343223
```

We find that after controlling for overload, the new-style bearings result in a lower hazard and therefore a longer survivor time.

Once an stcox model has been fit, typing stcox without arguments redisplays the previous results. Options that affect the display, such as nohr—which requests that coefficients rather than hazard ratios be displayed—can be specified upon estimation or when results are redisplayed:

```
. stcox, nohr
Cox regression -- Breslow method for ties
No. of subjects =           12                Number of obs   =         12
No. of failures =           12
Time at risk    =          896
                                              LR chi2(2)      =      23.39
Log likelihood  =    -8.577853                Prob > chi2     =     0.0000

          _t |      Coef.   Std. Err.      z    P>|z|     [95% Conf. Interval]
        load |   .4229578   .1433485     2.95   0.003     .1419999    .7039157
    bearings |  -2.754461   1.173115    -2.35   0.019    -5.053723   -.4551981
```

◁

❑ Technical note

stcox's iteration log looks like a standard Stata iteration log up to where it says "Refining estimates". The Cox proportional-hazards likelihood function is indeed a difficult function, both conceptually and numerically. Until Stata says "Refining estimates", it maximizes the Cox likelihood in the standard way by using double-precision arithmetic. Then just to be sure that the answers are accurate, Stata switches to quad-precision routines (double double precision) and completes the maximization procedure from its current location on the likelihood.

❑

Cox regression with censored data

▷ Example 2

We have data on 48 participants in a cancer drug trial. Of these 48, 28 receive treatment (drug = 1) and 20 receive a placebo (drug = 0). The participants range in age from 47 to 67 years. We wish to analyze time until death, measured in months. Our data include 1 observation for each patient. The variable studytime records either the month of their death or the last month that they were known to be alive. Some of the patients still live, so together with studytime is died, indicating their health status. Persons known to have died—"noncensored" in the jargon—have died = 1, whereas the patients who are still alive—"right-censored" in the jargon—have died = 0.

Here is an overview of our data:

```
. use http://www.stata-press.com/data/r11/drugtr
(Patient Survival in Drug Trial)
. st
-> stset studytime, failure(died)
       failure event:  died != 0 & died < .
obs. time interval:  (0, studytime]
 exit on or before:  failure
. summarize
```

Variable	Obs	Mean	Std. Dev.	Min	Max
studytime	48	15.5	10.25629	1	39
died	48	.6458333	.4833211	0	1
drug	48	.5833333	.4982238	0	1
age	48	55.875	5.659205	47	67
_st	48	1	0	1	1
_d	48	.6458333	.4833211	0	1
_t	48	15.5	10.25629	1	39
_t0	48	0	0	0	0

We typed stset studytime, failure(died) previously; that is how st knew about this dataset. To fit the Cox model, we type

```
. stcox drug age
         failure _d:  died
   analysis time _t:  studytime
Iteration 0:   log likelihood = -99.911448
Iteration 1:   log likelihood = -83.551879
Iteration 2:   log likelihood = -83.324009
Iteration 3:   log likelihood = -83.323546
Refining estimates:
Iteration 0:   log likelihood = -83.323546
Cox regression -- Breslow method for ties
No. of subjects =           48                   Number of obs   =         48
No. of failures =           31
Time at risk    =          744
                                                 LR chi2(2)      =      33.18
Log likelihood  =    -83.323546                  Prob > chi2     =     0.0000
```

_t	Haz. Ratio	Std. Err.	z	P>|z|	[95% Conf. Interval]
drug	.1048772	.0477017	-4.96	0.000	.0430057 .2557622
age	1.120325	.0417711	3.05	0.002	1.041375 1.20526

We find that the drug results in a lower hazard—and therefore a longer survivor time—controlling for age. Older patients are more likely to die. The model as a whole is statistically significant.

The hazard ratios reported correspond to a one-unit change in the corresponding variable. It is more typical to report relative risk for 5-year changes in age. To obtain such a hazard ratio, we create a new age variable such that a one-unit change indicates a 5-year change:

```
. replace age = age/5
age was int now float
(48 real changes made)

. stcox drug age, nolog
         failure _d:  died
   analysis time _t:  studytime

Cox regression -- Breslow method for ties

No. of subjects =          48                  Number of obs   =        48
No. of failures =          31
Time at risk    =         744
                                               LR chi2(2)      =     33.18
Log likelihood  =  -83.323544                  Prob > chi2     =    0.0000
```

_t	Haz. Ratio	Std. Err.	z	P>|z|	[95% Conf. Interval]	
drug	.1048772	.0477017	-4.96	0.000	.0430057	.2557622
age	1.764898	.3290196	3.05	0.002	1.224715	2.543338

Treatment of tied failure times

The proportional hazards model assumes that the hazard function is continuous and, thus, that there are no tied survival times. Because of the way that time is recorded, however, tied events do occur in survival data. In such cases, the partial likelihood must be modified. See *Methods and formulas* for more details on the methods described below.

Stata provides four methods for handling tied failures in calculating the Cox partial likelihood through the breslow, efron, exactm, and exactp options. If there are no ties in the data, the results are identical, regardless of the method selected.

Cox regression is a series of comparisons of those subjects who fail to those subjects at risk of failing; we refer to the latter set informally as a *risk pool*. When there are tied failure times, we must decide how to calculate the risk pools for these tied observations. Assume that there are 2 observations that fail in succession. In the calculation involving the second observation, the first observation is not in the risk pool because failure has already occurred. If the two observations have the same failure time, we must decide how to calculate the risk pool for the second observation and in which order to calculate the two observations.

There are two views of time. In the first, time is continuous, so ties should not occur. If they have occurred, the likelihood reflects the marginal probability that the tied-failure events occurred before the nonfailure events in the risk pool (the order that they occurred is not important). This is called the exact marginal likelihood (option exactm).

In the second view, time is discrete, so ties are expected. The likelihood is changed to reflect this discreteness and calculates the conditional probability that the observed failures are those that fail in the risk pool given the observed number of failures. This is called the exact partial likelihood (option exactp).

Let's assume that there are five subjects—e_1, e_2, e_3, e_4, and e_5—in the risk pool and that subjects e_1 and e_2 fail. Had we been able to observe the events at a better resolution, we might have seen that e_1 failed from risk pool $e_1 + e_2 + e_3 + e_4 + e_5$ and then e_2 failed from risk pool $e_2 + e_3 + e_4 + e_5$. Alternatively, e_2 might have failed first from risk pool $e_1 + e_2 + e_3 + e_4 + e_5$, and then e_1 failed from risk pool $e_1 + e_3 + e_4 + e_5$.

The Breslow method (option `breslow`) for handling tied values simply says that because we do not know the order, we will use the largest risk pool for each tied failure event. This method assumes that both e_1 and e_2 failed from risk pool $e_1 + e_2 + e_3 + e_4 + e_5$. This approximation is fast and is the default method for handling ties. If there are many ties in the dataset, this approximation will not be accurate because the risk pools include too many observations. The Breslow method is an approximation of the exact marginal likelihood.

The Efron method (option `efron`) for handling tied values assumes that the first risk pool is $e_1 + e_2 + e_3 + e_4 + e_5$ and the second risk pool is either $e_2 + e_3 + e_4 + e_5$ or $e_1 + e_3 + e_4 + e_5$. From this, Efron noted that the e_1 and e_2 terms were in the second risk pool with probability 1/2 and so used for the second risk pool $.5(e_1 + e_2) + e_3 + e_4 + e_5$. Efron's approximation is a more accurate approximation of the exact marginal likelihood than Breslow's but takes longer to calculate.

The exact marginal method (option `exactm`) is a misnomer in that the calculation performed is also an *approximation* of the exact marginal likelihood. It is an approximation because it evaluates the likelihood (and derivatives) by using 15-point Gauss–Laguerre quadrature. For small-to-moderate samples, this is slower than the Efron approximation, but the difference in execution time diminishes when samples become larger. You may want to consider the quadrature when deciding to use this method. If the number of tied deaths is large (on average), the quadrature approximation of the function is not well behaved. A little empirical checking suggests that if the number of tied deaths is larger (on average) than 30, the quadrature does not approximate the function well.

When we view time as discrete, the exact partial method (option `exactp`) is the final method available. This approach is equivalent to computing conditional logistic regression where the groups are defined by the risk sets and the outcome is given by the death variable. This is the slowest method to use and can take a significant amount of time if the the number of tied failures and the risk sets are large.

Cox regression with discrete time-varying covariates

▷ Example 3

In [ST] **stset**, we introduce the Stanford heart transplant data in which there are one or two records per patient depending on whether they received a new heart.

This dataset (Crowley and Hu 1977) consists of 103 patients admitted to the Stanford Heart Transplantation Program. Patients were admitted to the program after review by a committee and then waited for an available donor heart. While waiting, some patients died or were transferred out of the program, but 67% received a transplant. The dataset includes the year the patient was accepted into the program along with the patient's age, whether the patient had other heart surgery previously, and whether the patient received a transplant.

In the data, `posttran` becomes 1 when a patient receives a new heart, so it is a time-varying covariate. That does not, however, affect what we type to fit the model:

```
. use http://www.stata-press.com/data/r11/stan3, clear
(Heart transplant data)
. stset t1, failure(died) id(id)
 (output omitted)
```

```
. stcox age posttran surg year
         failure _d:  died
   analysis time _t:  t1
                 id:  id
Iteration 0:   log likelihood = -298.31514
Iteration 1:   log likelihood =  -289.7344
Iteration 2:   log likelihood = -289.53498
Iteration 3:   log likelihood = -289.53378
Iteration 4:   log likelihood = -289.53378
Refining estimates:
Iteration 0:   log likelihood = -289.53378

Cox regression -- Breslow method for ties

No. of subjects =         103                Number of obs   =       172
No. of failures =          75
Time at risk    =     31938.1
                                             LR chi2(4)      =     17.56
Log likelihood  =  -289.53378                Prob > chi2     =    0.0015
```

_t	Haz. Ratio	Std. Err.	z	P>|z|	[95% Conf. Interval]	
age	1.030224	.0143201	2.14	0.032	1.002536	1.058677
posttran	.9787243	.3032597	-0.07	0.945	.5332291	1.796416
surgery	.3738278	.163204	-2.25	0.024	.1588759	.8796
year	.8873107	.059808	-1.77	0.076	.7775022	1.012628

We find that older patients have higher hazards, that patients tend to do better over time, and that patients with prior surgery do better. Whether a patient ultimately receives a transplant does not seem to make much difference.

◁

Cox regression with continuous time-varying covariates

The basic proportional hazards regression assumes the relationship

$$h(t) = h_0(t) \exp(\beta_1 x_1 + \cdots + \beta_k x_k)$$

where $h_0(t)$ is the baseline hazard function. For most purposes, this model is sufficient, but sometimes we may wish to introduce variables of the form $z_i(t) = z_i g(t)$, which vary continuously with time so that

$$h(t) = h_0(t) \exp\left\{\beta_1 x_1 + \cdots + \beta_k x_k + g(t)(\gamma_1 z_1 + \cdots + \gamma_m z_m)\right\} \qquad (1)$$

where $z_1, \ldots, z_m$ are the time-varying covariates and where estimation has the net effect of estimating, say, a regression coefficient, γ_i, for a covariate, $g(t)z_i$, which is a function of the current time.

The time-varying covariates $z_1, \ldots, z_m$ are specified by using the tvc(*varlist*) option, and $g(t)$ is specified by using the texp(*exp*) option, where t in $g(t)$ is analysis time. For example, if we want $g(t) = \log(t)$, we would use texp(log(_t)) because _t stores the analysis time once the data are stset.

Because the calculations in Cox regression concern themselves only with the times at which failures occur, the above results could also be achieved by stsplitting the data at the observed failure times and manually generating the time-varying covariates. When this is feasible, tvc() merely represents a more convenient way to accomplish this. However, for large datasets with many distinct failure times, using stsplit may produce datasets that are too large to fit in memory, and even if this were not so, the estimation would take far longer to complete. For these reasons, the tvc() and texp() options described above were introduced.

▷ Example 4

Consider a dataset consisting of 45 observations on recovery time from walking pneumonia. Recovery time (in days) is recorded in the variable time, and there are measurements on the covariates age, drug1, and drug2, where drug1 and drug2 interact a choice of treatment with initial dosage level. The study was terminated after 30 days, so those who had not recovered by that time were censored (cured = 0).

```
. use http://www.stata-press.com/data/r11/drugtr2
. list age drug1 drug2 time cured in 1/12, sep(0)
```

	age	drug1	drug2	time	cured
1.	36	0	50	20.6	1
2.	14	0	50	6.8	1
3.	43	0	125	8.6	1
4.	25	100	0	10	1
5.	50	100	0	30	0
6.	26	0	100	13.6	1
7.	21	150	0	5.4	1
8.	25	0	100	15.4	1
9.	32	125	0	8.6	1
10.	28	150	0	8.5	1
11.	34	0	100	30	0
12.	40	0	50	30	0

Patient 1 took 50 mg of drug number 2 and was cured after 20.6 days, whereas patient 5 took 100 mg of drug number 1 and had yet to recover when the study ended and so was censored at 30 days.

We run a standard Cox regression after stsetting the data:

```
. stset time, failure(cured)
     failure event:  cured != 0 & cured < .
obs. time interval:  (0, time]
 exit on or before:  failure

       45  total obs.
        0  exclusions

       45  obs. remaining, representing
       36  failures in single record/single failure data
    677.9  total analysis time at risk, at risk from t =         0
                               earliest observed entry t =         0
                                    last observed exit t =        30
```

(Continued on next page)

```
. stcox age drug1 drug2
        failure _d:  cured
   analysis time _t:  time
Iteration 0:   log likelihood = -116.54385
Iteration 1:   log likelihood = -102.77311
Iteration 2:   log likelihood = -101.92794
Iteration 3:   log likelihood = -101.92504
Iteration 4:   log likelihood = -101.92504
Refining estimates:
Iteration 0:   log likelihood = -101.92504

Cox regression -- Breslow method for ties

No. of subjects =          45                Number of obs   =         45
No. of failures =          36
Time at risk    =  677.9000034
                                             LR chi2(3)      =      29.24
Log likelihood  =   -101.92504               Prob > chi2     =     0.0000
```

_t	Haz. Ratio	Std. Err.	z	P>|z|	[95% Conf. Interval]	
age	.8759449	.0253259	-4.58	0.000	.8276873	.9270162
drug1	1.008482	.0043249	1.97	0.049	1.000041	1.016994
drug2	1.00189	.0047971	0.39	0.693	.9925323	1.011337

The output includes p-values for the tests of the null hypotheses that each regression coefficient is 0 or, equivalently, that each hazard ratio is 1. That all hazard ratios are apparently close to 1 is a matter of scale; however, we can see that drug number 1 significantly increases the risk of being cured and so is an effective drug, whereas drug number 2 is ineffective (given the presence of age and drug number 1 in the model).

Suppose now that we wish to fit a model in which we account for the effect that as time goes by, the actual level of the drug remaining in the body diminishes, say, at an exponential rate. If it is known that the half-life of both drugs is close to 2 days, we can say that the actual concentration level of the drug in the patient's blood is proportional to the initial dosage times, $\exp(-0.35t)$, where t is analysis time. We now fit a model that reflects this change.

```
. stcox age, tvc(drug1 drug2) texp(exp(-0.35*_t)) nolog
        failure _d:  cured
   analysis time _t:  time
Cox regression -- Breslow method for ties

No. of subjects =          45                Number of obs   =         45
No. of failures =          36
Time at risk    =  677.9000034
                                             LR chi2(3)      =      36.98
Log likelihood  =   -98.052763               Prob > chi2     =     0.0000
```

_t	Haz. Ratio	Std. Err.	z	P>|z|	[95% Conf. Interval]	
main						
age	.8614636	.028558	-4.50	0.000	.8072706	.9192948
tvc						
drug1	1.304744	.1135967	3.06	0.002	1.100059	1.547514
drug2	1.200613	.1113218	1.97	0.049	1.001103	1.439882

Note: variables in tvc equation interacted with exp(-0.35*_t)

The first equation, rh, reports the results (hazard ratios) for the covariates that do not vary over time; the second equation, t, reports the results for the time-varying covariates.

As the level of drug in the blood system decreases, the drug's effectiveness diminishes. Accounting for this serves to unmask the effects of both drugs in that we now see increased effects on both. In fact, the effect on recovery time of drug number 2 now becomes significant.

❑ Technical note

The interpretation of hazard ratios requires careful consideration here. For the first model, the hazard ratio for, say, drug1 is interpreted as the proportional change in hazard when the dosage level of drug1 is increased by one unit. For the second model, the hazard ratio for drug1 is the proportional change in hazard when the blood concentration level—i.e., drug1*exp($-0.35t$)—increases by 1.

❑

Because the number of observations in our data is relatively small, for illustrative purposes we can stsplit the data at each recovery time, manually generate the blood concentration levels, and refit the second model.

```
. generate id=_n
. streset, id(id)
 (output omitted)
. stsplit, at(failures)
(31 failure times)
(812 observations (episodes) created)
. generate drug1emt = drug1*exp(-0.35*_t)
. generate drug2emt = drug2*exp(-0.35*_t)
. stcox age drug1emt drug2emt
        failure _d:  cured
   analysis time _t:  time
                id:  id
Iteration 0:   log likelihood = -116.54385
Iteration 1:   log likelihood = -99.321912
Iteration 2:   log likelihood =  -98.07369
Iteration 3:   log likelihood =  -98.05277
Iteration 4:   log likelihood = -98.052763
Refining estimates:
Iteration 0:   log likelihood = -98.052763

Cox regression -- Breslow method for ties

No. of subjects =           45              Number of obs   =        857
No. of failures =           36
Time at risk    =  677.9000034
                                            LR chi2(3)      =      36.98
Log likelihood  =   -98.052763              Prob > chi2     =     0.0000
```

_t	Haz. Ratio	Std. Err.	z	P>|z|	[95% Conf. Interval]
age	.8614636	.028558	-4.50	0.000	.8072706 .9192948
drug1emt	1.304744	.1135967	3.06	0.002	1.100059 1.547514
drug2emt	1.200613	.1113218	1.97	0.049	1.001103 1.439882

We get the same answer. However, this required more work both for Stata and for you.

◁

Above we used tvc() and texp() to demonstrate fitting models with time-varying covariates, but these options can also be used to fit models with *time-varying coefficients*. For simplicity, consider a version of (1) that contains only one fixed covariate, x_1, and sets $z_1 = x_1$:

$$h(t) = h_0(t) \exp\{\beta_1 x_1 + g(t)\gamma_1 x_1\}$$

Rearranging terms results in

$$h(t) = h_0(t) \exp\left[\{\beta_1 + \gamma_1 g(t)\} x_1\right]$$

Given this new arrangement, we consider that $\beta_1 + \gamma_1 g(t)$ is a (possibly) time-varying coefficient on the covariate x_1, for some specified function of time $g(t)$. The coefficient has a time-invariant component, β_1, with γ_1 determining the magnitude of the time-dependent deviations from β_1. As such, a test of $\gamma_1 = 0$ is a test of time invariance for the coefficient on x_1.

Confirming that a coefficient is time invariant is one way of testing the proportional-hazards assumption. Proportional hazards implies that the relative hazard (i.e., β) is fixed over time, and this assumption would be violated if a time interaction proved significant.

▷ Example 5

Returning to our cancer drug trial, we now include a time interaction on age as a way of testing the proportional-hazards assumption for that covariate:

```
. use http://www.stata-press.com/data/r11/drugtr, clear
(Patient Survival in Drug Trial)
. stcox drug age, tvc(age)
 (output omitted)
Cox regression -- Breslow method for ties

No. of subjects =         48              Number of obs   =         48
No. of failures =         31
Time at risk    =        744
                                          LR chi2(3)      =      33.63
Log likelihood  =  -83.095036             Prob > chi2     =     0.0000
```

_t	Haz. Ratio	Std. Err.	z	P>|z|	[95% Conf. Interval]
main						
drug	.1059862	.0478178	-4.97	0.000	.0437737	.2566171
age	1.156977	.07018	2.40	0.016	1.027288	1.303037
tvc						
age	.9970966	.0042415	-0.68	0.494	.988818	1.005445

Note: variables in tvc equation interacted with _t

We used the default function of time, $g(t) = t$, although we could have specified otherwise with the texp() option. The estimation results are presented in terms of hazard ratios, and so 0.9971 is an estimate of $\exp(\gamma_{\text{age}})$. Tests of hypotheses, however, are in terms of the original metric, and so 0.494 is the significance for the test of $H_0: \gamma_{\text{age}} = 0$ versus the two-sided alternative. With respect to this specific form of misspecification, there is not much evidence to dispute the proportionality of hazards when it comes to age.

◁

❑ Technical note

Finally, specifying $g(t)$ via the texp(*exp*) option is intended for functions of analysis time, _t only, with the default being texp(_t) if left unspecified. However, specifying any other valid Stata expression will not produce a syntax error, yet usually will not yield the anticipated output. For example, specifying texp(*varname*) will not generate interaction terms. This has to do mainly with how the calculations are carried out—by careful summations over risk pools at each failure time.

❑

Robust estimate of variance

By default, stcox produces the conventional estimate for the variance–covariance matrix of the coefficients (and hence the reported standard errors). If, however, you specify the vce(robust) option, stcox switches to the robust variance estimator (Lin and Wei 1989).

The key to the robust calculation is using the efficient score residual for each subject in the data for the variance calculation. Even in simple single-record, single-failure survival data, the same subjects appear repeatedly in the risk pools, and the robust calculation needs to account for that.

▷ Example 6

Refitting the Stanford heart transplant data model with robust standard errors, we obtain

```
. use http://www.stata-press.com/data/r11/stan3, clear
(Heart transplant data)
. stset t1, failure(died) id(id)
                id:  id
     failure event:  died != 0 & died < .
obs. time interval:  (t1[_n-1], t1]
 exit on or before:  failure

       172  total obs.
         0  exclusions

       172  obs. remaining, representing
       103  subjects
        75  failures in single failure-per-subject data
   31938.1  total analysis time at risk, at risk from t =         0
                                earliest observed entry t =         0
                                     last observed exit t =      1799

. stcox age posttran surg year, vce(robust)
         failure _d:  died
   analysis time _t:  t1
                 id:  id

Iteration 0:   log pseudolikelihood = -298.31514
Iteration 1:   log pseudolikelihood =  -289.7344
Iteration 2:   log pseudolikelihood = -289.53498
Iteration 3:   log pseudolikelihood = -289.53378
Iteration 4:   log pseudolikelihood = -289.53378
Refining estimates:
Iteration 0:   log pseudolikelihood = -289.53378

Cox regression -- Breslow method for ties

No. of subjects    =          103              Number of obs    =       172
No. of failures    =           75
Time at risk       =      31938.1
                                                Wald chi2(4)     =     19.68
Log pseudolikelihood =   -289.53378             Prob > chi2      =    0.0006

                         (Std. Err. adjusted for 103 clusters in id)
```

		Robust				
_t	Haz. Ratio	Std. Err.	z	P>|z|	[95% Conf.	Interval]
age	1.030224	.0148771	2.06	0.039	1.001474	1.059799
posttran	.9787243	.2961736	-0.07	0.943	.5408498	1.771104
surgery	.3738278	.1304912	-2.82	0.005	.1886013	.7409665
year	.8873107	.0613176	-1.73	0.084	.7749139	1.01601

Note the word Robust above Std. Err. in the table and the phrase "Std. Err. adjusted for 103 clusters in id" above the table.

The hazard ratio estimates are the same as before, but the standard errors are slightly different.

◁

❑ Technical note

In the previous example, stcox knew to specify vce(cluster id) for us when we specified vce(robust).

To see the importance of vce(cluster id), consider simple single-record, single-failure survival data, a piece of which is

t0	t	died	x
0	5	1	1
0	9	0	1
0	8	0	0

and then consider the absolutely equivalent multiple-record survival data:

id	t0	t	died	x
1	0	3	0	1
1	3	5	1	1
2	0	6	0	1
2	6	9	0	1
3	0	3	0	0
3	3	8	0	0

Both datasets record the same underlying data, and so both should produce the same numerical results. This should be true regardless of whether vce(robust) is specified.

In the second dataset, were we to ignore id, it would appear that there are 6 observations on 6 subjects. The key ingredients in the robust calculation are the efficient score residuals, and viewing the data as 6 observations on 6 subjects produces different score residuals. Let's call the 6 score residuals $s_1, s_2, \ldots, s_6$ and the 3 score residuals that would be generated by the first dataset S_1, S_2, and S_3. $S_1 = s_1 + s_2$, $S_2 = s_3 + s_4$, and $S_3 = s_5 + s_6$.

That residuals sum is the key to understanding the vce(cluster *clustvar*) option. When you specify vce(cluster id), Stata makes the robust calculation based not on the overly detailed s_1, $s_2, \ldots, s_6$ but on $S_1 + S_2$, $S_3 + S_4$, and $S_5 + S_6$. That is, Stata sums residuals within clusters before entering them into subsequent calculations (where they are squared), so results estimated from the second dataset are equal to those estimated from the first. In more complicated datasets with time-varying regressors, delayed entry, and gaps, this action of summing within cluster, in effect, treats the cluster (which is typically a subject) as a unified whole.

Because we had stset an id() variable, stcox knew to specify vce(cluster id) for us when we specified vce(robust). You may, however, override the default clustering by specifying vce(cluster *clustvar*) with a different variable from the one you used in stset, id(). This is useful in analyzing multiple-failure data, where you need to stset a pseudo-ID establishing the time from the last failure as the onset of risk.

❑

Cox regression with multiple-failure data

▷ Example 7

In [ST] **stsum**, we introduce a multiple-failure dataset:

```
. use http://www.stata-press.com/data/r11/mfail
. stdescribe
```

		├──────── per subject ────────┤			
Category	total	mean	min	median	max
no. of subjects	926				
no. of records	1734	1.87257	1	2	4
(first) entry time		0	0	0	0
(final) exit time		470.6857	1	477	960
subjects with gap	0				
time on gap if gap	0	.	.	.	.
time at risk	435855	470.6857	1	477	960
failures	808	.8725702	0	1	3

This dataset contains two variables—x1 and x2—which we believe affect the hazard of failure.

If we simply want to analyze these multiple-failure data as if the baseline hazard remains unchanged as events occur (that is, the hazard may change with time, but time is measured from 0 and is independent of when the last failure occurred), we can type

```
. stcox x1 x2, vce(robust)
Iteration 0:   log pseudolikelihood = -5034.9569
Iteration 1:   log pseudolikelihood = -4978.4198
Iteration 2:   log pseudolikelihood = -4978.1915
Iteration 3:   log pseudolikelihood = -4978.1914
Refining estimates:
Iteration 0:   log pseudolikelihood = -4978.1914
Cox regression -- Breslow method for ties
No. of subjects    =       926           Number of obs   =      1734
No. of failures    =       808
Time at risk       =    435855
                                         Wald chi2(2)    =    152.13
Log pseudolikelihood =  -4978.1914       Prob > chi2     =    0.0000
                        (Std. Err. adjusted for 926 clusters in id)
```

		Robust				
_t	Haz. Ratio	Std. Err.	z	P>|z|	[95% Conf. Interval]	
x1	2.273456	.1868211	9.99	0.000	1.935259	2.670755
x2	.329011	.0523425	-6.99	0.000	.2408754	.4493951

We chose to fit this model with robust standard errors—we specified vce(robust)—but you can estimate conventional standard errors if you wish.

In [ST] **stsum**, we discuss analyzing this dataset as the time since last failure. We wished to assume that the hazard function remained unchanged with failure, except that one restarted the same hazard function. To that end, we made the following changes to our data:

```
. stgen nf = nfailures()
. egen newid = group(id nf)
```

```
. sort newid t

. by newid: replace t = t - t0[1]
(808 real changes made)

. by newid: gen newt0 = t0 - t0[1]

. stset t, id(newid) failure(d) time0(newt0) noshow
               id:  newid
    failure event:  d != 0 & d < .
obs. time interval:  (newt0, t]
 exit on or before:  failure
```

```
       1734  total obs.
          0  exclusions

       1734  obs. remaining, representing
       1734  subjects
        808  failures in single failure-per-subject data
     435444  total analysis time at risk, at risk from t =         0
                                  earliest observed entry t =         0
                                    last observed exit t =       797
```

That is, we took each subject and made many newid subjects out of each, with each subject entering at time 0 (now meaning the time of the last failure). id still identifies a real subject, but Stata thinks the identifier variable is newid because we stset, id(newid). If we were to fit a model with vce(robust), we would get

```
. stcox x1 x2, vce(robust) nolog

Cox regression -- Breslow method for ties

No. of subjects    =       1734              Number of obs   =       1734
No. of failures    =        808
Time at risk       =     435444
                                             Wald chi2(2)    =      88.51
Log pseudolikelihood =  -5062.5815           Prob > chi2     =     0.0000

                       (Std. Err. adjusted for 1734 clusters in newid)
```

	Haz. Ratio	Robust Std. Err.	z	P>|z|	[95% Conf. Interval]	
_t						
x1	2.002547	.1936906	7.18	0.000	1.656733	2.420542
x2	.2946263	.0569167	-6.33	0.000	.2017595	.4302382

Note carefully the message concerning the clustering: standard errors have been adjusted for clustering on newid. We, however, want the standard errors adjusted for clustering on id, so we must specify the vce(cluster *clustvar*) option:

```
. stcox x1 x2, vce(cluster id) nolog
Cox regression -- Breslow method for ties
No. of subjects    =       1734              Number of obs   =      1734
No. of failures    =        808
Time at risk       =     435444
                                             Wald chi2(2)    =     93.66
Log pseudolikelihood = -5062.5815            Prob > chi2     =    0.0000
                           (Std. Err. adjusted for 926 clusters in id)
```

		Robust			
_t	Haz. Ratio	Std. Err.	z	P>|z|	[95% Conf. Interval]
x1	2.002547	.1920151	7.24	0.000	1.659452 2.416576
x2	.2946263	.0544625	-6.61	0.000	.2050806 .4232709

That is, if you are using vce(robust), you must remember to specify vce(cluster *clustvar*) for yourself when

1. you are analyzing multiple-failure data and

2. you have reset time to time since last failure, so what Stata considers the subjects are really subsubjects.

◁

Stratified estimation

When you type

```
. stcox xvars, strata(svars)
```

you are allowing the baseline hazard functions to differ for the groups identified by *svars*. This is equivalent to fitting separate Cox proportional hazards models under the constraint that the coefficients are equal but the baseline hazard functions are not.

▷ Example 8

Say that in the Stanford heart experiment data, there was a change in treatment for all patients, before and after transplant, in 1970 and then again in 1973. Further assume that the proportional-hazards assumption is not reasonable for these changes in treatment—perhaps the changes result in short-run benefit but little expected long-run benefit. Our interest in the data is not in the effect of these treatment changes but in the effect of transplantation, for which we still find the proportional-hazards assumption reasonable. We might fit our model to account for these fictional changes by typing

```
. use http://www.stata-press.com/data/r11/stan3, clear
(Heart transplant data)
. generate pgroup = year
. recode pgroup min/69=1 70/72=2 73/max=3
(pgroup: 172 changes made)
```

(Continued on next page)

```
. stcox age posttran surg year, strata(pgroup) nolog
        failure _d:  died
  analysis time _t:  t1
              id:   id
Stratified Cox regr. -- Breslow method for ties

No. of subjects =         103                Number of obs    =       172
No. of failures =          75
Time at risk    =     31938.1
                                             LR chi2(4)       =     20.67
Log likelihood  =   -213.35033               Prob > chi2      =    0.0004
```

_t	Haz. Ratio	Std. Err.	z	P>|z|	[95% Conf. Interval]	
age	1.027406	.0150188	1.85	0.064	.9983874	1.057268
posttran	1.075476	.3354669	0.23	0.816	.583567	1.982034
surgery	.2222415	.1218386	-2.74	0.006	.0758882	.6508429
year	.5523966	.1132688	-2.89	0.004	.3695832	.825638

Stratified by pgroup

Of course, we could obtain the robust estimate of variance by also including the vce(robust) option.

◁

Cox regression with shared frailty

A shared-frailty model is the survival-data analog to regression models with random effects. A *frailty* is a latent random effect that enters multiplicatively on the hazard function. In a Cox model, the data are organized as $i = 1, \ldots, n$ groups with $j = 1, \ldots, n_i$ observations in group i. For the jth observation in the ith group, the hazard is

$$h_{ij}(t) = h_0(t)\alpha_i \exp(\mathbf{x}_{ij}\boldsymbol{\beta})$$

where α_i is the group-level frailty. The frailties are unobservable positive quantities and are assumed to have mean 1 and variance θ, to be estimated from the data. You can fit a Cox shared-frailty model by specifying shared(*varname*), where *varname* defines the groups over which frailties are shared. stcox, shared() treats the frailties as being gamma distributed, but this is mainly an issue of computational convenience; see *Methods and formulas*. Theoretically, any distribution with positive support, mean 1, and finite variance may be used to model frailty.

Shared-frailty models are used to model within-group correlation; observations within a group are correlated because they share the same frailty. The estimate of θ is used to measure the degree of within-group correlation, and the shared-frailty model reduces to standard Cox when $\theta = 0$.

For $\nu_i = \log\alpha_i$, the hazard can also be expressed as

$$h_{ij}(t) = h_0(t) \exp(\mathbf{x}_{ij}\boldsymbol{\beta} + \nu_i)$$

and thus the log-frailties, ν_i, are analogous to random effects in standard linear models.

▷ Example 9

Consider the data from a study of 38 kidney dialysis patients, as described in McGilchrist and Aisbett (1991). The study is concerned with the prevalence of infection at the catheter insertion point. Two recurrence times (in days) are measured for each patient, and each recorded time is the time from initial insertion (onset of risk) to infection or censoring:

```
. use http://www.stata-press.com/data/r11/catheter, clear
(Kidney data, McGilchrist and Aisbett, Biometrics, 1991)
. list patient time infect age female in 1/10
```

	patient	time	infect	age	female
1.	1	16	1	28	0
2.	1	8	1	28	0
3.	2	13	0	48	1
4.	2	23	1	48	1
5.	3	22	1	32	0
6.	3	28	1	32	0
7.	4	318	1	31.5	1
8.	4	447	1	31.5	1
9.	5	30	1	10	0
10.	5	12	1	10	0

Each patient (patient) has two recurrence times (time) recorded, with each catheter insertion resulting in either infection (infect==1) or right-censoring (infect==0). Among the covariates measured are age and sex (female==1 if female, female==0 if male).

One subtlety to note concerns the use of the generic term *subjects*. In this example, the subjects are taken to be the individual catheter insertions, not the patients themselves. This is a function of how the data were recorded—the onset of risk occurs at catheter insertion (of which there are two for each patient), and not, say, at the time of admission of the patient into the study. We therefore have two subjects (insertions) within each group (patient).

It is reasonable to assume independence of patients but unreasonable to assume that recurrence times within each patient are independent. One solution would be to fit a standard Cox model, adjusting the standard errors of the estimated hazard ratios to account for the possible correlation by specifying vce(cluster patient).

We could instead model the correlation by assuming that the correlation is the result of a latent patient-level effect, or frailty. That is, rather than fitting a standard model and specifying vce(cluster patient), we could fit a frailty model by specifying shared(patient):

```
. stset time, fail(infect)
 (output omitted)
. stcox age female, shared(patient)
         failure _d:  infect
   analysis time _t:  time

Fitting comparison Cox model:

Estimating frailty variance:

Iteration 0:   log profile likelihood = -182.06713
Iteration 1:   log profile likelihood =  -181.9791
Iteration 2:   log profile likelihood = -181.97453
Iteration 3:   log profile likelihood = -181.97453
```

```
Fitting final Cox model:
Iteration 0:   log likelihood = -199.05599
Iteration 1:   log likelihood = -183.72296
Iteration 2:   log likelihood = -181.99509
Iteration 3:   log likelihood = -181.97455
Iteration 4:   log likelihood = -181.97453
Refining estimates:
Iteration 0:   log likelihood = -181.97453

Cox regression --
      Breslow method for ties            Number of obs    =       76
      Gamma shared frailty               Number of groups =       38
Group variable: patient

No. of subjects =         76             Obs per group: min =      2
No. of failures =         58                            avg =      2
Time at risk    =       7424                            max =      2
                                         Wald chi2(2)   =    11.66
Log likelihood  =   -181.97453           Prob > chi2    =   0.0029
```

_t	Haz. Ratio	Std. Err.	z	P>|z|	[95% Conf. Interval]	
age	1.006202	.0120965	0.51	0.607	.9827701	1.030192
female	.2068678	.095708	-3.41	0.001	.0835376	.5122756

theta	.4754497	.2673108

```
Likelihood-ratio test of theta=0: chibar2(01) =     6.27 Prob>=chibar2 = 0.006
Note: standard errors of hazard ratios are conditional on theta.
```

From the output, we obtain $\widehat{\theta} = 0.475$, and given the standard error of $\widehat{\theta}$ and likelihood-ratio test of H_0: $\theta = 0$, we find a significant frailty effect, meaning that the correlation within patient cannot be ignored. Contrast this with the analysis of the same data in [ST] **streg**, which considered both Weibull and lognormal shared-frailty models. For Weibull, there was significant frailty; for lognormal, there was not.

The estimated ν_i are not displayed in the coefficient table but may be retrieved postestimation by using `predict` with the `effects` option; see [ST] **stcox postestimation** for an example.

◁

In shared-frailty Cox models, the estimation consists of two steps. In the first step, the optimization is in terms of θ only. For fixed θ, the second step consists of fitting a standard Cox model via penalized log likelihood, with the ν_i introduced as estimable coefficients of dummy variables identifying the groups. The penalty term in the penalized log likelihood is a function of θ; see *Methods and formulas*. The final estimate of θ is taken to be the one that maximizes the penalized log likelihood. Once the optimal θ is obtained, it is held fixed, and a final penalized Cox model is fit. As a result, the standard errors of the main regression parameters (or hazard ratios, if displayed as such) are treated as conditional on θ fixed at its optimal value.

With gamma-distributed frailty, hazard ratios decay over time in favor of the *frailty effect* and thus the displayed "Haz. Ratio" in the above output is actually the hazard ratio only for $t = 0$. The degree of decay depends on θ. Should the estimated θ be close to 0, the hazard ratios do regain their usual interpretation; see Gutierrez (2002) for details.

❑ Technical note

The likelihood-ratio test of $\theta = 0$ is a boundary test and thus requires careful consideration concerning the calculation of its p-value. In particular, the null distribution of the likelihood-ratio test statistic is not the usual χ_1^2 but is rather a 50:50 mixture of a χ_0^2 (point mass at zero) and a χ_1^2, denoted as $\overline{\chi}_{01}^2$. See Gutierrez, Carter, and Drukker (2001) for more details.

❑

❑ Technical note

In [ST] **streg**, shared-frailty models are compared and contrasted with *unshared* frailty models. Unshared-frailty models are used to model heterogeneity, and the frailties are integrated out of the conditional survivor function to produce an unconditional survivor function, which serves as a basis for all likelihood calculations.

Given the nature of Cox regression (the baseline hazard remains unspecified), there is no Cox regression analog to the unshared parametric frailty model as fit using **streg**. That is not to say that you cannot fit a shared-frailty model with 1 observation per group; you can as long as you do not fit a null model. However, there are subtle differences in singleton-group data between shared- and unshared-frailty models; see Gutierrez (2002).

❑

Saved results

stcox saves the following in e():

Scalars

e(N)	number of observations
e(N_sub)	number of subjects
e(N_fail)	number of failures
e(N_g)	number of groups
e(df_m)	model degrees of freedom
e(r2_p)	pseudo-R-squared
e(ll)	log likelihood
e(ll_0)	log likelihood, constant-only model
e(ll_c)	log likelihood, comparison model
e(N_clust)	number of clusters
e(chi2)	χ^2
e(chi2_c)	χ^2, comparison model
e(risk)	total time at risk
e(g_min)	smallest group size
e(g_avg)	average group size
e(g_max)	largest group size
e(theta)	frailty parameter
e(se_theta)	standard error of θ
e(p_c)	significance, comparison model
e(rank)	rank of e(V)

(*Continued on next page*)

Macros
    e(cmd)   cox or stcox_fr
    e(cmd2)   stcox
    e(cmdline)   command as typed
    e(depvar)   _t
    e(t0)   _t0
    e(texp)   function used for time-varying covariates
    e(ties)   method used for handling ties
    e(shared)   frailty grouping variable
    e(clustvar)   name of cluster variable
    e(offset)   offset
    e(chi2type)   Wald or LR; type of model χ^2 test
    e(vce)   *vcetype* specified in vce()
    e(vcetype)   title used to label Std. Err.
    e(k_eform)   number of leading equations appropriate for eform output
    e(method)   requested estimation method
    e(crittype)   optimization criterion
    e(datasignature)   the checksum
    e(datasignaturevars)   variables used in calculation of checksum
    e(properties)   b V
    e(estat_cmd)   program used to implement estat
    e(predict)   program used to implement predict
    e(footnote)   program used to implement the footnote display
    e(asbalanced)   factor variables fvset as asbalanced
    e(asobserved)   factor variables fvset as asobserved

Matrices
    e(b)   coefficient vector
    e(V)   variance–covariance matrix of the
    e(V_modelbased)   model-based variance estimators

Functions
    e(sample)   marks estimation sample

Methods and formulas

stcox is implemented as an ado-file.

The proportional hazards model with explanatory variables was first suggested by Cox (1972). For an introductory explanation, see Hosmer Jr., Lemeshow, and May (2008, chap. 3, 4, and 7) Kahn and Sempos (1989, 193–198), and Selvin (2004, 412–442). For an introduction for the social scientist, see Box-Steffensmeier and Jones (2004, chap. 4). For a comprehensive review of the methods in this entry, see Klein and Moeschberger (2003). For a detailed development of these methods, see Kalbfleisch and Prentice (2002). For more Stata-specific insight, see Cleves et al. (2008), Dupont (2009), and Vittinghoff et al. (2005).

Let $\mathbf{x}_i$ be the row vector of covariates for the time interval $(t_{0i}, t_i]$ for the ith observation in the dataset $i = 1, \ldots, N$. stcox obtains parameter estimates, $\widehat{\boldsymbol{\beta}}$, by maximizing the partial log-likelihood function

$$\log L = \sum_{j=1}^{D} \left[\sum_{i \in D_j} \mathbf{x}_i \boldsymbol{\beta} - d_j \log \left\{ \sum_{k \in R_j} \exp(\mathbf{x}_k \boldsymbol{\beta}) \right\} \right]$$

where j indexes the ordered failure times $t_{(j)}$, $j = 1, \ldots, D$; D_j is the set of d_j observations that fail at $t_{(j)}$; d_j is the number of failures at $t_{(j)}$; and R_j is the set of observations k that are at risk at time $t_{(j)}$ (i.e., all k such that $t_{0k} < t_{(j)} \le t_k$). This formula for $\log L$ is for unweighted data and handles ties by using the Peto–Breslow approximation (Peto 1972; Breslow 1974), which is the default method of handling ties in stcox.

If strata(*varnames*) is specified, then the partial log likelihood is the sum of each stratum-specific partial log likelihood, obtained by forming the ordered failure times $t_{(j)}$, the failure sets D_j, and the risk sets R_j, using only those observations within that stratum.

The variance of $\widehat{\beta}$ is estimated by the conventional inverse matrix of (negative) second derivatives of $\log L$, unless vce(robust) is specified, in which case the method of Lin and Wei (1989) is used. That method treats efficient score residuals as analogs to the log-likelihood scores one would find in fully parametric models; see *Methods and formulas* in [ST] **stcox postestimation** for how to calculate efficient score residuals. If vce(cluster *clustvar*) is specified, the efficient score residuals are summed within cluster before the sandwich (robust) estimator is applied.

Tied values are handled using one of four approaches. The log likelihoods corresponding to the four approaches are given with weights (exactp and efron do not allow weights) and offsets by

$$\log L_{\text{breslow}} = \sum_{j=1}^{D} \sum_{i \in D_j} \left[w_i(\mathbf{x}_i\boldsymbol{\beta} + \text{offset}_i) - w_i \log \left\{ \sum_{\ell \in R_j} w_\ell \exp(\mathbf{x}_\ell\boldsymbol{\beta} + \text{offset}_\ell) \right\} \right]$$

$$\log L_{\text{efron}} = \sum_{j=1}^{D} \sum_{i \in D_j} \left[\mathbf{x}_i\boldsymbol{\beta} + \text{offset}_i - d_j^{-1} \sum_{k=0}^{d_j-1} \log \left\{ \sum_{\ell \in R_j} \exp(\mathbf{x}_\ell\boldsymbol{\beta} + \text{offset}_\ell) - kA_j \right\} \right]$$

$$A_j = d_j^{-1} \sum_{\ell \in D_j} \exp(\mathbf{x}_\ell\boldsymbol{\beta} + \text{offset}_\ell)$$

$$\log L_{\text{exactm}} = \sum_{j=1}^{D} \log \int_0^\infty \prod_{\ell \in D_j} \left\{ 1 - \exp\left(-\frac{e_\ell}{s}t\right) \right\}^{w_\ell} \exp(-t) dt$$

$$e_\ell = \exp(\mathbf{x}_\ell\boldsymbol{\beta} + \text{offset}_\ell)$$

$$s = \sum_{\substack{k \in R_j \\ k \notin D_j}} w_k \exp(\mathbf{x}_k\boldsymbol{\beta} + \text{offset}_k) = \text{sum of weighted nondeath risk scores}$$

$$\log L_{\text{exactp}} = \sum_{j=1}^{D} \left\{ \sum_{i \in R_j} \delta_{ij}(\mathbf{x}_i\boldsymbol{\beta} + \text{offset}_i) - \log f(r_j, d_j) \right\}$$

$$f(r, d) = f(r-1, d) + f(r-1, d-1) \exp(\mathbf{x}_k\boldsymbol{\beta} + \text{offset}_k)$$

$$k = r^{\text{th}} \text{ observation in the set } R_j$$

$$r_j = \text{cardinality of the set } R_j$$

$$f(r, d) = \begin{cases} 0 & \text{if } r < d \\ 1 & \text{if } d = 0 \end{cases}$$

where δ_{ij} is an indicator for failure of observation i at time $t_{(j)}$.

Calculations for the exact marginal log likelihood (and associated derivatives) are obtained with 15-point Gauss–Laguerre quadrature. The `breslow` and `efron` options both provide approximations of the exact marginal log likelihood. The `efron` approximation is a better (closer) approximation, but the `breslow` approximation is faster. The choice of the approximation to use in a given situation should generally be driven by the proportion of ties in the data.

For shared-frailty models, the data are organized into G groups with the ith group consisting of n_i observations, $i = 1, \ldots, G$. From Therneau and Grambsch (2000, 253–255), estimation of θ takes place via maximum profile log likelihood. For fixed θ, estimates of $\boldsymbol{\beta}$ and $\nu_1, \ldots, \nu_G$ are obtained by maximizing

$$\log L(\theta) = \log L_{\text{Cox}}(\boldsymbol{\beta}, \nu_1, \ldots, \nu_G) + \sum_{i=1}^{G} \left[\frac{1}{\theta} \{\nu_i - \exp(\nu_i)\} + \left(\frac{1}{\theta} + D_i\right) \left\{1 - \log\left(\frac{1}{\theta} + D_i\right)\right\} - \frac{\log \theta}{\theta} + \log \Gamma\left(\frac{1}{\theta} + D_i\right) - \log \Gamma\left(\frac{1}{\theta}\right) \right]$$

where D_i is the number of death events in group i, and $\log L_{\text{Cox}}(\boldsymbol{\beta}, \nu_1, \ldots, \nu_G)$ is the standard Cox partial log likelihood, with the ν_i treated as the coefficients of indicator variables identifying the groups. That is, the jth observation in the ith group has log relative hazard $\mathbf{x}_{ij}\boldsymbol{\beta} + \nu_i$. The estimate of the frailty parameter, $\widehat{\theta}$, is chosen as that which maximizes $\log L(\theta)$. The final estimates of $\boldsymbol{\beta}$ are obtained by maximizing $\log L(\widehat{\theta})$ in $\boldsymbol{\beta}$ and the ν_i. The ν_i are not reported in the coefficient table but are available via `predict`; see [ST] **stcox postestimation**. The estimated variance–covariance matrix of $\widehat{\boldsymbol{\beta}}$ is obtained as the appropriate submatrix of the variance matrix of $(\widehat{\boldsymbol{\beta}}, \widehat{\nu}_1, \ldots, \widehat{\nu}_G)$, and that matrix is obtained as the inverse of the negative Hessian of $\log L(\widehat{\theta})$. Therefore, standard errors and inference based on $\widehat{\boldsymbol{\beta}}$ should be treated as conditional on $\theta = \widehat{\theta}$.

The likelihood-ratio test statistic for testing H_0: $\theta = 0$ is calculated as minus twice the difference between the log-likelihood for a Cox model without shared frailty and $\log L(\widehat{\theta})$ evaluated at the final $(\widehat{\boldsymbol{\beta}}, \widehat{\nu}_1, \ldots, \widehat{\nu}_G)$.

David Roxbee Cox (1924–) was born in Birmingham, England. He earned PhD degrees in mathematics and statistics from the universities of Cambridge and Leeds, and he worked at the Royal Aircraft Establishment, the Wool Industries Research Association, and the universities of Cambridge, London (Birkbeck and Imperial Colleges), and Oxford. He was knighted in 1985. Sir David has worked on a wide range of theoretical and applied statistical problems, with outstanding contributions in areas such as experimental design, stochastic processes, binary data, survival analysis, asymptotic techniques, and multivariate dependencies.

Acknowledgment

We thank Peter Sasieni of Cancer Research UK for his statistical advice and guidance in implementing the robust variance estimator for this command.

References

Box-Steffensmeier, J. M., and B. S. Jones. 2004. *Event History Modeling: A Guide for Social Scientists.* Cambridge: Cambridge University Press.

Breslow, N. E. 1974. Covariance analysis of censored survival data. *Biometrics* 30: 89–99.

Cleves, M. A. 1999. ssa13: Analysis of multiple failure-time data with Stata. *Stata Technical Bulletin* 49: 30–39. Reprinted in *Stata Technical Bulletin Reprints*, vol. 9, pp. 338–349. College Station, TX: Stata Press.

Cleves, M. A., W. W. Gould, R. G. Gutierrez, and Y. Marchenko. 2008. *An Introduction to Survival Analysis Using Stata.* 2nd ed. College Station, TX: Stata Press.

Cox, D. R. 1972. Regression models and life-tables (with discussion). *Journal of the Royal Statistical Society, Series B* 34: 187–220.

———. 1975. Partial likelihood. *Biometrika* 62: 269–276.

Cox, D. R., and D. Oakes. 1984. *Analysis of Survival Data.* London: Chapman & Hall/CRC.

Cox, D. R., and E. J. Snell. 1968. A general definition of residuals (with discussion). *Journal of the Royal Statistical Society, Series B* 30: 248–275.

Crowley, J., and M. Hu. 1977. Covariance analysis of heart transplant survival data. *Journal of the American Statistical Association* 72: 27–36.

Cui, J. 2005. Buckley–James method for analyzing censored data, with an application to a cardiovascular disease and an HIV/AIDS study. *Stata Journal* 5: 517–526.

Dupont, W. D. 2009. *Statistical Modeling for Biomedical Researchers: A Simple Introduction to the Analysis of Complex Data.* 2nd ed. Cambridge: Cambridge University Press.

Fleming, T. R., and D. P. Harrington. 1991. *Counting Processes and Survival Analysis.* New York: Wiley.

Gutierrez, R. G. 2002. Parametric frailty and shared frailty survival models. *Stata Journal* 2: 22–44.

Gutierrez, R. G., S. Carter, and D. M. Drukker. 2001. sg160: On boundary-value likelihood-ratio tests. *Stata Technical Bulletin* 60: 15–18. Reprinted in *Stata Technical Bulletin Reprints*, vol. 10, pp. 269–273. College Station, TX: Stata Press.

Hills, M., and B. L. De Stavola. 2006. *A Short Introduction to Stata for Biostatistics.* London: Timberlake.

Hosmer Jr., D. W., S. Lemeshow, and S. May. 2008. *Applied Survival Analysis: Regression Modeling of Time to Event Data.* 2nd ed. New York: Wiley.

Jenkins, S. P. 1997. sbe17: Discrete time proportional hazards regression. *Stata Technical Bulletin* 39: 22–32. Reprinted in *Stata Technical Bulletin Reprints*, vol. 7, pp. 109–121. College Station, TX: Stata Press.

Kahn, H. A., and C. T. Sempos. 1989. *Statistical Methods in Epidemiology.* New York: Oxford University Press.

Kalbfleisch, J. D., and R. L. Prentice. 2002. *The Statistical Analysis of Failure Time Data.* 2nd ed. New York: Wiley.

Klein, J. P., and M. L. Moeschberger. 2003. *Survival Analysis: Techniques for Censored and Truncated Data.* 2nd ed. New York: Springer.

Lin, D. Y., and L. J. Wei. 1989. The robust inference for the Cox proportional hazards model. *Journal of the American Statistical Association* 84: 1074–1078.

McGilchrist, C. A., and C. W. Aisbett. 1991. Regression with frailty in survival analysis. *Biometrics* 47: 461–466.

Newman, S. C. 2001. *Biostatistical Methods in Epidemiology.* New York: Wiley.

Peto, R. 1972. Contribution to the discussion of paper by D. R. Cox. *Journal of the Royal Statistical Society, Series B* 34: 205–207.

Reid, N. 1994. A conversation with Sir David Cox. *Statistical Science* 9: 439–455.

Royston, J. P. 2001. Flexible parametric alternatives to the Cox model, and more. *Stata Journal* 1: 1–28.

———. 2006. Explained variation for survival models. *Stata Journal* 6: 83–96.

———. 2007. Profile likelihood for estimation and confidence intervals. *Stata Journal* 7: 376–387.

Schoenfeld, D. A. 1982. Partial residuals for the proportional hazards regression model. *Biometrika* 69: 239–241.

Selvin, S. 2004. *Statistical Analysis of Epidemiologic Data.* 3rd ed. New York: Oxford University Press.

Sterne, J. A. C., and K. Tilling. 2002. G-estimation of causal effects, allowing for time-varying confounding. *Stata Journal* 2: 164–182.

Therneau, T. M., and P. M. Grambsch. 2000. *Modeling Survival Data: Extending the Cox Model*. New York: Springer.

Vittinghoff, E., D. V. Glidden, S. C. Shiboski, and C. E. McCulloch. 2005. *Regression Methods in Biostatistics: Linear Logistic, Survival, and Repeated Measures Models*. New York: Springer.

Also see

[ST] **stcox postestimation** — Postestimation tools for stcox

[ST] **stcurve** — Plot survivor, hazard, cumulative hazard, or cumulative incidence function

[ST] **stcox PH-assumption tests** — Tests of proportional-hazards assumption

[ST] **stcrreg** — Competing-risks regression

[ST] **sts** — Generate, graph, list, and test the survivor and cumulative hazard functions

[ST] **stset** — Declare data to be survival-time data

[ST] **streg** — Parametric survival models

[SVY] **svy estimation** — Estimation commands for survey data

[U] **20 Estimation and postestimation commands**

Title

> **stcox PH-assumption tests** — Tests of proportional-hazards assumption

Syntax

Check proportional-hazards assumption:

   Log-log plot of survival

      stphplot [*if*] , {by(*varname*) | <u>str</u>ata(*varname*)} [*stphplot_options*]

   Kaplan–Meier and predicted survival plot

      stcoxkm [*if*] , by(*varname*) [*stcoxkm_options*]

   Using Schoenfeld residuals

      estat phtest [, *phtest_options*]

stphplot_options	description
Main	
* by(*varname*)	fit separate Cox models; the default
* <u>str</u>ata(*varname*)	fit stratified Cox model
<u>adj</u>ust(*varlist*)	adjust to average values of *varlist*
zero	adjust to zero values of *varlist*; use with adjust()
Options	
<u>noneg</u>ative	plot $\ln\{-\ln(\text{survival})\}$
<u>nolnt</u>ime	plot curves against analysis time
<u>nosh</u>ow	do not show st setting information
Plot	
plot#opts(*stphplot_plot_options*)	affect rendition of the #th connected line and #th plotted points
Add plots	
addplot(*plot*)	add other plots to the generated graph
Y axis, X axis, Titles, Legend, Overall	
twoway_options	any options other than by() documented in [G] ***twoway_options***

*Either by(*varname*) or strata(*varname*) is required with stphplot.

stphplot_plot_options	description
cline_options	change look of lines or connecting method
marker_options	change look of markers (color, size, etc.)

stcoxkm_options	description
Main	
*by(*varname*)	report the nominal or ordinal covariate
ties(breslow)	use Breslow method to handle tied failures
ties(efron)	use Efron method to handle tied failures
ties(exactm)	use exact marginal-likelihood method to handle tied failures
ties(exactp)	use exact partial-likelihood method to handle tied failures
separate	draw separate plot for predicted and observed curves
noshow	do not show st setting information
Observed plot	
obsopts(*stcoxkm_plot_options*)	affect rendition of the observed curve
obs#opts(*stcoxkm_plot_options*)	affect rendition of the #th observed curve; not allowed with separate
Predicted plot	
predopts(*stcoxkm_plot_options*)	affect rendition of the predicted curve
pred#opts(*stcoxkm_plot_options*)	affect rendition of the #th predicted curve; not allowed with separate
Add plots	
addplot(*plot*)	add other plots to the generated graph
Y axis, X axis, Titles, Legend, Overall	
twoway_options	any options documented in [G] *twoway_options*
byopts(*byopts*)	how subgraphs are combined, labeled, etc.

* by(*varname*) is required with stcoxkm.

stcoxkm_plot_options	description
connect_options	change look of connecting method
marker_options	change look of markers (color, size, etc.)

You must stset your data before using stphplot and stcoxkm; see [ST] **stset**.
fweights, iweights, and pweights may be specified using stset; see [ST] **stset**.

phtest_options	description
Main	
log	use natural logarithm time-scaling function
km	use 1 − KM product-limit estimate as the time-scaling function
rank	use rank of analysis time as the time-scaling function
time(*varname*)	use *varname* containing a monotone transformation of analysis time as the time-scaling function
plot(*varname*)	plot smoothed, scaled Schoenfeld residuals versus time
bwidth(#)	use bandwidth of #; default is bwidth(0.8)
detail	test proportional-hazards assumption separately for each covariate
Scatterplot	
marker_options	change look of markers (color, size, etc.)
marker_label_options	add marker labels; change look or position
Smoothed line	
lineopts(*cline_options*)	affect rendition of the smoothed line
Y axis, X axis, Titles, Legend, Overall	
twoway_options	any options other than by() documented in [G] *twoway_options*

estat phtest is not appropriate after estimation with svy.

Menu

stphplot

Statistics > Survival analysis > Regression models > Graphically assess proportional-hazards assumption

stcoxkm

Statistics > Survival analysis > Regression models > Kaplan-Meier versus predicted survival

estat phtest

Statistics > Survival analysis > Regression models > Test proportional-hazards assumption

Description

stphplot plots $-\ln\{-\ln(\text{survival})\}$ curves for each category of a nominal or ordinal covariate versus ln(analysis time). These are often referred to as "log-log" plots. Optionally, these estimates can be adjusted for covariates. The proportional-hazards assumption is not violated when the curves are parallel.

stcoxkm plots Kaplan–Meier observed survival curves and compares them with the Cox predicted curves for the same variable. The closer the observed values are to the predicted, the less likely it is that the proportional-hazards assumption has been violated. Do not run stcox before running this command; stcoxkm will execute stcox itself to fit the model and obtain predicted values.

estat phtest tests the proportional-hazards assumption on the basis of Schoenfeld residuals after fitting a model with stcox.

Options for stphplot

[Main]

by(*varname*) specifies the nominal or ordinal covariate. Either by() or strata() is required with stphplot.

strata(*varname*) is an alternative to by(). Rather than fitting separate Cox models for each value of *varname*, strata() fits one stratified Cox model. You must also specify adjust(*varlist*) with the strata(*varname*) option; see [ST] **sts graph**.

adjust(*varlist*) adjusts the estimates to that for the *average* values of the *varlist* specified. The estimates can also be adjusted to *zero* values of *varlist* by specifying the zero option. adjust(*varlist*) can be specified with by(); it is required with strata(*varname*).

zero is used with adjust() to specify that the estimates be adjusted to the 0 values of the *varlist* rather than to average values.

[Options]

nonegative specifies that ln{−ln(survival)} be plotted instead of −ln{−ln(survival)}.

nolntime specifies that curves be plotted against analysis time instead of against ln(analysis time).

noshow prevents stphplot from showing the key st variables. This option is seldom used because most people type stset, show or stset, noshow to set whether they want to see these variables mentioned at the top of the output of every st command; see [ST] **stset**.

[Plot]

plot#opts(*stphplot_plot_options*) affects the rendition of the #th connected line and #th plotted points; see [G] *cline_options* and [G] *marker_options*.

[Add plots]

addplot(*plot*) provides a way to add other plots to the generated graph; see [G] *addplot_option*.

[Y axis, X axis, Titles, Legend, Overall]

twoway_options are any of the options documented in [G] *twoway_options*, excluding by(). These include options for titling the graph (see [G] *title_options*) and for saving the graph to disk (see [G] *saving_option*).

Options for stcoxkm

[Main]

by(*varname*) specifies the nominal or ordinal covariate. by() is required.

ties(breslow | efron | exactm | exactp) specifies one of the methods available to stcox for handling tied failures. If none is specified, ties(breslow) is assumed; see [ST] **stcox**.

separate produces separate plots of predicted and observed values for each value of the variable specified with by().

noshow prevents stcoxkm from showing the key st variables. This option is seldom used because most people type stset, show or stset, noshow to set whether they want to see these variables mentioned at the top of the output of every st command; see [ST] **stset**.

stcox PH-assumption tests — Tests of proportional-hazards assumption

⌐ Observed plot ⌐

obsopts(*stcoxkm_plot_options*) affects the rendition of the observed curve; see [G] ***connect_options*** and [G] ***marker_options***.

obs#opts(*stcoxkm_plot_options*) affects the rendition of the #th observed curve; see [G] ***connect_options*** and [G] ***marker_options***. This option is not allowed with separate.

⌐ Predicted plot ⌐

predopts(*stcoxkm_connect_options*) affects the rendition of the predicted curve; see [G] ***connect_options*** and [G] ***marker_options***.

pred#opts(*stcoxkm_connect_options*) affects the rendition of the #th predicted curve; see [G] ***connect_options*** and [G] ***marker_options***. This option is not allowed with separate.

⌐ Add plots ⌐

addplot(*plot*) provides a way to add other plots to the generated graph; see [G] ***addplot_option***.

⌐ Y axis, X axis, Titles, Legend, Overall ⌐

twoway_options are any of the options documented in [G] ***twoway_options***, excluding by(). These include options for titling the graph (see [G] ***title_options***) and for saving the graph to disk (see [G] ***saving_option***).

byopts(*byopts*) affects the appearance of the combined graph when by() and separate are specified, including the overall graph title and the organization of subgraphs. See [G] ***by_option***.

Options for estat phtest

⌐ Main ⌐

log, km, rank, and time() are used to specify the time scaling function.

By default, estat phtest performs the tests using the identity function, i.e., analysis time itself.

log specifies that the natural log of analysis time be used.

km specifies that 1 minus the Kaplan–Meier product-limit estimate be used.

rank specifies that the rank of analysis time be used.

time(*varname*) specifies a variable containing an arbitrary monotonic transformation of analysis time. You must ensure that *varname* is a monotonic transform.

plot(*varname*) specifies that a scatterplot and smoothed plot of scaled Schoenfeld residuals versus time be produced for the covariate specified by *varname*. By default, the smoothing is performed using the running-mean method implemented in lowess, mean noweight; see [R] **lowess**.

bwidth(*#*) specifies the bandwidth. Centered subsets of bwidth() $\times N$ observations are used for calculating smoothed values for each point in the data except for endpoints, where smaller, uncentered subsets are used. The greater the bwidth(), the greater the smoothing. The default is bwidth(0.8).

detail specifies that a separate test of the proportional-hazards assumption be produced for each covariate in the Cox model. By default, estat phtest produces only the global test.

Scatterplot

marker_options affect the rendition of markers drawn at the plotted points, including their shape, size, color, and outline; see [G] ***marker_options***.

marker_label_options specify if and how the markers are to be labeled; see [G] ***marker_label_options***.

Smoothed line

lineopts(*cline_options*) affects the rendition of the smoothed line; see [G] ***cline_options***.

Y axis, X axis, Titles, Legend, Overall

twoway_options are any of the options documented in [G] ***twoway_options***, excluding by(). These include options for titling the graph (see [G] ***title_options***) and for saving the graph to disk (see [G] ***saving_option***).

Remarks

Cox proportional hazards models assume that the hazard ratio is constant over time. Suppose that a group of cancer patients on an experimental treatment is monitored for 10 years. If the hazard of dying for the nontreated group is twice the rate as that of the treated group (HR = 2.0), the proportional-hazards assumption implies that this ratio is the same at 1 year, at 2 years, or at any point on the time scale. Because the Cox model, by definition, is constrained to follow this assumption, it is important to evaluate its validity. If the assumption fails, alternative modeling choices would be more appropriate (e.g., a stratified Cox model, time-varying covariates).

stphplot and stcoxkm provide graphical methods for assessing violations of the proportional-hazards assumption. Although using graphs to assess the validity of the assumption is subjective, it can be a helpful tool.

stphplot plots $-\ln\{-\ln(\text{survival})\}$ curves for each category of a nominal or ordinal covariate versus ln(analysis time). These are often referred to as "log–log" plots. Optionally, these estimates can be adjusted for covariates. If the plotted lines are reasonably parallel, the proportional-hazards assumption has not been violated, and it would be appropriate to base the estimate for that variable on one baseline survivor function.

Another graphical method of evaluating the proportional-hazards assumption, though less common, is to plot the Kaplan–Meier observed survival curves and compare them with the Cox predicted curves for the same variable. This plot is produced with stcoxkm. When the predicted and observed curves are close together, the proportional-hazards assumption has not been violated. See Garrett (1997) for more details.

Many popular tests for proportional hazards are, in fact, tests of nonzero slope in a generalized linear regression of the scaled Schoenfeld residuals on time (see Grambsch and Therneau [1994]). The estat phtest command tests, for individual covariates and globally, the null hypothesis of zero slope, which is equivalent to testing that the log hazard-ratio function is constant over time. Thus rejection of the null hypothesis of a zero slope indicates deviation from the proportional-hazards assumption. The estat phtest command allows three common time-scaling options (log, km, and rank) and also allows you to specify a user-defined function of time through the time() option. When no option is specified, the tests are performed using analysis time without further transformation.

▷ Example 1

These examples use data from a leukemia remission study (Garrett 1997). The data consist of 42 patients who are monitored over time to see how long (weeks) it takes them to go out of remission (relapse: 1 = yes, 0 = no). Half the patients receive a new experimental drug, and the other half receive a standard drug (treatment1: 1 = drug A, 0 = standard). White blood cell count, a strong indicator of the presence of leukemia, is divided into three categories (wbc3cat: 1 = normal, 2 = moderate, 3 = high).

```
. use http://www.stata-press.com/data/r11/leukemia
(Leukemia Remission Study)
. describe
Contains data from http://www.stata-press.com/data/r11/leukemia.dta
  obs:            42                          Leukemia Remission Study
 vars:             8                          23 Mar 2009 10:39
 size:           504 (99.9% of memory free)
```

variable name	storage type	display format	value label	variable label
weeks	byte	%8.0g		Weeks in Remission
relapse	byte	%8.0g	yesno	Relapse
treatment1	byte	%8.0g	trt1lbl	Treatment I
treatment2	byte	%8.0g	trt2lbl	Treatment II
wbc3cat	byte	%9.0g	wbclbl	White Blood Cell Count
wbc1	byte	%8.0g		wbc3cat==Normal
wbc2	byte	%8.0g		wbc3cat==Moderate
wbc3	byte	%8.0g		wbc3cat==High

```
Sorted by: weeks
. stset weeks, failure(relapse) noshow
     failure event:  relapse != 0 & relapse < .
obs. time interval:  (0, weeks]
 exit on or before:  failure

       42  total obs.
        0  exclusions

       42  obs. remaining, representing
       30  failures in single record/single failure data
      541  total analysis time at risk, at risk from t =         0
                              earliest observed entry t =         0
                                   last observed exit t =        35
```

In this example, we examine whether the proportional-hazards assumption holds for drug A versus the standard drug (treatment1). First, we will use stphplot, followed by stcoxkm.

(Continued on next page)

```
. stphplot, by(treatment1)
```

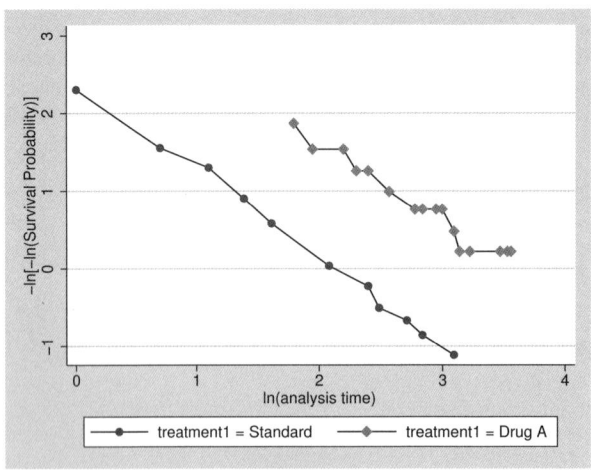

Figure 1.

```
. stcoxkm, by(treatment1)
```

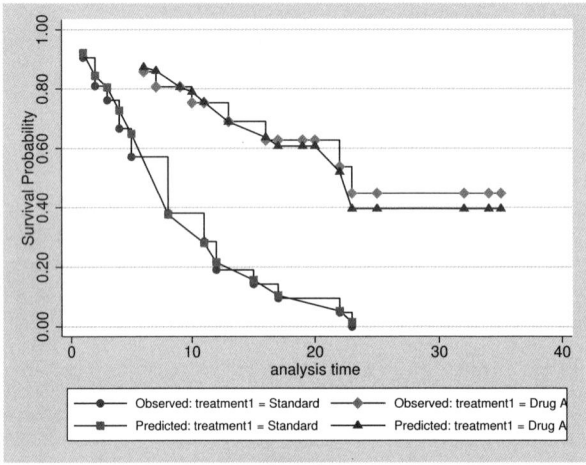

Figure 2.

Figure 1 (`stphplot`) displays lines that are parallel, implying that the proportional-hazards assumption for `treatment1` has not been violated. This is confirmed in figure 2 (`stcoxkm`), where the observed values and predicted values are close together.

The graph in figure 3 is the same as the one in figure 1, adjusted for white blood cell count (using two dummy variables). The adjustment variables were centered temporarily by `stphplot` before the adjustment was made.

stcox PH-assumption tests — Tests of proportional-hazards assumption

```
. stphplot, strata(treatment1) adj(wbc2 wbc3)
```

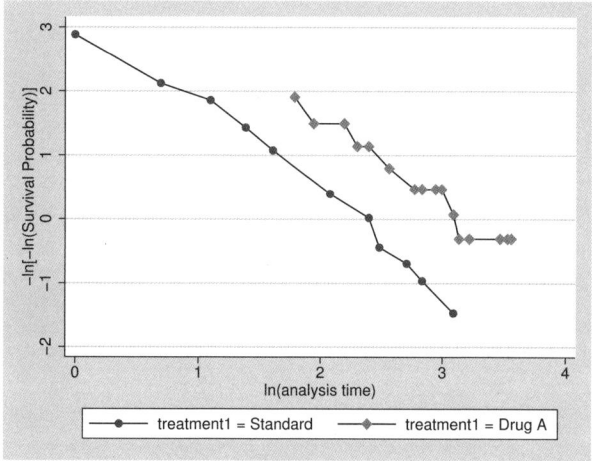

Figure 3.

The lines in figure 3 are still parallel, although they are somewhat closer together. Examining the proportional-hazards assumption on a variable without adjusting for covariates is usually adequate as a diagnostic tool before using the Cox model. However, if you know that adjustment for covariates in a final model is necessary, you may wish to reexamine whether the proportional-hazards assumption still holds.

Another variable in this dataset measures a different drug (`treatment2`: $1 =$ drug B, $0 =$ standard). We wish to examine the proportional-hazards assumption for this variable.

```
. stphplot, by(treatment2)
```

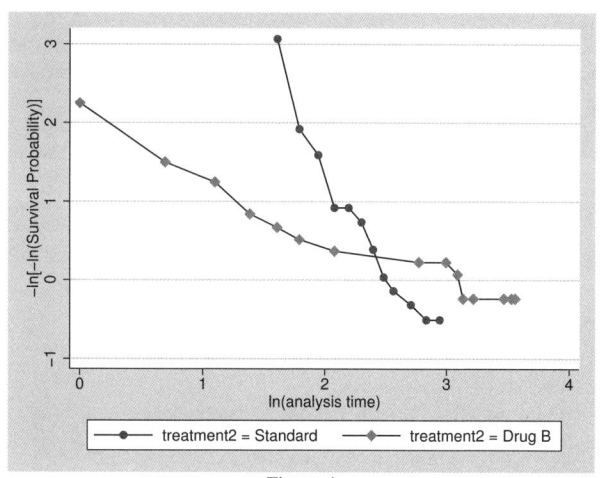

Figure 4.

```
. stcoxkm, by(treatment2) separate
```

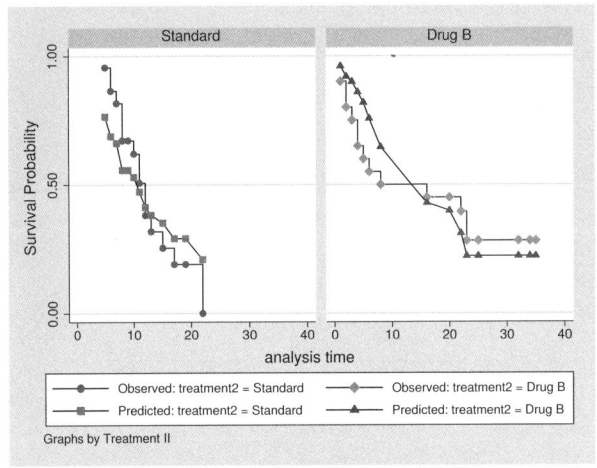

Figure 5.

This variable violates the proportional-hazards assumption. In figure 4, we see that the lines are not only nonparallel but also cross in the data region. In figure 5, we see that there are considerable differences between the observed and predicted values. We have overestimated the positive effect of drug B for the first half of the study and have underestimated it in the later weeks. One hazard ratio describing the effect of this drug would be inappropriate. We definitely would want to stratify on this variable in our Cox model.

◁

▷ Example 2: estat phtest

In this example, we use `estat phtest` to examine whether the proportional-hazards assumption holds for a model with covariates `wbc2`, `wbc1`, and `treatment1`. After stsetting the data, we first run `stcox` with these three variables as regressors. Then we use `estat phtest`:

```
. stset weeks, failure(relapse)

     failure event:  relapse != 0 & relapse < .
obs. time interval:  (0, weeks]
 exit on or before:  failure

─────────────────────────────────────────────────────────────────
        42  total obs.
         0  exclusions
─────────────────────────────────────────────────────────────────
        42  obs. remaining, representing
        30  failures in single record/single failure data
       541  total analysis time at risk, at risk from t =         0
                                 earliest observed entry t =      0
                                      last observed exit t =     35
```

```
. stcox treatment1 wbc2 wbc3, nolog
         failure _d:  relapse
   analysis time _t:  weeks

Cox regression -- Breslow method for ties

No. of subjects =           42                Number of obs   =         42
No. of failures =           30
Time at risk    =          541
                                              LR chi2(3)      =      33.02
Log likelihood  =   -77.476905                Prob > chi2     =     0.0000
```

_t	Haz. Ratio	Std. Err.	z	P>|z|	[95% Conf. Interval]	
treatment1	.2834551	.1229874	-2.91	0.004	.1211042	.6634517
wbc2	3.637825	2.201306	2.13	0.033	1.111134	11.91015
wbc3	10.92214	7.088783	3.68	0.000	3.06093	38.97284

```
. estat phtest, detail
      Test of proportional-hazards assumption

      Time:  Time
```

	rho	chi2	df	Prob>chi2
treatment1	-0.07019	0.15	1	0.6948
wbc2	-0.03223	0.03	1	0.8650
wbc3	0.01682	0.01	1	0.9237
global test		0.33	3	0.9551

Because we specified the `detail` option with the `estat phtest` command, both covariate-specific and global tests were produced. We can see that there is no evidence that the proportional-hazards assumption has been violated.

Another variable in this dataset measures a different drug (`treatment2`: 1 = drug B, 0 = standard). We now wish to examine the proportional-hazards assumption for the previous model by substituting `treatment2` for `treatment1`.

We fit a new Cox model and perform the test for proportional hazards:

```
. stcox treatment2 wbc2 wbc3, nolog
         failure _d:  relapse
   analysis time _t:  weeks

Cox regression -- Breslow method for ties

No. of subjects =           42                Number of obs   =         42
No. of failures =           30
Time at risk    =          541
                                              LR chi2(3)      =      23.93
Log likelihood  =   -82.019053                Prob > chi2     =     0.0000
```

_t	Haz. Ratio	Std. Err.	z	P>|z|	[95% Conf. Interval]	
treatment2	.8483777	.3469054	-0.40	0.688	.3806529	1.890816
wbc2	3.409628	2.050784	2.04	0.041	1.048905	11.08353
wbc3	14.0562	8.873693	4.19	0.000	4.078529	48.44314

```
. estat phtest, detail
    Test of proportional-hazards assumption

    Time:  Time
```

	rho	chi2	df	Prob>chi2
treatment2	-0.51672	10.19	1	0.0014
wbc2	-0.09860	0.29	1	0.5903
wbc3	-0.03559	0.04	1	0.8448
global test		10.24	3	0.0166

treatment2 violates the proportional-hazards assumption. A single hazard ratio describing the effect of this drug is inappropriate.

The test of the proportional-hazards assumption is based on the principle that, for a given regressor, the assumption restricts $\beta(t_j) = \beta$ for all t_j. This implies that a plot of $\beta(t_j)$ versus time will have a slope of zero. Grambsch and Therneau (1994) showed that $E(s_j^*) + \widehat{\beta} \approx \beta(t_j)$, where s_j^* is the scaled Schoenfeld residual at failure time t_j and $\widehat{\beta}$ is the estimated coefficient from the Cox model. Thus a plot of $s_j^* + \widehat{\beta}$ versus some function of time provides a graphical assessment of the assumption.

Continuing from above, if you type

```
. predict sch*, scaledsch
```

you obtain three variables—sch1, sch2, and sch3—corresponding to the three regressors, treatment2, wbc2, and wbc3. Given the utility of $s_j^* + \widehat{\beta}$, what is stored in variable sch1 is actually $s_{j1}^* + \widehat{\beta}_1$ and not just the scaled Schoenfeld residual for the first variable, s_{j1}^*, itself. The estimated coefficient, $\widehat{\beta}_1$, is added automatically. The same holds true for the second created variable representing the second regressor, sch2 = $s_{j2}^* + \widehat{\beta}_2$, and so on.

As such, a graphical assessment of the proportional-hazards assumption for the first regressor is as simple as

```
. scatter sch1 _t || lfit sch1 _t
```

which plots a scatter of $s_{j1}^* + \widehat{\beta}_1$ versus analysis time, _t, and overlays a linear fit. Is the slope zero? The answer is no for the first regressor, treatment2, and that agrees with our results from estat phtest.

◁

❑ Technical note

The tests of the proportional-hazards assumption assume homogeneity of variance across risk sets. This allows the use of the estimated overall (pooled) variance–covariance matrix in the equations. Although these tests have been shown by Grambsch and Therneau (1994) to be fairly robust to departures from this assumption, exercise care where this assumption may not hold, particularly when performing a stratified Cox analysis. In such cases, we recommend that you check the proportional-hazards assumption separately for each stratum.

❑

Saved results

estat phtest saves the following in r():

Scalars
r(df)   global test degrees of freedom   r(chi2)   global test χ^2

Methods and formulas

stphplot, stcoxkm, and estat phtest are implemented as ado-files.

For one covariate, x, the Cox proportional-hazards model reduces to

$$h(t; x) = h_0(t) \exp(x\beta)$$

where $h_0(t)$ is the baseline hazard function from the Cox model. Let $S_0(t)$ and $H_0(t)$ be the corresponding Cox baseline survivor and baseline cumulative hazard functions, respectively.

The proportional-hazards assumption implies that

$$H(t) = H_0(t) \exp(x\beta)$$

or

$$\ln H(t) = \ln H_0(t) + x\beta$$

where $H(t)$ is the cumulative hazard function. Thus, under the proportional-hazards assumption, the logs of the cumulative hazard functions at each level of the covariate have equal slope. This is the basis for the method implemented in stphplot.

The proportional-hazards assumption also implies that

$$S(t) = S_0(t)^{\exp(x\beta)}$$

Let $\widehat{S}(t)$ be the estimated survivor function based on the Cox model. This function is a step function like the Kaplan–Meier estimate and, in fact, reduces to the Kaplan–Meier estimate when $x = 0$. Thus for each level of the covariate of interest, we can assess violations of the proportional-hazards assumption by comparing these survival estimates with estimates calculated independently of the model. See Kalbfleisch and Prentice (2002) or Hess (1995).

stcoxkm plots Kaplan–Meier estimated curves for each level of the covariate together with the Cox model predicted baseline survival curve. The closer the observed values are to the predicted values, the less likely it is that the proportional-hazards assumption has been violated.

Grambsch and Therneau (1994) presented a scaled adjustment for the Schoenfeld residuals that permits the interpretation of the smoothed residuals as a nonparametric estimate of the log hazard-ratio function. These scaled Schoenfeld residuals, $\mathbf{r}^*_{S_i}$, can be obtained directly with predict's scaledsch option; see [ST] stcox postestimation.

Scaled Schoenfeld residuals are centered at $\widehat{\beta}$ for each covariate and, when there is no violation of proportional hazards, should have slope zero when plotted against functions of time. The estat phtest command uses these residuals, tests the null hypothesis that the slope is equal to zero for each covariate in the model, and performs the global test proposed by Grambsch and Therneau (1994). The test of zero slope is equivalent to testing that the log hazard-ratio function is constant over time.

For a specified function of time, $g(t)$, the statistic for testing the pth individual covariate is, for $\overline{g}(t) = d^{-1} \sum_{i=1}^{N} \delta_i g(t_i)$,

$$\chi_c^2 = \frac{\left[\sum_{i=1}^{N} \{\delta_i g(t_i) - \overline{g}(t)\} r_{S_{pi}}^* \right]^2}{d \text{ Var}(\widehat{\beta}_p) \sum_{i=1}^{N} \{\delta_i g(t_i) - \overline{g}(t)\}^2}$$

which is asymptotically distributed as χ^2 with 1 degree of freedom. $r_{S_{pi}}^*$ is the scaled Schoenfeld residual for observation i, and δ_i indicates failure for observation i, with $d = \sum \delta_i$.

The statistic for the global test is calculated as

$$\chi_g^2 = \left[\sum_{i=1}^{N} \{\delta_i g(t_i) - \overline{g}(t)\} \mathbf{r}_{S_i} \right]' \left[\frac{d \text{ Var}(\widehat{\boldsymbol{\beta}})}{\sum_{i=1}^{N} \{\delta_i g(t_i) - \overline{g}(t)\}^2} \right] \left[\sum_{i=1}^{N} \{\delta_i g(t_i) - \overline{g}(t)\} \mathbf{r}_{S_i} \right]$$

for $\mathbf{r}_{S_i}$, a vector of the m (unscaled) Schoenfeld residuals for the ith observation; see [ST] **stcox postestimation**. The global test statistic is asymptotically distributed as χ^2 with m degrees of freedom.

The equations for the scaled Schoenfeld residuals and the two test statistics just described assume homogeneity of variance across risk sets. Although these tests are fairly robust to deviations from this assumption, care must be exercised, particularly when dealing with a stratified Cox model.

Acknowledgment

The original versions of stphplot and stcoxkm were written by Joanne M. Garrett, University of North Carolina at Chapel Hill. We also thank Garrett for her contributions to the estat phtest command.

References

Barthel, F. M.-S., and J. P. Royston. 2006. Graphical representation of interactions. *Stata Journal* 6: 348–363.

Breslow, N. E. 1974. Covariance analysis of censored survival data. *Biometrics* 30: 89–99.

Cox, D. R. 1972. Regression models and life-tables (with discussion). *Journal of the Royal Statistical Society, Series B* 34: 187–220.

———. 1975. Partial likelihood. *Biometrika* 62: 269–276.

Cox, D. R., and D. Oakes. 1984. *Analysis of Survival Data*. London: Chapman & Hall/CRC.

Cox, D. R., and E. J. Snell. 1968. A general definition of residuals (with discussion). *Journal of the Royal Statistical Society, Series B* 30: 248–275.

Garrett, J. M. 1997. sbe14: Odds ratios and confidence intervals for logistic regression models with effect modification. *Stata Technical Bulletin* 36: 15–22. Reprinted in *Stata Technical Bulletin Reprints*, vol. 6, pp. 104–114. College Station, TX: Stata Press.

———. 1998. ssa12: Predicted survival curves for the Cox proportional hazards model. *Stata Technical Bulletin* 44: 37–41. Reprinted in *Stata Technical Bulletin Reprints*, vol. 8, pp. 285–290. College Station, TX: Stata Press.

Grambsch, P. M., and T. M. Therneau. 1994. Proportional hazards tests and diagnostics based on weighted residuals. *Biometrika* 81: 515–526.

Hess, K. R. 1995. Graphical methods for assessing violations of the proportional hazards assumption in Cox regression. *Statistics in Medicine* 14: 1707–1723.

Kalbfleisch, J. D., and R. L. Prentice. 2002. *The Statistical Analysis of Failure Time Data*. 2nd ed. New York: Wiley.

Rogers, W. H. 1994. ssa4: Ex post tests and diagnostics for a proportional hazards model. *Stata Technical Bulletin* 19: 23–27. Reprinted in *Stata Technical Bulletin Reprints*, vol. 4, pp. 186–191. College Station, TX: Stata Press.

Also see

[ST] **stcox** — Cox proportional hazards model

[ST] **sts** — Generate, graph, list, and test the survivor and cumulative hazard functions

[ST] **stset** — Declare data to be survival-time data

[U] **20 Estimation and postestimation commands**

Title

stcox postestimation — Postestimation tools for stcox

Description

The following postestimation commands are of special interest after `stcox`:

command	description
estat concordance	Harrell's C
stcurve	plot the survivor, hazard, and cumulative hazard functions

`estat concordance` is not appropriate after estimation with `svy`.

For information on `estat concordance`, see below. For information on `stcurve`, see [ST] **stcurve**.

The following standard postestimation commands are also available:

command	description
estat	AIC, BIC, VCE, and estimation sample summary
estat (svy)	postestimation statistics for survey data
estimates	cataloging estimation results
lincom	point estimates, standard errors, testing, and inference for linear combinations of coefficients
linktest	link test for model specification
lrtest[1]	likelihood-ratio test
margins	marginal means, predictive margins, marginal effects, and average marginal effects
nlcom	point estimates, standard errors, testing, and inference for nonlinear combinations of coefficients
predict	predictions, residuals, influence statistics, and other diagnostic measures
predictnl	point estimates, standard errors, testing, and inference for generalized predictions
test	Wald tests of simple and composite linear hypotheses
testnl	Wald tests of nonlinear hypotheses

[1] `lrtest` is not appropriate with `svy` estimation results.

See the corresponding entries in the *Base Reference Manual* for details, but see [SVY] **estat** for details about `estat (svy)`.

Special-interest postestimation commands

`estat concordance` calculates Harrell's C, which is defined as the proportion of all usable subject pairs in which the predictions and outcomes are concordant. `estat concordance` also reports the Somers' D rank correlation, which is obtained by calculating $2(C - 0.5)$.

Syntax for predict

> predict [*type*] *newvar* [*if*] [*in*] [, *sv_statistic* **nooff**set **parti**al]
>
> predict [*type*] { *stub** | *newvarlist* } [*if*] [*in*], *mv_statistic* [**parti**al]

sv_statistic	description
Main	
hr	predicted hazard ratio, also known as the relative hazard; the default
xb	linear prediction $\mathbf{x}_j \beta$
stdp	standard error of the linear prediction; $\text{SE}(\mathbf{x}_j \beta)$
*basesurv	baseline survivor function
*basechazard	baseline cumulative hazard function
*basehc	baseline hazard contributions
*mgale	martingale residuals
*csnell	Cox–Snell residuals
*deviance	deviance residuals
*ldisplace	likelihood displacement values
*lmax	LMAX measures of influence
*effects	log frailties

mv_statistic	description
Main	
*scores	efficient score residuals
*esr	synonym for scores
*dfbeta	DFBETA measures of influence
*schoenfeld	Schoenfeld residuals
*scaledsch	scaled Schoenfeld residuals

Unstarred statistics are available both in and out of sample; type predict ... if e(sample) ... if wanted only for the estimation sample. Starred statistics are calculated only for the estimation sample, even when e(sample) is not specified. nooffset is allowed only with unstarred statistics.
mgale, csnell, deviance, ldisplace, lmax, dfbeta, schoenfeld, and scaledsch are not allowed with svy estimation results.

Menu

Statistics > Postestimation > Predictions, residuals, etc.

Options for predict

Main

hr, the default, calculates the relative hazard (hazard ratio), that is, the exponentiated linear prediction, $\exp(\mathbf{x}_j \widehat{\beta})$.

xb calculates the linear prediction from the fitted model. That is, you fit the model by estimating a set of parameters, $\beta_0, \beta_1, \beta_2, \ldots, \beta_k$, and the linear prediction is $\widehat{\beta}_1 x_{1j} + \widehat{\beta}_2 x_{2j} + \cdots + \widehat{\beta}_k x_{kj}$, often written in matrix notation as $\mathbf{x}_j \widehat{\beta}$.

The $x_{1j}, x_{2j}, \ldots, x_{kj}$ used in the calculation are obtained from the data currently in memory and do not have to correspond to the data on the independent variables used in estimating β.

stdp calculates the standard error of the prediction, that is, the standard error of $x_j\widehat{\beta}$.

basesurv calculates the baseline survivor function. In the null model, this is equivalent to the Kaplan–Meier product-limit estimate. If stcox's strata() option was specified, baseline survivor functions for each stratum are provided.

basechazard calculates the cumulative baseline hazard. If stcox's strata() option was specified, cumulative baseline hazards for each stratum are provided.

basehc calculates the baseline hazard contributions. These are used to construct the product-limit type estimator for the baseline survivor function generated by basesurv. If stcox's strata() option was specified, baseline hazard contributions for each stratum are provided.

mgale calculates the martingale residuals. For multiple-record-per-subject data, by default only one value per subject is calculated, and it is placed on the last record for the subject.

Adding the partial option will produce partial martingale residuals, one for each record within subject; see partial below. Partial martingale residuals are the additive contributions to a subject's overall martingale residual. In single-record-per-subject data, the partial martingale residuals are the martingale residuals.

csnell calculates the Cox–Snell generalized residuals. For multiple-record data, by default only one value per subject is calculated and, it is placed on the last record for the subject.

Adding the partial option will produce partial Cox–Snell residuals, one for each record within subject; see partial below. Partial Cox–Snell residuals are the additive contributions to a subject's overall Cox–Snell residual. In single-record data, the partial Cox–Snell residuals are the Cox–Snell residuals.

deviance calculates the deviance residuals. Deviance residuals are martingale residuals that have been transformed to be more symmetric about zero. For multiple-record data, by default only one value per subject is calculated, and it is placed on the last record for the subject.

Adding the partial option will produce partial deviance residuals, one for each record within subject; see partial below. Partial deviance residuals are transformed partial martingale residuals. In single-record data, the partial deviance residuals are the deviance residuals.

ldisplace calculates the *likelihood displacement values*. A likelihood displacement value is an influence measure of the effect of deleting a subject on the overall coefficient vector. For multiple-record data, by default only one value per subject is calculated, and it is placed on the last record for the subject.

Adding the partial option will produce partial likelihood displacement values, one for each record within subject; see partial below. Partial displacement values are interpreted as effects due to deletion of individual records rather than deletion of individual subjects. In single-record data, the partial likelihood displacement values are the likelihood displacement values.

lmax calculates the LMAX measures of influence. LMAX values are related to likelihood displacement values because they also measure the effect of deleting a subject on the overall coefficient vector. For multiple-record data, by default only one LMAX value per subject is calculated, and it is placed on the last record for the subject.

Adding the partial option will produce partial LMAX values, one for each record within subject; see partial below. Partial LMAX values are interpreted as effects due to deletion of individual records rather than deletion of individual subjects. In single-record data, the partial LMAX values are the LMAX values.

effects is for use after stcox, shared() and provides estimates of the log frailty for each group. The log frailties are random group-specific offsets to the linear predictor that measure the group effect on the log relative-hazard.

scores calculates the efficient score residuals for each regressor in the model. For multiple-record data, by default only one score per subject is calculated, and it is placed on the last record for the subject.

Adding the partial option will produce partial efficient score residuals, one for each record within subject; see partial below. Partial efficient score residuals are the additive contributions to a subject's overall efficient score residual. In single-record data, the partial efficient score residuals are the efficient score residuals.

One efficient score residual variable is created for each regressor in the model; the first new variable corresponds to the first regressor, the second to the second, and so on.

esr is a synonym for scores.

dfbeta calculates the DFBETA measures of influence for each regressor in the model. The DFBETA value for a subject estimates the change in the regressor's coefficient due to deletion of that subject. For multiple-record data, by default only one value per subject is calculated, and it is placed on the last record for the subject.

Adding the partial option will produce partial DFBETAs, one for each record within subject; see partial below. Partial DFBETAs are interpreted as effects due to deletion of individual records rather than deletion of individual subjects. In single-record data, the partial DFBETAs are the DFBETAs.

One DFBETA variable is created for each regressor in the model; the first new variable corresponds to the first regressor, the second to the second, and so on.

schoenfeld calculates the Schoenfeld residuals. This option may not be used after stcox with the exactm or exactp option. Schoenfeld residuals are calculated and reported only at failure times.

One Schoenfeld residual variable is created for each regressor in the model; the first new variable corresponds to the first regressor, the second to the second, and so on.

scaledsch calculates the scaled Schoenfeld residuals. This option may not be used after stcox with the exactm or exactp option. Scaled Schoenfeld residuals are calculated and reported only at failure times.

One scaled Schoenfeld residual variable is created for each regressor in the model; the first new variable corresponds to the first regressor, the second to the second, and so on.

Note: The easiest way to use the preceding four options is, for example,

```
. predict double stub*, scores
```

where *stub* is a short name of your choosing. Stata then creates variables *stub*1, *stub*2, etc. You may also specify each variable explicitly, in which case there must be as many (and no more) variables specified as there are regressors in the model.

nooffset is allowed only with hr, xb, and stdp, and is relevant only if you specified offset(*varname*) for stcox. It modifies the calculations made by predict so that they ignore the offset variable; the linear prediction is treated as $x_j\widehat{\beta}$ rather than $x_j\widehat{\beta} + \text{offset}_j$.

partial is relevant only for multiple-record data and is valid with mgale, csnell, deviance, ldisplace, lmax, scores, esr, and dfbeta. Specifying partial will produce "partial" versions of these statistics, where one value is calculated for each record instead of one for each subject. The subjects are determined by the id() option to stset.

Specify partial if you wish to perform diagnostics on individual records rather than on individual subjects. For example, a partial DFBETA would be interpreted as the effect on a coefficient due to deletion of one record, rather than the effect due to deletion of all records for a given subject.

Syntax for estat concordance

estat concordance [*if*] [*in*] [, *concordance_options*]

concordance_options	description
Main	
all	compute statistic for all observations in the data
noshow	do not show st setting information

Menu

Statistics > Postestimation > Reports and statistics

Options for estat concordance

⌐ Main ⌐

all requests that the statistic be computed for all observations in the data. By default, estat concordance computes over the estimation subsample.

noshow prevents estat concordance from displaying the identities of the key st variables above its output.

Remarks

Remarks are presented under the following headings:

> Baseline functions
> Making baseline reasonable
> Residuals and diagnostic measures
> Multiple records per subject
> Predictions after stcox with the tvc() option
> Predictions after stcox with the shared() option
> estat concordance

Baseline functions

`predict` after `stcox` provides estimates of the baseline survivor and baseline cumulative hazard function, among other things. Here the term *baseline* means that these are the functions when all covariates are set to zero, that is, they reflect (perhaps hypothetical) individuals who have zero-valued measurements. When you specify `predict`'s `basechazard` option, you obtain the baseline cumulative hazard. When you specify `basesurv`, you obtain the baseline survivor function. Additionally, when you specify `predict`'s `basehc` option, you obtain estimates of the baseline hazard contribution at each failure time, which are factors used to develop the product-limit estimator for the survivor function generated by `basesurv`.

Although in theory $S_0(t) = \exp\{-H_0(t)\}$, where $S_0(t)$ is the baseline survivor function and $H_0(t)$ is the baseline cumulative hazard, the estimates produced by `basechazard` and `basesurv` do not exactly correspond in this manner, although they closely do. The reason is that `predict` after `stcox` uses different estimation schemes for each; the exact formulas are given in *Methods and formulas*.

When the Cox model is fit with the `strata()` option, you obtain estimates of the baseline functions for each stratum.

▷ Example 1: Baseline survivor function

Baseline functions refer to the values of the functions when all covariates are set to 0. Let's graph the survival curve for the Stanford heart transplant model that we fit in example 3 of [ST] **stcox**, and to make the baseline curve reasonable, let's do that at age = 40 and year = 70.

Thus we will begin by creating variables that, when 0, correspond to the baseline values we desire, and then we will fit our model with these variables instead. We then predict the baseline survivor function and graph it:

```
. use http://www.stata-press.com/data/r11/stan3
(Heart transplant data)
. generate age40 = age - 40
. generate year70 = year - 70
. stcox age40 posttran surg year70, nolog
         failure _d:  died
   analysis time _t:  t1
                 id:  id
Cox regression -- Breslow method for ties

No. of subjects =          103                     Number of obs   =        172
No. of failures =           75
Time at risk    =      31938.1
                                                   LR chi2(4)      =      17.56
Log likelihood  =    -289.53378                    Prob > chi2     =     0.0015
```

_t	Haz. Ratio	Std. Err.	z	P>|z|	[95% Conf. Interval]	
age40	1.030224	.0143201	2.14	0.032	1.002536	1.058677
posttran	.9787243	.3032597	-0.07	0.945	.5332291	1.796416
surgery	.3738278	.163204	-2.25	0.024	.1588759	.8796
year70	.8873107	.059808	-1.77	0.076	.7775022	1.012628

```
. predict s, basesurv
```

```
. summarize s

    Variable |       Obs        Mean    Std. Dev.       Min        Max
-------------+--------------------------------------------------------
           s |       172    .6291871    .2530009    .130666   .9908968
```

Our recentering of age and year did not affect the estimation, a fact you can verify by refitting the model with the original age and year variables.

To see how the values of the baseline survivor function are stored, we first sort according to analysis time and then list some observations.

```
. sort _t id
. list id _t0 _t _d s in 1/20

     +-------------------------------+
     | id   _t0   _t   _d          s |
     |-------------------------------|
  1. |  3     0    1    0    .9908968|
  2. | 15     0    1    1    .9908968|
  3. | 20     0    1    0    .9908968|
  4. | 45     0    1    0    .9908968|
  5. | 39     0    2    0    .9633915|
     |-------------------------------|
  6. | 43     0    2    1    .9633915|
  7. | 46     0    2    0    .9633915|
  8. | 61     0    2    1    .9633915|
  9. | 75     0    2    1    .9633915|
 10. | 95     0    2    0    .9633915|
     |-------------------------------|
 11. |  6     0    3    1    .9356873|
 12. | 23     0    3    0    .9356873|
 13. | 42     0    3    1    .9356873|
 14. | 54     0    3    1    .9356873|
 15. | 60     0    3    0    .9356873|
     |-------------------------------|
 16. | 68     0    3    0    .9356873|
 17. | 72     0    4    0    .9356873|
 18. | 94     0    4    0    .9356873|
 19. | 38     0    5    0    .9264087|
 20. | 70     0    5    0    .9264087|
     +-------------------------------+
```

At time $\_t = 2$, the baseline survivor function is 0.9634, or more precisely, $S_0(2 + \Delta t) = 0.9634$. What we mean by $S_0(t + \Delta t)$ is the probability of surviving just beyond t. This is done to clarify that the probability includes escaping failure at precisely time t.

The above also indicates that our estimate of $S_0(t)$ is a step function, and that the steps occur only at times when failure is observed—our estimated $S_0(t)$ does not change from $\_t = 3$ to $\_t = 4$ because no failure occurred at time 4. This behavior is analogous to that of the Kaplan–Meier estimate of the survivor function; see [ST] **sts**.

Here is a graph of the baseline survival curve:

```
. line s _t, sort c(J)
```

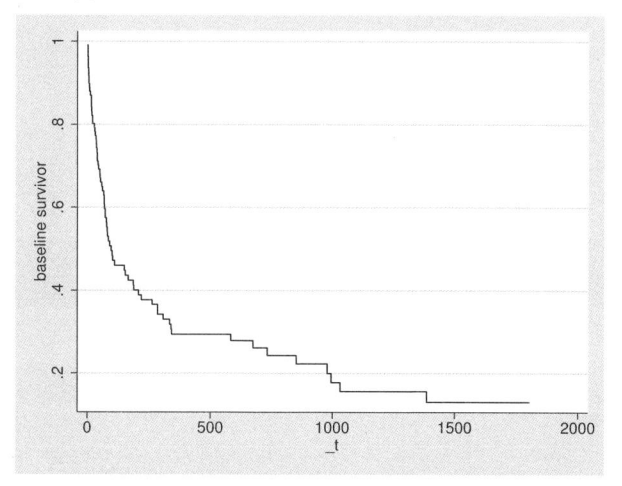

This graph was easy enough to produce because we wanted the survivor function at baseline. To graph survivor functions after stcox with covariates set to any value (baseline or otherwise), use stcurve; see [ST] **stcurve**.

◁

The similarity to Kaplan–Meier is not limited to the fact that both are step functions that change only when failure occurs. They are also calculated in much the same way, with predicting basesurv after stcox having the added benefit that the result is automatically adjusted for all the covariates in your Cox model. When you have no covariates, both methods are equivalent. If you continue from the previous example, you will find that

```
. sts generate s1 = s
```

and

```
. stcox, estimate
. predict double s2, basesurv
```

produce the identical variables s1 and s2, both containing estimates of the overall survivor function, unadjusted for covariates. We used type double for s2 to precisely match sts generate, which gives results in double precision.

If we had fit a stratified model by using the strata() option, the recorded survivor-function estimate on each observation would be for the stratum of that observation. That is, what you get is one variable that holds not an overall survivor curve, but instead a set of stratum-specific curves.

▷ Example 2: Baseline cumulative hazard

Obtaining estimates of the baseline cumulative hazard, $H_0(t)$, is just as easy as obtaining the baseline survivor function. Using the same data as previously,

```
. use http://www.stata-press.com/data/r11/stan3, clear
(Heart transplant data)
. generate age40 = age - 40
. generate year70 = year - 70
```

```
. stcox age40 posttran surg year70
  (output omitted)
. predict ch, basechazard
. line ch _t, sort c(J)
```

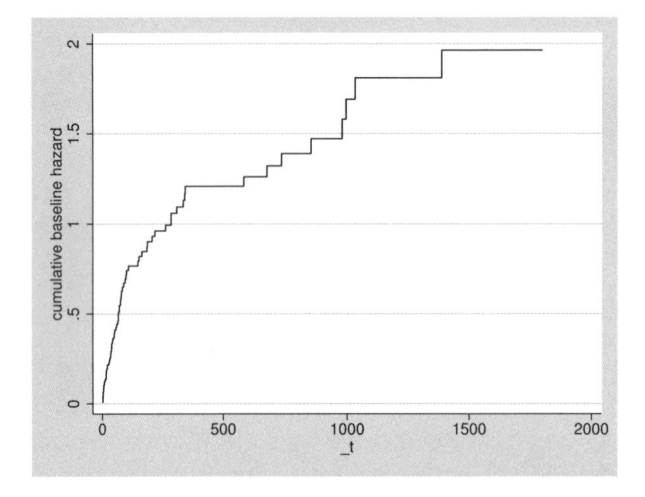

The estimated baseline cumulative hazard is also a step function with the steps occurring at the observed times of failure. When there are no covariates in your Cox model, what you obtain is equivalent to the Nelson–Aalen estimate of the cumulative hazard (see [ST] **sts**), but using `predict, basechazard` after `stcox` allows you to also adjust for covariates.

To obtain cumulative hazard curves at values other than baseline, you could either recenter your covariates—as we did previously with `age` and `year`—so that the values in which you are interested become baseline, or simply use `stcurve`; see [ST] **stcurve**.

◁

▷ Example 3: Baseline hazard contributions

Mathematically, a baseline hazard contribution, $h_i = (1 - \alpha_i)$ (see Kalbfleisch and Prentice 2002, 115), is defined at every analytic time t_i at which a failure occurs and is undefined at other times. Stata stores h_i in observations where a failure occurred and stores missing values in the other observations.

```
. use http://www.stata-press.com/data/r11/stan3, clear
(Heart transplant data)
. generate age40 = age - 40
. generate year70 = year - 70
. stcox age40 posttran surg year70
  (output omitted)
. predict double h, basehc
(97 missing values generated)
```

```
. list id _t0 _t _d h in 1/10
```

	id	_t0	_t	_d	h
1.	1	0	50	1	.01503465
2.	2	0	6	1	.02035303
3.	3	0	1	0	.
4.	3	1	16	1	.03339642
5.	4	0	36	0	.
6.	4	36	39	1	.01365406
7.	5	0	18	1	.01167142
8.	6	0	3	1	.02875689
9.	7	0	51	0	.
10.	7	51	675	1	.06215003

At time $\_t = 50$, the hazard contribution h_1 is 0.0150. At time $\_t = 6$, the hazard contribution h_2 is 0.0204. In observation 3, no hazard contribution is stored. Observation 3 contains a missing value because observation 3 did not fail at time 1. We also see that values of the hazard contributions are stored only in observations that are marked as failing.

Hazard contributions by themselves have no substantive interpretation, and in particular they should *not* be interpreted as estimating the hazard function at time t. Hazard contributions are simply mass points that are used as components to calculate the survivor function; see *Methods and formulas*. You can also use hazard contributions to estimate the hazard, but because they are only mass points, they need to be smoothed first. This smoothing is done automatically with stcurve; see [ST] **stcurve**. In summary, hazard contributions in their raw form serve no purpose other than to help replicate calculations done by Stata, and we demonstrate this below simply for illustrative purposes.

When we created the new variable h for holding the hazard contributions, we used type double because we plan on using h in some further calculations below and we wish to be as precise as possible.

In contrast with the baseline hazard contributions, the baseline survivor function, $S_0(t)$, is defined at all values of t: its estimate changes its value when failures occur, and at times when no failures occur, the estimated $S_0(t)$ is equal to its value at the time of the last failure.

Continuing with our example, we now predict the baseline survivor function:

```
. predict double s, basesurv
. list id _t0 _t _d h s in 1/10
```

	id	_t0	_t	_d	h	s
1.	1	0	50	1	.01503465	.68100303
2.	2	0	6	1	.02035303	.89846438
3.	3	0	1	0	.	.99089681
4.	3	1	16	1	.03339642	.84087361
5.	4	0	36	0	.	.7527663
6.	4	36	39	1	.01365406	.73259264
7.	5	0	18	1	.01167142	.82144038
8.	6	0	3	1	.02875689	.93568733
9.	7	0	51	0	.	.6705895
10.	7	51	675	1	.06215003	.26115633

In the above, we sorted by id, but it is easier to see how h and s are related if we sort by $\_t$ and put the failures on top:

```
. gsort +_t -_d
. list id _t0 _t _d h s in 1/18
```

	id	_t0	_t	_d	h	s
1.	15	0	1	1	.00910319	.99089681
2.	20	0	1	0	.	.99089681
3.	45	0	1	0	.	.99089681
4.	3	0	1	0	.	.99089681
5.	75	0	2	1	.02775802	.96339147
6.	61	0	2	1	.02775802	.96339147
7.	43	0	2	1	.02775802	.96339147
8.	39	0	2	0	.	.96339147
9.	95	0	2	0	.	.96339147
10.	46	0	2	0	.	.96339147
11.	6	0	3	1	.02875689	.93568733
12.	42	0	3	1	.02875689	.93568733
13.	54	0	3	1	.02875689	.93568733
14.	60	0	3	0	.	.93568733
15.	23	0	3	0	.	.93568733
16.	68	0	3	0	.	.93568733
17.	72	0	4	0	.	.93568733
18.	94	0	4	0	.	.93568733

The baseline hazard contribution is stored on every failure record—if multiple failures occur at a given time, the value of the hazard contribution is repeated—and the baseline survivor is stored on every record. (More correctly, baseline values are stored on records that meet the criterion and that were used in estimation. If some observations are explicitly or implicitly excluded from the estimation, their baseline values will be set to missing, no matter what.)

With this listing, we can better understand how the hazard contributions are used to calculate the survivor function. Because the patient with id = 15 died at time $t_1 = 1$, the hazard contribution for that patient is $h_{15} = 0.00910319$. Because that was the only death at $t_1 = 1$, the estimated survivor function at this time is $S_0(1) = 1 - h_{15} = 1 - 0.00910319 = 0.99089681$. The next death occurs at time $t_1 = 2$, and the hazard contribution at this time for patient 43 (or patient 61 or patient 75, it does not matter) is $h_{43} = 0.02775802$. Multiplying the previous survivor function value by $1 - h_{43}$ gives the new survivor function at $t_1 = 2$ as $S_0(2) = 0.96339147$. The other survivor function values are then calculated in succession, using this method at each failure time. At times when no failures occur, the survivor function remains unchanged.

◁

❑ Technical note

Consider manually obtaining the estimate of $S_0(t)$ from the h_i:

```
. sort _t _d
. by _t: keep if _d & _n==_N
. generate double s2 = 1-h
. replace s2 = s2[_n-1]*s2 if _n>1
```

s2 will be equivalent to s as produced above. If you had obtained stratified estimates, the code would be

```
. sort group _t _d
. by group _t: keep if _d & _n==_N
. generate double s2 = 1-h
. by group: replace s2 = s2[_n-1]*s2 if _n>1
```
❑

Making baseline reasonable

When predicting with basesurv or basechazard, for numerical accuracy reasons, the baseline functions must correspond to something reasonable in your data. Remember, the baseline functions correspond to all covariates equal to 0 in your Cox model.

Consider, for instance, a Cox model that includes the variable calendar year among the covariates. Say that year varies between 1980 and 1996. The baseline functions would correspond to year 0, almost 2,000 years in the past. Say that the estimated coefficient on year is -0.2, meaning that the hazard ratio for one year to the next is a reasonable 0.82.

Think carefully about the contribution to the predicted log cumulative hazard: it would be approximately $-0.2 \times 2,000 = -400$. Now $e^{-400} \approx 10^{-173}$, which on a digital computer is so close to 0 that there is simply no hope that $H_0(t)e^{-400}$ will produce an accurate estimate of $H(t)$.

Even with less extreme numbers, problems arise, even in the calculation of the baseline survivor function. Baseline hazard contributions near 1 produce baseline survivor functions with steps differing by many orders of magnitude because the calculation of the survivor function is cumulative. Producing a meaningful graph of such a survivor function is hopeless, and adjusting the survivor function to other values of the covariates is too much work.

For these reasons, covariate values of 0 must be meaningful if you are going to specify the basechazard or basesurv option. As the baseline values move to absurdity, the first problem you will encounter is a baseline survivor function that is too hard to interpret, even though the baseline hazard contributions are estimated accurately. Further out, the procedure Stata uses to estimate the baseline hazard contributions will break down—it will produce results that are exactly 1. Hazard contributions that are exactly 1 produce survivor functions that are uniformly 0, and they will remain 0 even after adjusting for covariates.

This, in fact, occurs with the Stanford heart transplant data:

```
. use http://www.stata-press.com/data/r11/stan3, clear
(Heart transplant data)
. stcox age posttran surg year
 (output omitted)
. predict ch, basechazard
. predict s, basesurv
. summarize ch s
```

Variable	Obs	Mean	Std. Dev.	Min	Max
ch	172	745.1134	682.8671	11.88239	2573.637
s	172	1.45e-07	9.43e-07	0	6.24e-06

The hint that there are problems is that the values of ch are huge and the values of s are close to 0. In this dataset, age (which ranges from 8 to 64 with a mean value of 45) and year (which ranges from 67 to 74) are the problems. The baseline functions correspond to a newborn at the turn of the century on the waiting list for a heart transplant!

To obtain accurate estimates of the baseline functions, type

```
. drop ch s
. generate age40 = age - 40
. generate year70 = year - 70
. stcox age40 posttran surg year70
 (output omitted)
. predict ch, basechazard
. predict s, basesurv
. summarize ch s
```

Variable	Obs	Mean	Std. Dev.	Min	Max
ch	172	.5685743	.521076	.0090671	1.963868
s	172	.6291871	.2530009	.130666	.9908968

Adjusting the variables does not affect the coefficient (and, hence, hazard-ratio) estimates, but it changes the values at which the baseline functions are estimated to be within the range of the data.

❑ **Technical note**

Above we demonstrated what can happen to predicted baseline functions when baseline values represent a departure from what was observed in the data. In the above example, the Cox model fit was fine and only the baseline functions lacked accuracy. As baseline values move even further toward absurdity, the risk-set accumulations required to fit the Cox model will also break down. If you are having difficulty getting stcox to converge or you obtain missing coefficients, one possible solution is to recenter your covariates just as we did above.

❑

Residuals and diagnostic measures

Stata can calculate Cox–Snell residuals, martingale residuals, deviance residuals, efficient score residuals (esr), Schoenfeld residuals, scaled Schoenfeld residuals, likelihood displacement values, LMAX values, and DFBETA influence measures.

Although the uses of residuals vary and depend on the data and user preferences, traditional and suggested uses are the following: Cox–Snell residuals are useful in assessing overall model fit. Martingale residuals are useful in determining the functional form of covariates to be included in the model and are occasionally useful in identifying outliers. Deviance residuals are useful in examining model accuracy and identifying outliers. Schoenfeld and scaled Schoenfeld residuals are useful for checking and testing the proportional-hazards assumption. Likelihood displacement values and LMAX values are useful in identifying influential subjects. DFBETAs also measure influence, but they do so on a coefficient-by-coefficient basis. Likelihood displacement values, LMAX values, and DFBETAs are all based on efficient score residuals.

▷ **Example 4: Cox–Snell residuals**

Let's first examine the use of Cox–Snell residuals. Using the cancer data introduced in example 2 in [ST] **stcox**, we first perform a Cox regression and then predict the Cox–Snell residuals.

```
. use http://www.stata-press.com/data/r11/drugtr, clear
(Patient Survival in Drug Trial)
. stset studytime, failure(died)
 (output omitted)
```

```
. stcox age drug, nolog
        failure _d:  died
   analysis time _t:  studytime

Cox regression -- Breslow method for ties

No. of subjects =           48              Number of obs    =         48
No. of failures =           31
Time at risk    =          744
                                            LR chi2(2)       =      33.18
Log likelihood  =   -83.323546              Prob > chi2      =     0.0000
```

_t	Haz. Ratio	Std. Err.	z	P>|z|	[95% Conf. Interval]	
age	1.120325	.0417711	3.05	0.002	1.041375	1.20526
drug	.1048772	.0477017	-4.96	0.000	.0430057	.2557622

```
. predict cs, csnell
```

The `csnell` option tells `predict` to output the Cox–Snell residuals to a new variable, cs. If the Cox regression model fits the data, these residuals should have a standard censored exponential distribution with hazard ratio 1. We can verify the model's fit by calculating—based, for example, on the Kaplan–Meier estimated survivor function or the Nelson–Aalen estimator—an empirical estimate of the cumulative hazard function, using the Cox–Snell residuals as the time variable and the data's original censoring variable. If the model fits the data, the plot of the cumulative hazard versus cs should approximate a straight line with slope 1.

To do this, we first re-`stset` the data, specifying cs as our new failure-time variable and died as the failure/censoring indicator. We then use the `sts generate` command to generate the km variable containing the Kaplan–Meier survivor estimates. Finally, we generate the cumulative hazard, H, by using the relationship $H = -\ln(km)$ and plot it against cs.

```
. stset cs, failure(died)
 (output omitted)
. sts generate km = s
. generate H = -ln(km)
(1 missing value generated)
. line H cs cs, sort ytitle("") clstyle(. refline)
```

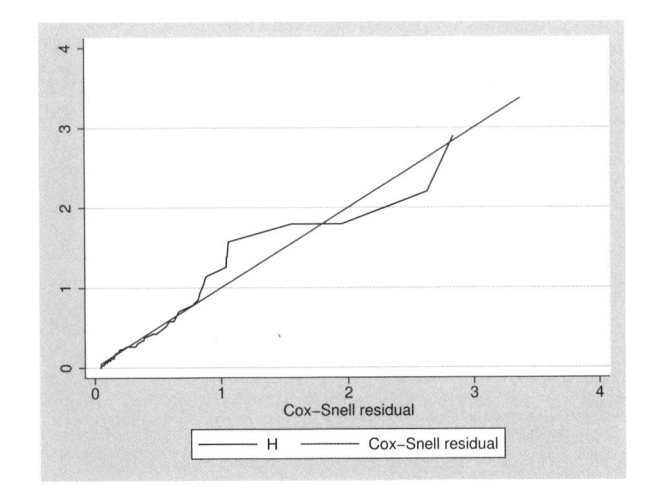

We specified cs twice in the graph command above so that a reference 45° line is plotted. Comparing the jagged line with the reference line, we observe that the Cox model does not fit these data too badly.

◁

❑ Technical note

The statement that "if the Cox regression model fits the data, the Cox–Snell residuals have a standard censored exponential distribution with hazard ratio 1" holds only if the true parameters, β, and the true cumulative baseline hazard function, $H_0(t)$, are used in calculating the residuals. Because we use estimates $\widehat{\beta}$ and $\widehat{H}_0(t)$, deviations from the 45° line in the above plots could be due in part to uncertainty about these estimates. This is particularly important for small sample sizes and in the right-hand tail of the distribution, where the baseline hazard is more variable because of the reduced effective sample caused by prior failures and censoring.

❑

▷ Example 5: Martingale residuals

Let's now examine the martingale residuals. Martingale residuals are useful in assessing the functional form of a covariate to be entered into a Cox model. Sometimes the covariate may need transforming so that the transformed variable will satisfy the assumptions of the proportional hazards model. To find the appropriate functional form of a variable, we fit a Cox model excluding the variable and then plot a lowess smooth of the martingale residuals against some transformation of the variable in question. If the transformation is appropriate, then the smooth should be approximately linear.

We apply this procedure to our cancer data to find an appropriate transformation of age (or to verify that age need not be transformed).

```
. use http://www.stata-press.com/data/r11/drugtr, clear
(Patient Survival in Drug Trial)
. stset studytime, failure(died)
 (output omitted)
. stcox drug
 (output omitted)
. predict mg, mgale
. lowess mg age, mean noweight title("") note("") m(o)
```

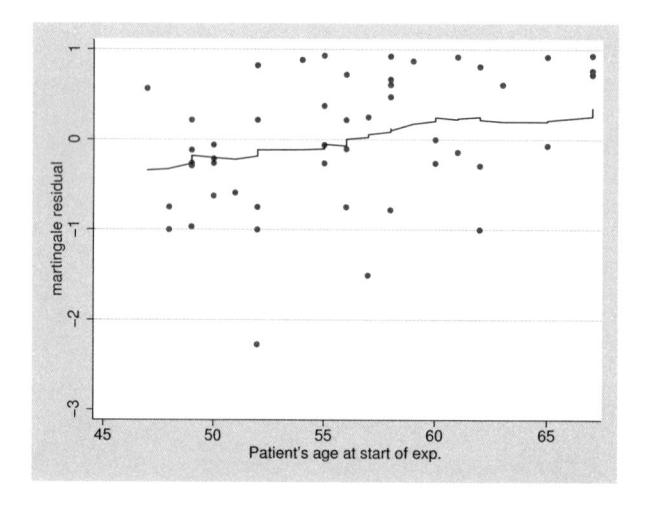

We used the `lowess` command with the `mean` and `noweight` options to obtain a plot of the running-mean smoother to ease interpretation. A lowess smoother or other smoother could also be used; see [R] **lowess**. The smooth appears nearly linear, supporting the inclusion of the untransformed version of `age` in our Cox model. Had the smooth not been linear, we would have tried smoothing the martingale residuals against various transformations of `age` until we found one that produced a near-linear smooth.

◁

Martingale residuals can also be interpreted as the difference over time of the observed number of failures minus the difference predicted by the model. Thus a plot of the martingale residuals versus the linear predictor may be used to detect outliers.

Plots of martingale residuals are sometimes difficult to interpret, however, because these residuals are skewed, taking values in $(-\infty, 1)$. For this reason, deviance residuals are preferred for examining model accuracy and identifying outliers.

▷ Example 6: Deviance residuals

Deviance residuals are a rescaling of the martingale residuals so that they are symmetric about 0 and thus are more like residuals obtained from linear regression. Plots of these residuals against the linear predictor, survival time, rank order of survival, or observation number can be useful in identifying aberrant observations and assessing model fit. We continue from the previous example, but we need to first refit the Cox model with `age` included:

```
. drop mg
. stcox drug age
 (output omitted)
. predict mg, mgale
. predict xb, xb
. scatter mg xb
```

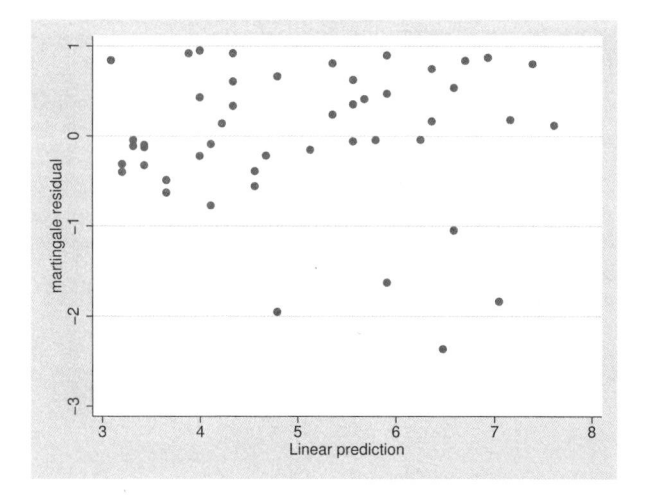

```
. predict dev, deviance

. scatter dev xb
```

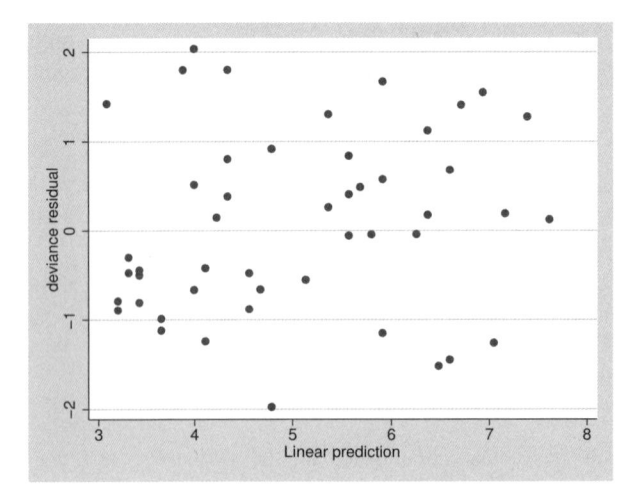

We first plotted the martingale residuals versus the linear predictor and then plotted the deviance residuals versus the linear predictor. Given their symmetry about 0, deviance residuals are easier to interpret, although both graphs yield the same information. With uncensored data, deviance residuals should resemble white noise if the fit is adequate. Censored observations would be represented as clumps of deviance residuals near 0 (Klein and Moeschberger 2003, 381). Given what we see above, there do not appear to be any outliers.

◁

In evaluating the adequacy of the fitted model, we must determine if any one subject has a disproportionate influence on the estimated parameters. This is known as influence or leverage analysis. The preferred method of performing influence or leverage analysis is to compare the estimated parameter, $\widehat{\beta}$, obtained from the full data, with estimated parameters $\widehat{\beta}_i$, obtained by fitting the model to the $N-1$ subjects remaining after the ith subject is removed. If $\widehat{\beta} - \widehat{\beta}_i$ is close to 0, the ith subject has little influence on the estimate. The process is repeated for all subjects included in the original model. To compute these differences for a dataset with N subjects, we would have to execute stcox N additional times, which could be impractical for large datasets.

To avoid fitting N additional Cox models, an approximation to $\widehat{\beta} - \widehat{\beta}_i$ can be made based on the efficient score residuals; see *Methods and formulas*. The difference $\widehat{\beta} - \widehat{\beta}_i$ is commonly referred to as DFBETA in the literature; see [R] **regress postestimation**.

▷ Example 7: DFBETAs

You obtain DFBETAs by using predict's dfbeta option:

```
. use http://www.stata-press.com/data/r11/drugtr, clear
(Patient Survival in Drug Trial)
. stset studytime, failure(died)
(output omitted)
. stcox age drug
(output omitted)
. predict df*, dfbeta
```

The last command saves the estimates of $\text{DFBETA}_i = \widehat{\beta} - \widehat{\beta}_i$ for $i = 1, \ldots, N$ in the variables df1 and df2. We can now plot these versus either time or subject (observation) number to identify subjects with disproportionate influence. To maximize the available information, we plot versus time and label the points by their subject numbers.

. generate obs = _n
. scatter df1 studytime, yline(0) mlabel(obs)

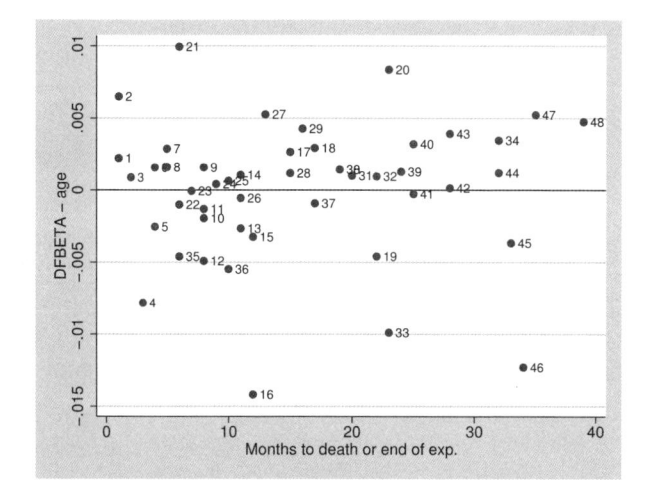

. scatter df2 studytime, yline(0) mlabel(obs)

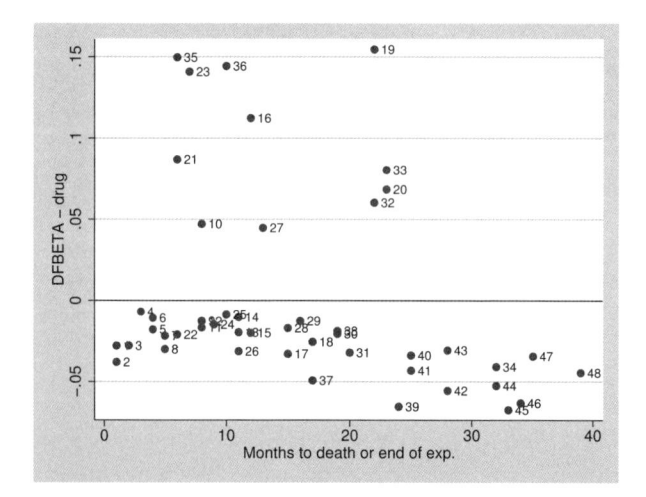

From the second graph we see that observation 35, if removed, would decrease the coefficient on drug by approximately 0.15 or, equivalently, decrease the hazard ratio for drug by a factor of approximately $\exp(-0.15) = 0.861$.

◁

DFBETAs as measures of influence have a straightforward interpretation. Their only disadvantage is that the number of values to examine grows both with sample size and with the number of regressors.

Two alternative measures of influence are *likelihood displacement* values and LMAX values, and both measure each subject's influence on the coefficient vector as a whole. Thus, for each, you have only one value per subject regardless of the number of regressors. As was the case with DFBETAs, likelihood displacement and LMAX calculations are also based on efficient score residuals; see *Methods and formulas*.

Likelihood displacement values measure influence by approximating what happens to the model log likelihood (more precisely, twice the log likelihood) when you omit subject i. Formally, the likelihood displacement value for subject i approximates the quantity

$$2\left\{\log L\left(\widehat{\beta}\right) - \log L\left(\widehat{\beta}_i\right)\right\}$$

where $\widehat{\beta}$ and $\widehat{\beta}_i$ are defined as previously and $L(\cdot)$ is the partial likelihood for the Cox model estimated from all the data. In other words, when you calculate $L(\cdot)$, you use all the data, but you evaluate at the parameter estimates $\widehat{\beta}_i$ obtained by omitting the ith subject. Note that because $\widehat{\beta}$ represents an optimal solution, likelihood displacement values will always be nonnegative.

That likelihood displacements measure influence can be seen through the following logic: if subject i is influential, then the vector $\widehat{\beta}_i$ will differ substantially from $\widehat{\beta}$. When that occurs, evaluating the log likelihood at such a suboptimal solution will give you a very different log likelihood.

LMAX values are closely related to likelihood displacements and are derived from an eigensystem analysis of the matrix of efficient score residuals; see *Methods and formulas* for details.

Both likelihood displacement and LMAX values measure each subject's overall influence, but they are not directly comparable with each other. Likelihood displacement values should be compared only with other likelihood displacement values, and LMAX values only with other LMAX values.

▷ Example 8: Likelihood displacement and LMAX values

You obtain likelihood displacement values with `predict`'s `ldisplace` option, and you obtain LMAX values with the `lmax` option. Continuing from the previous example:

```
. predict ld, ldisplace
. predict lmax, lmax
. list _t0 _t _d ld lmax in 1/10
```

	_t0	_t	_d	ld	lmax
1.	0	1	1	.0059511	.0735375
2.	0	1	1	.032366	.1124505
3.	0	2	1	.0038388	.0686295
4.	0	3	1	.0481942	.0113989
5.	0	4	1	.0078195	.0331513
6.	0	4	1	.0019887	.0308102
7.	0	5	1	.0069245	.0614247
8.	0	5	1	.0051647	.0763283
9.	0	8	1	.0021315	.0353402
10.	0	8	0	.0116187	.1179539

We can plot the likelihood displacement values versus time and label the points by observation number:

    . scatter ld studytime, mlabel(obs)

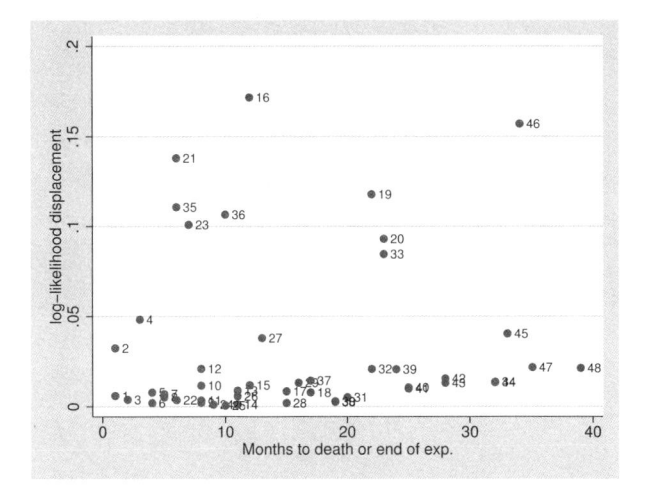

[Scatter plot of log-likelihood displacement vs. Months to death or end of exp., with observations labeled. Observations 16 and 46 appear as outliers.]

The above shows subjects 16 and 46 to be somewhat influential. A plot of LMAX values will show subject 16 as influential but not subject 46, a fact we leave to you to verify.

◁

Schoenfeld residuals and scaled Schoenfeld residuals are most often used to test the proportional-hazards assumption, as described in [ST] **stcox PH-assumption tests**.

Multiple records per subject

In the previous section, we analyzed data from a cancer study, and in doing so we were very loose in differentiating "observations" versus "subjects". In fact, we used both terms interchangeably. We were able to get away with that because in that dataset each subject (patient) was represented by only one observation—the subjects were the observations.

Oftentimes, however, subjects need representation by multiple observations, or records. For example, if a patient leaves the study for some time only to return later, at least one additional record will be needed to denote the subject's return to the study and the gap in their history. If the covariates of interest for a subject change during the study (for example, transitioning from smoking to nonsmoking), then this will also require representation by multiple records.

Multiple records per subject are not a problem for Stata; you simply specify an `id()` variable when stsetting your data, and this `id()` variable tells Stata which records belong to which subjects. The other commands in Stata's st suite know how to then incorporate this information into your analysis.

For `predict` after `stcox`, by default Stata handles diagnostic measures as always being at the *subject level*, regardless of whether that subject comprises one observation or multiple ones.

▷ Example 9: Stanford heart transplant data

As an example, consider, as we did previously, data from the Stanford heart transplant study:

```
. use http://www.stata-press.com/data/r11/stan3, clear
(Heart transplant data)
. stset
-> stset t1, id(id) failure(died)
                id:  id
     failure event:  died != 0 & died < .
obs. time interval:  (t1[_n-1], t1]
 exit on or before:  failure

       172  total obs.
         0  exclusions

       172  obs. remaining, representing
       103  subjects
        75  failures in single failure-per-subject data
   31938.1  total analysis time at risk, at risk from t =         0
                                    earliest observed entry t =  0
                                      last observed exit t =  1799

. list id _t0 _t _d age posttran surgery year in 1/10
```

	id	_t0	_t	_d	age	posttran	surgery	year
1.	1	0	50	1	30	0	0	67
2.	2	0	6	1	51	0	0	68
3.	3	0	1	0	54	0	0	68
4.	3	1	16	1	54	1	0	68
5.	4	0	36	0	40	0	0	68
6.	4	36	39	1	40	1	0	68
7.	5	0	18	1	20	0	0	68
8.	6	0	3	1	54	0	0	68
9.	7	0	51	0	50	0	0	68
10.	7	51	675	1	50	1	0	68

The data come to us already `stset`, and we type `stset` without arguments to examine the current settings. We verify that the `id` variable has been set as the patient id. We also see that we have 172 records representing 103 subjects, implying multiple records for some subjects. From our listing, we see that multiple records are necessary to accommodate changes in patients' heart-transplant status (pretransplant versus posttransplant).

Residuals and other diagnostic measures, where applicable, will by default take place at the subject level, meaning that (for example) there will be 103 likelihood displacement values for detecting influential subjects (not observations, but subjects).

```
. stcox age posttran surg year
  (output omitted)
. predict ld, ldisplace
(69 missing values generated)
```

```
. list id _t0 _t _d age posttran surgery year ld in 1/10
```

	id	_t0	_t	_d	age	posttran	surgery	year	ld
1.	1	0	50	1	30	0	0	67	.0596877
2.	2	0	6	1	51	0	0	68	.0154667
3.	3	0	1	0	54	0	0	68	.
4.	3	1	16	1	54	1	0	68	.0298421
5.	4	0	36	0	40	0	0	68	.
6.	4	36	39	1	40	1	0	68	.0359712
7.	5	0	18	1	20	0	0	68	.1260891
8.	6	0	3	1	54	0	0	68	.0199614
9.	7	0	51	0	50	0	0	68	.
10.	7	51	675	1	50	1	0	68	.0659499

Because here we are not interested in predicting any baseline functions, it is perfectly safe to leave age and year uncentered. The "(69 missing values generated)" message after predict tells us that only 103 out of the 172 observations of ld were filled in; that is, we received only one likelihood displacement per subject. Regardless of the current sorting of the data, the ld value for a subject is stored in the last chronological record for that subject as determined by analysis time, _t.

Patient 4 has two records in the data, one pretransplant and one posttransplant. As such, the ld value for that patient is interpreted as the change in twice the log likelihood due to deletion of both of these observations, i.e., the deletion of patient 4 from the study. The interpretation is at the patient level, not the record level.

◁

If, instead, you want likelihood displacement values that you can interpret at the observation level (i.e., changes in twice the log likelihood due to deleting one record), you simply add the partial option to the predict command above:

```
. predict ld, ldisplace partial
```

We do not think these kinds of observation-level diagnostics are generally what you would want, but they are available.

In the above, we discussed likelihood displacement values, but the same issue concerning subject-level versus observation-level interpretation also exists with Cox–Snell residuals, martingale residuals, deviance residuals, efficient score residuals, LMAX values, and DFBETAs. Regardless of which diagnostic you examine, this issue of interpretation is the same.

There is one situation where you do want to use the partial option. If you are using martingale residuals to determine functional form and the variable you are thinking of adding varies within subject, then you want to graph the partial martingale residuals against that new variable. Because the variable changes within subject, the martingale residuals should also change accordingly.

Predictions after stcox with the tvc() option

The residuals and diagnostics discussed previously are not available after estimation with stcox with the tvc() option, which is a convenience option for handling time-varying covariates:

```
. use http://www.stata-press.com/data/r11/drugtr, clear
(Patient Survival in Drug Trial)

. stcox drug age, tvc(age) nolog
         failure _d:  died
   analysis time _t:  studytime

Cox regression -- Breslow method for ties

No. of subjects =         48                Number of obs   =        48
No. of failures =         31
Time at risk    =        744
                                            LR chi2(3)      =     33.63
Log likelihood  =  -83.095036               Prob > chi2     =    0.0000
```

_t	Haz. Ratio	Std. Err.	z	P>|z|	[95% Conf. Interval]	
main						
drug	.1059862	.0478178	-4.97	0.000	.0437737	.2566171
age	1.156977	.07018	2.40	0.016	1.027288	1.303037
tvc						
age	.9970966	.0042415	-0.68	0.494	.988818	1.005445

```
Note: variables in tvc equation interacted with _t

. predict dev, deviance
this prediction is not allowed after estimation with tvc();
see tvc note for an alternative to the tvc() option
r(198);
```

The above fits a Cox model to the cancer data and includes an interaction of age with analysis time, _t. Such interactions are useful for testing the proportional-hazards assumption: significant interactions are violations of the proportional-hazards assumption for the variable being interacted with analysis time (or some function of analysis time). That is not the situation here.

In any case, models with tvc() interactions do not allow predicting the residuals and diagnostics discussed thus far. The solution in such situations is to forgo the use of tvc(), expand the data, and use factor variables to specify the interaction:

```
. generate id = _n

. streset, id(id)
 (output omitted)

. stsplit, at(failures)
(21 failure times)
(534 observations (episodes) created)
```

```
. stcox drug age c.age#c._t, nolog
        failure _d:  died
   analysis time _t:  studytime
              id:  id
Cox regression -- Breslow method for ties

No. of subjects =          48                Number of obs   =         582
No. of failures =          31
Time at risk    =         744
                                             LR chi2(3)      =       33.63
Log likelihood  =   -83.095036               Prob > chi2     =      0.0000
```

_t	Haz. Ratio	Std. Err.	z	P>|z|	[95% Conf. Interval]	
drug	.1059862	.0478178	-4.97	0.000	.0437737	.2566171
age	1.156977	.07018	2.40	0.016	1.027288	1.303037
c.age#c._t	.9970966	.0042415	-0.68	0.494	.988818	1.005445

```
. predict dev, deviance
(534 missing values generated)

. summarize dev
```

Variable	Obs	Mean	Std. Dev.	Min	Max
dev	48	.0658485	1.020993	-1.804876	2.065424

We split the observations, currently one per subject, so that the interaction term is allowed to vary over time. Splitting the observations requires that we first establish a subject id variable. Once that is done, we split the observations with stsplit and the at(failures) option, which splits the records only at the observed failure times. This amount of splitting is the minimal amount required to reproduce our previous Cox model. We then include the interaction term c.age#c._t in our model, verify that our Cox model is the same as before, and obtain our 48 deviance residuals, one for each subject.

Predictions after stcox with the shared() option

A Cox shared frailty model is a Cox model with added group-level random effects such that

$$h_{ij}(t) = h_0(t) \exp(\mathbf{x}_{ij}\boldsymbol{\beta} + \nu_i)$$

with ν_i representing the added effect due to being in group i; see *Cox regression with shared frailty* in [ST] **stcox** for more details. You fit this kind of model by specifying the shared(*varname*) option with stcox, where *varname* identifies the groups. stcox will produce an estimate of $\boldsymbol{\beta}$, its covariance matrix, and an estimate of the variance of the ν_i. What it will not produce are estimates of the ν_i themselves. These you can obtain postestimation with predict.

▷ Example 10: Shared frailty models

In example 9 of [ST] **stcox**, we fit a shared frailty model to data from 38 kidney dialysis patients, measuring the time to infection at the catheter insertion point. Two recurrence times (in days) were measured for each patient.

The estimated ν_i are not displayed in the stcox coefficient table but may be retrieved postestimation by using predict with the effects option:

```
. use http://www.stata-press.com/data/r11/catheter, clear
(Kidney data, McGilchrist and Aisbett, Biometrics, 1991)

. qui stcox age female, shared(patient)

. predict nu, effects

. sort nu

. list patient nu in 1/2
```

	patient	nu
1.	21	-2.448707
2.	21	-2.448707

```
. list patient nu in 75/L
```

	patient	nu
75.	7	.5187159
76.	7	.5187159

From the results above, we estimate that the least frail patient is patient 21, with $\widehat{\nu}_{21} = -2.45$, and that the frailest patient is patient 7, with $\widehat{\nu}_7 = 0.52$.

◁

❏ Technical note

When used with shared-frailty models, predict's basehc, basesurv, and basechazard options produce estimates of baseline quantities that are based on the last-step penalized Cox model fit. Therefore, the term *baseline* means that not only are the covariates set to 0 but the ν_i are as well.

Other predictions, such as martingale residuals, are conditional on the estimated frailty variance being fixed and known at the onset.

❏

estat concordance

estat concordance calculates Harrell's C, which is defined as the proportion of all usable subject pairs in which the predictions and outcomes are concordant. estat concordance also reports the Somers' D rank correlation, which is derived by calculating $2(C - 0.5)$. estat concordance may not be used after a Cox regression model with time-varying covariates and may not be applied to weighted data or to data with delayed entries.

▷ Example 11: Harrell's C

Using our cancer data, we wish to evaluate the predictive value of the measurement of drug and age. After fitting a Cox regression model, we use estat concordance to calculate Harrell's C index.

```
. use http://www.stata-press.com/data/r11/drugtr, clear
(Patient Survival in Drug Trial)

. stcox drug age
        failure _d:  died
   analysis time _t:  studytime

Iteration 0:   log likelihood = -99.911448
Iteration 1:   log likelihood = -83.551879
Iteration 2:   log likelihood = -83.324009
Iteration 3:   log likelihood = -83.323546
Refining estimates:
Iteration 0:   log likelihood = -83.323546

Cox regression -- Breslow method for ties

No. of subjects =         48                Number of obs    =         48
No. of failures =         31
Time at risk    =        744
                                            LR chi2(2)       =      33.18
Log likelihood  =  -83.323546               Prob > chi2      =     0.0000
```

_t	Haz. Ratio	Std. Err.	z	P>|z|	[95% Conf. Interval]	
drug	.1048772	.0477017	-4.96	0.000	.0430057	.2557622
age	1.120325	.0417711	3.05	0.002	1.041375	1.20526

```
. estat concordance, noshow
  Harrell's C concordance statistic
  Number of subjects (N)                =       48
  Number of comparison pairs (P)        =      849
  Number of orderings as expected (E)   =      679
  Number of tied predictions (T)        =       15

          Harrell's C = (E + T/2) / P    =    .8086
                        Somers' D       =    .6172
```

The result of stcox shows that the drug results in a lower hazard and therefore a longer survival time, controlling for age and older patients being more likely to die. The value of Harrell's C is 0.8086, which indicates that we can correctly order survival times for pairs of patients 81% of the time on the basis of measurement of drug and age. See *Methods and formulas* for the full definition of concordance.

◁

❑ Technical note

estat concordance does not work after a Cox regression model with time-varying covariates. When the covariates are varying with time, the prognostic score, PS = $x\beta$, will not capture or condense the information in given measurements, in which case it does not make sense to calculate the rank correlation between PS and survival time.

❑

Saved results

estat concordance saves the following in r():

Scalars
r(N)	number of observations	r(n_T)	number of tied predictions
r(n_P)	number of comparison pairs	r(D)	Somers' D coefficient
r(n_E)	number of orderings as expected	r(C)	Harrell's C coefficient

r(n_P), r(n_E), and r(n_T) are returned only when strata are not specified.

Methods and formulas

All methods presented in this entry have been implemented as ado-files that use Mata.

Let $\mathbf{x}_i$ be the row vector of covariates for the time interval $(t_{0i}, t_i]$ for the ith observation in the dataset ($i = 1, \ldots, N$). The Cox partial log-likelihood function, using the default Peto–Breslow method for tied failures is

$$\log L_{\text{breslow}} = \sum_{j=1}^{D} \sum_{i \in D_j} \left[w_i(\mathbf{x}_i \boldsymbol{\beta} + \text{offset}_i) - w_i \log \left\{ \sum_{\ell \in R_j} w_\ell \exp(\mathbf{x}_\ell \boldsymbol{\beta} + \text{offset}_\ell) \right\} \right]$$

where j indexes the ordered failure times t_j ($j = 1, \ldots, D$), D_j is the set of d_j observations that fail at t_j, d_j is the number of failures at t_j, and R_j is the set of observations k that are at risk at time t_j (i.e., all k such that $t_{0k} < t_j \leq t_k$). w_i and offset_i are, respectively, the weight and linear offset for observation i, if specified.

If the Efron method for ties is specified at estimation, the partial log likelihood is

$$\log L_{\text{efron}} = \sum_{j=1}^{D} \sum_{i \in D_j} \left[\mathbf{x}_i \boldsymbol{\beta} + \text{offset}_i - d_j^{-1} \sum_{k=0}^{d_j - 1} \log \left\{ \sum_{\ell \in R_j} \exp(\mathbf{x}_\ell \boldsymbol{\beta} + \text{offset}_\ell) - k A_j \right\} \right]$$

for $A_j = d_j^{-1} \sum_{\ell \in D_j} \exp(\mathbf{x}_\ell \boldsymbol{\beta} + \text{offset}_\ell)$. Weights are not supported with the Efron method.

At estimation, Stata also supports the exact marginal and exact partial methods for handling ties, but only the Peto–Breslow and Efron methods are supported in regard to the calculation of residuals, diagnostics, and other predictions. As such, only the partial log-likelihood formulas for those two methods are presented above, for easier reference in what follows.

If you specified efron at estimation, all predictions are carried out using the Efron method; that is, the handling of tied failures is done analogously to the way it was done when calculating $\log L_{\text{efron}}$. If you specified breslow (or nothing, because breslow is the default), exactm, or exactp, all predictions are carried out using the Peto–Breslow method. That is not to say that if you specify exactm at estimation, your predictions will be the same as if you had specified breslow. The formulas used will be the same, but the parameter estimates at which they are evaluated will differ because those were based on different ways of handling ties.

Define $z_i = \mathbf{x}_i \widehat{\boldsymbol{\beta}} + \text{offset}_i$. Schoenfeld residuals for the pth variable using the Peto–Breslow method are given by

$$r_{S_{pi}} = \delta_i (x_{pi} - a_{pi})$$

where

$$a_{pi} = \frac{\sum_{\ell \in R_i} w_\ell x_{p\ell} \exp(z_\ell)}{\sum_{\ell \in R_i} w_\ell \exp(z_\ell)}$$

δ_i indicates failure for observation i, and x_{pi} is the pth element of $\mathbf{x}_i$. For the Efron method, Schoenfeld residuals are

$$r_{S_{pi}} = \delta_i (x_{pi} - b_{pi})$$

where

$$b_{pi} = d_i^{-1} \sum_{k=0}^{d_i - 1} \frac{\sum_{\ell \in R_i} x_{p\ell} \exp(z_\ell) - k d_i^{-1} \sum_{\ell \in D_i} x_{p\ell} \exp(z_\ell)}{\sum_{\ell \in R_i} \exp(z_\ell) - k d_i^{-1} \sum_{\ell \in D_i} \exp(z_\ell)}$$

Schoenfeld residuals are derived from the first derivative of the log likelihood, with

$$\left.\frac{\partial \log L}{\partial \beta_p}\right|_{\widehat{\beta}} = \sum_{i=1}^{N} r_{S_{pi}} = 0$$

and only those observations that fail ($\delta_i = 1$) contribute a Schoenfeld residual to the derivative.

For censored observations, Stata stores a missing value for the Schoenfeld residual even though the above implies a value of 0. This is to emphasize that no calculation takes place when the observation is censored.

Scaled Schoenfeld residuals are given by

$$\mathbf{r}^*_{S_i} = \widehat{\boldsymbol{\beta}} + d\ \text{Var}(\widehat{\boldsymbol{\beta}}) \mathbf{r}_{S_i}$$

where $\mathbf{r}_{S_i} = (r_{S_{1i}}, \ldots, r_{S_{mi}})'$, m is the number of regressors, and d is the total number of failures.

In what follows, we assume the Peto–Breslow method for handling ties. Formulas for the Efron method, while tedious, can be obtained by applying similar principles of averaging across risk sets, as demonstrated above with Schoenfeld residuals.

Efficient score residuals are obtained by

$$r_{E_{pi}} = r_{S_{pi}} - \exp(z_i) \sum_{j: t_{0i} < t_j \leq t_i} \frac{\delta_j w_j (x_{pi} - a_{pj})}{\sum_{\ell \in R_j} w_\ell \exp(z_\ell)}$$

Like Schoenfeld residuals, efficient score residuals are also additive components of the first derivative of the log likelihood. Whereas Schoenfeld residuals are the contributions of each failure, efficient score residuals are the contributions of each observation. Censored observations contribute to the log likelihood (and its derivative) because they belong to risk sets at times when other observations fail. As such, an observation's contribution is twofold: 1) If the observation ends in failure, a risk assessment is triggered, i.e., a term in the log likelihood is computed. 2) Whether failed or censored, an observation contributes to risk sets for other observations that do fail. Efficient score residuals reflect both contributions.

The above computes efficient score residuals at the observation level. If you have multiple records per subject and do not specify the `partial` option, then the efficient score residual for a given subject is calculated by summing the efficient scores over the observations within that subject.

Martingale residuals are

$$r_{M_i} = \delta_i - \exp(z_i) \sum_{j: t_{0i} < t_j \leq t_i} \frac{w_j \delta_j}{\sum_{\ell \in R_j} w_\ell \exp(z_\ell)}$$

The above computes martingale residuals at the observation level. If you have multiple records per subject and do not specify the `partial` option, then the martingale residual for a given subject is calculated by summing r_{M_i} over the observations within that subject.

Martingale residuals are in the range $(-\infty, 1)$. Deviance residuals are transformations of martingale residuals designed to have a distribution that is more symmetric about zero. Deviance residuals are calculated using

$$r_{D_i} = \text{sign}(r_{M_i}) \left[-2 \{ r_{M_i} + \delta_i \log(\delta_i - r_{M_i}) \} \right]^{1/2}$$

These residuals are expected to be symmetric about zero but do not necessarily sum to zero.

The above computes deviance residuals at the observation level. If you have multiple records per subject and do not specify the `partial` option, then the deviance residual for a given subject is calculated by applying the above transformation to the *subject-level* martingale residual.

The estimated baseline hazard contribution is obtained at each failure time as $h_j = 1 - \widehat{\alpha}_j$, where $\widehat{\alpha}_j$ is the solution to

$$\sum_{k \in D_j} \frac{\exp(z_k)}{1 - \widehat{\alpha}_j^{\exp(z_k)}} = \sum_{\ell \in R_j} \exp(z_\ell)$$

(Kalbfleisch and Prentice 2002, eq. 4.34, 115).

The estimated baseline survivor function is

$$\widehat{S}_0(t) = \prod_{j: t_j \leq t} \widehat{\alpha}_j$$

When estimated with no covariates, $\widehat{S}_0(t)$ is the Kaplan–Meier estimate of the survivor function.

The estimated baseline cumulative hazard function, if requested, is related to the baseline survivor function calculation, yet the values of $\widehat{\alpha}_j$ are set at their starting values and are not iterated. Equivalently,

$$\widehat{H}_0(t) = \sum_{j: t_j \leq t} \frac{d_j}{\sum_{\ell \in R_j} \exp(z_\ell)}$$

When estimated with no covariates, $\widehat{H}_0(t)$ is the Nelson–Aalen estimate of the cumulative hazard.

Cox–Snell residuals are calculated with

$$r_{C_i} = \delta_i - r_{M_i}$$

where r_{M_i} are the martingale residuals. Equivalently, Cox–Snell residuals can be obtained with

$$r_{C_i} = \exp(z_i) \widehat{H}_0(t_i)$$

The above computes Cox–Snell residuals at the observation level. If you have multiple records per subject and do not specify the `partial` option, then the Cox–Snell residual for a given subject is calculated by summing r_{C_i} over the observations within that subject.

DFBETAs are calculated with

$$\text{DFBETA}_i = \mathbf{r}_{E_i} \widetilde{\text{Var}}(\widehat{\beta})$$

where $\mathbf{r}_{E_i} = (r_{E_{1i}}, \ldots, r_{E_{mi}})$ is a row vector of efficient score residuals with one entry for each regressor, and $\widetilde{\text{Var}}(\widehat{\beta})$ is the model-based variance matrix of $\widehat{\beta}$.

Likelihood displacement values are calculated with

$$\mathrm{LD}_i = \mathbf{r}_{E_i} \mathrm{Var}(\widehat{\boldsymbol{\beta}}) \mathbf{r}'_{E_i}$$

(Collett 2003, 136). In both of the above, $\mathbf{r}_{E_i}$ can represent either one observation or, in multiple-record data, the cumulative efficient score for an entire subject. For the former, the interpretation is that due to deletion of one record; for the latter, the interpretation is that due to deletion of all a subject's records.

Following Collett (2003, 137), LMAX values are obtained from an eigensystem analysis of

$$\mathbf{B} = \boldsymbol{\Theta}\,\mathrm{Var}(\widehat{\boldsymbol{\beta}})\,\boldsymbol{\Theta}'$$

where $\boldsymbol{\Theta}$ is the $N \times m$ matrix of efficient score residuals, with element (i, j) representing the jth regressor and the ith observation (or subject). LMAX values are then the absolute values of the elements of the unit-length eigenvector associated with the largest eigenvalue of the $N \times N$ matrix $\mathbf{B}$.

For shared-frailty models, the data are organized into G groups, with the ith group consisting of n_i observations, $i = 1, \ldots, G$. From Therneau and Grambsch (2000, 253–255), for fixed θ, estimates of $\boldsymbol{\beta}$ and $\nu_1, \ldots, \nu_G$ are obtained by maximizing

$$\log L(\theta) = \log L_{\mathrm{Cox}}(\boldsymbol{\beta}, \nu_1, \ldots, \nu_G) + \sum_{i=1}^{G} \left[\frac{1}{\theta} \{\nu_i - \exp(\nu_i)\} + \right.$$

$$\left. \left(\frac{1}{\theta} + D_i\right) \left\{1 - \log\left(\frac{1}{\theta} + D_i\right)\right\} - \frac{\log\theta}{\theta} + \log\Gamma\left(\frac{1}{\theta} + D_i\right) - \log\Gamma\left(\frac{1}{\theta}\right) \right]$$

where D_i is the number of death events in group i, and $\log L_{\mathrm{Cox}}(\boldsymbol{\beta}, \nu_1, \ldots, \nu_G)$ is the standard Cox partial log likelihood, with the ν_i treated as the coefficients of indicator variables identifying the groups. That is, the jth observation in the ith group has log relative-hazard $\mathbf{x}_{ij}\boldsymbol{\beta} + \nu_i$.

You obtain the estimates of $\nu_1, \ldots, \nu_G$ with `predict`'s `effects` option after `stcox, shared()`.

estat concordance

Harrell's C was proposed by Harrell Jr. et al. (1982) and was developed to evaluate the results of a medical test. The C index is defined as the proportion of all usable subject pairs in which the predictions and outcomes are concordant. The C index may be applied to ordinary continuous outcomes, dichotomous diagnostic outcomes, ordinal outcomes, and censored time-until-event response variables.

In predicting the time until death, C is calculated by considering all comparable patient pairs. A pair of patients is comparable if either 1) the two have different values on the time variable, and the one with the lowest value presents a failure, or 2) the two have the same value on the time variable, and exactly one of them presents a failure. If the predicted survival time is larger for the patient who lived longer, the predictions for the pair are said to be concordant with the outcomes. From Fibrinogen Studies Collaboration (2009), Harrell's C is defined as $\sum_k (E_k + T_k/2) / \sum_k (D_k)$, where D_k is the total number of pairs usable for comparison in stratum k, E_k is the number of pairs for which the predictions are concordant with the outcomes and the predictions are not identical in stratum k, and T_k is the number of usable pairs for which the predictions are identical in stratum k. If there are no strata specified, then the formula for Harrell's C reduces to $(E + T/2)/D$.

For a Cox proportional hazards model, the probability that the patient survives past time t is given by $S_0(t)$ raised to the $\exp(\mathbf{x}\beta)$ power, where $S_0(t)$ is the baseline survivor function, $\mathbf{x}$ denotes a set of measurements for the patient, and β is the vector of coefficients. A Cox regression model is fit by the stcox command. The hazard ratio, $\exp(\mathbf{x}\beta)$, is obtained by predict after stcox. Because the predicted survivor time and the predicted survivor function are one-to-one functions of each other, the predicted survivor function can be used to calculate C instead of the predicted survival time. The predicted survivor function decreases when the predicted hazard ratio increases; therefore, Harrell's C can be calculated by computing E, T, and D, based on the observed outcomes and the predicted hazard ratios.

C takes a value between 0 and 1. A value of 0.5 indicates no predictive discrimination, and values of 0 or 1.0 indicate perfect separation of subjects with different outcomes. See Harrell Jr., Lee, and Mark (1996) for more details. Somers' D rank correlation is calculated by $2C - 1$; see Newson (2002) for a discussion of Somers' D.

References

Collett, D. 2003. *Modelling Survival Data in Medical Research*. 2nd ed. London: Chapman & Hall/CRC.

Fibrinogen Studies Collaboration. 2009. Measures to assess the prognostic ability of the stratified Cox proportional hazards model. *Statistics in Medicine* 28: 389–411.

Harrell Jr., F. E., R. M. Califf, D. B. Pryor, K. L. Lee, and R. A. Rosati. 1982. Evaluating the yield of medical tests. *Journal of the American Medical Association* 247: 2543–2546.

Harrell Jr., F. E., K. L. Lee, and D. B. Mark. 1996. Multivariable prognostic models: Issues in developing models, evaluating assumptions and adequacy, and measuring and reducing errors. *Statistics in Medicine* 15: 361–387.

Kalbfleisch, J. D., and R. L. Prentice. 2002. *The Statistical Analysis of Failure Time Data*. 2nd ed. New York: Wiley.

Klein, J. P., and M. L. Moeschberger. 2003. *Survival Analysis: Techniques for Censored and Truncated Data*. 2nd ed. New York: Springer.

Newson, R. 2002. Parameters behind "nonparametric" statistics: Kendall's tau, Somers' D and median differences. *Stata Journal* 2: 45–64.

———. 2006. Confidence intervals for rank statistics: Somers' D and extensions. *Stata Journal* 6: 309–334.

Rogers, W. H. 1994. ssa4: Ex post tests and diagnostics for a proportional hazards model. *Stata Technical Bulletin* 19: 23–27. Reprinted in *Stata Technical Bulletin Reprints*, vol. 4, pp. 186–191. College Station, TX: Stata Press.

Schoenfeld, D. A. 1982. Partial residuals for the proportional hazards regression model. *Biometrika* 69: 239–241.

Therneau, T. M., and P. M. Grambsch. 2000. *Modeling Survival Data: Extending the Cox Model*. New York: Springer.

Also see

[ST] **stcox** — Cox proportional hazards model

[ST] **stcurve** — Plot survivor, hazard, cumulative hazard, or cumulative incidence function

[U] **20 Estimation and postestimation commands**

Title

stcrreg — Competing-risks regression

Syntax

<u>stcrreg</u> [*varlist*] [*if*] [*in*] , <u>compete</u>(*crvar*[==*numlist*]) [*options*]

options	description
Model	
* <u>compete</u>(*crvar*[==*numlist*])	specify competing-risks event(s)
<u>tvc</u>(*tvarlist*)	time-varying covariates
<u>texp</u>(*exp*)	multiplier for time-varying covariates; default is texp(_t)
<u>off</u>set(*varname*)	include *varname* in model with coefficient constrained to 1
<u>constraints</u>(*constraints*)	apply specified linear constraints
<u>coll</u>inear	keep collinear variables
SE/Robust	
<u>vce</u>(*vcetype*)	*vcetype* may be <u>robust</u>, <u>cl</u>uster *clustvar*, <u>boot</u>strap, or <u>jackknife</u>
<u>noadj</u>ust	do not use standard degree-of-freedom adjustment
Reporting	
<u>level</u>(#)	set confidence level; default is level(95)
<u>noshr</u>	report coefficients, not subhazard ratios
<u>nosh</u>ow	do not show st setting information
<u>nohead</u>er	suppress header from coefficient table
<u>notable</u>	suppress coefficient table
<u>nodisplay</u>	suppress output; iteration log is still displayed
<u>nocns</u>report	do not display constraints
display_options	control spacing and display of omitted variables and base and empty cells
Maximization	
maximize_options	control the maximization process; seldom used
† <u>coeflegend</u>	display coefficients' legend instead of coefficient table

*compete(*crvar*[==*numlist*]) is required.

†coeflegend does not appear in the dialog box.

You must stset your data before using stcrreg; see [ST] **stset**.

varlist and *tvarlist* may contain factor variables; see [U] **11.4.3 Factor variables**.

bootstrap, by, jackknife, mi estimate, nestreg, statsby, and stepwise are allowed; see [U] **11.1.10 Prefix commands**.

vce(bootstrap) and vce(jackknife) are not allowed with the mi estimate prefix.

fweights, iweights, and pweights may be specified using stset; see [ST] **stset**. In multiple-record data, weights are applied to subjects as a whole, not to individual observations. iweights are treated as fweights that can be noninteger, but not negative. Weights are not allowed with the bootstrap prefix.

See [U] **20 Estimation and postestimation commands** for more capabilities of estimation commands.

Menu

Statistics > Survival analysis > Regression models > Competing-risks regression

Description

stcrreg fits, via maximum likelihood, competing-risks regression models on st data, according to the method of Fine and Gray (1999). Competing-risks regression posits a model for the subhazard function of a failure event of primary interest. In the presence of competing failure events that impede the event of interest, a standard analysis using Cox regression (see [ST] **stcox**) is able to produce incidence-rate curves that either 1) are appropriate only for a hypothetical universe where competing events do not occur or 2) are appropriate for the data at hand, yet the effects of covariates on these curves are not easily quantified. Competing-risks regression, as performed using stcrreg, provides an alternative model that can produce incidence curves that represent the observed data and for which describing covariate effects is straightforward.

stcrreg can be used with single- or multiple-record data. stcrreg cannot be used when you have multiple failures per subject.

Options

___Model___

compete(*crvar*[==*numlist*]) is required and specifies the events that are associated with failure due to competing risks.

If compete(*crvar*) is specified, *crvar* is interpreted as an indicator variable; any nonzero, nonmissing values are interpreted as representing competing events.

If compete(*crvar*==*numlist*) is specified, records with *crvar* taking on any of the values in *numlist* are assumed to be competing events.

The syntax for compete() is the same as that for stset's failure() option. Use stset, failure() to specify the failure event of interest, that is, the failure event you wish to model using stcox, streg, stcrreg, or whatever. Use stcrreg, compete() to specify the event or events that compete with the failure event of interest. Competing events, because they are not the failure event of primary interest, must be stset as censored.

If you have multiple records per subject, only the value of *crvar* for the last chronological record for each subject is used to determine the event type for that subject.

tvc(*tvarlist*) specifies those variables that vary continuously with respect to time, i.e., time-varying covariates. These variables are multiplied by the function of time specified in texp().

texp(*exp*) is used in conjunction with tvc(*tvarlist*) to specify the function of analysis time that should be multiplied by the time-varying covariates. For example, specifying texp(ln(_t)) would cause the time-varying covariates to be multiplied by the logarithm of analysis time. If tvc(*tvarlist*) is used without texp(*exp*), Stata understands that you mean texp(_t), and thus multiplies the time-varying covariates by the analysis time.

Both tvc(*tvarlist*) and texp(*exp*) are explained more in *Option tvc() and testing the proportional-subhazards assumption* below.

offset(*varname*), constraints(*constraints*), collinear; see [R] **estimation options**.

⌐ SE/Robust ⌐

vce(*vcetype*) specifies the type of standard error reported, which includes types that are robust to some kinds of misspecification, that allow for intragroup correlation, and that use bootstrap or jackknife methods; see [R] ***vce_option***. vce(robust) is the default in single-record-per-subject st data. For multiple-record st data, vce(cluster *idvar*) is the default, where *idvar* is the ID variable previously stset.

Standard Hessian-based standard errors—*vcetype* oim—are not statistically appropriate for this model and thus are not allowed.

noadjust is for use with vce(robust) or vce(cluster *clustvar*). noadjust prevents the estimated variance matrix from being multiplied by $N/(N-1)$ or $g/(g-1)$, where g is the number of clusters. The default adjustment is somewhat arbitrary because it is not always clear how to count observations or clusters. In such cases, however, the adjustment is likely to be biased toward 1, so we would still recommend making it.

⌐ Reporting ⌐

level(*#*); see [R] **estimation options**.

noshr specifies that coefficients be displayed rather than exponentiated coefficients or subhazard ratios. This option affects only how results are displayed and not how they are estimated. noshr may be specified at estimation time or when redisplaying previously estimated results (which you do by typing stcrreg without a variable list).

noshow prevents stcrreg from showing the key st variables. This option is seldom used because most people type stset, show or stset, noshow to set whether they want to see these variables mentioned at the top of the output of every st command; see [ST] **stset**.

noheader suppresses the header information from the output. The coefficient table is still displayed. noheader may be specified at estimation time or when redisplaying previously estimated results.

notable suppresses the table of coefficients from the output. The header information is still displayed. notable may be specified at estimation time or when redisplaying previously estimated results.

nodisplay suppresses the output. The iteration log is still displayed.

nocnsreport; see [R] **estimation options**.

display_options: noomitted, vsquish, noemptycells, baselevels, allbaselevels; see [R] **estimation options**.

⌐ Maximization ⌐

maximize_options: difficult, technique(*algorithm_spec*), iterate(*#*), [no]log, trace, gradient, showstep, hessian, showtolerance, tolerance(*#*), ltolerance(*#*), nrtolerance(*#*), nonrtolerance, from(*init_specs*); see [R] **maximize**. These options are seldom used.

The following option is available with stcrreg but is not shown in the dialog box:

coeflegend; see [R] **estimation options**.

(*Continued on next page*)

Remarks

Remarks are presented under the following headings:

> The case for competing-risks regression
> Using stcrreg
> Multiple competing-event types
> stcrreg as an alternative to stcox
> Multiple records per subject
> Option tvc() and testing the proportional-subhazards assumption

The case for competing-risks regression

In this section, we provide a brief history and literature review of competing-risks analysis, and provide the motivation behind the stcrreg model. If you know you want to use stcrreg and are anxious to get started, you can safely skip this section.

Based on the method of Fine and Gray (1999), competing-risks regression provides a useful alternative to Cox regression (Cox 1972) for survival data in the presence of competing risks. Consider the usual survival analysis where one measures time-to-failure as a function of experimental or observed factors. For example, we may be interested in measuring time from initial treatment to recurrence of breast cancer in relation to factors such as treatment type and smoking status. The term *competing risk* refers to the chance that instead of cancer recurrence, you will observe a *competing event*, for example, death. The competing event, death, impedes the occurrence of the *event of interest*, breast cancer. This is not to be confused with the usual right-censoring found in survival data, such as censoring due to loss to follow-up. When subjects are lost to follow-up, they are still considered at risk of recurrent breast cancer—it is just that the researcher is not in a position to record the precise time that it happens. In contrast, death is a permanent condition that prevents future breast cancer. While censoring merely obstructs you from observing the event of interest, a competing event prevents the event of interest from occurring altogether. Because competing events are distinct from standard censorings, a competing-risks analysis requires some new methodology and some caution when interpreting the results from the old methodology.

Putter, Fiocco, and Geskus (2007) and Gichangi and Vach (2005) provide excellent tutorials covering the problem of competing risks, nonparametric estimators and tests, competing-risks regression, and the more general multistate models. Textbook treatments of competing-risks analysis can be found within Andersen et al. (1993), Klein and Moeschberger (2003), Therneau and Grambsch (2000), and Marubini and Valsecchi (1997). The texts by Crowder (2001) and Pintilie (2006) are devoted entirely to the topic. In what follows, we assume that you are familiar with the basic concepts of survival analysis, e.g., hazard functions and Kaplan–Meier curves. For such an introduction to survival analysis aimed at Stata users, see Cleves et al. (2008).

Without loss of generality, assume a situation where there is only one event that competes with the failure event of interest. Before analyzing the problem posed by competing-risks data—the problem stcrreg proposes to solve—we first formalize the mechanism behind it. Ignoring censoring for the moment, recording a failure time in a competing-risks scenario can be represented as observing the minimum of two potential failures times: the time to the event of interest, T_1, and the time to the competing event, T_2. The problem of competing risks then becomes one of understanding the nature of the bivariate distribution of (T_1, T_2), and in particular the correlation therein. Although conceptually simple, unfortunately this joint distribution cannot be identified by the data (Pepe and Mori 1993; Tsiatis 1975; Gail 1975). If you get to observe only the minimum, you are getting only half the picture.

An alternate representation of the competing-risks scenario that relies on quantities that are data-identifiable is described by Beyersman et al. (2009). In that formulation, we consider the hazard for the event of interest, $h_1(t)$, and that for the competing event, $h_2(t)$. Both hazards can be estimated from available data and when combined form a total hazard that any event will occur equal to $h(t) = h_1(t) + h_2(t)$. As risk accumulates according to $h(t)$, event times T are observed. Whether these events turn out to be failures of interest (type 1) or competing events (type 2) is determined by the two component hazards at that precise time. The event will be a failure of interest with probability $h_1(T)/\{h_1(T) + h_2(T)\}$, or a competing event with probability one minus that.

Instead of focusing on the survivor function for the event of interest, $P(T > t$ and event type 1), when competing risks are present you want to focus on the failure function, $P(T \leq t$ and event type 1), also known as the *cumulative incidence function* (CIF). That is because you will not know what type of event will occur until after it has occurred. It makes more sense to ask "What is the probability of breast cancer within 5 months?" than to ask "What is probability that nothing happens before 5 months, and that when something does happen, it will be breast cancer and not death?"

Much of the literature on competing risks focuses on the inadequacy of the Kaplan–Meier (1958) estimator (which we refer to as KM) as a measure of prevalence for the event of interest. Among others, Gooley et al. (1999) point out that 1−KM is a biased estimate of the CIF. The bias results from KM treating competing events as if they were censored. That is, subjects that experience competing events are treated as if they could later experience the event of interest, even though that is impossible. Although you could interpret 1−KM as the probability of a type 1 failure in a hypothetical setting where type 2 failures do not occur, this requires you to assume that $h_1(t)$ remains unchanged given that $h_2(t) = 0$, a rather strong and untestable assumption. Regardless of whether the independence assumption holds, 1−KM is still not representative of the data at hand, under which competing events do take place.

As such, 1−KM should be rejected in favor of the *cumulative incidence estimator* of the CIF; see Coviello and Boggess (2004) for a Stata-specific presentation. The cumulative incidence estimator is superior to 1−KM because it acknowledges that cumulative incidence is a function of both cause-specific hazards, $h_1(t)$ and $h_2(t)$. Conversely, 1−KM treats the CIF as a function solely of $h_1(t)$.

When you have covariates, you can use `stcox` to perform regression on $h_1(t)$ by treating failures of type 2 as censored, on $h_2(t)$ by treating failures of type 1 as censored, or on $h_1(t)$ and $h_2(t)$ simultaneously by using the method of data duplication described by Lunn and McNeil (1995) and Cleves (1999). Because cause-specific hazards are identified by the data, all three of the above analyses are suitable for estimating how covariates affect the mechanism behind a given type of failure. For example, if you are interested in how smoking affects breast cancer in general terms (competing death notwithstanding), then a Cox model for $h_1(t)$ that treats death as censored is perfectly valid; see Pintilie (2007).

If you are interested in the incidence of breast cancer, however, you want to use a Cox model that models both $h_1(t)$ and $h_2(t)$, because the CIF for breast cancer will likely depend on both. Based on the fitted model, you will have a hard time spotting the effects of covariates on cumulative incidence, because the covariates can affect $h_1(t)$ and $h_2(t)$ differently, and the CIF is a nonlinear function of these effects and of the baseline hazards. Whether increasing a covariate increases or decreases the cumulative incidence depends on time and on the nominal value of that covariate, as well as on the values of the other covariates. There is no way to determine the full effects of the covariates by just looking at the model coefficients. You would have to estimate and graph the CIF for various sets of covariate values, and this requires a bit of programming; see example 4.

An alternative model for the CIF that does make it easy to see the effects of covariates is that due to Fine and Gray (1999). They specify a model for the *hazard of the subdistribution* (Gray 1988), formally defined for failure type 1 as

$$\overline{h}_1(t) = \lim_{\delta \to 0} \left\{ \frac{P(t < T \leq t+\delta \text{ and event type 1}) \mid T > t \text{ or } (T \leq t \text{ and not event type 1})}{\delta} \right\}$$

Less formally, think of this hazard as that which generates failure events of interest while keeping subjects who experience competing events "at risk" so that they can be adequately counted as not having any chance of failing. The advantage of modeling the subdistribution hazard, or *subhazard*, is that you can readily calculate the CIF from it;

$$\text{CIF}_1(t) = 1 - \exp\{-\overline{H}_1(t)\}$$

where $\overline{H}_1(t) = \int_0^t \overline{h}_1(t)dt$ is the *cumulative subhazard*.

Competing-risks regression performed in this manner using stcrreg is quite similar to Cox regression performed using stcox. The model is semiparametric in that the baseline subhazard $\overline{h}_{1,0}(t)$ (that for covariates set to zero) is left unspecified, while the effects of the covariates $\mathbf{x}$ are assumed to be proportional:

$$\overline{h}_1(t|\mathbf{x}) = \overline{h}_{1,0}(t) \exp(\mathbf{x}\boldsymbol{\beta})$$

Estimation with stcrreg will produce estimates of β, or exponentiated coefficients known as *subhazard ratios*. A positive (negative) coefficient means that the effect of increasing that covariate is to increase (decrease) the subhazard and thus increase (decrease) the CIF across the board.

Estimates of the baseline cumulative subhazard and of the baseline CIF are available via predict after stcrreg; see [ST] **stcrreg postestimation**. Because proportionality holds for cumulative subhazards as well, adjusting the baseline cumulative hazard and baseline CIF for a given set of covariate values is quite easy and, in fact, done automatically for you by stcurve; see [ST] **stcurve**.

Using stcrreg

If you have used stcox before, stcrreg will look very familiar.

▷ Example 1: Cervical cancer study

Pintilie (2006, sec. 1.6.2) describes data from 109 cervical cancer patients that were treated at a cancer center between 1994 and 2000. The patients were treated and then the time in years until relapse or loss to follow-up was recorded. Relapses were recorded as either "local" if cancer relapsed in the pelvis, or "distant" if cancer recurred elsewhere but not in the pelvis. Patients who did not respond to the initial treatment were considered to have relapsed locally after one day.

```
. use http://www.stata-press.com/data/r11/hypoxia
(Hypoxia study)

. describe

Contains data from http://www.stata-press.com/data/r11/hypoxia.dta
  obs:            109                          Hypoxia study
 vars:             16                          7 Apr 2009 09:44
 size:          4,142 (99.9% of memory free)   (_dta has notes)
```

variable name	storage type	display format	value label	variable label
stnum	int	%8.0g		Patient ID
age	byte	%8.0g		Age (years)
hgb	int	%8.0g		Hemoglobin (g/l)
tumsize	float	%9.0g		Tumor size (cm)
ifp	float	%9.0g		Interstitial fluid pressure (marker, mmHg)
hp5	float	%9.0g		Hypoxia marker (percentage of meas. < 5 mmHg)
pelvicln	str1	%9s		Pelvic node involvement: N=Negative, E=Equivocal, Y=Positive
resp	str2	%9s		Response after treatment: CR=Complete response, NR=No response
pelrec	byte	%9.0g	yesno	Pelvic disease observed
disrec	byte	%9.0g	yesno	Distant disease observed
survtime	float	%9.0g		Time from diagnosis to death or last follow-up time (yrs)
stat	byte	%8.0g		Status at last follow-up: 0=Alive, 1=Dead
dftime	float	%9.0g		Time from diagnosis to first failure or last follow-up (yrs)
dfcens	byte	%8.0g		Censoring variable: 1=Failure, 0=Censored
failtype	byte	%8.0g		Failure type: 1 if pelrec, 2 if disrec & not pelrec, 0 otherwise
pelnode	byte	%8.0g		1 if pelvic nodes negative or equivocal

Sorted by:

The `dftime` variable records analysis time in years and the `failtype` variable records the type of event observed: 0 for loss to follow-up (censored), 1 for a local relapse, and 2 for a distant relapse. Among the covariates used in the analysis were a hypoxia marker (`hp5`) that measures the degree of oxygenation in the tumor, interstitial fluid pressure (`ifp`), tumor size (`tumsize`), and an indicator of pelvic node involvement (`pelnode == 0` if positive involvement and `pelnode == 1` otherwise). The main goal of the study was to determine whether `ifp` and `hp5` influence the outcome, controlling for the other covariates. Following Pintilie (2006), we focus on `ifp` and not on `hp5`. For more details regarding this study and the process behind the measured data, see Fyles et al. (2002) and Milosevic et al. (2001).

We wish to fit a competing-risks model that treats a local relapse as the event of interest and a distant relapse as the competing event. Although a distant relapse does not strictly prevent a future local relapse, presumably, the treatment protocol changed based on which event was first observed. As such, both events can be treated as competing with one another because the conditions of the study ended once any relapse was observed. Because no deaths occurred before first relapse, death is not considered a competing event in this analysis.

To fit the model, we first `stset` the data and specify that a local relapse, `failtype == 1`, is the event of interest. We then specify to `stcrreg` the covariates and that a distant relapse (`failtype == 2`) is a competing event.

```
. stset dftime, failure(failtype == 1)
 (output omitted)

. stcrreg ifp tumsize pelnode, compete(failtype == 2)

        failure _d:  failtype == 1
   analysis time _t:  dftime

Iteration 0:   log pseudolikelihood = -138.67925
Iteration 1:   log pseudolikelihood = -138.53082
Iteration 2:   log pseudolikelihood = -138.5308
Iteration 3:   log pseudolikelihood = -138.5308

Competing-risks regression                      No. of obs      =        109
                                                No. of subjects =        109
Failure event  : failtype == 1                  No. failed      =         33
Competing event: failtype == 2                  No. competing   =         17
                                                No. censored    =         59

                                                Wald chi2(3)    =      33.21
Log pseudolikelihood = -138.5308                Prob > chi2     =     0.0000

                             Robust
         _t         SHR    Std. Err.       z    P>|z|     [95% Conf. Interval]

        ifp    1.033206    .0178938     1.89    0.059     .9987231    1.068879
    tumsize    1.297332    .1271191     2.66    0.008     1.070646    1.572013
     pelnode    .4588123    .1972067    -1.81    0.070     .1975931    1.065365
```

From the above we point out the following:

- When we `stset` the data, distant relapses were set as censored because they are not the event of interest and any standard, noncompeting-risks analysis would want to treat them as censored. `stcrreg` option `compete()` tells Stata which of these "censored" events are actually competing events that require special consideration in a competing-risks regression. Because competing events are not the event of interest, `stcrreg` will issue an error if competing events are not `stset` as censored.

- `stcrreg` lists the event code(s) for the event of interest under "Failure event(s):" and the competing event code(s) under "Competing event(s):". The syntax for `stset` and `stcrreg` allows you to have multiple codes for both. For competing events, multiple event codes can be devoted entirely to one competing event type, many competing event types, or some combination of both. The methodology behind `stcrreg` extends to more than one competing event type and is concerned only with whether events are competing events, not with their exact type. The focus is on the event of interest.

- We see that out of the 109 patients, 33 experienced a local relapse, 17 experienced a distant relapse, and the remaining 59 were lost to follow-up before any relapse.

- In the column labeled "SHR" are the estimated subhazard ratios, and you interpret these similarly to hazard ratios in Cox regression. Because the estimated subhazard ratio for `ifp` is greater than 1, higher interstitial fluid pressures are associated with higher incidence of local relapses controlling for tumor size, pelvic node involvement, and the fact that distant relapses can also occur. However, this effect is not highly significant.

- To see the estimated coefficients instead of subhazard ratios, use the `noshr` option either when fitting the model or when replaying results.

- Standard errors are listed as "Robust", even though we did not specify any sampling weights, vce(robust), or vce(cluster *clustvar*). As mentioned in the previous section, competing-risks regression works by keeping subjects who experience competing events at risk so that they can be adequately counted as having no chance of failing. Doing so requires a form of sample weighting that invalidates the usual model-based standard errors; see *Methods and formulas*. Robust standard errors are conventional in stcrreg.

- The output lists a "log pseudolikelihood" rather than the standard log likelihood. This is also a consequence of the inherent sample weighting explained in the previous bullet. The log pseudolikelihood is used as a maximization criterion to obtain parameter estimates, but is not representative of the distribution of the data. For this reason, likelihood-ratio (LR) tests (the lrtest command) are not valid after stcrreg. Use Wald tests (the test command) instead.

As mentioned above, you can use the noshr option to obtain coefficients instead of subhazard ratios.

```
. stcrreg, noshr
Competing-risks regression              No. of obs       =         109
                                        No. of subjects  =         109
Failure event  : failtype == 1          No. failed       =          33
Competing event: failtype == 2          No. competing    =          17
                                        No. censored     =          59

                                        Wald chi2(3)     =       33.21
Log pseudolikelihood = -138.5308        Prob > chi2      =      0.0000
```

	Coef.	Robust Std. Err.	z	P>|z|	[95% Conf.	Interval]
ifp	.0326664	.0173188	1.89	0.059	-.0012777	.0666105
tumsize	.2603096	.0979851	2.66	0.008	.0682623	.4523568
pelnode	-.779114	.4298199	-1.81	0.070	-1.621546	.0633175

Just as with stcox, this model has no constant term. It is absorbed as part of the baseline subhazard, which is not directly estimated.

◁

▷ Example 2: CIF curves after stcrreg

In the above analysis, we stated that with increased interstitial fluid pressure comes an increase in the incidence of local relapses in the presence of possible distant relapses. To demonstrate this visually, we use stcurve to compare two CIF curves: one for ifp == 5 and one for ifp == 20. For both curves, we assume positive pelvic node involvement (pelnode==0) and tumor size set at the mean over the data.

(*Continued on next page*)

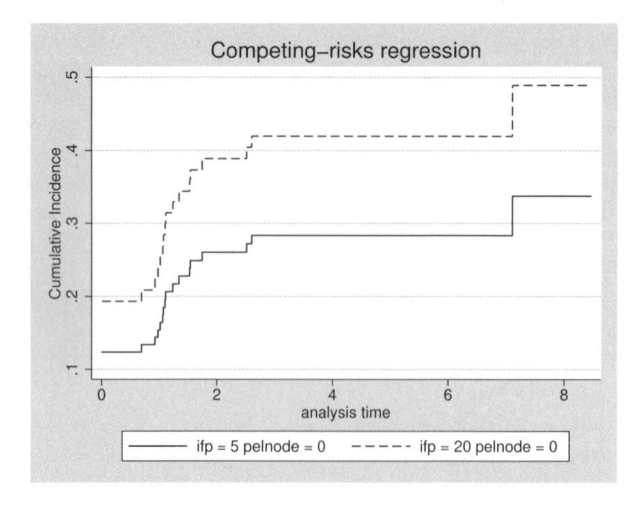

```
. stcurve, cif at1(ifp = 5 pelnode = 0) at2(ifp = 20 pelnode = 0)
```

For positive pelvic node involvement and mean tumor size, the probability of local relapse within 2 years is roughly 26% when the interstitial fluid pressure is 5 mmHg and near 40% when this is increased to 20 mmHg. Both probabilities take into account the possibility that a distant relapse could occur instead.

◁

Multiple competing-event types

Competing-risks regression generalizes to the case where more than one type of event competes with the event of interest. If you have such data, after you stset the failure event of interest, you can lump together all competing event codes into the compete() option of stcrreg. It does not matter whether multiple codes represent the same competing-event type, or if they represent multiple types. The results will be the same.

▷ Example 3: UDCA in patients with PBC

Therneau and Grambsch (2000, sec. 8.4.3) analyze data from patients with primary biliary cirrhosis (PBC), a chronic liver disease characterized by progressive destruction of the bile ducts. Data were obtained from 170 patients in a randomized double-blind trial conducted at the Mayo Clinic from 1988 to 1992. The trial was for a new treatment, ursodeoxycholic acid (UDCA; Lindor et al. [1994]).

stcrreg — Competing-risks regression 205

```
. use http://www.stata-press.com/data/r11/udca, clear
(Randomized trial of UDCA in PBC)
. describe
Contains data from http://www.stata-press.com/data/r11/udca.dta
  obs:           188                          Randomized trial of UDCA in PBC
 vars:             8                          3 Apr 2009 09:37
 size:         6,016 (99.9% of memory free)   (_dta has notes)

              storage  display    value
variable name   type   format     label      variable label

id              int    %9.0g                 Patient ID
entry           float  %td                   Date of enrollment
eventtime       float  %td                   Date of first event or loss to
                                                follow-up
treat           byte   %9.0g                 0=placebo 1=UDCA
stage           byte   %9.0g                 histologic stage: 0=stage 1/2 at
                                                entry 1=stage 3/4
lbili           float  %9.0g                 log(bilirubin value)
etype           float  %9.0g      event      Event type (see notes)
wt              double %4.2f                 Observation weight

Sorted by: id
```

The etype variable is coded as any of eight distinct event types (or no event) according to table 1.

Table 1. Event codes for the etype variable

Event Code	Event type
0	No event (censored)
1	Death
2	Transplant
3	Histologic progression
4	Development of varices
5	Development of ascites
6	Development of encephalopathy
7	Doubling of bilirubin
8	Worsening of symptoms

Cleves (1999) analyzed these data by estimating the cause-specific hazards for each of the eight events. In the version of the data used there, the time at which any adverse event occurred was recorded, but here we record only the time of the first adverse event for each patient. We do so because we wish to perform a competing-risks analysis where we are interested in the time to the first adverse event and the type of that event. The events compete because only one can be first.

We are interested in whether treatment will decrease the incidence of histologic progression (etype == 3) as the first adverse outcome, in reference to treatment (treat), the logarithm of bilirubin level (lbili), and histologic stage at entry (stage). Because the patients entered the study at different times (entry), when stsetting the data we must specify this variable as the origin, or onset of risk.

The competing-risks analysis described above could thus proceed as follows:

```
. stset eventtime, failure(etype == 3) origin(entry)
. stcrreg treat lbili stage, compete(etype == 1 2 4 5 6 7 8)
```

except for one minor complication. Some patients experienced multiple "first events", and thus ties exist. For example, consider patient 8 who experienced four adverse events at the same time:

```
. list if id == 8
```

	id	entry	eventtime	treat	stage	lbili	etype	wt
8.	8	25may1988	02jul1990	0	1	1.629241	ascites	0.25
9.	8	25may1988	02jul1990	0	1	1.629241	ence	0.25
10.	8	25may1988	02jul1990	0	1	1.629241	bili_2	0.25
11.	8	25may1988	02jul1990	0	1	1.629241	worse	0.25

While most patients are represented by one record each, patients with multiple first events are represented by multiple records. Rather than break ties arbitrarily, we take advantage of how importance weights (iweights) are handled by stcrreg. Importance weights are treated like frequency weights, but they are allowed to be noninteger. As such, we define the weight variable (wt) to equal one for single-record patients and to equal one divided by the number of tied events for multiple-record patients. In this way, each patient contributes a total weight of one observation.

The only further modification we need is to specify vce(cluster id) so that our standard errors account for the correlation within multiple records on the same patient.

```
. stset eventtime [iw=wt], failure(etype == 3) origin(entry)
 (output omitted)
. stcrreg treat lbili stage, compete(etype == 1 2 4 5 6 7 8) vce(cluster id)
        failure _d:  etype == 3
  analysis time _t:  (eventtime-origin)
            origin:  time entry
            weight:  [iweight=wt]

Iteration 0:   log pseudolikelihood = -62.158461
Iteration 1:   log pseudolikelihood = -61.671367
Iteration 2:   log pseudolikelihood = -61.669225
Iteration 3:   log pseudolikelihood = -61.669225

Competing-risks regression                         No. of obs      =        170
                                                   No. of subjects =        170
Failure event   : etype == 3                       No. failed      =   12.66667
Competing events: etype == 1 2 4 5 6 7 8           No. competing   =   59.33333
                                                   No. censored    =         98

                                                   Wald chi2(3)    =       1.89
Log pseudolikelihood = -61.669225                  Prob > chi2     =     0.5955

                              (Std. Err. adjusted for 170 clusters in id)
```

		Robust				
_t	SHR	Std. Err.	z	P>|z|	[95% Conf. Interval]	
treat	.5785214	.3238038	-0.98	0.328	.1931497	1.732786
lbili	1.012415	.367095	0.03	0.973	.4974143	2.060623
stage	.5537101	.3305371	-0.99	0.322	.1718534	1.78405

In the above, we clustered on id but we did not stset it as an id() variable. That was because we wanted stcrreg to treat each observation within patient as its own distinct spell, not as a set of overlapping spells.

Treatment with UDCA seems to decrease the incidence of histologic progression as a first adverse event. However, the effect is not significant, most likely as a result of observing so few failures.

◁

stcrreg as an alternative to stcox

In this section, we demonstrate that you may also use stcox to perform a cumulative-incidence analysis, and we compare that approach with one that uses stcrreg.

▷ Example 4: HIV and SI as competing events

Geskus (2000) and Putter, Fiocco, and Geskus (2007) analyzed data from 324 homosexual men from the Amsterdam Cohort Studies on HIV infection and AIDS. During the course of infection, the syncytium inducing (SI) HIV phenotype appeared in many of these individuals. The appearance of the SI phenotype strongly impairs prognosis. Thus the time to SI appearance in the absence of an AIDS diagnosis is of interest. In this context, a diagnosis of AIDS acts as a competing event.

```
. use http://www.stata-press.com/data/r11/hiv_si, clear
(HIV and SI as competing risks)
. describe
Contains data from http://www.stata-press.com/data/r11/hiv_si.dta
  obs:           324                          HIV and SI as competing risks
 vars:             4                          3 Apr 2009 13:40
 size:         3,888 (99.9% of memory free)   (_dta has notes)

              storage  display     value
variable name   type   format      label      variable label

patnr           int    %8.0g                  ID
time            float  %9.0g                  Years from HIV infection
status          byte   %10.0g      stat       1 = AIDS, 2 = SI, 0 = event-free
ccr5            byte   %9.0g       ccr5       1 if WM (deletion in C-C
                                              chemokine receptor 5 gene)

Sorted by:
```

In what follows, we recreate the analysis performed by Putter, Fiocco, and Geskus (2007), treating AIDS and SI as competing events and modeling cumulative incidence in relation to covariate ccr5. ccr5 equals 1 if a specific deletion in the C-C chemokine receptor 5 gene is present and equals zero otherwise (wild type).

We can model the cumulative incidence of SI on ccr5 directly with stcrreg:

```
. stset time, failure(status == 2)          // SI is the event of interest
 (output omitted)
. stcrreg ccr5, compete(status == 1)        // AIDS is the competing event
 (output omitted)
Competing-risks regression                      No. of obs       =      324
                                                No. of subjects  =      324
Failure event  : status == 2                    No. failed       =      107
Competing event: status == 1                    No. competing    =      113
                                                No. censored     =      104

                                                Wald chi2(1)     =     0.01
Log pseudolikelihood = -579.06241               Prob > chi2      =   0.9172
```

		Robust				
_t	SHR	Std. Err.	z	P>|z|	[95% Conf.	Interval]
ccr5	1.023865	.2324119	0.10	0.917	.6561827	1.597574

It seems that this particular genetic mutation has little relation with the incidence of SI, a point we emphasize further with a graph:

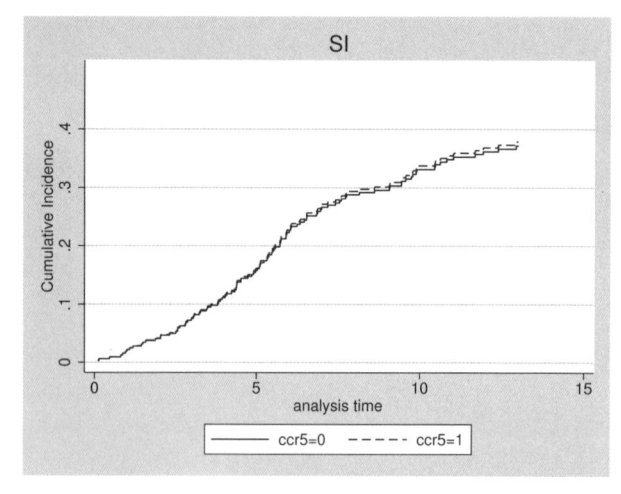

```
. stcurve, cif at1(ccr5=0) at2(ccr5=1) title(SI) range(0 13) yscale(range(0 0.5))
```

The above analysis compared SI incidence curves under the assumption that the subhazard for SI, that which generates SI events in the presence of AIDS, was proportional with respect to ccr5. Because we modeled the subhazard and not the cause-specific hazard, obtaining estimates of cumulative incidence was straightforward and depended only on the subhazard for SI and not on that for AIDS.

As explained in *The case for competing-risks regression*, the cumulative incidence of SI is a function of both the cause-specific hazard for SI, $h_1(t)$, and that for AIDS, $h_2(t)$, because SI and AIDS are competing events. Suppose for the moment that we are not interested in the incidence of SI in the presence of AIDS, but instead in the biological mechanism that causes SI in general. We can model this mechanism with stcox by treating AIDS events as censored.

```
. stcox ccr5
(output omitted)
Cox regression -- no ties

No. of subjects =          324              Number of obs   =        324
No. of failures =          107
Time at risk    =  2261.959996
                                            LR chi2(1)      =       1.19
Log likelihood  =   -549.73443              Prob > chi2     =     0.2748
```

_t	Haz. Ratio	Std. Err.	z	P>|z|	[95% Conf. Interval]
ccr5	.7755334	.1846031	-1.07	0.286	.4863914 1.23656

Because we initially stset our data with SI as the event of interest, AIDS events are treated as censored by stcox (but not by stcrreg). In any case, the ccr5 mutation somewhat decreases the risk of SI, but this effect is not significant.

We make the above interpretation with no regard to AIDS as a competing risk because we are interested only in the biological mechanism behind SI. To estimate the cumulative incidence of SI, we first need to make a choice. Either we can pretend a diagnosis of AIDS does not exist as a competing risk and use stcurve to plot survivor curves for SI based on the Cox model above, or we can acknowledge AIDS as a competing risk and model that cause-specific hazard also.

We choose the latter. Before fitting the model, however, we need to re-stset the data with AIDS as the event of interest.

```
. stset time, failure(status == 1)         // AIDS is the event of interest
  (output omitted )
. stcox ccr5
  (output omitted )
Cox regression -- Breslow method for ties
No. of subjects =          324                   Number of obs   =        324
No. of failures =          113
Time at risk    =   2261.959996
                                                 LR chi2(1)      =      21.98
Log likelihood  =    -555.37301                  Prob > chi2     =     0.0000
```

_t	Haz. Ratio	Std. Err.	z	P>|z|	[95% Conf. Interval]	
ccr5	.2906087	.0892503	-4.02	0.000	.1591812	.530549

Patients with the ccr5 mutation have a significantly lower risk of AIDS.

We have now modeled both cause-specific hazards separately. Cleves (1999); Lunn and McNeil (1995); and Putter, Fiocco, and Geskus (2007) (among others) describe an approach based on data duplication where both hazards can be modeled simultaneously. Such an approach has the advantage of being able to set the effects of ccr5 on both hazards as equal and to test that hypothesis. Also, you can model the baseline hazards as proportional rather than entirely distinct. However, for the least parsimonious model with event-specific covariate effects and event-specific baseline hazards, the data duplication method is no different than fitting separate models for each event type, just as we have done above. Because data duplication will reveal no simpler model for these data, we do not describe it further.

We can derive estimates of cumulative incidence for SI based on the above cause-specific hazard models, but the process is a bit more complicated than before. The cumulative incidence of SI (event type 1) in the presence of AIDS (event type 2) is calculated as

$$\widehat{\mathrm{CIF}}_1(t) = \sum_{j:t_j \leq t} \widehat{h}_1(t_j) \widehat{S}(t_{j-1})$$

with

$$\widehat{S}(t) = \prod_{j:t_j \leq t} \left\{ 1 - \widehat{h}_1(t_j) - \widehat{h}_2(t_j) \right\}$$

The t_j index the times at which events (of any type) occur, and $\widehat{h}_1(t_j)$ and $\widehat{h}_2(t_j)$ are the cause-specific *hazard contributions* for SI and AIDS respectively. Baseline hazard contributions can be obtained with predict after stcox, and they can be transformed to hazard contributions for any covariate pattern by multiplying them by the exponentiated linear predictor for that pattern. Hazard contributions represent the increments of the cumulative hazards at each event time. $\widehat{S}(t)$ estimates the probability that you are event free at time t.

We begin by refitting both models and predicting the hazard contributions.

```
. stset time, failure(status == 2)         // SI
  (output omitted )
. stcox ccr5
  (output omitted )
. predict h_si_0, basehc
(217 missing values generated)
. gen h_si_1 = h_si_0*exp(_b[ccr5])
(217 missing values generated)
```

```
. stset time, failure(status == 1)            // AIDS
  (output omitted)
. stcox ccr5
  (output omitted)
. predict h_aids_0, basehc
(211 missing values generated)
. gsort _t -_d
. by _t: replace h_aids_0 = . if _n > 1
(1 real change made, 1 to missing)
. gen h_aids_1 = h_aids_0*exp(_b[ccr5])
(212 missing values generated)
```

Variables h_si_0 and h_aids_0 hold the baseline hazard contributions, those for ccr5 == 0. Variables h_si_1 and h_aids_1 hold the hazard contributions for ccr5 == 1, and they were obtained by multiplying the baseline contributions by the exponentiated coefficient for ccr5. When we ran stcox with AIDS as the event of interest, the output indicated that we had tied failure times (the analysis for SI had no ties). As such, this required the extra step of setting any duplicated hazard contributions to missing. As it turned out, this affected only one observation.

Hazard contributions are generated only at times when events are observed and are set to missing otherwise. Because we will be summing and multiplying over event times, we next drop the observations that contribute nothing and then replace missing with zero for those observations that have some hazard contributions missing and some nonmissing.

```
. drop if missing(h_si_0) & missing(h_aids_0)
(105 observations deleted)
. replace h_aids_0 = 0 if missing(h_aids_0)
(107 real changes made)
. replace h_aids_1 = 0 if missing(h_aids_1)
(107 real changes made)
. replace h_si_0 = 0 if missing(h_si_0)
(112 real changes made)
. replace h_si_1 = 0 if missing(h_si_1)
(112 real changes made)
```

We can now sort by analysis time and calculate the estimated event-free survivor functions. Recall that you can express a product as an exponentiated sum of logarithms, which allows us to take advantage of Stata's sum() function for obtaining running sums.

```
. sort _t
. gen S_0 = exp(sum(log(1- h_aids_0 - h_si_0)))
. gen S_1 = exp(sum(log(1- h_aids_1 - h_si_1)))
```

Finally, we calculate the estimated CIFs and graph:

```
. gen cif_si_0 = sum(S_0[_n-1]*h_si_0)
. label var cif_si_0 "ccr5 = 0"
. gen cif_si_1 = sum(S_1[_n-1]*h_si_1)
. label var cif_si_1 "ccr5 = 1"
```

```
. twoway line cif_si* _t if _t<13, connect(J J) sort yscale(range(0 0.5))
> title(SI) ytitle(Cumulative Incidence) xtitle(analysis time)
```

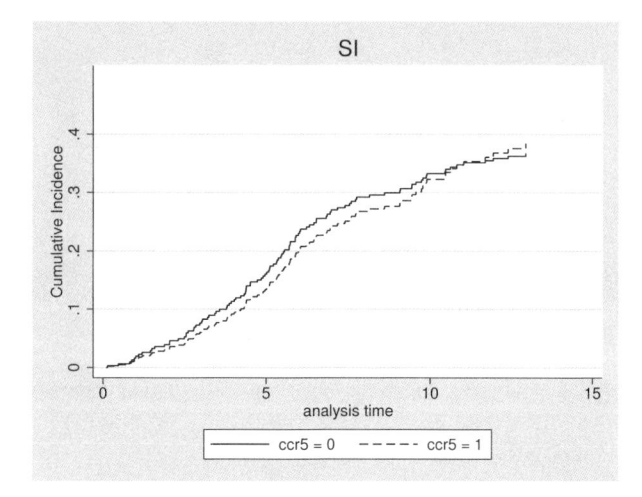

[Figure: Line plot titled "SI" showing Cumulative Incidence vs analysis time, with two curves for ccr5 = 0 (solid) and ccr5 = 1 (dashed).]

This model formulation shows ccr5 to have more of an effect on the incidence of SI, although the effect is still small. Note that under this formulation, the effect of ccr5 is not constrained to be overall increasing or overall decreasing. In fact, when $t > 11$ years or so, those with the ccr5 mutation actually have an increased SI incidence. That is due to time-accumulated reduced competition from AIDS, the risk of which is significantly lower when the ccr5 mutation is present.

Putter, Fiocco, and Geskus (2007) also performed the same analysis using AIDS as the event of interest, something we leave to you as an exercise.

◁

We have described two different modeling approaches for estimating the cumulative incidence of SI. Although you may prefer the stcrreg approach because it is much simpler, that does not mean it is a better model than the one based on stcox. The better model is the one whose assumptions more closely fit the data. The stcrreg model assumes that the effect of ccr5 is proportional on the subhazard for SI. The stcox model assumes proportionality on the cause-specific hazards for both SI and AIDS. Because our analysis uses only one binary covariate, we can compare both models with a nonparametric estimator of the CIF to see which fits the data more closely; see [ST] **stcrreg postestimation**.

Multiple records per subject

stcrreg can be used with data where you have multiple records per subject, as long as 1) you stset an ID variable that identifies the subjects and 2) you carefully consider the role played by time-varying covariates in subjects who fail because of competing events. We explain both issues below.

Stata's st suite of commands allows for multiple records per subject. Having multiple records allows you to record gaps in subjects' histories and to keep track of time-varying covariates. If you have multiple records per subject, you identify which records belong to which subjects by specifying an ID variable to stset option id().

Consider the sample data listed below:

```
. list if id == 18

     id   _t0    _t   _d     x
1.   18    3     5    0    5.1
2.   18    5     8    0    7.8
3.   18   11    12    0    6.7
4.   18   12    20    1    8.9
```

These data reflect the following:

- Subject 18 first became at risk at analysis time 3 (delayed entry) with covariate value x equal to 5.1.
- At time 5, subject 18's x value changed to 7.8.
- Subject 18 left the study at time 8 only to return at time 11 (gap), with x equal to 6.7 at that time.
- At time 12, x changed to 8.9.
- Subject 18 failed at time 20 with x equal to 8.9 at that time.

An analysis of these data with Cox regression using stcox is capable of processing all this information. Intermittent records are treated as censored (_d==0), and either failure or censoring occurs on the last record (here failure with _d==1). When subjects are not under observation, they are simply not considered at risk of failure. Time-varying covariates are also processed correctly. For example, if some other subject failed at time 7, then the risk calculations would count subject 18 at risk with x equal to 7.8 at that time.

stcox will give the same results for the above data whether or not you stset the ID variable, id. Whether you treat the above data as four distinct subjects (three censored and one failed) or as one subject with a four-record history is immaterial. The only difference you may encounter concerns robust and replication-based standard errors, in which case if you stset an ID variable, then stcox will automatically cluster on this variable.

Such a distinction, however, is of vital importance to stcrreg. While stcox is concerned only about detecting one type of failure, stcrreg relies on precise accounting of the number of subjects who fail because of the event of interest, those who fail because to competing events, and those who are censored. In particular, the weighting mechanism behind stcrreg depends on an accurate estimate of the probability a subject will be censored; see *Methods and formulas*. As such, it makes a difference whether you want to treat the above as four distinct subjects or as one subject. If you have multiple records per subject, you must stset your ID variable before using stcrreg. When counting the number failed, number competing, and number censored, stcrreg only considers what happened at the end of a subject's history. Intermittent records are treated simply as temporary entries to and exits from the analysis, and the exits are not counted as censored in the strict sense.

Furthermore, when using stcrreg with covariates that change over multiple records (time-varying covariates), you need to carefully consider what happens when subjects experience competing failures. For the above sample data, subject 18 failed because of the event interest (_d==1). Consider, however, what would have happened had this subject failed because of a competing event instead. Competing-risks regression keeps such subjects "at risk" of failure from the event of interest even after they fail from competing events; see *Methods and formulas*. Because these subjects will be used in future risk calculations for which they have no data, stcrreg will use the last available covariate values for these calculations. For the above example, if subject 18 experiences a competing event at time 20, then the last available value of x, 8.9, will be used in all subsequent risk calculations. If the last available values are as good a guess as any as to what future values would have been—for example, a binary covariate recording pretransplant versus posttransplant status—then this is not an issue. If, however, you have reason to believe that a subject's covariates would have been much different had the subject remained under observation, then the results from stcrreg could be biased.

▷ Example 5: Hospital-acquired pneumonia

Consider the following simulated data from a competing-risks analysis studying the effects of pneumonia.

```
. use http://www.stata-press.com/data/r11/pneumonia, clear
(Hospital-acquired pneumonia)
. describe
Contains data from http://www.stata-press.com/data/r11/pneumonia.dta
  obs:           957                          Hospital-acquired pneumonia
 vars:             7                          7 Apr 2009 15:35
 size:        12,441 (99.9% of memory free)

              storage  display     value
variable name   type   format      label      variable label

id              int    %9.0g                  Patient id
age             byte   %9.0g                  Age at admission
ndays           int    %9.0g                  Days in ICU
died            byte   %9.0g                  1 if died
censored        byte   %9.0g                  1 if alive and in ICU at the end
                                                of the study
discharged      byte   %9.0g                  1 if discharged
pneumonia       byte   %9.0g                  1 if pneumonia

Sorted by:  id
```

The above data are for 855 ICU patients. One hundred twenty-three patients contracted pneumonia, of which 21 did before admission and 102 during their stay. Those patients who contracted pneumonia during their stay are represented by two records with the time-varying covariate pneumonia recording the change in status.

We perform a competing-risks regression for the cumulative incidence of death during ICU stay with age and pneumonia as covariates. We also treat hospital discharge as a competing event.

```
. stset ndays, id(id) failure(died)
 (output omitted)
. stcrr age pneumonia, compete(discharged) noshow nolog
Competing-risks regression                       No. of obs       =      957
                                                 No. of subjects  =      855
Failure events   : died nonzero, nonmissing      No. failed       =      178
Competing events: discharged nonzero, nonmissing No. competing    =      641
                                                 No. censored     =       36

                                                 Wald chi2(2)     =   121.21
Log pseudolikelihood = -1128.6096                Prob > chi2      =   0.0000

                             (Std. Err. adjusted for 855 clusters in id)
```

		Robust				
_t	SHR	Std. Err.	z	P>|z|	[95% Conf. Interval]	
age	1.021612	.0076443	2.86	0.004	1.006739	1.036705
pneumonia	5.587052	.9641271	9.97	0.000	3.983782	7.835558

Both increased age and contracting pneumonia are associated with an increased incidence of death in the ICU.

◁

Option tvc() and testing the proportional-subhazards assumption

In the previous section, we considered data with multiple records per subject. Such data makes it possible to record discrete time-varying covariates, those whose values change at discrete points in time. Each change is captured by a new record.

Consider instead what happens when you have covariates that vary continuously with respect to time. Competing-risks regression assumes the following relationship between subhazard and baseline subhazard

$$\overline{h}_1(t) = \overline{h}_{1,0}(t) \exp(\beta_1 x_1 + \cdots + \beta_k x_k)$$

where $\overline{h}_{1,0}(t)$ is the baseline subhazard function. For most purposes, this model is sufficient, but sometimes we may wish to introduce variables of the form $z_i(t) = z_i g(t)$, which vary continuously with time so that

$$\overline{h}_1(t) = \overline{h}_{1,0}(t) \exp\left\{\beta_1 x_1 + \cdots + \beta_k x_k + g(t)(\gamma_1 z_1 + \cdots + \gamma_m z_m)\right\} \quad (1)$$

where $(z_1, \ldots, z_m)$ are the time-varying covariates. Fitting this model has the net effect of estimating the regression coefficient, γ_i, for the covariate $g(t)z_i$, which is a function of analysis time.

The time-varying covariates $(z_1, \ldots, z_m)$ are specified using the tvc(*tvarlist*) option, and $g(t)$ is specified using the texp(*exp*) option, where t in $g(t)$ is analysis time. For example, if we want $g(t) = \log(t)$, we would use texp(log(_t)) because _t stores the analysis time once the data are stset.

When subjects fail because of competing events, covariate values for these subjects continue to be used in subsequent risk calculations; see the previous section for details. When this occurs, any time-varying covariates specified using tvc() will continue to respect their time interactions even after these subjects fail. Because such behavior is unlikely to reflect any real data situation, we do not recommend using tvc() for this purpose.

We do, however, recommend using tvc() to model *time-varying coefficients*, because these can be used to test the proportionality assumption behind competing-risks regression. Consider a version of (1) that contains only one fixed covariate, x_1, and sets $z_1 = x_1$:

$$\overline{h}_1(t) = \overline{h}_{1,0}(t) \exp\left[\{\beta_1 + \gamma_1 g(t)\} x_1\right]$$

Given this new arrangement, we consider that $\beta_1 + \gamma_1 g(t)$ is a (possibly) time-varying coefficient on the covariate x_1, for some specified function of time $g(t)$. The coefficient has a time-invariant component β_1, with γ_1 determining the magnitude of the time-dependent deviations from β_1. As such, a test of $\gamma_1 = 0$ is a test of time invariance for the coefficient on x_1.

Confirming that a coefficient is time invariant is one way of testing the proportional-subhazards assumption. Proportional subhazards implies that the relative subhazard (i.e., β) is fixed over time, and this assumption would be violated if a time interaction proved significant.

▷ Example 6: Testing proportionality of subhazards

Returning to our cervical cancer study, we now include time interactions on all three covariates as a way of testing the proportional-subhazards assumption for each:

```
. use http://www.stata-press.com/data/r11/hypoxia, clear
(Hypoxia study)
. stset dftime, failure(failtype == 1)
 (output omitted)
. stcrreg ifp tumsize pelnode, compete(failtype == 2) tvc(ifp tumsize pelnode)
> noshr
 (output omitted)
```

Competing-risks regression					No. of obs	=	109
					No. of subjects	=	109
Failure event : failtype == 1					No. failed	=	33
Competing event: failtype == 2					No. competing	=	17
					No. censored	=	59
					Wald chi2(6)	=	44.93
Log pseudolikelihood = -136.79					Prob > chi2	=	0.0000

_t	Coef.	Robust Std. Err.	z	P>|z|	[95% Conf. Interval]	
main						
ifp	.0262093	.0174458	1.50	0.133	-.0079838	.0604025
tumsize	.37897	.1096628	3.46	0.001	.1640348	.5939052
pelnode	-.766362	.473674	-1.62	0.106	-1.694746	.162022
tvc						
ifp	.0055901	.0081809	0.68	0.494	-.0104441	.0216243
tumsize	-.1415204	.0908955	-1.56	0.119	-.3196722	.0366314
pelnode	.0610457	.5676173	0.11	0.914	-1.051464	1.173555

Note: variables in tvc equation interacted with _t

We used the default function of time $g(t) = t$, although we could have specified otherwise with the texp() option. After looking at the significance levels in the equation labeled "tvc", we find no indication that the proportionality assumption has been violated.

◁

When you use tvc() in this manner, there is no issue of postfailure covariate values for subjects who fail from competing events. The covariate values are assumed constant—the *coefficients* change with time.

(*Continued on next page*)

Saved results

stcrreg saves the following in e():

Scalars
e(N)	number of observations
e(N_sub)	number of subjects
e(N_fail)	number of failures
e(N_compete)	number of competing events
e(N_censor)	number of censored subjects
e(k)	number of parameters
e(k_eq)	number of equations in e(b)
e(k_eq_model)	number of equations in model Wald test
e(k_autoCns)	number of base, empty, and omitted constraints
e(k_dv)	number of dependent variables
e(df_m)	model degrees of freedom
e(ll)	log pseudolikelihood
e(N_clust)	number of clusters
e(chi2)	χ^2
e(p)	significance
e(k_eform)	number of leading equations appropriate for eform output
e(rank)	rank of e(V)
e(fmult)	1 if > 1 failure events, 0 otherwise
e(crmult)	1 if > 1 competing events, 0 otherwise
e(fnz)	1 if nonzero indicates failure, 0 otherwise
e(crnz)	1 if nonzero indicates competing, 0 otherwise
e(ic)	number of iterations
e(rc)	return code
e(converged)	1 if converged, 0 otherwise

Macros
    e(cmd)   stcrreg
    e(cmdline)   command as typed
    e(depvar)   name of dependent variable
    e(mainvars)   variables in main equation
    e(tvc)   time-varying covariates
    e(texp)   function used for time-varying covariates
    e(fevent)   failure event(s) in estimation output
    e(crevent)   competing event(s) in estimation output
    e(compete)   competing event(s) as typed
    e(wtype)   weight type
    e(wexp)   weight expression
    e(title)   title in estimation output
    e(clustvar)   name of cluster variable
    e(offset1)   offset
    e(chi2type)   Wald; type of model χ^2 test
    e(vce)   *vcetype* specified in vce()
    e(vcetype)   title used to label Std. Err.
    e(opt)   type of optimization
    e(which)   max or min; whether optimizer is to perform maximization or minimization
    e(ml_method)   type of ml method
    e(user)   name of likelihood-evaluator program
    e(technique)   maximization technique
    e(singularHmethod)   m-marquardt or hybrid; method used when Hessian is singular
    e(crittype)   optimization criterion
    e(properties)   b V
    e(predict)   program used to implement predict
    e(asbalanced)   factor variables fvset as asbalanced
    e(asobserved)   factor variables fvset as asobserved

Matrices
    e(b)   coefficient vector
    e(Cns)   constraints matrix
    e(ilog)   iteration log
    e(gradient)   gradient vector
    e(V)   variance–covariance matrix of the estimators
    e(V_modelbased)   model-based variance

Functions
    e(sample)   marks estimation sample

Methods and formulas

    stcrreg is implemented as an ado-file that makes use of Mata.

    In what follows, we assume single-record data and time-invariant covariates or coefficients. Extensions to both multiple-record data and continuous time-varying covariates are achieved by treating the mechanisms that generate censorings, competing events, and failure events of interest as counting processes; see Fine and Gray (1999) and Andersen et al. (1993) for further details.

Let $\mathbf{x}_i$ be the row vector of m covariates for the time interval $(t_{0i}, t_i]$ for the ith observation in the dataset ($i = 1, \ldots, n$). stcrreg obtains parameter estimates $\widehat{\boldsymbol{\beta}}$ by maximizing the log-pseudolikelihood function

$$\log L = \sum_{i=1}^{n} \delta_i w_i \left[\mathbf{x}_i \boldsymbol{\beta} + \text{offset}_i - \log \left\{ \sum_{j \in R_i} w_j \pi_{ji} \exp(\mathbf{x}_j \boldsymbol{\beta} + \text{offset}_j) \right\} \right]$$

where δ_i indicates a failure of interest for observation i and R_i is the set of observations, j, that are at risk at time t_i (i.e., all j such that $t_{0j} < t_i \leq t_j$). w_i and offset_i are the usual observation weights and linear offsets, if specified.

The log likelihood given above is identical to that for standard Cox regression (Breslow method for ties) with the exception of the weights π_{ji}. These weights are used to keep subjects who fail because of competing events in subsequent risk sets, and to decrease their weight over time as their likelihood of being otherwise censored increases.

Formally, extend R_i above not only to include those at risk of failure at time t_i, but also to include those subjects already having experienced a competing-risks event. Also, define

$$\pi_{ji} = \frac{\widehat{S}_c(t_i)}{\widehat{S}_c\{\min(t_j, t_i)\}}$$

if subject j experiences a competing event; $\pi_{ji} = 1$ otherwise. $\widehat{S}_c(t)$ is the Kaplan–Meier estimate of the survivor function for the censoring distribution—that which treats censorings as the events of interest—evaluated at time t, and t_j is the time at which subject j experienced his or her competing-failure event. As a matter of convention, $\widehat{S}_c(t)$ is treated as the probability of being censored up to but *not including* time t.

Because of the sample weighting inherent to this estimator, the standard Hessian-based estimate of variance is not statistically appropriate and is thus rejected in favor of a robust, sandwich-type estimator, as derived by Fine and Gray (1999).

Define $z_i = \mathbf{x}_i \widehat{\boldsymbol{\beta}} + \text{offset}_i$. (Pseudo)likelihood scores are given by

$$\widehat{\mathbf{u}}_i = \widehat{\boldsymbol{\eta}}_i + \widehat{\boldsymbol{\psi}}_i$$

where $\widehat{\boldsymbol{\eta}}_i = (\widehat{\eta}_{1i}, \ldots, \widehat{\eta}_{mi})'$, and

$$\widehat{\eta}_{ki} = \delta_i (x_{ki} - a_{ki}) - \exp(z_i) \sum_{j: t_{0i} < t_j \leq t_i} \frac{\delta_j w_j \pi_{ij}(x_{ki} - a_{kj})}{\sum_{\ell \in R_j} w_\ell \pi_{\ell j} \exp(z_\ell)}$$

for

$$a_{ki} = \frac{\sum_{\ell \in R_i} w_\ell \pi_{\ell i} x_{k\ell} \exp(z_\ell)}{\sum_{\ell \in R_i} w_\ell \pi_{\ell i} \exp(z_\ell)}$$

The $\widehat{\boldsymbol{\psi}}_i$ are variance contributions due to data estimation of the weights π_{ji}, with

$$\widehat{\boldsymbol{\psi}}_i = \frac{\gamma_i \widehat{\mathbf{q}}(t_i)}{r(t_i)} - \sum_{j: t_{0i} < t_j \leq t_i} \frac{\gamma_j \widehat{h}_c(t_j) \widehat{\mathbf{q}}(t_j)}{r(t_j)}$$

γ_i indicates censoring for observation i, $r(t)$ is the number at risk of failure (or censoring) at time t,

$$\widehat{h}_c(t) = \frac{\sum_{i=1}^{n} \gamma_i I(t_i = t)}{r(t)}$$

and the kth component of $\widehat{\mathbf{q}}(t)$ is

$$\widehat{q}_k(t) = \sum_{i \in C(t)} w_i \exp(z_i) \sum_{j:t_{0i}<t_j\leq t_i} \frac{\delta_j w_j \pi_{ij}(x_{ki} - a_{kj})}{\sum_{\ell \in R_j} w_\ell \pi_{\ell j} \exp(z_\ell)} I(t_j \geq t)$$

where $C(t)$ is the set of observations that experienced a competing event prior to time t.

By default, stcrreg calculates the Huber/White/sandwich estimator of the variance and calculates its clustered version if either the vce(cluster *clustvar*) option is specified or an ID variable has been stset. See *Maximum likelihood estimators* and *Methods and formulas* in [P] _robust for details on how the pseudolikelihood scores defined above are used to calculate this variance estimator.

Acknowledgment

We thank Jason Fine of the Gillings School of Global Public Health, University of North Carolina at Chapel Hill, for answering our technical questions.

References

Andersen, P. K., Ø. Borgan, R. D. Gill, and N. Keiding. 1993. *Statistical Models Based on Counting Processes*. New York: Springer.

Beyersman, J., A. Latouche, A. Buchholz, and M. Schumacher. 2009. Simulating competing risks data in survival analysis. *Statistics in Medicine* 28: 956–971.

Beyersman, J., and M. Schumacher. 2008. Time-dependent covariates in the proportional subdistribution hazards model for competing risks. *Biostatistics* 9: 765–776.

Cleves, M. A. 1999. ssa13: Analysis of multiple failure-time data with Stata. *Stata Technical Bulletin* 49: 30–39. Reprinted in *Stata Technical Bulletin Reprints*, vol. 9, pp. 338–349. College Station, TX: Stata Press.

Cleves, M. A., W. W. Gould, R. G. Gutierrez, and Y. Marchenko. 2008. *An Introduction to Survival Analysis Using Stata*. 2nd ed. College Station, TX: Stata Press.

Coviello, V., and M. Boggess. 2004. Cumulative incidence estimation in the presence of competing risks. *Stata Journal* 4: 103–112.

Cox, D. R. 1972. Regression models and life-tables (with discussion). *Journal of the Royal Statistical Society, Series B* 34: 187–220.

Crowder, M. J. 2001. *Classical Competing Risks*. Boca Raton, FL: Chapman & Hall/CRC.

Fine, J. P., and R. J. Gray. 1999. A proportional hazards model for the subdistribution of a competing risk. *Journal of the American Statistical Association* 94: 496–509.

Fyles, A., M. Milosevic, D. Hedley, M. Pintilie, W. Levin, L. Manchul, and R. P. Hill. 2002. Tumor hypoxia has independent predictor impact only in patients with node-negative cervix cancer. *Journal of Clinical Oncology* 20: 680–687.

Gail, M. H. 1975. A review and critique of some models used in competing risk analysis. *Biometrics* 31: 209–222.

Geskus, R. B. 2000. On the inclusion of prevalent cases in HIV/AIDS natural history studies through a marker-based estimate of time since seroconversion. *Statistics in Medicine* 19: 1753–1769.

Gichangi, A., and W. Vach. 2005. The analysis of competing risks data: A guided tour. Preprint series, Department of Statistics, University of Southern Denmark.
http://www.stat.sdu.dk/publications/preprints/pp009/Anthony%20Gichangi%20Competing%20Risk%20Tutorial.pdf.

Gooley, T. A., W. Leisenring, J. Crowley, and B. E. Storer. 1999. Estimation of failure probabilities in the presence of competing risks: New representations of old estimators. *Statistics in Medicine* 18: 695–706.

Gray, R. J. 1988. A class of k-sample tests for comparing the cumulative incidence of a competing risk. *Annals of Statistics* 16: 1141–1154.

Kaplan, E. L., and P. Meier. 1958. Nonparametric estimation from incomplete observations. *Journal of the American Statistical Association* 53: 457–481.

Klein, J. P., and M. L. Moeschberger. 2003. *Survival Analysis: Techniques for Censored and Truncated Data.* 2nd ed. New York: Springer.

Lambert, P. C. 2007. Modeling of the cure fraction in survival studies. *Stata Journal* 7: 351–375.

Lin, D. Y., and L. J. Wei. 1989. The robust inference for the Cox proportional hazards model. *Journal of the American Statistical Association* 84: 1074–1078.

Lindor, K. D., E. R. Dickson, W. P. Baldus, R. A. Jorgensen, J. Ludwig, P. A. Murtaugh, J. M. Harrison, R. H. Wiesner, M. L. Anderson, S. M. Lange, G. LeSage, S. S. Rossi, and A. F. Hofman. 1994. Ursodeoxycholic acid in the treatment of primary biliary cirrhosis. *Gastroenterology* 106: 1284–1290.

Lunn, M., and D. McNeil. 1995. Applying Cox regression to competing risks. *Biometrics* 51: 524–532.

Marubini, E., and M. G. Valsecchi. 1997. *Analysing Survival Data from Clinical Trials and Observational Studies.* Chichester, UK: Wiley.

Milosevic, M., A. Fyles, D. Hedley, M. Pintilie, W. Levin, L. Manchul, and R. P. Hill. 2001. Interstitial fluid pressure predicts survival in patients with cervix cancer independent of clinical prognostic factors and tumor oxygen measurements. *Cancer Research* 61: 6400–6405.

Pepe, M. S., and M. Mori. 1993. Kaplan–Meier, marginal or conditional probability curves in summarizing competing risks failure time data? *Statistics in Medicine* 12: 737–751.

Pintilie, M. 2006. *Competing Risks: A Practical Perspective.* Chichester, UK: Wiley.

———. 2007. Analysing and interpreting competing risk data. *Statistics in Medicine* 26: 1360–1367.

Putter, H., M. Fiocco, and R. B. Geskus. 2007. Tutorial in biostatistics: Competing risks and multi-state models. *Statistics in Medicine* 26: 2389–2430.

Therneau, T. M., and P. M. Grambsch. 2000. *Modeling Survival Data: Extending the Cox Model.* New York: Springer.

Tsiatis, A. 1975. A nonidentifiability aspect of the problem of competing risks. *Proceedings of the National Academy of Sciences* 72: 20–22.

Also see

[ST] **stcrreg postestimation** — Postestimation tools for stcrreg

[ST] **stcox postestimation** — Postestimation tools for stcox

[ST] **stcurve** — Plot survivor, hazard, cumulative hazard, or cumulative incidence function

[ST] **stcox PH-assumption tests** — Tests of proportional-hazards assumption

[ST] **sts** — Generate, graph, list, and test the survivor and cumulative hazard functions

[ST] **stset** — Declare data to be survival-time data

[ST] **streg** — Parametric survival models

[U] **20 Estimation and postestimation commands**

Title

stcrreg postestimation — Postestimation tools for stcrreg

Description

The following postestimation command is of special interest after `stcrreg`:

command	description
stcurve	plot the cumulative subhazard and cumulative incidence functions

For information on `stcurve`, see [ST] **stcurve**.

The following standard postestimation commands are also available:

command	description
estat	AIC, BIC, VCE, and estimation sample summary
estimates	cataloging estimation results
lincom	point estimates, standard errors, testing, and inference for linear combinations of coefficients
margins	marginal means, predictive margins, marginal effects, and average marginal effects
nlcom	point estimates, standard errors, testing, and inference for nonlinear combinations of coefficients
predict	predictions, residuals, influence statistics, and other diagnostic measures
predictnl	point estimates, standard errors, testing, and inference for generalized predictions
test	Wald tests of simple and composite linear hypotheses
testnl	Wald tests of nonlinear hypotheses

See the corresponding entries in the *Base Reference Manual* for details.

Syntax for predict

predict [*type*] *newvar* [*if*] [*in*] [, *sv_statistic* <u>nooff</u>set]

predict [*type*] { *stub** | *newvarlist* } [*if*] [*in*], *mv_statistic* [<u>partial</u>]

sv_statistic	description
Main	
shr	predicted subhazard ratio, also known as the relative subhazard; the default
xb	linear prediction $x_j\beta$
stdp	standard error of the linear prediction; $\text{SE}(x_j\beta)$
*basecif	baseline cumulative incidence function (CIF)
*basecshazard	baseline cumulative subhazard function
*kmcensor	Kaplan–Meier survivor curve for the censoring distribution

mv_statistic	description
Main	
*scores	pseudolikelihood scores
*esr	efficient score residuals
*dfbeta	DFBETA measures of influence
*schoenfeld	Schoenfeld residuals

Unstarred statistics are available both in and out of sample; type predict ... if e(sample) ... if wanted only for the estimation sample. Starred statistics are calculated only for the estimation sample, even when e(sample) is not specified.

nooffset is allowed only with unstarred statistics.

Menu

Statistics > Postestimation > Predictions, residuals, etc.

Options for predict

Main

shr, the default, calculates the relative subhazard (subhazard ratio), that is, the exponentiated linear prediction, $\exp(x_j\widehat{\beta})$.

xb calculates the linear prediction from the fitted model. That is, you fit the model by estimating a set of parameters, $\beta_1, \beta_2, \ldots, \beta_k$, and the linear prediction is $\widehat{\beta}_1 x_{1j} + \widehat{\beta}_2 x_{2j} + \cdots + \widehat{\beta}_k x_{kj}$, often written in matrix notation as $x_j\widehat{\beta}$.

The $x_{1j}, x_{2j}, \ldots, x_{kj}$ used in the calculation are obtained from the data currently in memory and do not have to correspond to the data on the independent variables used in estimating β.

stdp calculates the standard error of the prediction, that is, the standard error of $x_j\widehat{\beta}$.

basecif calculates the baseline CIF. This is the CIF of the subdistribution for the cause-specific failure process.

basecshazard calculates the baseline cumulative subhazard function. This is the cumulative hazard function of the subdistribution for the cause-specific failure process.

kmcensor calculates the Kaplan–Meier survivor function for the censoring distribution. These estimates are used to weight within risk pools observations that have experienced a competing event. As such, these values are not predictions or diagnostics in the strict sense, but are provided for those who wish to reproduce the pseudolikelihood calculations performed by stcrreg.

nooffset is allowed only with shr, xb, and stdp, and is relevant only if you specified offset(*varname*) for stcrreg. It modifies the calculations made by predict so that they ignore the offset variable; the linear prediction is treated as $x_j\widehat{\beta}$ rather than $x_j\widehat{\beta} + \text{offset}_j$.

scores calculates the pseudolikelihood scores for each regressor in the model. These scores are components of the robust estimate of variance. For multiple-record data, by default only one score per subject is calculated and it is placed on the last record for the subject.

Adding the partial option will produce partial scores, one for each record within subject; see partial below. Partial pseudolikelihood scores are the additive contributions to a subject's overall pseudolikelihood score. In single-record data, the partial pseudolikelihood scores are the pseudolikelihood scores.

One score variable is created for each regressor in the model; the first new variable corresponds to the first regressor, the second to the second, and so on.

esr calculates the efficient score residuals for each regressor in the model. Efficient score residuals are diagnostic measures equivalent to pseudolikelihood scores, with the exception that efficient score residuals treat the censoring distribution (that used for weighting) as known rather than estimated. For multiple-record data, by default only one score per subject is calculated and it is placed on the last record for the subject.

Adding the partial option will produce partial efficient score residuals, one for each record within subject; see partial below. Partial efficient score residuals are the additive contributions to a subject's overall efficient score residual. In single-record data, the partial efficient scores are the efficient scores.

One efficient variable is created for each regressor in the model; the first new variable corresponds to the first regressor, the second to the second, and so on.

dfbeta calculates the DFBETA measures of influence for each regressor of in the model. The DFBETA value for a subject estimates the change in the regressor's coefficient due to deletion of that subject. For multiple-record data, by default only one value per subject is calculated and it is placed on the last record for the subject.

Adding the partial option will produce partial DFBETAs, one for each record within subject; see partial below. Partial DFBETAs are interpreted as effects due to deletion of individual records rather than deletion of individual subjects. In single-record data, the partial DFBETAs are the DFBETAs.

One DFBETA variable is created for each regressor in the model; the first new variable corresponds to the first regressor, the second to the second, and so on.

schoenfeld calculates the Schoenfeld-like residuals. Schoenfeld-like residuals are diagnostic measures analogous to Schoenfeld residuals in Cox regression. They compare a failed observation's covariate values to the (weighted) average covariate values for all those at risk at the time of failure. Schoenfeld-like residuals are calculated only for those observations that end in failure; missing values are produced otherwise.

One Schoenfeld residual variable is created for each regressor in the model; the first new variable corresponds to the first regressor, the second to the second, and so on.

Note: The easiest way to use the preceding four options is, for example,

. predict double *stub**, scores

where *stub* is a short name of your choosing. Stata then creates variables *stub*1, *stub*2, etc. You may also specify each variable name explicitly, in which case there must be as many (and no more) variables specified as there are regressors in the model.

partial is relevant only for multiple-record data and is valid with scores, esr, and dfbeta. Specifying partial will produce "partial" versions of these statistics, where one value is calculated for each record instead of one for each subject. The subjects are determined by the id() option to stset.

Specify partial if you wish to perform diagnostics on individual records rather than on individual subjects. For example, a partial DFBETA would be interpreted as the effect on a coefficient due to deletion of one record, rather than the effect due to deletion of all records for a given subject.

Remarks

Remarks are presented under the following headings:

> *Baseline functions*
> *Null models*
> *Measures of influence*

Baseline functions

▷ Example 1: Cervical cancer study

In example 1 of [ST] **stcrreg**, we fit a proportional subhazards model on data from a cervical cancer study.

```
. use http://www.stata-press.com/data/r11/hypoxia
(Hypoxia study)
. stset dftime, failure(failtype == 1)
 (output omitted)
. stcrreg ifp tumsize pelnode, compete(failtype == 2)
 (output omitted)
Competing-risks regression                      No. of obs      =        109
                                                No. of subjects =        109
Failure event  : failtype == 1                  No. failed      =         33
Competing event: failtype == 2                  No. competing   =         17
                                                No. censored    =         59

                                                Wald chi2(3)    =      33.21
Log pseudolikelihood =   -138.5308              Prob > chi2     =     0.0000

------------------------------------------------------------------------------
             |               Robust
          _t |        SHR   Std. Err.      z    P>|z|     [95% Conf. Interval]
-------------+----------------------------------------------------------------
         ifp |   1.033206   .0178938     1.89   0.059     .9987231    1.068879
     tumsize |   1.297332   .1271191     2.66   0.008     1.070646    1.572013
      pelnode|   .4588123   .1972067    -1.81   0.070     .1975931    1.065365
------------------------------------------------------------------------------
```

After fitting the model, we can predict the baseline cumulative subhazard, $\overline{H}_{1,0}(t)$, and the baseline CIF, $\text{CIF}_{1,0}(t)$:

```
. predict bch, basecsh
. predict bcif, basecif
```

```
. list dftime failtype ifp tumsize pelnode bch bcif in 1/15
```

	dftime	failtype	ifp	tumsize	pelnode	bch	bcif
1.	6.152	0	8	7	1	.0658792	.063756
2.	8.008	0	8.2	2	1	.0813224	.0781036
3.	.003	1	8.6	10	1	.0260186	.025683
4.	1.073	1	3.3	8	1	.0379107	.0372011
5.	.003	1	18.5	8	0	.0260186	.025683
6.	7.929	0	20	8	1	.0813224	.0781036
7.	8.454	0	21.8	4	1	.0813224	.0781036
8.	7.107	1	31.6	5	1	.0813224	.0781036
9.	8.378	0	16.5	5	1	.0813224	.0781036
10.	8.178	0	31.5	3	1	.0813224	.0781036
11.	3.395	0	18.5	4	1	.0658792	.063756
12.	.003	1	12.8	5	0	.0260186	.025683
13.	1.35	1	18.4	4	1	.051079	.0497964
14.	.003	1	18.5	8	1	.0260186	.025683
15.	.512	2	21	10	0	.0260186	.025683

The baseline functions are for subjects who have zero-valued covariates, which in this example are not representative of the data. If baseline is an extreme departure from the covariate patterns in your data, then we recommend recentering your covariates to avoid numerical overflows when predicting baseline functions; see *Making baseline reasonable* in [ST] **stcox postestimation** for more details.

For our data, baseline is close enough to not cause any numerical problems, but far enough to not be of scientific interest (zero tumor size?). You can transform the baseline functions to those for other covariate patterns according to the relationships

$$\overline{H}_1(t) = \exp(\mathbf{x}\boldsymbol{\beta})\overline{H}_{1,0}(t)$$

and

$$\text{CIF}_1(t) = 1 - \exp\{-\exp(\mathbf{x}\boldsymbol{\beta})\overline{H}_{1,0}(t)\}$$

but it is rare that you will ever have to do that. **stcurve** will predict, transform, and graph these functions for you. When you use **stcurve**, you specify the covariate settings, and any you leave unspecified are set at the mean over the data used in the estimation.

(*Continued on next page*)

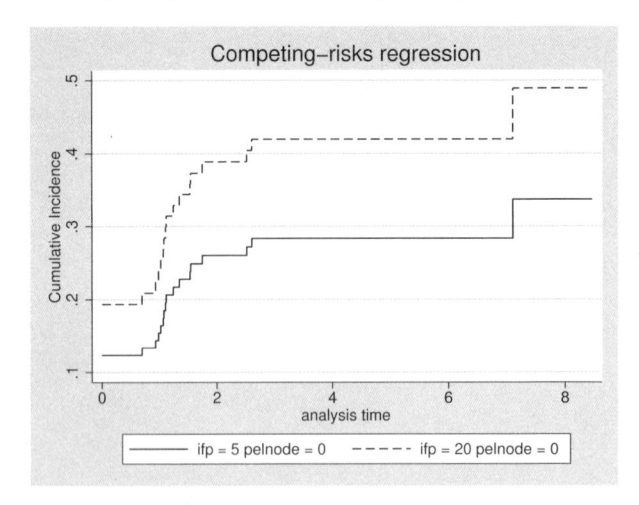

```
. stcurve, cif at1(ifp = 5 pelnode = 0) at2(ifp = 20 pelnode = 0)
```

Because they were left unspecified, the cumulative incidence curves are for mean tumor size. If you wish to graph cumulative subhazards instead of CIFs, use the `stcurve` option `cumhaz` in place of `cif`.

◁

Null models

Predicting baseline functions after fitting a null model (one without covariates) yields nonparametric estimates of the cumulative subhazard and the CIF.

▷ Example 2: HIV and SI as competing events

In example 4 of [ST] **stcrreg**, we analyzed the incidence of appearance of the SI HIV phenotype, where a diagnosis of AIDS is a competing event. We modeled SI incidence in reference to a genetic mutation indicated by the covariate ccr5. We compared two approaches: one that used `stcrreg` and assumed that the subhazard of SI was proportional with respect to ccr5 versus one that used `stcox` and assumed that the cause-specific hazards for both SI and AIDS were each proportional with respect to ccr5. For both approaches, we produced cumulative incidence curves for SI comparing those who did not have the mutation (ccr5==0) to those who did (ccr5==1).

To see which approach better fits these data, we now produce cumulative incidence curves that make no model assumption about the effect of ccr5. We do this by fitting null models on the two subsets of the data defined by ccr5 and predicting the baseline CIF for each. Because the models have no covariates, the estimated baseline CIFs are nonparametric estimators.

```
. use http://www.stata-press.com/data/r11/hiv_si, clear
(HIV and SI as competing risks)
. stset time, failure(status == 2)                  // SI is the event of interest
 (output omitted )
```

```
. stcrreg if !ccr5, compete(status == 1) noshow    // AIDS is the competing event
Competing-risks regression                          No. of obs       =        259
                                                    No. of subjects  =        259
Failure event  : status == 2                        No. failed       =         84
Competing event: status == 1                        No. competing    =        101
                                                    No. censored     =         74

                                                    Wald chi2(0)     =       0.00
Log pseudolikelihood = -435.80148                   Prob > chi2      =          .

                      |             Robust
                  _t  |    SHR    Std. Err.      z    P>|z|     [95% Conf. Interval]


. predict cif_si_0, basecif
(65 missing values generated)
. label var cif_si_0 "ccr5 = 0"
. stcrreg if ccr5, compete(status == 1) noshow
Competing-risks regression                          No. of obs       =         65
                                                    No. of subjects  =         65
Failure event  : status == 2                        No. failed       =         23
Competing event: status == 1                        No. competing    =         12
                                                    No. censored     =         30

                                                    Wald chi2(0)     =       0.00
Log pseudolikelihood = -88.306665                   Prob > chi2      =          .

                      |             Robust
                  _t  |    SHR    Std. Err.      z    P>|z|     [95% Conf. Interval]


. predict cif_si_1, basecif
(259 missing values generated)
. label var cif_si_1 "ccr5 = 1"
. twoway line cif_si* _t if _t<13, connect(J J) sort yscale(range(0 0.5))
> title(SI) ytitle(Cumulative Incidence) xtitle(analysis time)
```

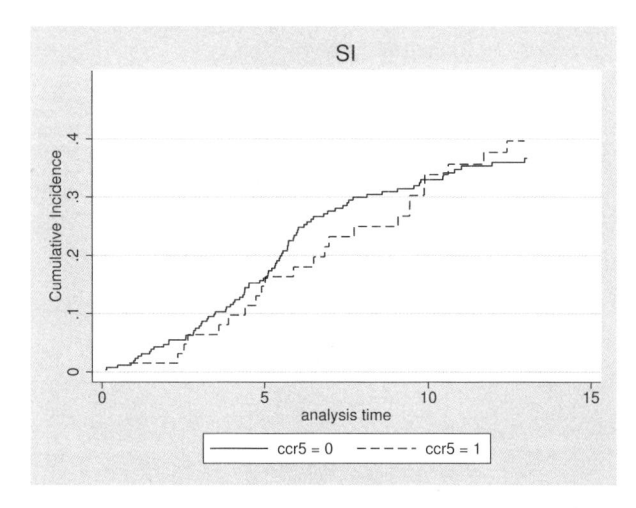

After comparing with the graphs produced in [ST] **stcrreg**, we find that the nonparametric analysis favors the `stcox` approach over the `stcrreg` approach.

◁

Technical note

Predicting the baseline CIF after fitting a null model with `stcrreg` produces a nonparametric CIF estimator that is asymptotically equivalent, but not exactly equal, to an alternate estimator that is often used; see Coviello and Boggess (2004) for the details of that estimator. The estimator used by `predict` after `stcrreg` is a competing-risks extension of the Nelson–Aalen estimator (Nelson 1972; Aalen 1978); see *Methods and formulas*. The other is a competing-risks extension of the Kaplan–Meier (1958) estimator.

In large samples with many failures, the difference is negligible.

Measures of influence

With `predict` after `stcrreg`, you can obtain pseudolikelihood scores that are used to calculate robust estimates of variance, Schoenfeld residuals that reflect each failure's contribution to the gradient of the log pseudolikelihood, efficient score residuals that represent each subject's (observation's) contribution to the gradient, and DFBETAs that measure the change in coefficients due to deletion of a subject or observation.

▷ Example 3: DFBETAs

Returning to our cervical cancer study, we obtain DFBETAs for each of the three coefficients in the model and graph those for the first with respect to analysis time.

```
. use http://www.stata-press.com/data/r11/hypoxia, clear
(Hypoxia study)
. stset dftime, failure(failtype == 1)
 (output omitted)
. stcrreg ifp tumsize pelnode, compete(failtype == 2)
 (output omitted)
. predict df*, dfbeta
. generate obs = _n
. twoway scatter df1 dftime, yline(0) mlabel(obs)
```

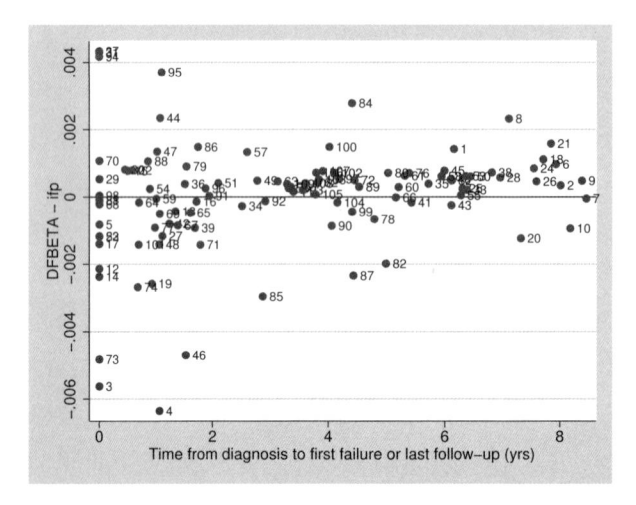

predict created the variables df1, df2, and df3, holding DFBETA values for variables ifp, tumsize, and pelnode, respectively. Based on the graph, we see that subject 4 is the most influential on the coefficient for ifp, the first covariate in the model.

◁

In the previous example, we had single-record data. If you have data with multiple records per subject, then by default DFBETAs will be calculated at the subject level, with one value representing each subject and measuring the effect of deleting all records for that subject. If you instead want record-level DFBETAs that measure the change due to deleting single records within subjects, add the partial option; see [ST] **stcox postestimation** for further details.

Methods and formulas

All methods presented in this entry have been implemented as ado-files that use Mata.

Continuing the discussion from *Methods and formulas* in [ST] **stcrreg**, the baseline cumulative subhazard function is calculated as

$$\widehat{H}_{1,0}(t) = \sum_{j:t_j \leq t} \frac{\delta_j}{\sum_{\ell \in R_j} w_\ell \pi_{\ell j} \exp(z_\ell)}$$

The baseline CIF is $\widehat{\text{CIF}}_{1,0}(t) = 1 - \exp\{-\widehat{H}_{1,0}(t)\}$.

The Kaplan–Meier survivor curve for the censoring distribution is

$$\widehat{S}_c(t) = \prod_{t_{(j)} < t} \left\{ 1 - \frac{\sum_i \gamma_i I(t_i = t_{(j)})}{r(t_{(j)})} \right\}$$

where $t_{(j)}$ indexes the times at which censorings occur.

Both the pseudolikelihood scores, $\widehat{\mathbf{u}}_i$, and the efficient score residuals, $\widehat{\eta}_i$, are as defined previously. DFBETAs are calculated according to Collett (2003):

$$\text{DFBETA}_i = \widehat{\eta}_i' \text{Var}^*(\widehat{\beta})$$

where $\text{Var}^*(\widehat{\beta})$ is the model-based variance estimator, that is, the inverse of the negative Hessian.

Schoenfeld residuals are $\mathbf{r}_i = (\widehat{r}_{1i}, \ldots, \widehat{r}_{mi})$ with

$$\widehat{r}_{ki} = \delta_i (x_{ki} - a_{ki})$$

References

Aalen, O. O. 1978. Nonparametric inference for a family of counting processes. *Annals of Statistics* 6: 701–726.

Collett, D. 2003. *Modelling Survival Data in Medical Research*. 2nd ed. London: Chapman & Hall/CRC.

Coviello, V., and M. Boggess. 2004. Cumulative incidence estimation in the presence of competing risks. *Stata Journal* 4: 103–112.

Fyles, A., M. Milosevic, D. Hedley, M. Pintilie, W. Levin, L. Manchul, and R. P. Hill. 2002. Tumor hypoxia has independent predictor impact only in patients with node-negative cervix cancer. *Journal of Clinical Oncology* 20: 680–687.

Geskus, R. B. 2000. On the inclusion of prevalent cases in HIV/AIDS natural history studies through a marker-based estimate of time since seroconversion. *Statistics in Medicine* 19: 1753–1769.

Kaplan, E. L., and P. Meier. 1958. Nonparametric estimation from incomplete observations. *Journal of the American Statistical Association* 53: 457–481.

Milosevic, M., A. Fyles, D. Hedley, M. Pintilie, W. Levin, L. Manchul, and R. P. Hill. 2001. Interstitial fluid pressure predicts survival in patients with cervix cancer independent of clinical prognostic factors and tumor oxygen measurements. *Cancer Research* 61: 6400–6405.

Nelson, W. 1972. Theory and applications of hazard plotting for censored failure data. *Technometrics* 14: 945–966.

Pintilie, M. 2006. *Competing Risks: A Practical Perspective*. Chichester, UK: Wiley.

Putter, H., M. Fiocco, and R. B. Geskus. 2007. Tutorial in biostatistics: Competing risks and multi-state models. *Statistics in Medicine* 26: 2389–2430.

Also see

[ST] **stcrreg** — Competing-risks regression

[ST] **stcurve** — Plot survivor, hazard, cumulative hazard, or cumulative incidence function

[U] **20 Estimation and postestimation commands**

Title

> **stcurve** — Plot survivor, hazard, cumulative hazard, or cumulative incidence function

Syntax

stcurve [, *stcurve_options*]

stcurve_options	description
Main	
*<u>s</u>urvival	plot survivor function
*<u>haz</u>ard	plot hazard function
*<u>cumh</u>az	plot cumulative hazard function
*cif	plot cumulative incidence function
at(*varname*=# [*varname*=# ...])	value of the specified covariates and mean of unspecified covariates
[at1(*varname*=# [*varname*=# ...])	
[at2(*varname*=# [*varname*=# ...])	
[...]]]	
Options	
<u>alpha1</u>	conditional frailty model
<u>uncon</u>ditional	unconditional frailty model
<u>range</u>(# #)	range of analysis time
<u>outf</u>ile(*filename* [, replace])	save values used to plot the curves
width(#)	override "optimal" width; use with hazard
<u>kernel</u>(*kernel*)	kernel function; use with hazard
<u>nob</u>oundary	no boundary correction; use with hazard
Plot	
connect_options	affect rendition of plotted survivor, hazard, or cumulative hazard function
Add plots	
addplot(*plot*)	add other plots to the generated graph
Y axis, X axis, Titles, Legend, Overall	
twoway_options	any options other than by() documented in [G] ***twoway_options***

*One of survival, hazard, cumhaz, or cif must be specified.
 survival and hazard are not allowed after estimation with stcrreg.
 cif is allowed only after estimation with stcrreg.

Menu

Statistics > Survival analysis > Regression models > Plot survivor, hazard, cumulative hazard, or cumulative incidence function

Description

stcurve plots the survivor, hazard, or cumulative hazard function after stcox or streg and plots the cumulative subhazard or cumulative incidence function (CIF) after stcrreg.

Options

      ⌈ Main ⌉

survival requests that the survivor function be plotted. survival is not allowed after estimation with stcrreg.

hazard requests that the hazard function be plotted. hazard is not allowed after estimation with stcrreg.

cumhaz requests that the cumulative hazard function be plotted when used after stcox or streg and requests that the cumulative subhazard function be plotted when used after stcrreg.

cif requests that the cumulative incidence function be plotted. This option is available only after estimation with stcrreg.

at(*varname*=# ...) requests that the covariates specified by *varname* be set to #. By default, stcurve evaluates the function by setting each covariate to its mean value. This option causes the function to be evaluated at the value of the covariates listed in at() and at the mean of all unlisted covariates.

at1(*varname*=# ...), at2(*varname*=# ...), ..., at10(*varname*=# ...) specify that multiple curves (up to 10) be plotted on the same graph. at1(), at2(), ..., at10() work like the at() option. They request that the function be evaluated at the value of the covariates specified and at the mean of all unlisted covariates. at1() specifies the values of the covariates for the first curve, at2() specifies the values of the covariates for the second curve, and so on.

      ⌈ Options ⌉

alpha1, when used after fitting a frailty model, plots curves that are conditional on a frailty value of one. This is the default for shared-frailty models.

unconditional, when used after fitting a frailty model, plots curves that are unconditional on the frailty; i.e., the curve is "averaged" over the frailty distribution. This is the default for unshared-frailty models.

range(# #) specifies the range of the time axis to be plotted. If this option is not specified, stcurve plots the desired curve on an interval expanding from the earliest to the latest time in the data.

outfile(*filename* [, replace]) saves in *filename*.dta the values used to plot the curve(s).

width(#) is for use with hazard (and applies only after stcox) and is used to specify the bandwidth to be used in the kernel smooth used to plot the estimated hazard function. If left unspecified, a default bandwidth is used, as described in [R] **kdensity**.

kernel(*kernel*) is for use with hazard and is for use only after stcox because, for Cox regression, an estimate of the hazard function is obtained by smoothing the estimated *hazard contributions*. kernel() specifies the kernel function for use in calculating the weighted kernel-density estimate required to produce a smoothed hazard-function estimator. The default is kernel(Epanechnikov), yet *kernel* may be any of the kernels supported by kdensity; see [R] **kdensity**.

stcurve — Plot survivor, hazard, cumulative hazard, or cumulative incidence function   233

noboundary is for use with hazard and applies only to the plotting of smoothed hazard functions after stcox. It specifies that no boundary-bias adjustments are to be made when calculating the smoothed hazard-function estimator. By default, the smoothed hazards are adjusted near the boundaries; see [ST] **sts graph**. If the epan2, biweight, or rectangular kernel is used, the bias correction near the boundary is performed using boundary kernels. For other kernels, the plotted range of the smoothed hazard function is restricted to be inside of one bandwidth from each endpoint. For these other kernels, specifying noboundary merely removes this range restriction.

⎡ Plot ⎤

connect_options affect the rendition of the plotted survivor, hazard, or cumulative hazard function; see [G] *connect_options*.

⎡ Add plots ⎤

addplot(*plot*) provides a way to add other plots to the generated graph; see [G] *addplot_option*.

⎡ Y axis, X axis, Titles, Legend, Overall ⎤

twoway_options are any of the options documented in [G] *twoway_options*, excluding by(). These include options for titling the graph (see [G] *title_options*) and for saving the graph to disk (see [G] *saving_option*).

Remarks

Remarks are presented under the following headings:

> stcurve after stcox
> stcurve after streg
> stcurve after stcrreg

stcurve after stcox

After fitting a Cox model, stcurve can be used to plot the estimated hazard, cumulative hazard, and survivor functions.

▷ Example 1

```
. use http://www.stata-press.com/data/r11/drugtr
(Patient Survival in Drug Trial)
. stcox age drug
 (output omitted )
. stcurve, survival
```

(*Continued on next page*)

234   stcurve — Plot survivor, hazard, cumulative hazard, or cumulative incidence function

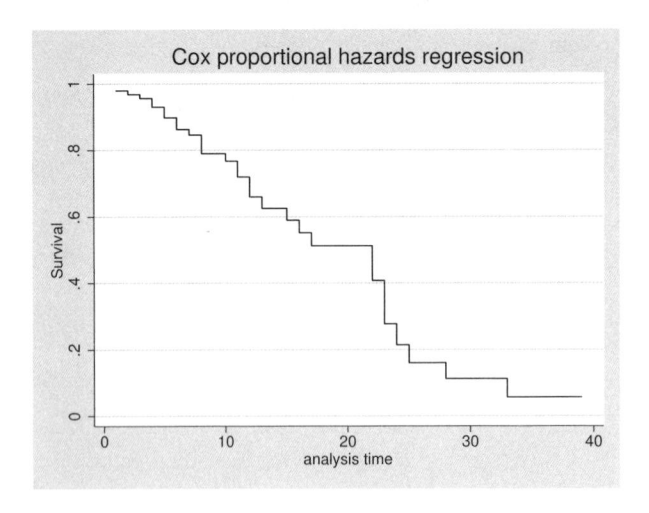

By default, the curve is evaluated at the mean values of all the predictors, but we can specify other values if we wish.

        . stcurve, survival at1(drug=0) at2(drug=1)

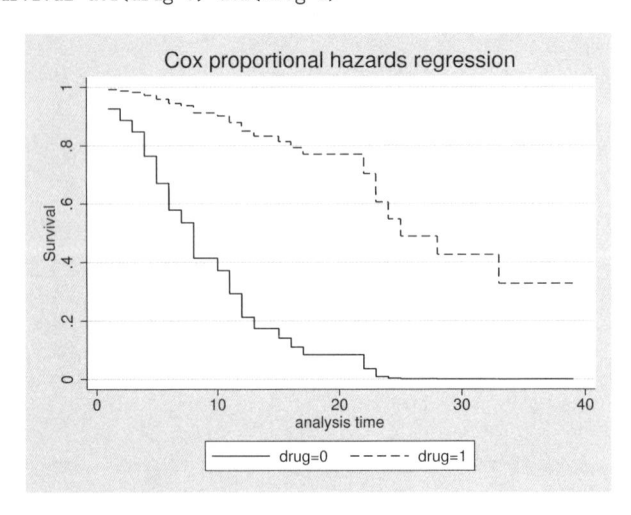

In this example, we asked for two plots, one for the placebo group and one for the treatment group. For both groups, the value of age was held at its mean value for the overall estimation sample.

◁

▷ Example 2

stcurve can also be used to plot estimated hazard functions. The hazard function is estimated by a kernel smooth of the estimated hazard contributions; see [ST] **sts graph** for details. We can thus customize the smooth as we would any other; see [R] **kdensity** for details.

        . stcurve, hazard at1(drug=0) at2(drug=1) kernel(gauss) yscale(log)

[Figure: Cox proportional hazards regression — Smoothed hazard function vs analysis time, with curves for drug=0 (solid) and drug=1 (dashed).]

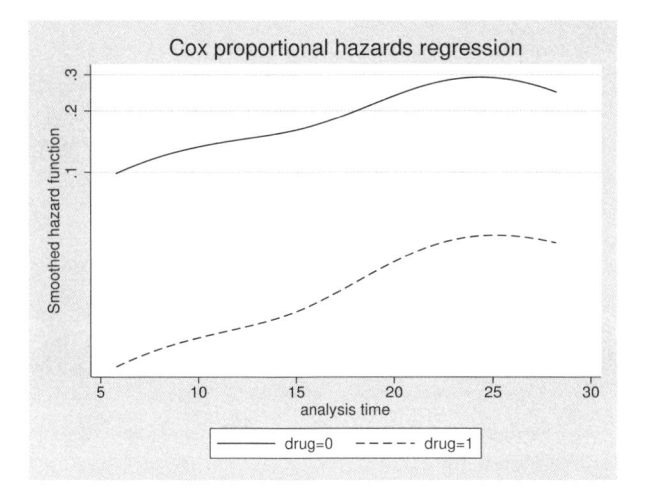

For the hazard plot, we plotted on a log scale to demonstrate the proportionality of hazards under this model; see the technical note below on smoothed hazards.

◁

❑ Technical note

For survivor or cumulative hazard estimation, stcurve works by first estimating the baseline function and then modifying it to adhere to the specified (or by default, mean) covariate patterns. As mentioned previously, *baseline* (when all covariates are equal to zero) must correspond to something that is meaningful and preferably in the range of your data. Otherwise, stcurve could encounter numerical difficulties. We ignored our own advice above and left age unchanged. Had we encountered numerical problems, or funny-looking graphs, we would have known to try shifting age so that age==0 was in the range of our data.

For hazard estimation, stcurve works by first transforming the estimated hazard contributions to adhere to the necessary covariate pattern and then applying the smooth. When you plot multiple curves, each is smoothed independently, although the same bandwidth is used for each.

The smoothing takes place in the hazard scale and not in the log-hazard scale. As a result, the resulting curves will look nearly, but not exactly, parallel when plotted on a log scale. This inexactitude is a product of the smoothing and should not be interpreted as a deviation from the proportional-hazards assumption; stcurve (after stcox) assumes proportionality of hazards and will reflect this in the produced plots. If smoothing were a perfect science, the curves would be parallel when plotted on a log scale. If you encounter estimated hazards exhibiting severe disproportionality, this may signal a numerical problem as described above. Try recentering your covariates so that baseline is more reasonable.

❑

stcurve after streg

stcurve is used after streg to plot the fitted survivor, hazard, and cumulative hazard functions. By default, stcurve computes the means of the covariates and evaluates the fitted model at each time in the data, censored or uncensored. The resulting plot is therefore the survival experience of a subject with a covariate pattern equal to the average covariate pattern in the study. You can produce the plot at other values of the covariates by using the at() option or specify a time range by using the range() option.

▷ Example 3

We pick up where [ST] **streg** left off. The cancer dataset we are using has three values for variable drug: 1 corresponds to placebo, and 2 and 3 correspond to two alternative treatments. Using the cancer data with drug remapped to form an indicator of treatment, let's fit a loglogistic regression model and plot its survival curves. We can perform a loglogistic regression by issuing the following commands:

```
. use http://www.stata-press.com/data/r11/cancer, clear
(Patient Survival in Drug Trial)
. replace drug = drug==2 | drug==3        // 0, placebo : 1, nonplacebo
(48 real changes made)
. stset studytime, failure(died)
 (output omitted)
. streg age drug, distribution(llogistic) nolog

        failure _d:  died
   analysis time _t: studytime

Loglogistic regression -- accelerated failure-time form
No. of subjects =         48                  Number of obs   =        48
No. of failures =         31
Time at risk    =        744
                                              LR chi2(2)      =     35.14
Log likelihood  =  -43.21698                  Prob > chi2     =    0.0000
```

_t	Coef.	Std. Err.	z	P>|z|	[95% Conf. Interval]	
age	-.0803289	.0221598	-3.62	0.000	-.1237614	-.0368964
drug	1.420237	.2502148	5.68	0.000	.9298251	1.910649
_cons	6.446711	1.231914	5.23	0.000	4.032204	8.861218
/ln_gam	-.8456552	.1479337	-5.72	0.000	-1.1356	-.5557105
gamma	.429276	.0635044			.3212293	.5736646

Now we wish to plot the survivor and the hazard functions:

```
. stcurve, survival ylabels(0 .5 1)
```

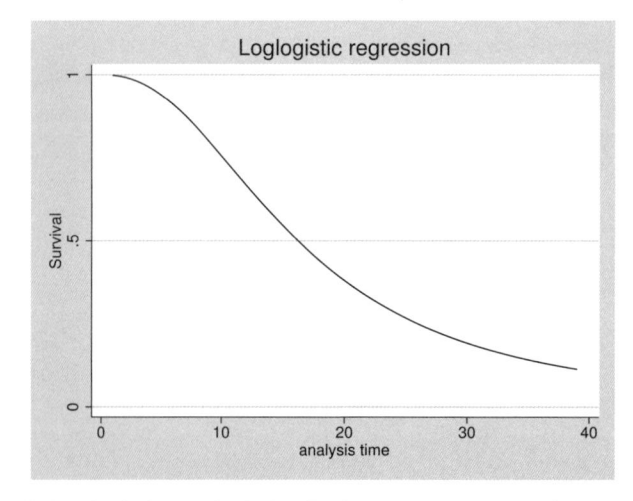

Figure 3. Loglogistic survival distribution at mean value of all covariates

stcurve — Plot survivor, hazard, cumulative hazard, or cumulative incidence function

. stcurve, hazard

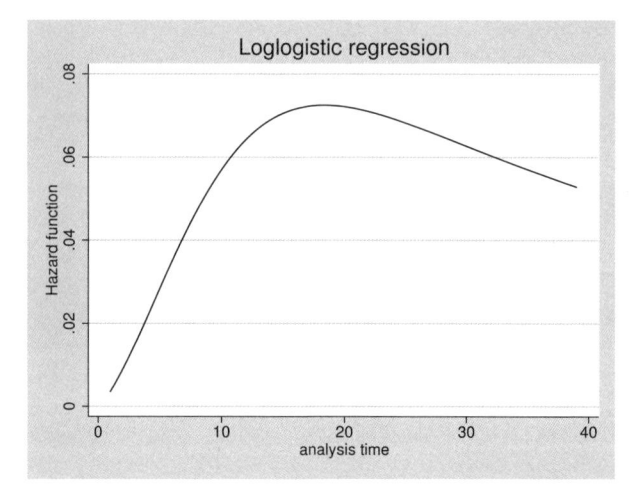

Figure 4. Loglogistic hazard distribution at mean value of all covariates

These plots show the fitted survivor and hazard functions evaluated for a cancer patient of average age receiving the average drug. Of course, the "average drug" has no meaning here because drug is an indicator variable. It makes more sense to plot the curves at a fixed value (level) of the drug. We can do this with the at option. For example, we may want to compare the average-age patient's survival curve under placebo (drug==0) and under treatment (drug==1).

We can plot both curves on the same graph:

. stcurve, surv at1(drug = 0) at2(drug = 1) ylabels(0 .5 1)

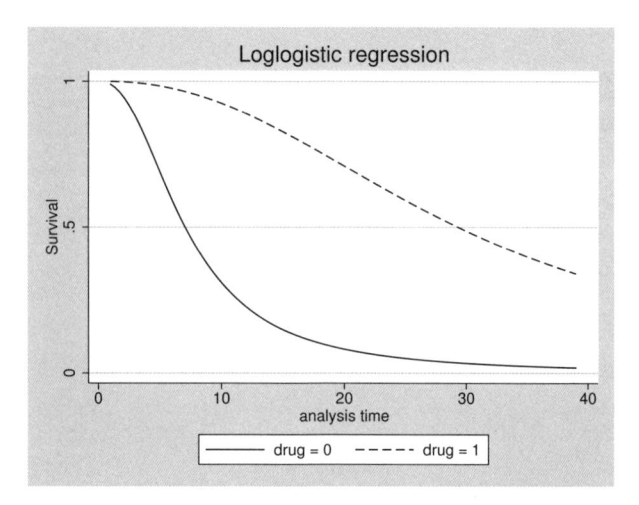

Figure 5. Loglogistic survival distribution at mean age for placebo

In the plot, we can see from the loglogistic model that the survival experience of an average-age patient receiving the placebo is worse than the survival experience of that same patient receiving treatment. We can also see the accelerated-failure-time feature of the loglogistic model. The survivor function for treatment is a time-decelerated (stretched-out) version of the survivor function for placebo.

◁

▷ Example 4

In our discussion of frailty models in [ST] **streg**, we emphasize the distinction between the individual hazard (or survivor) function and the hazard (survivor) function for the population. When significant frailty is present, the population hazard will tend to begin falling past a certain point, regardless of the shape of the individual hazard. This is due to the frailty effect—as time passes, the frailer individuals will fail, leaving a more homogeneous population comprising only the most robust individuals.

The frailty effect may be demonstrated using stcurve to plot the estimated hazard (both individual and population) after fitting a frailty model. Use the alpha1 option to specify the individual hazard ($\alpha = 1$) and the unconditional option to specify the population hazard. Applying this to the Weibull/inverse-Gaussian shared-frailty model on the kidney data of example 11 of [ST] **streg**,

```
. use http://www.stata-press.com/data/r11/catheter, clear
(Kidney data, McGilchrist and Aisbett, Biometrics, 1991)
. stset time infect
 (output omitted)
. qui streg age female, d(weibull) frailty(invgauss) shared(patient)
. stcurve, hazard at(female = 1) alpha1
```

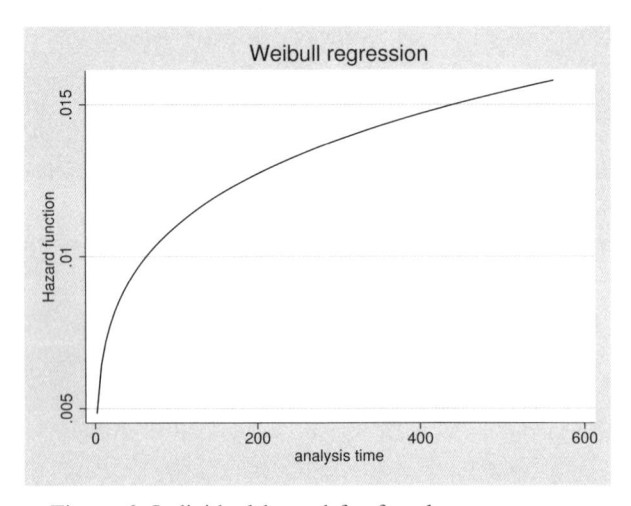

Figure 6. Individual hazard for females at mean age

Compare with

```
. stcurve, hazard at(female = 1) unconditional
```

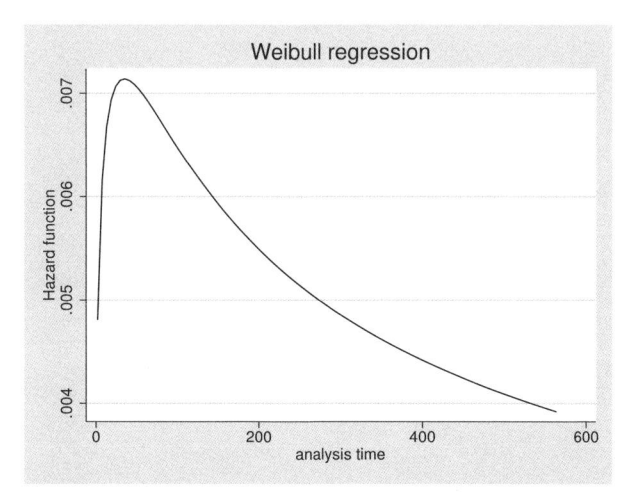

Figure 7. Population hazard for females at mean age

◁

stcurve after stcrreg

▷ Example 5

In [ST] **stcrreg**, we analyzed data from 109 patients with primary cervical cancer, treated at a cancer center between 1994 and 2000. We fit a competing-risks regression model where local relapse was the failure event of interest (`failtype == 1`), distant relapse with no local relapse was the competing risk event (`failtype == 2`), and we were interested primarily in the effect of insterstitial fluid pressure (`ifp`) while controlling for tumor size and pelvic node involvement.

After fitting the competing-risks regression model, we can use `stcurve` to plot the estimated cumulative incidence of local relapses in the presence of the competing risk. We wish to compare the cumulative incidence curves for `ifp == 5` versus `ifp == 20`, assuming positive pelvic node involvement (`pelnode == 0`) and a tumor size that is the average over the data.

```
. use http://www.stata-press.com/data/r11/hypoxia
(Hypoxia study)
. stset dftime, fail(failtype==1)
  (output omitted)
. stcrreg ifp tumsize pelnode, compete(failtype==2)
  (output omitted)
. stcurve, cif at1(ifp=5 pelnode=0) at2(ifp=20 pelnode=0)
```

(*Continued on next page*)

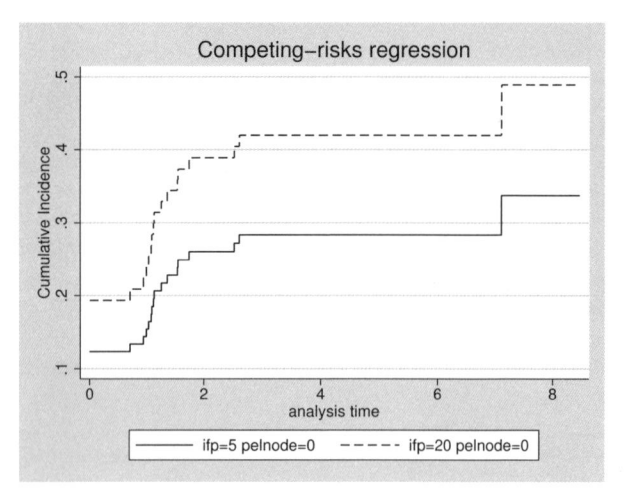

Figure 8. Comparative cumulative incidence functions

Methods and formulas

stcurve is implemented as an ado-file.

Reference

Cleves, M. A. 2000. stata54: Multiple curves plotted with stcurv command. *Stata Technical Bulletin* 54: 2–4. Reprinted in *Stata Technical Bulletin Reprints*, vol. 9, pp. 7–10. College Station, TX: Stata Press.

Also see

[ST] **stcox** — Cox proportional hazards model

[ST] **stcox postestimation** — Postestimation tools for stcox

[ST] **stcrreg** — Competing-risks regression

[ST] **stcrreg postestimation** — Postestimation tools for stcrreg

[ST] **streg** — Parametric survival models

[ST] **streg postestimation** — Postestimation tools for streg

[ST] **sts** — Generate, graph, list, and test the survivor and cumulative hazard functions

[ST] **stset** — Declare data to be survival-time data

Title

> **stdescribe** — Describe survival-time data

Syntax

> stdescribe [*if*] [*in*] [, <u>w</u>eight <u>nosh</u>ow]

You must stset your data before using stdescribe; see [ST] **stset**.

fweights, iweights, and pweights may be specified using stset; see [ST] **stset**.

by is allowed; see [D] **by**.

Menu

Statistics > Survival analysis > Summary statistics, tests, and tables > Describe survival-time data

Description

stdescribe produces a summary of the st data in a computer or data-based sense rather than in an analytical or statistical sense.

stdescribe can be used with single- or multiple-record or single- or multiple-failure st data.

Options

⌐ Main ┐

weight specifies that the summary use weighted rather than unweighted statistics. weight does nothing unless you specified a weight when you stset the data. The weight option and the ability to ignore weights are unique to stdescribe. The purpose of stdescribe is to describe the data in a computer sense—the number of records, etc.—and for that purpose, the weights are best ignored.

noshow prevents stdescribe from showing the key st variables. This option is seldom used because most people type stset, show or stset, noshow to set whether they want to see these variables mentioned at the top of the output of every st command; see [ST] **stset**.

Remarks

Here is an example of stdescribe with single-record survival data:

```
. use http://www.stata-press.com/data/r11/page2
. stdescribe
         failure _d:  dead
   analysis time _t:  time
```

			per subject		
Category	total	mean	min	median	max
no. of subjects	40				
no. of records	40	1	1	1	1
(first) entry time		0	0	0	0
(final) exit time		227.95	142	231	344
subjects with gap	0				
time on gap if gap	0				
time at risk	9118	227.95	142	231	344
failures	36	.9	0	1	1

There is one record per subject. The purpose of this summary is not analysis—it is to describe how the data are arranged. We can quickly see that there is one record per subject (the number of subjects equals the number of records, but if there is any doubt, the minimum and maximum number of records per subject is 1), that all the subjects entered at time 0, that the subjects exited between times 142 and 344 (median 231), that there are no gaps (as there could not be if there is only one record per subject), that the total time at risk is 9,118 (distributed reasonably evenly across the subjects), and that the total number of failures is 36 (with a maximum of 1 failure per subject).

Here is a description of the multiple-record Stanford heart transplant data that we introduced in [ST] **stset**:

```
. use http://www.stata-press.com/data/r11/stan3
(Heart transplant data)
. stdescribe
         failure _d:  died
   analysis time _t:  t1
                 id:  id
```

			per subject		
Category	total	mean	min	median	max
no. of subjects	103				
no. of records	172	1.669903	1	2	2
(first) entry time		0	0	0	0
(final) exit time		310.0786	1	90	1799
subjects with gap	0				
time on gap if gap	0	.	.	.	.
time at risk	31938.1	310.0786	1	90	1799
failures	75	.7281553	0	1	1

Here patients have one or two records. Although this is not revealed by the output, a patient has one record if the patient never received a heart transplant and two if the patient did receive a transplant; the first reflects the patient's survival up to the time of transplantation and the second their subsequent survival:

```
. stset, noshow                    /* to not show the st marker variables */
. stdescribe if !transplant
```

		per subject
Category	total	mean	min	median	max
no. of subjects	34				
no. of records	34	1	1	1	1
(first) entry time		0	0	0	0
(final) exit time		96.61765	1	21	1400
subjects with gap	0				
time on gap if gap	0	.	.	.	.
time at risk	3285	96.61765	1	21	1400
failures	30	.8823529	0	1	1

```
. stdescribe if transplant
```

		per subject
Category	total	mean	min	median	max
no. of subjects	69				
no. of records	138	2	2	2	2
(first) entry time		0	0	0	0
(final) exit time		415.2623	5.1	207	1799
subjects with gap	0				
time on gap if gap	0	.	.	.	.
time at risk	28653.1	415.2623	5.1	207	1799
failures	45	.6521739	0	1	1

Finally, here are the results of stdescribe from multiple-failure data:

```
. use http://www.stata-press.com/data/r11/mfail2
. stdescribe
```

		per subject
Category	total	mean	min	median	max
no. of subjects	926				
no. of records	1734	1.87257	1	2	4
(first) entry time		0	0	0	0
(final) exit time		470.6857	1	477	960
subjects with gap	6				
time on gap if gap	411	68.5	16	57.5	133
time at risk	435444	470.2419	1	477	960
failures	808	.8725702	0	1	3

The maximum number of failures per subject observed is three, although 50% had just one failure, and six subjects have gaps in their histories.

Saved results

stdescribe saves the following in r():

Scalars
r(N_sub)	number of subjects	r(gap)	total gap, if gap
r(N_total)	number of records	r(gap_min)	minimum gap, if gap
r(N_min)	minimum number of records	r(gap_mean)	mean gap, if gap
r(N_mean)	mean number of records	r(gap_med)	median gap, if gap
r(N_med)	median number of records	r(gap_max)	maximum gap, if gap
r(N_max)	maximum number of records	r(tr)	total time at risk
r(t0_min)	minimum first entry time	r(tr_min)	minimum time at risk
r(t0_mean)	mean first entry time	r(tr_mean)	mean time at risk
r(t0_med)	median first entry time	r(tr_med)	median time at risk
r(t0_max)	maximum first entry time	r(tr_max)	maximum time at risk
r(t1_min)	minimum final exit time	r(N_fail)	number of failures
r(t1_mean)	mean final exit time	r(f_min)	minimum number of failures
r(t1_med)	median final exit time	r(f_mean)	mean number of failures
r(t1_max)	maximum final exit time	r(f_med)	median number of failures
r(N_gap)	number of subjects with gap	r(f_max)	maximum number of failures

Methods and formulas

stdescribe is implemented as an ado-file.

Reference

Cleves, M. A., W. W. Gould, R. G. Gutierrez, and Y. Marchenko. 2008. *An Introduction to Survival Analysis Using Stata*. 2nd ed. College Station, TX: Stata Press.

Also see

[ST] **stset** — Declare data to be survival-time data

[ST] **stsum** — Summarize survival-time data

[ST] **stvary** — Report whether variables vary over time

Title

> **stfill** — Fill in by carrying forward values of covariates

Syntax

> stfill *varlist* [*if*] [*in*] , { <u>b</u>aseline | <u>f</u>orward } [*options*]

options	description
Main	
* <u>b</u>aseline	replace with values at baseline
* <u>f</u>orward	carry forward values
<u>nosh</u>ow	do not show st setting information

* Either baseline or forward is required.

You must stset your data before using stfill; see [ST] **stset**.

fweights, iweights, and pweights may be specified using stset; see [ST] **stset**.

Menu

Statistics > Survival analysis > Setup and utilities > Fill forward with values of covariates

Description

stfill is intended for use with multiple-record st data for which id() has been stset. stfill may be used with single-record data, but it does nothing. That is, stfill can be used with multiple-record or single- or multiple-failure st data.

stfill, baseline changes variables to contain the value at the earliest time each subject was observed, making the variable constant over time. stfill, baseline changes all subsequent values of the specified variables to equal the first value, whether they originally contained missing or not.

stfill, forward fills in missing values of each variable with that of the most recent time at which the variable was last observed. stfill, forward changes only missing values.

You must specify either the baseline or the forward option.

if *exp* and in *range* operate slightly differently from their usual definitions to work as you would expect. if and in restrict where changes can be made to the data, but no matter what, all stset observations are used to provide the values to be carried forward.

Options

> **Main**

baseline specifies that values be replaced with the values at baseline, the earliest time at which the subject was observed. All values of the specified variables are replaced, missing and nonmissing.

forward specifies that values be carried forward and that previously observed, nonmissing values be used to fill in later values that are missing in the specified variables.

noshow prevents stfill from showing the key st variables. This option is seldom used because most people type stset, show or stset, noshow to set whether they want to see these variables mentioned at the top of the output of every st command; see [ST] **stset**.

Remarks

stfill assists in fixing data errors and makes baseline analyses easier.

▷ Example 1

Let's begin by repairing broken data.

You have a multiple-record st dataset that, because of how it was constructed, has a problem with the gender variable:

```
. use http://www.stata-press.com/data/r11/mrecord
. stvary sex
        failure _d:  myopic
   analysis time _t:  t
                 id:  id
```

	subjects for whom the variable is				
variable	constant	varying	never missing	always missing	sometimes missing
sex	131	1	22	0	110

For 110 subjects, sex is sometimes missing, and for one more subject, the value of sex changes over time! The sex change is an error, but the missing values occurred because sometimes the subject's sex was not filled in on the revisit forms. We will assume that you have checked the changing-sex subject and determined that the baseline record is correct in that case, too.

```
. stfill sex, baseline
        failure _d:  myopic
   analysis time _t:  t
                 id:  id
replace all values with value at earliest observed:
          sex:  221 real changes made
. stvary sex
        failure _d:  myopic
   analysis time _t:  t
                 id:  id
```

	subjects for whom the variable is				
variable	constant	varying	never missing	always missing	sometimes missing
sex	132	0	132	0	0

The sex variable is now completely filled in.

In this same dataset, there is another variable—bp, blood pressure—that is not always filled in because readings were not always taken.

```
. stvary bp
        failure _d:  myopic
  analysis time _t:  t
               id:  id
```

	subjects for whom the variable is		never	always	sometimes
variable	constant	varying	missing	missing	missing
bp	18	114	9	0	123

(bp is constant for 18 patients because it was taken only once—at baseline.) Anyway, you decide that it will be good enough when bp is missing to use the previous value of bp:

```
. stfill bp, forward noshow
replace missing values with previously observed values:
        bp:  263 real changes made
. stvary bp, noshow
```

	subjects for whom the variable is		never	always	sometimes
variable	constant	varying	missing	missing	missing
bp	18	114	132	0	0

So much for data repair and fabrication.

◁

▷ Example 2

Much later, deep in analysis, you are concerned about the bp variable and decide to compare results with a model that simply includes blood pressure at baseline. You are undecided on the issue and want to have both variables in your data:

```
. stset, noshow
. gen bp0 = bp
. stfill bp0, baseline
replace all values with value at earliest observed:
       bp0:  406 real changes made
. stvary bp bp0
```

	subjects for whom the variable is		never	always	sometimes
variable	constant	varying	missing	missing	missing
bp	18	114	132	0	0
bp0	132	0	132	0	0

◁

Methods and formulas

stfill is implemented as an ado-file.

Also see

[ST] **stbase** — Form baseline dataset

[ST] **stgen** — Generate variables reflecting entire histories

[ST] **stset** — Declare data to be survival-time data

[ST] **stvary** — Report whether variables vary over time

Title

stgen — Generate variables reflecting entire histories

Syntax

stgen [*type*] *newvar* = *function*

where *function* is

ever(*exp*)
never(*exp*)
always(*exp*)
min(*exp*)
max(*exp*)
when(*exp*)
when0(*exp*)
count(*exp*)
count0(*exp*)
minage(*exp*)
maxage(*exp*)
avgage(*exp*)
nfailures()
ngaps()
gaplen()
hasgap()

You must stset your data before using stgen; see [ST] **stset**.

Menu

Statistics > Survival analysis > Setup and utilities > Generate variable reflecting entire histories

Description

stgen provides a convenient way to generate new variables reflecting entire histories—variables you could create for yourself by using generate (and especially, generate with the by *varlist*: prefix), but that would require too much thought, and there would be too much chance of making a mistake.

These functions are intended for use with multiple-record survival data but may be used with single-record data. With single-record data, each function reduces to one generate, and generate would be a more natural way to approach the problem.

stgen can be used with multiple-record or single- or multiple-failure st data.

If you want to generate calculated values, such as the survivor function, see [ST] **sts**.

Functions

In the description of the functions below, time units refer to the same units as *timevar* from stset *timevar*, ... For instance, if *timevar* is the number of days since 01 January 1960 (a Stata date), time units are days. If *timevar* is in years—years since 1960, years since diagnosis, or whatever—time units are years.

When we say variable X records a "time", we mean a variable that records when something occurred in the same units and with the same base as *timevar*. If *timevar* is a Stata date, "time" is correspondingly a Stata date.

t units, or analysis-time units, refer to a variable in the units *timevar*/scale() from stset *timevar*, scale(...) .... If you did not specify a scale(), *t* units are the same as time units. Alternatively, say that *timevar* is recorded as a Stata date and you specified scale(365.25). Then *t* units are years. If you specified a nonconstant scale—scale(myvar), where myvar varies from subject to subject—*t* units are different for every subject.

"An analysis time" refers to the time something occurred, recorded in the units (*timevar*−origin())/scale(). We speak about analysis time only in terms of the beginning and end of each time-span record.

Although in *Description* above we said that stgen creates variables reflecting entire histories, stgen restricts itself to the stset observations, so "entire history" means the entire history as it is currently stset. If you really want to use entire histories as recorded in the data, type streset, past or streset, past future before using stgen. Then type streset to reset to the original analysis sample.

The following functions are available:

ever(*exp*) creates *newvar* containing 1 (true) if the expression is ever true (nonzero) and 0 otherwise. For instance,

. stgen everlow = ever(bp<100)

would create everlow containing, for each subject, uniformly 1 or 0. Every record for a subject would contain everlow = 1 if, on any stset record for the subject, bp < 100; otherwise, everlow would be 0.

never(*exp*) is the reverse of ever(); it creates *newvar* containing 1 (true) if the expression is always false (0) and 0 otherwise. For instance,

. stgen neverlow = never(bp<100)

would create neverlow containing, for each subject, uniformly 1 or 0. Every record for a subject would contain neverlow = 1 if, on every stset record for the subject, bp < 100 is false.

always(*exp*) creates *newvar* containing 1 (true) if the expression is always true (nonzero) and 0 otherwise. For instance,

. stgen lowlow = always(bp<100)

would create lowlow containing, for each subject, uniformly 1 or 0. Every record for a subject would contain lowlow = 1 if, on every stset record for a subject, bp < 100.

min(*exp*) and max(*exp*) create *newvar* containing the minimum or maximum nonmissing value of *exp* within id(). min() and max() are often used with variables recording a time (see definition above), such as min(visitdat).

when(*exp*) and when0(*exp*) create *newvar* containing the time when *exp* first became true within the previously stset id(). The result is in time, not *t* units; see the definition above.

when() and when0() differ about when the *exp* became true. Records record time spans (*time0*, *time1*]. when() assumes that the expression became true at the end of the time span, *time1*. when0() assumes that the expression became true at the beginning of the time span, *time0*.

Assume that you previously stset myt, failure(*eventvar*=...) .... when() would be appropriate for use with *eventvar*, and, presumably, when0() would be appropriate for use with the remaining variables.

count(*exp*) and count0(*exp*) create *newvar* containing the number of occurrences when *exp* is true within id().

count() and count0() differ in when they assume that *exp* occurs. count() assumes that *exp* corresponds to the end of the time-span record. Thus even if *exp* is true in this record, the count would remain unchanged until the next record.

count0() assumes that *exp* corresponds to the beginning of the time-span record. Thus if *exp* is true in this record, the count is immediately updated.

For example, assume that you previously stset myt, failure(*eventvar*=...) .... count() would be appropriate for use with *eventvar*, and, presumably, count0() would be appropriate for use with the remaining variables.

minage(*exp*), maxage(*exp*), and avgage(*exp*) return the elapsed time, in time units, because *exp* is at the beginning, end, or middle of the record, respectively. *exp* is expected to evaluate to a time in time units. minage(), maxage(), and avgage() would be appropriate for use with the result of when(), when0(), min(), and max(), for instance.

Also see [ST] **stsplit**; stsplit will divide the time-span records into new time-span records that record specified intervals of ages.

nfailures() creates *newvar* containing the cumulative number of failures for each subject as of the entry time for the observation. nfailures() is intended for use with multiple-failure data; with single-failure data, nfailures() is always 0. In multiple-failure data,

． stgen nfail = nfailures()

might create, for a particular subject, the following:

id	time0	time1	fail	x	nfail
93	0	20	0	1	0
93	20	30	1	1	0
93	30	40	1	2	1
93	40	60	0	1	2
93	60	70	0	2	2
93	70	80	1	1	2

The total number of failures for this subject is 3, and yet the maximum of the new variable nfail is 2. At time 70, the beginning of the last record, there had been two failures previously, and there were two failures up to but not including time 80.

ngaps() creates *newvar* containing the cumulative number of gaps for each subject as of the entry time for the record. Delayed entry (an opening gap) is not considered a gap. For example,

． stgen ngap = ngaps()

might create, for a particular subject, the following:

id	time0	time1	fail	x	ngap
94	10	30	0	1	0
94	30	40	0	2	0
94	50	60	0	1	1
94	60	70	0	2	1
94	82	90	1	1	2

gaplen() creates *newvar* containing the time on gap, measured in analysis-time units, for each subject as of the entry time for the observation. Delayed entry (an opening gap) is not considered a gap. Continuing with the previous example,

```
. stgen gl = gaplen()
```

would produce

id	time0	time1	fail	x	ngap	gl
94	10	30	0	1	0	0
94	30	40	0	2	0	0
94	50	60	0	1	1	10
94	60	70	0	2	1	0
94	82	90	1	1	2	12

hasgap() creates *newvar* containing uniformly 1 if the subject ever has a gap and 0 otherwise. Delayed entry (an opening gap) is not considered a gap.

Remarks

stgen does nothing you cannot do in other ways, but it is convenient.

Consider how you would obtain results like those created by stgen should you need something that stgen will not create for you. Say that we have an st dataset for which we have previously

```
. stset t, failure(d) id(id)
```

Assume that these are some of the data:

id	t	d	bp
27	30	0	90
27	50	0	110
27	60	1	85
28	11	0	120
28	40	1	130

If we were to type

```
. stgen everlow = ever(bp<100)
```

the new variable, everlow, would contain for these two subjects

id	t	d	bp	everlow
27	30	0	90	1
27	50	0	110	1
27	60	1	85	1
28	11	0	120	0
28	40	1	130	0

Variable everlow is 1 for subject 27 because, in two of the three observations, bp < 100, and everlow is 0 for subject 28 because everlow is never less than 100 in either observation.

Here is one way we could have created everlow for ourselves:

```
. generate islow = bp<100
. sort id
. by id: generate sumislow = sum(islow)
. by id: generate everlow = sumislow[_N]>0
. drop islow sumislow
```

The generic term for code like this is explicit subscripting; see [U] **13.7 Explicit subscripting**.

Anyway, that is what stgen did for us, although, internally, stgen used denser code that was equivalent to

```
. by id, sort: generate everlow=sum(bp<100)
. by id: replace everlow = everlow[_N]>0
```

Obtaining things like the time on gap is no more difficult. When we stset the data, stset created variable _t0 to record the entry time. stgen's gaplen() function is equivalent to

```
. sort id _t
. by id: generate gaplen = _t0-_t[_n-1]
. by id: replace gaplen = 0 if _n == 1
```

Seeing this, you should realize that if all you wanted was the cumulative length of the gap before the current record, you could type

```
. sort id _t
. by id: generate curgap = sum(_t0-_t[_n-1])
```

If, instead, you wanted a variable that was 1 if there were a gap just before this record and 0 otherwise, you could type

```
. sort id _t
. by id: generate iscurgap = (_t0-_t[_n-1])>0
```

▷ Example 1

Let's use the stgen commands to real effect. We have a multiple-record, multiple-failure dataset.

```
. use http://www.stata-press.com/data/r11/mrmf
. st
-> stset t, id(id) failure(d) time0(t0) exit(time .) noshow
             id:  id
  failure event:  d != 0 & d < .
obs. time interval:  (t0, t]
 exit on or before:  time .
. stdescribe
```

		├──────── per subject ────────┤			
Category	total	mean	min	median	max
no. of subjects	926				
no. of records	1734	1.87257	1	2	4
(first) entry time		0	0	0	0
(final) exit time		470.6857	1	477	960
subjects with gap	6				
time on gap if gap	411	68.5	16	57.5	133
time at risk	435444	470.2419	1	477	960
failures	808	.8725702	0	1	3

Also in this dataset are two covariates, x1 and x2. We wish to fit a Cox model on these data but wish to assume that the baseline hazard for first failures is different from that for second and later failures.

Our data contain six subjects with gaps. Because failures might have occurred during the gap, we begin by dropping those six subjects:

```
. stgen hg = hasgap()
. drop if hg
(14 observations deleted)
```

The six subjects had 14 records among them. We can now create variable `nf` containing the number of failures and, from that, create variable `group`, which will be 0 when subjects have experienced no previous failures and 1 thereafter:

```
. stgen nf = nfailures()
. generate byte group = nf>0
```

We can now fit our stratified model:

```
. stcox x1 x2, strata(group) vce(robust)
Iteration 0:   log pseudolikelihood = -4499.9966
Iteration 1:   log pseudolikelihood = -4444.7797
Iteration 2:   log pseudolikelihood = -4444.4596
Iteration 3:   log pseudolikelihood = -4444.4596
Refining estimates:
Iteration 0:   log pseudolikelihood = -4444.4596

Stratified Cox regr. -- Breslow method for ties

No. of subjects    =       920              Number of obs   =      1720
No. of failures    =       800
Time at risk       =    432153
                                            Wald chi2(2)    =    102.78
Log pseudolikelihood  =  -4444.4596         Prob > chi2     =    0.0000

                       (Std. Err. adjusted for 920 clusters in id)
```

		Robust				
_t	Haz. Ratio	Std. Err.	z	P>|z|	[95% Conf. Interval]	
x1	2.087903	.1961725	7.84	0.000	1.736738	2.510074
x2	.2765613	.052277	-6.80	0.000	.1909383	.4005806

Stratified by group

Methods and formulas

`stgen` is implemented as an ado-file.

Also see

[ST] **stci** — Confidence intervals for means and percentiles of survival time

[ST] **sts** — Generate, graph, list, and test the survivor and cumulative hazard functions

[ST] **stset** — Declare data to be survival-time data

[ST] **stvary** — Report whether variables vary over time

Title

stir — Report incidence-rate comparison

Syntax

stir *exposedvar* [*if*] [*in*] [, *options*]

options	description
Main	
<u>strata</u>(*varname*)	stratify on *varname*
<u>nosh</u>ow	do not show st setting information
Options	
ird	report incidence-rate difference rather than ratio
<u>es</u>tandard	combine external weights with within-stratum statistics
<u>is</u>tandard	combine internal weights with within-stratum statistics
<u>s</u>tandard(*varname*)	combine user-specified weights with within-stratum statistics
<u>po</u>ol	display pooled estimate
<u>noc</u>rude	do not display crude estimate
<u>noh</u>om	do not display homogeneity test
tb	calculate test-based confidence intervals
<u>l</u>evel(*#*)	set confidence level; default is level(95)

Options except noshow, tb, and level(*#*) are relevant only if strata() is specified.

You must stset your data before using stir; see [ST] **stset**.

by is allowed; see [D] **by**.

fweights and iweights may be specified using stset; see [ST] **stset**. stir may not be used with pweighted data.

Menu

Statistics > Survival analysis > Summary statistics, tests, and tables > Report incidence-rate comparison

Description

stir reports point estimates and confidence intervals for the incidence-rate ratio and difference. stir is an interface to the ir command; see [ST] **epitab**.

By the logic of ir, *exposedvar* should be a 0/1 variable, with 0 meaning unexposed and 1 meaning exposed. stir, however, allows any two-valued coding and even allows *exposedvar* to be a string variable.

stir may not be used with pweighted data.

stir can be used with single- or multiple-record or single- or multiple-failure st data.

Options

[Main]

strata(*varname*) specifies that the calculation be stratified on *varname*, which may be a numeric or string variable. Within-stratum statistics are shown and then combined with Mantel–Haenszel weights.

noshow prevents stir from showing the key st variables. This option is seldom used because most people type stset, show or stset, noshow to set whether they want to see these variables mentioned at the top of the output of every st command; see [ST] **stset**.

[Options]

ird, estandard, istandard, standard(*varname*), pool, nocrude, and nohom are relevant only if strata() is specified; see [ST] **epitab**.

tb and level(*#*) are relevant in all cases; see [ST] **epitab**.

Remarks

stir examines the incidence rate and time at risk.

```
. use http://www.stata-press.com/data/r11/page2
. stir group, noshow
note: Exposed <-> group==2 and Unexposed <-> group==1
```

	group Exposed	Unexposed	Total		
Failure	19	17	36		
Time	5023	4095	9118		
Incidence rate	.0037826	.0041514	.0039482		
	Point estimate		[95% Conf. Interval]		
Inc. rate diff.	-.0003688		-.002974	.0022364	
Inc. rate ratio	.9111616		.4484366	1.866047	(exact)
Prev. frac. ex.	.0888384		-.8660469	.5515634	(exact)
Prev. frac. pop	.04894				
	(midp) Pr(k<=19) =			0.3900	(exact)
	(midp) 2*Pr(k<=19) =			0.7799	(exact)

Saved results

stir saves the following in r():

Scalars
    r(p)   one-sided p-value   r(ub_irr) upper bound of CI for irr
    r(ird)   incidence-rate difference   r(afe)   attributable (prev.) fraction among exposed
    r(lb_ird)   lower bound of CI for ird   r(lb_afe) lower bound of CI for afe
    r(ub_ird)   upper bound of CI for ird   r(ub_afe) upper bound of CI for afe
    r(irr)   incidence-rate ratio   r(afp)   attributable fraction for the population
    r(lb_irr)   lower bound of CI for irr

Methods and formulas

stir is implemented as an ado-file.

stir simply accumulates numbers of failures and time at risk by exposed and unexposed (by strata, if necessary) and passes the calculation to ir; see [ST] **epitab**.

Reference

Dupont, W. D. 2009. *Statistical Modeling for Biomedical Researchers: A Simple Introduction to the Analysis of Complex Data*. 2nd ed. Cambridge: Cambridge University Press.

Also see

[ST] **stset** — Declare data to be survival-time data

[ST] **stsum** — Summarize survival-time data

[ST] **epitab** — Tables for epidemiologists

Title

> **stpower** — Sample-size, power, and effect-size determination for survival analysis

Syntax

Sample-size determination

>   stpower cox [...] [, ...]
>
>   stpower logrank [...] [, ...]
>
>   stpower exponential [...] [, ...]

Power determination

>   stpower cox [...], n(*numlist*) [...]
>
>   stpower logrank [...], n(*numlist*) [...]
>
>   stpower exponential [...], n(*numlist*) [...]

Effect-size determination

>   stpower cox, n(*numlist*) { power(*numlist*) | beta(*numlist*) } [...]
>
>   stpower logrank [...], n(*numlist*) { power(*numlist*) | beta(*numlist*) } [...]

See [ST] **stpower cox**, [ST] **stpower logrank**, and [ST] **stpower exponential**.

Description

stpower computes sample size and power for survival analysis comparing two survivor functions using the log-rank test or the *exponential test* (to be defined later), as well as for more general survival analysis investigating the effect of a single covariate in a Cox proportional hazards regression model, possibly in the presence of other covariates. It provides the estimate of the number of events required to be observed (or the expected number of events) in a study. The minimal effect size (minimal detectable difference, expressed as the hazard ratio or the log hazard-ratio) may also be obtained for the log-rank test and for the Wald test on a single coefficient from the Cox model.

This entry provides an overview of the relevant terminology, theory, and a few examples. For more details, see the entries specific to each stpower subcommand.

Remarks

Remarks are presented under the following headings:

>   *Theory and terminology*
>   *Introduction to stpower subcommands*
>   *Sample-size determination for survival studies*
>   *Creating output tables*
>   *Power curves*

Theory and terminology

The prominent feature of survival data is that the outcome is the time from an origin to the occurrence of a given event (failure), often referred to as the analysis time. Analyses of such data use the information from all subjects in a study, both those who experience an event by the end of the study and those who do not. However, inference about the survival experience of subjects is based on the event times and therefore depends on the number of events observed in a study. Indeed, if none of the subjects fails in a study, then the survival rate cannot be estimated and survivor functions of subjects from different groups cannot be compared. Therefore, power depends on the number of events observed in a study and not directly on the number of subjects recruited to the study. As a result, to obtain the estimate of the required number of subjects, the probability that a subject experiences an event during the course of the study needs to be estimated in addition to the required number of events. This distinguishes sample-size determination for survival studies from that for other studies in which the endpoint is not measured as a time to failure.

All the above leads us to consider the following two types of survival studies. The first type (a *type I study*) is a study in which all subjects experience an event by the end of the study (no censoring), and the second type (a *type II study*) is a study that terminates after a fixed period regardless of whether all subjects experienced an event by that time. For a type II study, subjects who did not experience an event at the end of the study are known to be right-censored. For a type I study, when all subjects fail by the end of the study, the estimate of the probability of a failure in a study is one and the required number of subjects is equal to the required number of failures. For a type II study, the probability of a failure needs to be estimated and therefore various aspects that affect this probability (and usually do not come into play at the analysis stage) must be taken into account for the computation of the sample size.

Under the assumption of random censoring (Lachin 2000, 355; Lawless 2003, 52; Chow and Liu 2004, 388), the type of censoring pattern is irrelevant to the analysis of survival data in which the goal is to make inferences about the survival distribution of subjects. It becomes important, however, for sample-size determination because the probability that a subject experiences an event in a study depends on the censoring distribution. We consider the following two types of random censoring: administrative censoring and loss to follow-up.

Under administrative censoring, a subject is known to have experienced either of the two outcomes at the end of a study: survival or failure. The probability of a subject failing in a study depends on the duration of the study. Often in practice, subjects may withdraw from a study, say, because of severe side effects from a treatment or may be lost to follow-up because of moving to a different location. Here the information about the outcome that subject would have experienced at the end of the study had he completed the course of the study is unavailable, and the probability of experiencing an event by the end of the study is affected by the process governing withdrawal of subjects from the study. In the literature, this type of censoring is often referred to as subject loss to follow-up, subject withdrawal, or sometimes subject dropout (Freedman 1982, Machin and Campbell 2005). Generally, great care must be taken when using this terminology because it may have slightly different meanings in different contexts. `stpower logrank` and `stpower cox` apply a conservative adjustment to the estimate of the sample size for withdrawal. `stpower exponential` assumes that losses to follow-up are exponentially distributed.

Another important component of sample-size and power determination that affects the estimate of the probability of a failure is the pattern of accrual of subjects into the study. The duration of a study is often divided into two phases: an accrual phase, during which subjects are recruited to the study, and a follow-up phase, during which subjects are followed up until the end of the study and no new subjects enter the study. For a fixed-duration study, fast accrual increases the average analysis time (average follow-up time) and increases the chance of a subject failing in a study, whereas slow accrual decreases the average analysis time and consequently decreases this probability. `stpower logrank` and `stpower exponential` provide facilities to take into account uniform accrual, and for `stpower exponential` only, truncated exponential accrual.

All sample-size formulas used by `stpower` rely on the proportional-hazards assumption, that is, the assumption that the hazard ratio does not depend on time. See the documentation entry of each subcommand for the additional assumptions imposed by the methods it uses. In the case when the proportional-hazards assumption is suspect, or in the presence of other complexities associated with the nature of the trial (for example, lagged effect of a treatment, more than two treatment groups, clustered data) and with the behavior of participants (for example, noncompliance of subjects with the assigned treatment, competing risks), one may consider obtaining required sample size or power by simulation. Feiveson (2002) demonstrates an example of such simulation for clustered survival data. Barthel et al. (2006); Barthel, Royston, and Babiker (2005); and Royston and Babiker (2002) present sample-size and power computation for multiarm trials under more flexible design conditions.

Introduction to stpower subcommands

`stpower cox` provides estimates of sample size, power, or the minimal detectable value of the coefficient when an effect of a single covariate on subject survival is to be explored using Cox proportional hazards regression. It is assumed that the effect is to be tested using the partial likelihood from the Cox model (for example, score or Wald test) on the coefficient of the covariate of interest.

`stpower logrank` reports estimates of sample size, power, or minimal detectable value of the hazard ratio in the case when the two survivor functions are to be compared using the log-rank test. The only requirement about the distribution of the survivor functions is that the two survivor functions must satisfy the proportional-hazards assumption.

`stpower exponential` reports estimates of sample size or power when the disparity in the two exponential survivor functions is to be tested using the *exponential test*, the parametric test comparing the two exponential hazard rates. In particular, we refer to (exponential) *hazard-difference test* as the exponential test for the difference between hazards and (exponential) *log hazard-ratio test* as the exponential test for the log of the hazard ratio or, equivalently, for the difference between log hazards.

All subcommands share a common syntax. Sample-size determination with a power of 80% or, equivalently, a probability of a *type II error*, a failure to reject the null hypothesis when the alternative hypothesis is true, of 20% is the default. Other values of power or type II error probability may be supplied via the `power()` or `beta()` options, respectively. If power determination is desired, sample size `n()` must be specified. If the minimal detectable difference is of interest, both sample size `n()` and `power()` (or type II error probability `beta()`) must be specified.

For sample-size and power computations, the default effect size corresponds to a value of the hazard ratio of 0.5 and may be changed by specifying the `hratio()` option. The hazard ratio is defined as a ratio of hazards of the experimental group to the control group (or the less favorable of the two groups). Other ways of specifying the effect size are available, and these are particular to each subcommand.

The default probability of a *type I error*, a rejection of the null hypothesis when the null hypothesis is true, of a test is 0.05 but may be changed by using the `alpha()` option. Results for one-sided tests may be requested by using the `onesided` option. To change the default setting of equal-sized groups in `stpower logrank` and `stpower exponential`, one of the `p1()` or `nratio()` options must be specified.

By default, all subcommands assume a type I study, that is, perform computations for uncensored survival data. The censoring information may be taken into account by specifying the appropriate arguments or options. See [ST] **stpower cox**, [ST] **stpower logrank**, and [ST] **stpower exponential** for details.

All subcommands can report results in a table. Results may be tabulated for various values of input parameters. See *Creating output tables* for examples. An example of how to produce a power curve is given in *Power curves*.

Sample-size determination for survival studies

Here we demonstrate using `stpower` to obtain an estimate of the sample size for three different survival studies.

▷ Example 1: Sample size for the test of the effect of a covariate in the Cox model

Consider a hypothetical study for which the goal is to investigate the effect of the expression of one gene on subject survival with the Cox proportional hazards regression model. Suppose that the Wald test is to be used to test the coefficient on the gene after fitting the Cox model. Gene expression values measure the level of activity of the gene. Consider the scenario described in Simon, Radmacher, and Dobbin (2002) in which the hazard ratio of 3 associated with a one-unit change in the $\log_2$ intensity of a gene (or, respectively, with a twofold change in gene expression level) is desired to be detected with 95% power using a two-sided, 0.001-level test. The estimate of the standard deviation of the $\log_2$-intensity level of the gene over the entire set of samples is assumed to be 0.75.

```
. stpower cox, hratio(3) sd(0.75) power(0.95) alpha(0.001)
Estimated sample size for Cox PH regression
Wald test, log-hazard metric
Ho: [b1, b2, ..., bp] = [0, b2, ..., bp]

Input parameters:
        alpha =     0.0010  (two sided)
           b1 =     1.0986
           sd =     0.7500
        power =     0.9500

Estimated number of events and sample size:
            E =         36
            N =         36
```

◁

Provided that all subjects experience an event in this study, a total of 36 events is required to be observed in the study to ensure the specified power.

See [ST] **stpower cox** for more details.

▷ Example 2: Sample size for the log-rank test

Consider an example from Machin and Campbell (2005) of a study comparing two forms of surgical resection for patients with gastric cancer. From a prestudy survey, the baseline 5-year survival rate was expected to be 20% and an anticipated increase in survival in the experimental group expressed as a hazard ratio of 0.6667 (corresponding to a 5-year survival rate of approximately 34%) was desired to be detected with 90% power using a two-sided, 0.05 level, log-rank test under 1:1 randomization. To obtain the estimate of the sample size for this study, we use `stpower logrank` with survival proportion in the control group 0.2 supplied as an argument, the `hratio(0.6667)` option to request a hazard ratio of 0.6667, and the `power(0.9)` option to request 90% power.

```
. stpower logrank 0.2, hratio(0.6667) power(0.9)
```

Estimated sample sizes for two-sample comparison of survivor functions
Log-rank test, Freedman method
Ho: S1(t) = S2(t)

Input parameters:

```
       alpha =   0.0500   (two sided)
          s1 =   0.2000
          s2 =   0.3420
      hratio =   0.6667
       power =   0.9000
          p1 =   0.5000
```

Estimated number of events and sample sizes:

```
           E =   264
           N =   362
          N1 =   181
          N2 =   181
```

From the output, 264 events (failures) are required to be observed in this study to ensure 90% power to detect a hazard ratio of 0.6667 by using the log-rank test. The respective estimate of the total number of subjects required to observe 264 events in a 5-year study is 362 with 181 subjects per surgical group. Our estimate, 181, of each group's sample size is close to the manually computed estimate of 180 from Machin and Campbell (2005). This study is an example of a type II study as previously described, because 20% of subjects were expected to survive (be censored) by the end of the study.

◁

See [ST] **stpower logrank** for more detailed examples and other available methods of sample-size computation for this type of analysis.

▷ Example 3: Sample size for two-sample test of exponential survivor functions

Consider an example from Lachin (2000, 412) of a study comparing two therapies, the combination of a new therapy with the standard one versus the standard alone, in the treatment of lupus nephritis patients. From previous studies, the survivor function of the control group treated with the standard therapy was log linear with a constant yearly hazard rate of 0.3. The number of events (failures) required to ensure 90% power to detect a 50% risk reduction, $\Delta = 0.5$, (or, respectively, the log hazard-ratio of $\ln(0.5) = -0.6931$) with a one-sided test at a 0.05 significance level was obtained to be 72 under equal-group allocation. In the absence of censoring, Lachin (2000) determined that a total of 72 subjects (36 per group) would have to be recruited to the study. To obtain this same estimate with stpower exponential, we supply the control hazard rate 0.3 as an argument and specify the power(0.9), onesided, and loghazard options to request 90% power, a one-sided test, and sample-size determination for the exponential log hazard-ratio test (or, test for the log-hazard difference), respectively.

```
. stpower exponential 0.3, power(0.9) onesided loghazard
Note: input parameters are hazard rates.

Estimated sample sizes for two-sample comparison of survivor functions
Exponential test, log-hazard difference, conditional
Ho: ln(h2/h1) = 0

Input parameters:

         alpha =    0.0500   (one sided)
            h1 =    0.3000
            h2 =    0.1500
     ln(h2/h1) =   -0.6931
         power =    0.9000
            p1 =    0.5000

Estimated sample sizes:
             N =        72
            N1 =        36
            N2 =        36
```

Further, the study was planned to continue for 6 years with a recruitment period of 4 years. Subjects who did not experience an event by the end of 6 years were censored. For this fixed-duration study with uniform entry (recruitment), the estimate of the sample size increases from 72 to 128. We specify the length of the accrual and the follow-up periods in the aperiod() and fperiod() options, respectively. We also request to display the expected number of events by using the detail option.

```
. stpower exponential 0.3, power(0.9) onesided loghazard aperiod(4) fperiod(2) detail
Note: input parameters are hazard rates.

Estimated sample sizes for two-sample comparison of survivor functions
Exponential test, log-hazard difference, conditional
Ho: ln(h2/h1) = 0

Input parameters:

         alpha =    0.0500   (one sided)
            h1 =    0.3000
            h2 =    0.1500
     ln(h2/h1) =   -0.6931
         power =    0.9000
            p1 =    0.5000

Accrual and follow-up information:
      duration =    6.0000
     follow-up =    2.0000
       accrual =    4.0000   (uniform)

Estimated sample sizes:
             N =       128
            N1 =        64
            N2 =        64

Estimated expected number of events:
         E|Ha =        72         E|Ho =        74
        E1|Ha =        44        E1|Ho =        37
        E2|Ha =        28        E2|Ho =        37
```

Under the alternative hypothesis of H_a: $\ln(\Delta) = -0.6931$, we expect to observe 44 events in the control group and 28 events in the experimental group. A total of 128 subjects (64 per group) is required to be enrolled into the study to observe an expected total of 72 events under the alternative.

See [ST] **stpower exponential** for more examples.

Creating output tables

stpower subcommands offer the `table` and `columns()` options to display results in a table. All tables in the examples below are produced for a default screen width of 79 characters. You may need to resize your Results window if you wish to clearly reproduce these tables.

▷ Example 4: Displaying results in a table with default columns

Continuing example 1, we display results in a table with the default columns.

```
. stpower cox, hratio(3) sd(0.75) power(0.95) alpha(0.001) table
Estimated sample size for Cox PH regression
Wald test, log-hazard metric
Ho: [b1, b2, ..., bp] = [0, b2, ..., bp]
```

Power	N	E	B1	SD	Alpha*
.95	36	36	1.09861	.75	.001

* two sided

Suppose we now believe that only 90% of subjects will experience an event by the end of the study. We account for this by using the `failprob(0.9)` option.

```
. stpower cox, hratio(3) sd(0.75) power(0.95) alpha(0.001) failprob(0.9) table
Estimated sample size for Cox PH regression
Wald test, log-hazard metric
Ho: [b1, b2, ..., bp] = [0, b2, ..., bp]
```

Power	N	E	B1	SD	Alpha*	Pr(E)
.95	40	36	1.09861	.75	.001	.9

* two sided

The specified options determine what table columns are displayed by default. See the documentation entry of each command for the details on the default columns.

◁

▷ Example 5: Producing tables when options contain multiple values

Recall example 2. Suppose that we would like to tabulate values of power for sample-size values of 200, 250, and 300 and for three different values of the survival probability in the experimental group, 0.34, 0.5, and 0.65. To fit the table on a screen of width 79, we specify the `colwidth(7)` option, requesting a width of 7 for all table columns.

```
. stpower logrank 0.2 (0.34 0.5 0.65), n(200 250 300) colwidth(7)
Estimated power for two-sample comparison of survivor functions
Log-rank test, Freedman method
Ho: S1(t) = S2(t)
```

Power	N	N1	N2	E	S1	S2	HR	Alpha*
.6646	200	100	100	146	.2	.34	.6703	.05
.7601	250	125	125	183	.2	.34	.6703	.05
.8318	300	150	150	219	.2	.34	.6703	.05
.995	200	100	100	130	.2	.5	.4307	.05
.9991	250	125	125	163	.2	.5	.4307	.05
.9998	300	150	150	195	.2	.5	.4307	.05
1	200	100	100	115	.2	.65	.2677	.05
1	250	125	125	144	.2	.65	.2677	.05
1	300	150	150	173	.2	.65	.2677	.05

* two sided

Optionally, the width of each column may be changed by specifying respective column widths in colwidth(), as we demonstrate in example 6. Using colwidth(7) is equivalent to colwidth(7 7 7 7 7 7 7 7) in the above.

By default, the results are displayed for all possible combinations of values of n() and the second argument list. If a table is desired instead with results for sequential pairs of values (0.34, 200), (0.5, 250), and (0.65, 300), one can specify the parallel option.

```
. stpower logrank 0.2 (0.34 0.5 0.65), n(200 250 300) colwidth(7) parallel
Estimated power for two-sample comparison of survivor functions
Log-rank test, Freedman method
Ho: S1(t) = S2(t)
```

Power	N	N1	N2	E	S1	S2	HR	Alpha*
.6646	200	100	100	146	.2	.34	.6703	.05
.9991	250	125	125	163	.2	.5	.4307	.05
1	300	150	150	173	.2	.65	.2677	.05

* two sided

If the parallel option is specified, options with multiple values must each contain the same number of values.

◁

▷ Example 6: Customized tables

In example 4, we used the table option to display the default columns. One can construct a customized table by specifying *colnames* within columns(). The columns will be displayed in the same order as they are specified in columns(). Continuing example 3, we display the total and per-group expected number of events under the null and under the alternative hypotheses for two values of the hazard ratio, 0.5 and 0.6 (Lachin 2000, 413). To obtain this table, we also specify the hratio(0.5 0.6) and columns(power ea ea1 ea2 eo eo1 eo2 hr alpha) options and omit the detail option in the earlier syntax. We request that the column widths of the first six columns be as specified in colwidth(7 6 7 7 6 7). Remaining columns will be displayed with the width of 7 (the last specified column width). If the widths of the remaining columns are desired to be unchanged, a missing (.), denoting the default column width of 9, may be specified as the last column width.

```
. stpower exponential 0.3, power(0.9) hratio(0.5 0.6) onesided loghazard
> aperiod(4) fperiod(2) columns(power ea ea1 ea2 eo eo1 eo2 hr alpha)
> colwidth(7 6 7 7 6 7)
Note: input parameters are hazard rates.

Estimated sample sizes for two-sample comparison of survivor functions
Exponential test, log-hazard difference, conditional
Ho: ln(h2/h1) = 0
```

Power	E|Ha	E1|Ha	E2|Ha	E|Ho	E1|Ho	E2|Ho	HR	Alpha*
.9	72	44	28	74	37	37	.5	.05
.9	132	76	56	134	67	67	.6	.05

* one sided

◁

The displayed table values may be saved to a Stata data file by using the saving() option. Also table and columns() in conjunction with noheader and continue may be used to produce tables for various values in options that do not allow specifying a number list. This task may be done by issuing an stpower subcommand repeatedly within a forvalues loop. See example 5 in [ST] **stpower exponential**.

Power curves

Here we demonstrate how to produce a graph of a power curve as a function of sample size.

▷ Example 7: Plotting a simple power curve

Continuing example 2, we plot power for sample-size values in a range from 200 to 400 for this study. We supply the integers in this range with a step size of 1 to the n() option and use the saving() option to save table values in a Stata dataset named mypower.dta.

```
. quietly stpower logrank 0.2, hratio(0.6667) n(200(1)400) saving(mypower)
```

We specify quietly to avoid displaying the resulting table. The values of columns of the table are saved in mypower.dta. We generate the power graph by plotting the values of power and sample size saved in variables power and n:

```
. use mypower

. line power n,
> title("Power vs sample size for the log-rank test")
> ytitle("Power") xtitle("Number of subjects")
> yline(.9, lpattern("-.")) xline(362, lpattern("-."))
> xlabel(200(20)400, grid) text(.915 345 "(.9, 362)")
> ylabel(.65(.05).95, grid)
> note("S1(5) = .2, S2(5) = .342, HR = .6667, alpha = .05 (two sided)")
```

![Power vs sample size for the log-rank test. S1(5) = .2, S2(5) = .342, HR = .6667, alpha = .05 (two-sided). Point marked at (.9, 362).]

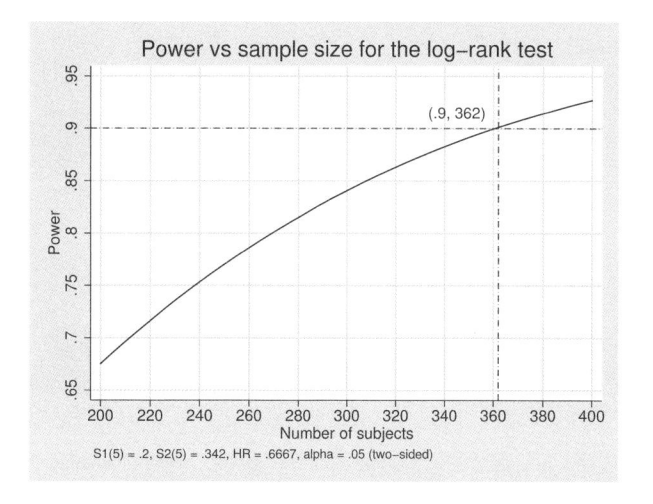

Power curves may also be plotted for combinations of values of different options. For an example of plotting two power curves, see example 7 in [ST] **stpower logrank**.

Methods and formulas

stpower is implemented as an ado-file.

stpower cox adopts the method of Hsieh and Lavori (2000) to compute sample size and power for the test of a covariate obtained after the Cox model fit.

stpower logrank uses the approach of Freedman (1982) and Schoenfeld (1981) for sample-size and power computation. The approach of Schoenfeld (1983) is used to obtain the estimates in the presence of uniform accrual.

stpower exponential implements methods of Lachin (1981); Lachin and Foulkes (1986); George and Desu (1974); and Rubinstein, Gail, and Santner (1981) for the two-sample test of exponential survivor functions. The explicit sample-size formula for the last method was given in Lakatos and Lan (1992).

See *Methods and formulas* in [ST] **stpower cox**, [ST] **stpower logrank**, and [ST] **stpower exponential** for more details.

References

Barthel, F. M.-S., A. Babiker, J. P. Royston, and M. K. B. Parmar. 2006. Evaluation of sample size and power for multi-arm survival trials allowing for non-uniform accrual, non-proportional hazards, loss to follow-up and cross-over. *Statistics in Medicine* 25: 2521–2542.

Barthel, F. M.-S., J. P. Royston, and A. Babiker. 2005. A menu-driven facility for complex sample size calculation in randomized controlled trials with a survival or a binary outcome: Update. *Stata Journal* 5: 123–129.

Chow, S.-C., and J.-P. Liu. 2004. *Design and Analysis of Clinical Trials: Concepts and Methodologies*. 2nd ed. Hoboken, NJ: Wiley.

Chow, S.-C., J. Shao, and H. Wang. 2003. *Sample Size Calculations in Clinical Research*. New York: Marcel Dekker.

Collett, D. 2003. *Modelling Survival Data in Medical Research*. 2nd ed. London: Chapman & Hall/CRC.

Cox, D. R. 1972. Regression models and life-tables (with discussion). *Journal of the Royal Statistical Society, Series B* 34: 187–220.

Cox, D. R., and D. Oakes. 1984. *Analysis of Survival Data*. London: Chapman & Hall/CRC.

Feiveson, A. H. 2002. Power by simulation. *Stata Journal* 2: 107–124.

Freedman, L. S. 1982. Tables of the number of patients required in clinical trials using the logrank test. *Statistics in Medicine* 1: 121–129.

George, S. L., and M. M. Desu. 1974. Planning the size and duration of a clinical trial studying the time to some critical event. *Journal of Chronic Diseases* 27: 15–24.

Hosmer Jr., D. W., S. Lemeshow, and S. May. 2008. *Applied Survival Analysis: Regression Modeling of Time to Event Data*. 2nd ed. New York: Wiley.

Hsieh, F. Y. 1992. Comparing sample size formulae for trials with unbalanced allocation using the logrank test. *Statistics in Medicine* 11: 1091–1098.

Hsieh, F. Y., and P. W. Lavori. 2000. Sample-size calculations for the Cox proportional hazards regression model with nonbinary covariates. *Controlled Clinical Trials* 21: 552–560.

Klein, J. P., and M. L. Moeschberger. 2003. *Survival Analysis: Techniques for Censored and Truncated Data*. 2nd ed. New York: Springer.

Lachin, J. M. 1981. Introduction to sample size determination and power analysis for clinical trials. *Controlled Clinical Trials* 2: 93–113.

——. 2000. *Biostatistical Methods: The Assessment of Relative Risks*. New York: Wiley.

Lachin, J. M., and M. A. Foulkes. 1986. Evaluation of sample size and power for analyses of survival with allowance for nonuniform patient entry, losses to follow-up, noncompliance, and stratification. *Biometrics* 42: 507–519.

Lakatos, E., and K. K. G. Lan. 1992. A comparison of sample size methods for the logrank statistic. *Statistics in Medicine* 11: 179–191.

Lawless, J. F. 2003. *Statistical Models and Methods for Lifetime Data*. 2nd ed. New York: Wiley.

Machin, D. 2004. On the evolution of statistical methods as applied to clinical trials. *Journal of Internal Medicine* 255: 521–528.

Machin, D., and M. J. Campbell. 2005. *Design of Studies for Medical Research*. Chichester, UK: Wiley.

Machin, D., M. J. Campbell, S. B. Tan, and S. H. Tan. 2009. *Sample Size Tables for Clinical Studies*. 3rd ed. Chichester, UK: Wiley–Blackwell.

Marubini, E., and M. G. Valsecchi. 1997. *Analysing Survival Data from Clinical Trials and Observational Studies*. Chichester, UK: Wiley.

Royston, J. P., and A. Babiker. 2002. A menu-driven facility for complex sample size calculation in randomized controlled trials with a survival or a binary outcome. *Stata Journal* 2: 151–163.

Rubinstein, L. V., M. H. Gail, and T. J. Santner. 1981. Planning the duration of a comparative clinical trial with loss to follow-up and a period of continued observation. *Journal of Chronic Diseases* 34: 469–479.

Schoenfeld, D. A. 1981. The asymptotic properties of nonparametric tests for comparing survival distributions. *Biometrika* 68: 316–319.

——. 1983. Sample-size formula for the proportional-hazards regression model. *Biometrics* 39: 499–503.

Schoenfeld, D. A., and J. R. Richter. 1982. Nomograms for calculating the number of patients needed for a clinical trial with survival as an endpoint. *Biometrics* 38: 163–170.

Simon, R., R. D. Radmacher, and K. Dobbin. 2002. Design of studies using DNA microarrays. *Genetic Epidemiology* 23: 21–36.

Væth, M., and E. Skovlund. 2004. A simple approach to power and sample size calculations in logistic regression and Cox regression models. *Statistics in Medicine* 23: 1781–1792.

Wittes, J. 2002. Sample size calculations for randomized control trials. *Epidemiologic Reviews* 24: 39–53.

Also see

[ST] **stpower cox** — Sample size, power, and effect size for the Cox proportional hazards model

[ST] **stpower exponential** — Sample size and power for the exponential test

[ST] **stpower logrank** — Sample size, power, and effect size for the log-rank test

[R] **sampsi** — Sample size and power determination

[ST] **Glossary**

Title

stpower cox — Sample size, power, and effect size for the Cox proportional hazards model

Syntax

Sample-size determination

    stpower cox [*coef*] [, *options*]

Power determination

    stpower cox [*coef*] , n(*numlist*) [*options*]

Effect-size determination

    stpower cox , n(*numlist*) { power(*numlist*) | beta(*numlist*) } [*options*]

where *coef* is the regression coefficient (effect size) of a covariate of interest, in a Cox proportional hazards model, desired to be detected by a test with a prespecified power. *coef* may be specified either as one number or as a list of values (see [U] **11.1.8 numlist**) enclosed in parentheses.

options	description
Main	
*alpha(*numlist*)	significance level; default is alpha(0.05)
*power(*numlist*)	power; default is power(0.8)
*beta(*numlist*)	probability of type II error; default is beta(0.2)
*n(*numlist*)	sample size; required to compute power or effect size
*hratio(*numlist*)	hazard ratio (effect size) associated with a one-unit increase in covariate of interest; default is hratio(0.5)
onesided	one-sided test; default is two sided
sd(*#*)	standard deviation of covariate of interest; default is sd(0.5)
r2(*#*)	squared coefficient of multiple correlation with other covariates; default is r2(0)
failprob(*#*)	probability of an event (failure) of interest; default is failprob(1), meaning no censoring
wdprob(*#*)	the proportion of subjects anticipated to withdraw from the study; default is wdprob(0)
parallel	treat number lists in starred options as parallel (do not enumerate all possible combinations of values) when multiple values per option are specified

Reporting

hr	report hazard ratio, not coefficient
table	display results in a table with default columns
columns(*colnames*)	display results in a table with specified *colnames* columns
notitle	suppress table title
nolegend	suppress table legend
colwidth(# [# ...])	column widths; default is colwidth(9)
separator(#)	draw a horizontal separator line every # lines; default is separator(0) meaning no separator lines
saving(*filename* [, replace])	save the table data to *filename*; use replace to overwrite existing *filename*
† noheader	suppress table header; seldom used
† continue	draw a continuation border in the table output; seldom used

*Starred options may be specified either as one number or as a list of values (see [U] **11.1.8 numlist**).
† noheader and continue are not shown in the dialog box.

colnames	description
alpha	significance level
power	power
beta	type II error probability
n	total number of subjects
e	total number of events (failures)
hr	hazard ratio
coef	coefficient (log hazard-ratio)
sd	standard deviation
r2	squared multiple-correlation coefficient
pr	overall probability of an event (failure)
w	proportion of withdrawals

By default, the following *colnames* are displayed:
  power, n, e, sd, and alpha are always displayed;
  coef is displayed, unless the hr option is specified, in which case hr is displayed;
  pr if overall probability of an event (failprob()) is specified;
  r2 if squared multiple-correlation coefficient (r2()) is specified; and
  w if withdrawal proportion (wdprob()) is specified.

Menu

Statistics > Survival analysis > Power and sample size > Cox proportional hazards model

Description

stpower cox estimates required sample size, power, and effect size for survival analyses that use Cox proportional hazards (PH) models. It also reports the number of events (failures) required to be observed in a study. The estimates of sample size or power are obtained for the test of the effect of one covariate, x_1 (binary or continuous), on time to failure adjusted for other predictors, $x_2, \ldots, x_p$, in a PH model. The command provides options to account for possible correlation between a covariate of interest and other predictors and for withdrawal of subjects from the study. Optionally, the minimal effect size (minimal detectable difference in a regression coefficient, β_1, or hazard ratio) may be obtained for given sample size and power.

You can use stpower cox to

- calculate required number of events and sample size when you know power and effect size expressed as a hazard ratio or a coefficient (log hazard-ratio),
- calculate power when you know sample size (number of events) and effect size expressed as a hazard ratio or a coefficient (log hazard-ratio), and
- calculate effect size and display it as a coefficient (log hazard-ratio) or a hazard ratio when you know sample size (number of events) and power.

stpower cox's input parameter, *coef*, is the value β_{1a} of the regression coefficient, β_1, of a covariate of interest, x_1, from a Cox PH model, which is desired to be detected by a test with prespecified power.

Options

Main

alpha(*numlist*) sets the significance level of the test. The default is alpha(0.05).

power(*numlist*) sets the power of the test. The default is power(0.8). If beta() is specified, this value is set to be 1−beta(). Only one of power() or beta() may be specified.

beta(*numlist*) sets the probability of a type II error of the test. The default is beta(0.2). If power() is specified, this value is set to be 1−power(). Only one of beta() or power() may be specified.

n(*numlist*) specifies the number of subjects in the study to be used to compute the power of the test or the minimal effect size (minimal detectable value of the regression coefficient, β_1, or hazard ratio) if power() or beta() is also specified.

hratio(*numlist*) specifies the hazard ratio associated with a one-unit increase in the covariate of interest, x_1, when other covariates are held constant. The default is hratio(0.5). This value defines the minimal clinically significant effect of a covariate on the response to be detected by a test with a certain power, specified in power(), in a Cox PH model. If *coef* is specified, hratio() is not allowed and the hazard ratio is instead computed as $\exp(coef)$.

onesided indicates a one-sided test. The default is two sided.

sd(*#*) specifies the standard deviation of the covariate of interest, x_1. The default is sd(0.5).

r2(*#*) specifies the squared multiple-correlation coefficient between x_1 and other predictors $x_2, \ldots, x_p$ in a Cox PH model. The default is r2(0), meaning that x_1 is independent of other covariates. This option defines the proportion of variance explained by the regression of x_1 on $x_2, \ldots, x_p$ (see [R] **regress**).

failprob(*#*) specifies the overall probability of a subject failing (or experiencing an event of interest, or not being censored) in the study. The default is failprob(1), meaning that all subjects experience an event (or fail) in the study; that is, no censoring of subjects occurs.

wdprob(*#*) specifies the proportion of subjects anticipated to withdraw from a study. The default is wdprob(0). wdprob() may not be combined with n().

parallel reports results sequentially (in parallel) over the list of numbers supplied to options allowing *numlist*. By default, results are computed over all combinations of the number lists in the following order of nesting: alpha(), hratio() or list of coefficients *coef*, power() or beta(), and n(). This option requires that options with multiple values each contain the same number of elements.

Reporting

hr specifies that the hazard ratio be displayed rather than the regression coefficient. This option affects how results are displayed and not how they are estimated.

table displays results in a tabular format and is implied if any number list contains more than one element. This option is useful if you are producing results one case at a time and wish to construct your own custom table by using a forvalues loop.

columns(*colnames*) specifies results in a table with specified *colnames* columns. The order of columns in the output table is the same as the order of *colnames* specified in columns(). Column names in columns() must be space-separated.

notitle prevents the table title from displaying.

nolegend prevents the table legend from displaying and column headers from being marked.

colwidth(# [# ...]) specifies column widths. The default is 9 for all columns. The number of specified values may not exceed the number of columns in the table. A missing value (.) may be specified for any column to indicate the default width (9). If fewer widths are specified than the number of columns in the table, the last width specified is used for the remaining columns.

separator(#) specifies how often separator lines should be drawn between rows of the table. The default is separator(0), meaning that no separator lines should be displayed.

saving(*filename* [, replace]) creates a Stata data file (.dta file) containing the table values with variable names corresponding to the displayed *colnames*. replace indicates that *filename* be overwritten, if it exists. saving() is appropriate only with tabular output.

The following options are available with stpower cox but are not shown in the dialog box:

noheader prevents the table header from displaying. This option is useful when the command is issued repeatedly, such as within a loop. noheader implies notitle.

continue draws a continuation border at the bottom of the table. This option is useful when the command is issued repeatedly within a loop.

Remarks

Remarks are presented under the following headings:

> *Introduction*
> *Computing sample size in the absence of censoring*
> *Computing sample size in the presence of censoring*
> *Link to the sample-size and power computation for the log-rank test*
> *Power and effect-size determination*
> *Performing the analysis with the Cox PH model*

Introduction

Consider a survival study for which the goal is to investigate the effect of a covariate of interest, x_1, on time to failure, possibly adjusted for other predictors, $x_2, \ldots, x_p$, using the Cox proportional hazards model (Cox 1972). For continuous x_1, the effect is measured as a hazard ratio, Δ, or a log hazard-ratio, $\ln(\Delta)$, associated with a one-unit increase in x_1 when the other covariates, $x_2, \ldots, x_p$, are held constant. For a binary predictor, the effect is a ratio of hazards or log hazards corresponding to the two categories of x_1 when other covariates are held constant. In both cases, to measure the effect of a covariate, a test of a hazard or log hazard-ratio is performed.

In a Cox PH model, the hazard function is assumed to be

$$h(t) = h_0(t) \exp(\beta_1 x_1 + \ldots + \beta_p x_p)$$

where no distributional assumption is made about the baseline hazard, $h_0(t)$. Under this assumption, the regression coefficient, β_1, is the log hazard-ratio, $\ln(\Delta)$, associated with a one-unit increase in x_1 when the other predictors are held constant, and the exponentiated regression coefficient, $\exp(\beta_1)$, is the hazard ratio, Δ. Therefore, the effect of x_1 on time to failure can be investigated by performing an appropriate test based on the partial likelihood (Hosmer Jr., Lemeshow, and May 2008; Klein and Moeschberger 2003) for the regression coefficient, β_1, from a Cox model. Negative values of β_1 correspond to the reduction in hazard for a one-unit increase in x_1, and, conversely, positive values correspond to the increase in hazard for a one-unit increase in x_1.

stpower cox provides the estimates of sample size or power for a test of the regression coefficient, β_1, with the null hypothesis $H_0: (\beta_1, \beta_2, \ldots, \beta_p) = (0, \beta_2, \ldots, \beta_p)$ against the alternative $H_a: (\beta_1, \beta_2, \ldots, \beta_p) = (\beta_{1a}, \beta_2, \ldots, \beta_p)$. The methods used are derived for the score test of H_0 versus H_a. In practice, however, the obtained results may be used in the context of the Wald test as well because the two tests usually lead to the same conclusions about the significance of the regression coefficient. Refer to section *The conditional versus unconditional approaches* in [ST] **stpower exponential** for more details about the results based on conditional and unconditional tests. From now on, we will refer to H_a as $H_a: \beta_1 = \beta_{1a}$ for simplicity.

stpower cox implements the method of Hsieh and Lavori (2000) for the sample-size and power computation, which reduces to the method of Schoenfeld (1983) for a binary covariate. The sample size is related to the power of a test through the number of events observed in the study; that is, for a fixed number of events, the power of a test is independent of the sample size. As a result, the sample size is estimated as the number of events divided by the overall probability of a subject failing in the study.

The argument *coef* or `hratio()` may be used to specify the effect size desired to be detected by a test. If argument *coef* is omitted, the value of the log of the hazard ratio specified in option `hratio()` or the log of the default hazard-ratio value of 0.5 is used to compute β_{1a}. If argument *coef* is specified, then `hratio()` is not allowed and the hazard ratio is computed as $\exp(coef)$.

If power determination is desired, then sample size `n()` must be specified. Otherwise, sample-size determination is assumed with `power(0.8)` (or, equivalently, `beta(0.2)`). The default setting for power or, alternatively, the probability of a *type II error*, a failure to reject the null hypothesis when the alternative hypothesis is true, may be changed by using `power()` or `beta()`, respectively. If both `n()` and `power()` (or `beta()`) are specified, then the value of the regression coefficient, β_{1a} (or hazard ratio if the `hr` option is specified), which can be detected by a test with requested `power()` for fixed sample size `n()`, is computed.

The default probability of a *type I error*, a rejection of the null hypothesis when the null hypothesis is true, of a test is 0.05 but may be changed by using the `alpha()` option. One-sided tests may be requested by using `onesided`. By default, no censoring, no correlation between x_1 and other predictors, and no withdrawal of subjects from the study are assumed. This may be changed by specifying `failprob()`, `r2()`, and `wdprob()`, respectively.

Optionally, the results may be displayed in a table by using `table` or `columns()`, as demonstrated in [ST] **stpower**. Refer to [ST] **stpower** and example 7 in *Power and effect-size determination* of [ST] **stpower logrank** to see how to obtain a graph of a power curve.

Computing sample size in the absence of censoring

First, consider a *type I study* in which all subjects fail by the end of the study (no censoring). Then the required sample size is the same as the number of events required to be observed in a study.

▷ Example 1: Sample size for a model with a binary covariate of interest

Consider a survival study for which the goal is to investigate the effect of a treatment on survival times of subjects. The covariate of interest is binary with levels defining whether a subject receives the treatment (the experimental group) or a placebo (the control or placebo group). Prior to conducting the study, investigators need an estimate of the sample size that ensures that a ratio of hazards of the experimental group to the control group of 0.5 ($\beta_{1a} = \ln(0.5) = -0.6931$) can be detected with a power of 80% with a two-sided, 0.05-level test. Under 1:1 randomization, a subject has a 50% chance of receiving the treatment. The corresponding binary covariate follows a Bernoulli distribution with the probability of a subject receiving a treatment, p, equal to 0.5. As such, the standard deviation of the covariate is $\{p(1-p)\}^{1/2} = 0.5$. Because these study parameters correspond to default values of `stpower cox`, to obtain the sample size for the above study we simply type

```
. stpower cox
Estimated sample size for Cox PH regression
Wald test, log-hazard metric
Ho: [b1, b2, ..., bp] = [0, b2, ..., bp]
Input parameters:
        alpha =    0.0500  (two sided)
           b1 =   -0.6931
           sd =    0.5000
        power =    0.8000
Estimated number of events and sample size:
            E =        66
            N =        66
```

Recall that if argument *coef* is omitted, a default value of $\ln(0.5) = -0.6931$ is assumed. From the output, we see that 66 events (failures) are required to be observed in the study to ensure a power of 80% to detect an alternative $H_a: \beta_1 = -0.6931$ using a two-sided test with a 0.05 significance level. Because we have no censoring, a total of 66 subjects is needed in the study to observe 66 events.

One can also request that the results be displayed in the hazard metric by specifying the `hr` option:

```
. stpower cox, hr
Estimated sample size for Cox PH regression
Wald test, hazard metric
Ho: [b1, b2, ..., bp] = [0, b2, ..., bp]
Input parameters:
        alpha =    0.0500  (two sided)
       hratio =    0.5000
           sd =    0.5000
        power =    0.8000
Estimated number of events and sample size:
            E =        66
            N =        66
```

◁

Suppose now that the covariate of interest, x_1, is continuous. Hsieh and Lavori (2000) extend the formula of Schoenfeld (1983) for the number of events to the case when a covariate is continuous. They also relax the assumption of Schoenfeld (1983) about the independence of x_1 of other covariates and provide an adjustment to the estimate of the number of events for possible correlation.

▷ Example 2: Sample size for a model with a continuous covariate of interest

Consider an example from Hsieh and Lavori (2000) of a study of multiple-myeloma patients treated with alkylating agents (Krall, Uthoff, and Harley 1975). Although in the original study of multiple-myeloma patients, 17 of a total of 65 patients are censored; here we assume that all patients die by the end of the study (a type I study, no censoring). Suppose that the covariate of interest, x_1, is the log of the amount of blood urea nitrogen (BUN) measured in a patient. The sample size for a one-sided, 0.05-level test to detect a coefficient (log hazard-ratio) of 1 for a unit increase in x_1 with a power of 80% is required. The standard deviation of x_1 is 0.3126. To obtain an estimate of the sample size, we supply *coef*, 1, as an argument, the sd(0.3126) option, and the onesided option for a one-sided test.

```
. stpower cox 1, sd(0.3126) onesided
Estimated sample size for Cox PH regression
Wald test, log-hazard metric
Ho: [b1, b2, ..., bp] = [0, b2, ..., bp]

Input parameters:
        alpha =    0.0500  (one sided)
           b1 =    1.0000
           sd =    0.3126
        power =    0.8000

Estimated number of events and sample size:
            E =        64
            N =        64
```

The estimate of the required number of events and the sample size is 64.

◁

Based on the derivation in Schoenfeld (1983) and Hsieh and Lavori (2000), sample-size estimates in the above examples may be used if other covariates are also present in the model as long as these covariates are independent of the covariate of interest. The independence assumption holds for randomized studies, but it is not true for nonrandomized studies, often encountered in practice. Also, in many studies, the main covariate of interest will often be correlated with other covariates. For example, age and gender will often be confounded with the covariate of interest, such as smoking. Below we investigate the effect of the confounding factor on the estimate of the required number of events.

▷ Example 3: Sample size when covariates are not independent

Continuing example 2, the effect of a covariate BUN is desired to be adjusted for eight other covariates in the model. Hsieh and Lavori (2000) report the coefficient of determination of $R^2 = 0.1837$ from regression of the log of BUN, x_1, on the eight other covariates.

```
. stpower cox 1, sd(0.3126) onesided r2(0.1837)
Estimated sample size for Cox PH regression
Wald test, log-hazard metric
Ho: [b1, b2, ..., bp] = [0, b2, ..., bp]

Input parameters:

        alpha =   0.0500  (one sided)
           b1 =   1.0000
           sd =   0.3126
        power =   0.8000
           R2 =   0.1837

Estimated number of events and sample size:
            E =        78
            N =        78
```

The number of events required to be observed in a study and, respectively, the number of subjects increase from 64 to 78 after adjusting for the inflation of the variance of the estimate of β_1 because of the correlation with other covariates. The variance of x_1 decreases by the factor $1 - R^2$, so the estimate of the number of events must also be adjusted by a variance inflation factor $\text{VIF} = 1/(1 - R^2)$.

◁

Computing sample size in the presence of censoring

In the previous section, we assumed that all subjects fail by the end of the study. In practice, the study often terminates after a fixed time, T. As a result, some subjects may not experience an event by the end of the study (a *type II study*). These subjects are censored. To obtain an estimate of the sample size in the presence of censoring, an estimate of the overall probability of a subject not being censored is required. The investigator may already have such an estimate from previous studies, or this probability may be computed as suggested in the literature (Schoenfeld 1983; Lachin and Foulkes 1986; Barthel et al. 2006; Barthel, Royston, and Babiker 2005; also see [ST] **stpower logrank** and [ST] **stpower exponential**).

▷ Example 4: Sample size in the presence of censoring

Consider the study from example 2. In reality, as mentioned earlier, 17 of a total of 65 patients survived until the end of the study. The overall death rate is estimated as $1 - 17/65 = 0.738$.

```
. stpower cox 1, sd(0.3126) onesided failprob(0.738)
Estimated sample size for Cox PH regression
Wald test, log-hazard metric
Ho: [b1, b2, ..., bp] = [0, b2, ..., bp]

Input parameters:

        alpha =   0.0500  (one sided)
           b1 =   1.0000
           sd =   0.3126
        power =   0.8000
    Pr(event) =   0.7380

Estimated number of events and sample size:
            E =        64
            N =        86
```

In the presence of censoring, the number of subjects required in the study increases from 64 to 86. The number of events remains the same (64) because the only change in the study is the presence of censoring, and censoring is assumed to be independent of failure (event) times.

If we also adjust for the correlation between the log of BUN and other covariates, we obtain the estimate of the sample size to be 106:

```
. stpower cox, hratio(2.7182) sd(0.3126) onesided r2(0.1837) failprob(0.738)
Estimated sample size for Cox PH regression
Wald test, log-hazard metric
Ho: [b1, b2, ..., bp] = [0, b2, ..., bp]

Input parameters:
            alpha =    0.0500   (one sided)
               b1 =    1.0000
               sd =    0.3126
            power =    0.8000
        Pr(event) =    0.7380
               R2 =    0.1837

Estimated number of events and sample size:
                E =        78
                N =       106
```

In the above example, we also demonstrate the alternative syntax. Rather than supplying the coefficient (log hazard-ratio) of 1, we use the `hratio()` option to specify the size of the effect expressed as the hazard ratio $\exp(1) = 2.7182$.

◁

❑ Technical note

Supplying the coefficient (log hazard-ratio) of 1 or -1 (or, respectively, a hazard ratio of $\exp(1) = 2.7182$ or $\exp(-1) = 1/2.7182 = 0.36788$) is irrelevant for sample-size and power determination because it results in the same estimates of sample size and power. However, the sign of the coefficient (or the value of the hazard ratio being larger or smaller than one) is important at the analysis stage because it determines the direction of the effect associated with a one-unit increase of a covariate value.

❑

Often, in practice, subjects may withdraw from a study before it terminates. As a result, the information about the subjects' response is lost. The proportion of subjects anticipated to withdraw from a study may be specified by using `wdprob()`. Refer to [ST] **stpower** and *Withdrawal of subjects from the study* in [ST] **stpower logrank** for a more detailed description and an example.

Link to the sample-size and power computation for the log-rank test

When there are no other covariates in a Cox regression model, the score test of the regression coefficient of a binary covariate is the same as the log-rank test (in the absence of tied observations). Powers of the two tests are the same and therefore so are the formulas for the number of events. The formula for the total number of events for a test of a binary covariate in the context of a PH model in the presence of other covariates is derived in Schoenfeld (1983) under the assumption that the covariate of interest is independent of the other covariates. This formula is also the same as the formula for the number of events when the log-rank test is used to compare survivor functions of two groups without covariates (Schoenfeld 1981). Indeed, using `stpower logrank` for the study described in example 1,

```
. stpower logrank, schoenfeld
Estimated sample sizes for two-sample comparison of survivor functions
Log-rank test, Schoenfeld method
Ho: S1(t) = S2(t)

Input parameters:

        alpha =    0.0500   (two sided)
    ln(hratio) =  -0.6931
        power =    0.8000
           p1 =    0.5000

Estimated number of events and sample sizes:
            E =       66
            N =       66
           N1 =       33
           N2 =       33
```

yields the same estimates of 66 for both the required number of events and the required sample size.

Schoenfeld (1983) notes that although the formulas for the number of events are the same for the two approaches (based on the log-rank test and based on the score test of a regression coefficient from a Cox regression model adjusting for other covariates), the powers are different. Suppose that the two groups defined by a binary covariate follow the PH model. Then, if covariates are ignored, the ratio of hazards will be nonproportional at every time t and the power of the log-rank test will be smaller than the power of the test based on a Cox PH model.

If a covariate of interest is binary, either stpower cox or stpower logrank with the schoenfeld option can be used to obtain the estimate of the sample size or power regardless of the presence of other covariates. However, if covariates are present, it is important to use the appropriate test that adjusts for other covariates when analyzing the data.

Væth and Skovlund (2004) demonstrate that for a continuous covariate, the sample-size formula for the log-rank test (assuming the equal-group allocation) may be used to obtain the sample size or power with the value of the hazard ratio equal to $\exp(2\beta_{1a}\sigma)$. By typing this expression into the sample-size formula for the log-rank test, one obtains the formula derived in Hsieh and Lavori (2000).

For example, we obtain the same estimate of the total number of events as computed in example 2 using stpower logrank with the schoenfeld option and with the value of the hazard ratio equal to $\exp(2\beta_{1a}\sigma) = \exp(2 \times 1 \times 0.3126) = 1.8686$.

```
. stpower log, hratio(1.8686) onesided schoenfeld
Estimated sample sizes for two-sample comparison of survivor functions
Log-rank test, Schoenfeld method
Ho: S1(t) = S2(t)

Input parameters:

        alpha =    0.0500   (one sided)
    ln(hratio) =   0.6252
        power =    0.8000
           p1 =    0.5000

Estimated number of events and sample sizes:
            E =       64
            N =       64
           N1 =       32
           N2 =       32
```

Power and effect-size determination

Suppose that, for some reason, the number of subjects required to ensure a certain power of a test to detect a specified effect size is not achieved by the end of the recruitment phase of a study. Investigators may want to know by how much the power of a test is decreased for the obtained sample size. If the decrease is significant, then what is the minimal effect size that can be detected with an acceptable level of power for this sample size?

▷ Example 5: Power determination

Consider the data of Krall, Uthoff, and Harley (1975) from the study described in example 2. Suppose that we want to test the effect of the log of BUN on patients' survival times adjusted for eight other covariates. In example 4, the required number of patients is estimated to be 106 to ensure a power of 80% of a 0.05 one-sided test to detect a value of 1 in the regression coefficient. This study, however, had only 65 patients. How does this reduction in sample size affect the power of the test to detect the alternative H_a: $\beta_1 = 1$?

```
. stpower cox 1, sd(0.3126) onesided r2(0.1837) failprob(0.738) n(65)
Estimated power for Cox PH regression
Wald test, log-hazard metric
Ho: [b1, b2, ..., bp] = [0, b2, ..., bp]

Input parameters:

            alpha =    0.0500  (one sided)
               b1 =    1.0000
               sd =    0.3126
                N =        65
        Pr(event) =    0.7380
               R2 =    0.1837

Estimated number of events and power:
                E =        48
            power =    0.6222
```

When the sample size decreases from 106 to 65, power decreases from 80% to 62%.

◁

▷ Example 6: Effect-size determination

Continuing the above example: if a power of 62% is unacceptable to investigators, they may want to find out what is the smallest value of the regression coefficient that can be detected with a preserved power of 80%. To obtain this estimate, we specify both the n() and power() options.

```
. stpower cox, sd(0.3126) onesided r2(0.1837) failprob(0.738) n(65) power(0.8)
Estimated coefficient for Cox PH regression
Wald test, log-hazard metric
Ho: [b1, b2, ..., bp] = [0, b2, ..., bp]

Input parameters:

            alpha =    0.0500  (one sided)
               sd =    0.3126
                N =        65
            power =    0.8000
        Pr(event) =    0.7380
               R2 =    0.1837

Estimated number of events and coefficient:
                E =        48
               b1 =   -1.2711
```

Stata reports the estimate of the regression coefficient of −1.2711. We can disregard the sign because, as mentioned earlier, it is irrelevant in the context of sample-size or power determination. Refer to *Methods and formulas* for details.

With only 65 subjects, the smallest change in log hazards for a one-unit increase in the log of BUN, which can be detected with a preserved 80% power, is roughly 1.27, corresponding to a 27% increase in the log hazard-ratio of 1 desired to be detected originally in example 4.

◁

Performing the analysis with the Cox PH model

After the data are collected, one can use `stcox` and `test` to fit the Cox PH model and perform a Wald test, as we demonstrate below.

▷ Example 7: Performing a Wald test

We demonstrate how to perform a Wald test for the regression coefficient of the log of BUN from a Cox model using the data from Krall, Uthoff, and Harley (1975) described in example 2. The dataset `myeloma.dta` consists of 11 variables, described below.

```
. use http://www.stata-press.com/data/r11/myeloma
(Multiple myeloma patients)

. describe

Contains data from http://www.stata-press.com/data/r11/myeloma.dta
  obs:            65                          Multiple myeloma patients
 vars:            11                          11 Feb 2009 19:26
 size:         1,950 (99.9% of memory free)

              storage  display     value
variable name   type   format      label       variable label

time            float  %9.0g                   Survival time from diagnosis to
                                                 nearest month + 1
died            byte   %9.0g                   0 - Alive, 1 - Dead
lnbun           float  %9.0g                   log BUN at diagnosis
hemo            float  %9.0g                   Hemoglobin at diagnosis
platelet        byte   %9.0g       normal      Platelets at diagnosis
age             byte   %9.0g                   Age (complete years)
lnwbc           float  %9.0g                   Log WBC at diagnosis
fracture        byte   %9.0g       present     Fractures at diagnosis
lnbm            float  %9.0g                   log % of plasma cells in bone
                                                 marrow
protein         byte   %9.0g                   Proteinuria at diagnosis
scalcium        byte   %9.0g                   Serum calcium (mgm%)

Sorted by:
```

Prior to using `stcox` to fit a Cox model, we need to set up the data by using `stset` (see [ST] **stset**). The analysis-time variable is `time` and the failure variable is `died`.

```
. stset time, failure(died)

     failure event:  died != 0 & died < .
obs. time interval:  (0, time]
 exit on or before:  failure

      65  total obs.
       0  exclusions

      65  obs. remaining, representing
      48  failures in single record/single failure data
  1560.5  total analysis time at risk, at risk from t =          0
                                earliest observed entry t =      0
                                 last observed exit t =         92
```

We include all nine covariates in the model and perform a fit by using `stcox`. Then we perform a Wald test of $H_0: \beta_1 = 1$ for the coefficient of `lnbun` using `test`.

```
. stcox lnbun hemo platelet age lnwbc fracture lnbm protein scalcium, nohr
          failure _d:  died
    analysis time _t:  time

Iteration 0:    log likelihood = -154.85799
Iteration 1:    log likelihood = -146.68114
Iteration 2:    log likelihood = -146.29446
Iteration 3:    log likelihood = -146.29404
Refining estimates:
Iteration 0:    log likelihood = -146.29404

Cox regression -- Breslow method for ties

No. of subjects =       65                     Number of obs   =        65
No. of failures =       48
Time at risk    =   1560.5
                                               LR chi2(9)      =     17.13
Log likelihood  =  -146.29404                  Prob > chi2     =    0.0468
```

_t	Coef.	Std. Err.	z	P>|z|	[95% Conf. Interval]	
lnbun	1.798354	.6483293	2.77	0.006	.5276519	3.069056
hemo	-.1263119	.0718333	-1.76	0.079	-.2671026	.0144789
platelet	-.2505915	.5074656	-0.49	0.621	-1.245206	.7440228
age	-.0127949	.019475	-0.66	0.511	-.0509653	.0253755
lnwbc	.3537259	.7131935	0.50	0.620	-1.044108	1.75156
fracture	.3378767	.4072774	0.83	0.407	-.4603722	1.136126
lnbm	.3589346	.4860298	0.74	0.460	-.5936663	1.311535
protein	.0130672	.0261696	0.50	0.618	-.0382243	.0643587
scalcium	.1259479	.1034015	1.22	0.223	-.0767153	.3286112

```
. test lnbun = 1

 ( 1)  lnbun = 1

           chi2(  1) =     1.52
         Prob > chi2 =    0.2182
```

By default, `stcox` reports estimates of hazard ratios and the two-sided tests of the equality of a coefficient to zero. We use the `nohr` option to request estimates of coefficients. From the output table, a one-sided test of $H_0: \beta_1 = 0$ versus $H_a: \beta_1 > 0$ is rejected at a 0.05 level (one-sided p-value is $0.006/2 = 0.003 < 0.05$). The estimate of the log-hazard difference associated with a one-unit increase of `lnbun` is $\widehat{\beta}_1 = 1.8$. From the `test` output, we cannot reject the hypothesis of $H_0: \beta_1 = 1$.

For these data, the observed effect size (coefficient) of 1.8 is large enough for the sample size of 65 to be sufficient to reject the null hypothesis of no effect of the BUN on the survival of subjects

($H_0: \beta_1 = 0$). However, if the goal of the study were to ensure that the test detects the effect size corresponding to the coefficient of at least 1 with 80% power, a sample of approximately 106 subjects would have been required.

◁

Saved results

stpower cox saves the following in r():

Scalars
- r(N)   total number of subjects
- r(E)   total number of events (failures)
- r(power)   power of test
- r(alpha)   significance level of test
- r(hratio)   hazard ratio
- r(onesided)   1 if one-sided test, 0 otherwise
- r(sd)   standard deviation
- r(Pr_E)   probability of an event (failure), if specified
- r(r2)   squared multiple correlation, if specified
- r(w)   proportion of withdrawals, if specified

Macros
- r(metric)   displayed metric (log-hazard or hazard)

Methods and formulas

stpower cox is implemented as an ado-file.

Let β_1 denote the regression coefficient of the covariate of interest, x_1, from a Cox PH model, possibly in the presence of other covariates, $x_2, \ldots, x_p$; and let Δ denote the hazard ratio associated with a one-unit increase of x_1 when other covariates are held constant. Under the PH model, $\beta_1 = \ln(\Delta)$, where $\ln(\Delta)$ is the change in log hazards associated with a one-unit increase in x_1 when other covariates are held constant.

Define E and N to be the total number of events (failures) and the total number of subjects required in the study; σ to be the standard deviation of x_1; p_E to be the overall probability of an event (failure); R^2 to be the proportion of variance explained by the regression of x_1 on $x_2, \ldots, x_p$ (or squared multiple-correlation coefficient); w to be the proportion of subjects withdrawn from a study (lost to follow-up); α to be the significance level; β to be the probability of a type II error; and $z_{(1-\alpha/k)}$ and $z_{(1-\beta)}$ to be the $(1-\alpha/k)$th and the $(1-\beta)$th quantiles of the standard normal distribution, with $k = 1$ for the one-sided test and $k = 2$ for the two-sided test.

The total number of events required to be observed in a study to ensure a power of $1 - \beta$ of a test to detect the regression coefficient, β_1, with a significance level α, according to Hsieh and Lavori (2000), is

$$E = \frac{(z_{1-\alpha/k} + z_{1-\beta})^2}{\sigma^2 \beta_1^2 (1 - R^2)}$$

For the case of randomized study and a binary covariate x_1, this formula was derived in Schoenfeld (1983). The formula is an approximation and relies on a set of assumptions such as distinct failure times, all subjects completing the course of the study (no withdrawal), and a local alternative under which $\ln(\Delta)$ is assumed to be of order $O(N^{-1/2})$. The formula is derived for the score test but may be applied to other tests (Wald, for example) that are based on the partial likelihood of a Cox model because all these tests are asymptotically equivalent (Schoenfeld 1983; Hosmer Jr., Lemeshow, and May 2008; Klein and Moeschberger 2003).

The total sample size required to observe the total number of events, E, is given by

$$N = \frac{E}{p_E}$$

The estimate of the sample size is rounded up to the nearest integer.

To account for a proportion of subjects, w, withdrawn from a study, a conservative adjustment to the total sample size suggested in the literature (Freedman 1982; Machin and Campbell 2005) is applied as follows:

$$N_w = \frac{N}{1 - w}$$

Withdrawal is assumed to be independent of administrative censoring and failure (event) times.

Power is estimated using the formula

$$1 - \beta = \Phi\left[|\beta_1|\sigma\{Np_E(1 - R^2)\}^{1/2} - z_{1-\alpha/k}\right]$$

where $\Phi(\cdot)$ is the standard normal cumulative distribution function.

The estimate of the regression coefficient for a fixed power, $1 - \beta$, and a sample size, N, is computed as

$$\beta_1^2 = \frac{(z_{1-\alpha/k} + z_{1-\beta})^2}{\sigma^2 N p_E(1 - R^2)}$$

Either of the two values $|\beta_1|$ and $-|\beta_1|$ satisfy the above equation. `stpower cox` reports the negative of the two values, which corresponds to the reduction in a hazard of a failure for a one-unit increase in x_1. Similarly, if the `hr` option is used, the corresponding value of the hazard ratio less than 1 is reported to reflect the reduction in hazard for a one-unit increase in x_1.

References

Barthel, F. M.-S., A. Babiker, J. P. Royston, and M. K. B. Parmar. 2006. Evaluation of sample size and power for multi-arm survival trials allowing for non-uniform accrual, non-proportional hazards, loss to follow-up and cross-over. *Statistics in Medicine* 25: 2521–2542.

Barthel, F. M.-S., J. P. Royston, and A. Babiker. 2005. A menu-driven facility for complex sample size calculation in randomized controlled trials with a survival or a binary outcome: Update. *Stata Journal* 5: 123–129.

Cox, D. R. 1972. Regression models and life-tables (with discussion). *Journal of the Royal Statistical Society, Series B* 34: 187–220.

Freedman, L. S. 1982. Tables of the number of patients required in clinical trials using the logrank test. *Statistics in Medicine* 1: 121–129.

Hosmer Jr., D. W., S. Lemeshow, and S. May. 2008. *Applied Survival Analysis: Regression Modeling of Time to Event Data*. 2nd ed. New York: Wiley.

Hsieh, F. Y., and P. W. Lavori. 2000. Sample-size calculations for the Cox proportional hazards regression model with nonbinary covariates. *Controlled Clinical Trials* 21: 552–560.

Klein, J. P., and M. L. Moeschberger. 2003. *Survival Analysis: Techniques for Censored and Truncated Data*. 2nd ed. New York: Springer.

Krall, J. M., V. A. Uthoff, and J. B. Harley. 1975. A step-up procedure for selecting variables associated with survival. *Biometrics* 31: 49–57.

Lachin, J. M., and M. A. Foulkes. 1986. Evaluation of sample size and power for analyses of survival with allowance for nonuniform patient entry, losses to follow-up, noncompliance, and stratification. *Biometrics* 42: 507–519.

Machin, D., and M. J. Campbell. 2005. *Design of Studies for Medical Research*. Chichester, UK: Wiley.

Schoenfeld, D. A. 1981. The asymptotic properties of nonparametric tests for comparing survival distributions. *Biometrika* 68: 316–319.

———. 1983. Sample-size formula for the proportional-hazards regression model. *Biometrics* 39: 499–503.

Væth, M., and E. Skovlund. 2004. A simple approach to power and sample size calculations in logistic regression and Cox regression models. *Statistics in Medicine* 23: 1781–1792.

Also see [ST] **stpower** for more references.

Also see

[ST] **stpower** — Sample-size, power, and effect-size determination for survival analysis

[ST] **stpower logrank** — Sample size, power, and effect size for the log-rank test

[ST] **stpower exponential** — Sample size and power for the exponential test

[R] **sampsi** — Sample size and power determination

[ST] **stcox** — Cox proportional hazards model

[R] **test** — Test linear hypotheses after estimation

[ST] **sts test** — Test equality of survivor functions

[ST] **Glossary**

Title

stpower exponential — Sample size and power for the exponential test

Syntax

Sample-size determination

   Specifying hazard rates

      <u>stpower</u> <u>exp</u>onential $\left[\, h_1 \, \left[\, h_2 \,\right]\,\right]$ [, *options*]

   Specifying survival probabilities

      <u>stpower</u> <u>exp</u>onential s_1 $\left[\, s_2 \,\right]$, t(#) [*options*]

Power determination

   Specifying hazard rates

      <u>stpower</u> <u>exp</u>onential $\left[\, h_1 \, \left[\, h_2 \,\right]\,\right]$, n(*numlist*) [*options*]

   Specifying survival probabilities

      <u>stpower</u> <u>exp</u>onential s_1 $\left[\, s_2 \,\right]$, t(#) n(*numlist*) [*options*]

where

   h_1 is the hazard rate in the control group;

   h_2 is the hazard rate in the experimental group;

   s_1 is the survival probability in the control group at reference (base) time t; and

   s_2 is the survival probability in the experimental group at reference (base) time t. h_1, h_2 and s_1, s_2 may each be specified either as one number or as a list of values (see [U] **11.1.8 numlist**) enclosed in parentheses.

stpower exponential — Sample size and power for the exponential test

options	description
Main	
t(#)	reference time t for survival probabilities s_1 and s_2
*alpha(numlist)	significance level; default is alpha(0.05)
*power(numlist)	power; default is power(0.8)
*beta(numlist)	probability of type II error; default is beta(0.2)
*n(numlist)	sample size; required to compute power
*hratio(numlist)	hazard ratio of the experimental group to the control group, h_2/h_1 or $\ln(s_2)/\ln(s_1)$; default is hratio(0.5)
onesided	one-sided test; default is two sided
*p1(numlist)	the proportion of subjects in the control group; default is p1(0.5), meaning equal group sizes
*nratio(numlist)	ratio of sample sizes, N_2/N_1; default is nratio(1), meaning equal group sizes
loghazard	power or sample-size computation for the test of the difference between log hazards; default is the test of the difference between hazards
unconditional	power or sample-size computation using the unconditional approach
parallel	treat number lists in starred options as parallel (do not enumerate all possible combinations of values) when multiple values per option are specified
Accrual/Follow-up	
fperiod(#)	length of the follow-up period; if not specified, the study is assumed to continue until all subjects experience an event (fail)
aperiod(#)	length of the accrual period; default is aperiod(0), meaning no accrual
aprob(#)	proportion of subjects accrued by time t^* under truncated exponential accrual; default is aprob(0.5)
aptime(#)	proportion of the accrual period, $t^*/$aperiod(), by which proportion of subjects in aprob() is accrued; default is aptime(0.5)
atime(#)	reference accrual time t^* by which the proportion of subjects in aprob() is accrued; default value is $0.5 \times$aperiod()
ashape(#)	shape of the truncated exponential accrual distribution; default is ashape(0), meaning uniform accrual
lossprob(# #)	proportion of subjects lost to follow-up by time losstime() in the control and the experimental groups; default is lossprob(0 0), meaning no losses to follow-up
losstime(#)	(reference) time by which the proportion of subjects specified in lossprob() is lost to follow-up; default is losstime(1)
losshaz(# #)	loss hazard rates in the control and the experimental groups; default is losshaz(0 0), meaning no losses to follow-up

Reporting	
detail	more detailed output
table	display results in a table with default columns
columns(*colnames*)	display results in a table with specified *colnames* columns
notitle	suppress table title
nolegend	suppress table legend
colwidth(# [# ...])	column widths; default is colwidth(9)
separator(#)	draw a horizontal separator line every # lines; default is separator(0), meaning no separator lines
saving(*filename* [, replace])	save the table data to *filename*; use replace to overwrite existing *filename*
†noheader	suppress table header; seldom used
†continue	draw a continuation border in the table output; seldom used

*Starred options may be specified either as one number or as a list of values (see [U] **11.1.8 numlist**).

†noheader and continue are not shown in the dialog box.

colnames	description
alpha	significance level
power	power
beta	type II error probability
n	total number of subjects
n1	number of subjects in the control group
n2	number of subjects in the experimental group
hr	hazard ratio
loghr	log of the hazard ratio (difference between log hazards)
diff	difference between hazards
h1	hazard rate in the control group
h2	hazard rate in the experimental group
s1	survival probability in the control group
s2	survival probability in the experimental group
t	reference survival time
p1	proportion of subjects in the control group
nratio	ratio of sample sizes, experimental to control
fperiod	follow-up period
aperiod	accrual period
aprob	% of subjects accrued by time atime (or by aptime % of accrual period)
aptime	% of an accrual period by which aprob % of subjects are accrued
atime	reference accrual time
ashape	shape of the accrual distribution
lpr1	proportion of subjects lost to follow-up in the control group
lpr2	proportion of subjects lost to follow-up in the experimental group
losstime	reference loss to follow-up time

lh1	loss hazard rate in the control group
lh2	loss hazard rate in the experimental group
eo	total expected number of events (failures) under the null
eo1	number of events in the control group under the null
eo2	number of events in the experimental group under the null
ea	total expected number of events (failures) under the alternative
ea1	number of events in the control group under the alternative
ea2	number of events in the experimental group under the alternative
lo	total expected number of losses to follow-up under the null
lo1	number of losses in the control group under the null
lo2	number of losses in the experimental group under the null
la	total expected number of losses to follow-up under the alternative
la1	number of losses in the control group under the alternative
la2	number of losses in the experimental group under the alternative

By default, the following *colnames* are displayed:

power, n, n1, n2, and alpha are always displayed;

h1 and h2 are displayed if hazard rates are specified, or s1 and s2 if survival probabilities are specified;

diff if hazard difference test is specified, or loghr if log-hazard difference test is specified;

aperiod if accrual period (aperiod()) is specified;

fperiod if follow-up period (fperiod()) is specified; and

lh1 and lh2 if losshaz() or lpr1 and lpr2 if lossprob() is specified.

Menu

Statistics > Survival analysis > Power and sample size > Exponential test

Description

stpower exponential estimates required sample size and power for survival analysis comparing two exponential survivor functions by using parametric tests for the difference between hazards or, optionally, for the difference between log hazards. It accommodates unequal allocation between the two groups, flexible accrual of subjects into the study, and group-specific losses to follow-up. The accrual distribution may be chosen to be uniform over the fixed accrual period, R, or truncated exponential over the period $[0, R]$. Losses to follow-up are assumed to be exponentially distributed. Also the computations may be carried out using the conditional or the unconditional approach.

You can use stpower exponential to

- calculate expected number of events and required sample size when you know power and effect size (supplied as hazard rates, survival probabilities, or hazard ratio) for studies with or without follow-up and accrual periods allowing for different accrual patterns and in the presence of losses to follow-up and

- calculate power when you know sample size and effect size (supplied as hazard rates, survival probabilities, or hazard ratio) for studies with or without follow-up and accrual periods allowing for different accrual patterns and in the presence of losses to follow-up.

If the t() option is specified, the command's input parameters are the values of survival probabilities in the control (or the less favorable) group, $S_1(t)$, and in the experimental group, $S_2(t)$, at a fixed time, t (reference survival time), specified in t(), given as s_1 and s_2, respectively. Otherwise, the input parameters are assumed to be the values of the hazard rates in the control group, λ_1, and in the experimental group, λ_2, given as h_1 and h_2, respectively. If survival probabilities are specified, they are converted to hazard rates by using the formula for the exponential survivor function and the value of time t in t().

Options

[Main]

t(#) specifies a fixed time t (reference survival time) such that the proportions of subjects in the control and experimental groups still alive past this time point are as specified in s_1 and s_2. If this option is specified, the input parameters, s_1 and s_2, are the survival probabilities $S_1(t)$ and $S_2(t)$. Otherwise, the input parameters are assumed to be hazard rates, λ_1 and λ_2, given as h_1 and h_2, respectively.

alpha(*numlist*) sets the significance level of the test. The default is alpha(0.05).

power(*numlist*) sets the power of the test. The default is power(0.8). If beta() is specified, this value is set to be 1−beta(). Only one of power() or beta() may be specified.

beta(*numlist*) sets the probability of a type II error of the test. The default is beta(0.2). If power() is specified, this value is set to be 1−power(). Only one of beta() or power() may be specified.

n(*numlist*) specifies the number of subjects in the study to be used to compute the power of the test. By default, the sample-size calculation is assumed. This option may not be combined with beta() or power().

hratio(*numlist*) specifies the hazard ratio of the experimental group to the control group. The default is hratio(0.5). This value defines the clinically significant improvement of the experimental procedure over the control desired to be detected by a test, with a certain power specified in power(). If h_1 and h_2 (or s_1 and s_2) are given, hratio() is not allowed and the hazard ratio is computed as h_2/h_1 (or $\ln(s_2)/\ln(s_1)$).

onesided indicates a one-sided test. The default is two sided.

p1(*numlist*) specifies the proportion of subjects in the control group. The default is p1(0.5), meaning equal allocation of subjects to the control and the experimental groups. Only one of p1() or nratio() may be specified.

nratio(*numlist*) specifies the sample-size ratio of the experimental group relative to the control group, N_2/N_1. The default is nratio(1), meaning equal allocation between the two groups. Only one of nratio() or p1() may be specified.

loghazard requests sample-size or power computation for the test of the difference between log hazards (or the test of the log of the hazard ratio). This option implies uniform accrual. By default, the test of the difference between hazards is assumed.

unconditional requests that the unconditional approach be used for sample-size or power computation; see *The conditional versus unconditional approaches* and *Methods and formulas* for details.

parallel reports results sequentially (in parallel) over the list of numbers supplied to options allowing *numlist*. By default, results are computed over all combinations of the number lists in the following order of nesting: alpha(); p1() or nratio(); list of hazard rates h_1 and h_2 or survival probabilities s_1 and s_2; hratio(); power() or beta(); and n(). This option requires that options with multiple values each contain the same number of elements.

Accrual/Follow-up

fperiod(#) specifies the follow-up period of the study, f. By default, it is assumed that subjects are followed up until the last subject experiences an event (fails). The (minimal) follow-up period is defined as the length of the period after the recruitment of the last subject to the study until the end of the study. If T is the duration of a study and R is the length of an accrual period, then the follow-up period is $f = T - R$.

aperiod(#) specifies the accrual period, R, during which subjects are to be recruited into the study. The default is aperiod(0), meaning no accrual.

aprob(#) specifies the proportion of subjects expected to be accrued by time t^* according to the truncated exponential distribution. The default is aprob(0.5). This option is useful when the shape parameter is unknown but the proportion of accrued subjects at a certain time is known. aprob() is often used in conjunction with aptime() or atime(). This option may not be specified with ashape() or loghazard and requires specifying a nonzero accrual period in aperiod().

aptime(#) specifies the proportion of the accrual period, t^*/R, by which the proportion of subjects specified in aprob() is expected to be accrued according to the truncated exponential distribution. The default is aptime(0.5). This option may not be combined with atime(), ashape(), or loghazard and requires specifying a nonzero accrual period in aperiod().

atime(#) specifies the time point t^*, reference accrual time, by which the proportion of subjects specified in aprob() is expected to be accrued according to the truncated exponential distribution. The default value is $0.5 \times R$. This option may not be combined with aptime(), ashape(), or loghazard and requires specifying a nonzero accrual period in aperiod(). The value in atime() may not exceed the value in aperiod().

ashape(#) specifies the shape, γ, of the truncated exponential accrual distribution. The default is ashape(0), meaning uniform accrual. This option is not allowed in conjunction with loghazard and requires specifying a nonzero accrual period in aperiod().

lossprob(# #) specifies the proportion of subjects lost to follow-up by time losstime() in the control and the experimental groups, respectively. The default is lossprob(0 0), meaning no losses to follow-up. This option requires specifying aperiod() or fperiod() and may not be combined with losshaz().

losstime(#) specifies the time at which the proportion of subjects specified in lossprob() is lost to follow-up, also referred to as the reference loss to follow-up time. The default is losstime(1). This option requires specifying lossprob().

losshaz(# #) specifies exponential hazard rates of losses to follow-up, η_1 and η_2, in the control and the experimental groups, respectively. The default is losshaz(0 0), meaning no losses to follow-up. This option requires specifying aperiod() or fperiod() and may not be combined with lossprob().

Reporting

detail displays more detailed output; the expected number of events (failures) and losses to follow-up under the null and alternative hypotheses are displayed. This option is not appropriate with tabular output.

table displays results in a tabular format and is implied if any number list contains more than one element. This option is useful if you are producing results one case at a time and wish to construct your own custom table by using a forvalues loop.

columns(*colnames*) specifies results in a table with specified *colnames* columns. The order of the columns in the output table is the same as the order of *colnames* specified in columns(). Column names in columns() must be space-separated.

notitle prevents the table title from displaying.

nolegend prevents the table legend from displaying and column headers from being marked.

colwidth(# [# ...]) specifies column widths. The default is 9 for all columns. The number of specified values may not exceed the number of columns in the table. A missing value (.) may be specified for any column to indicate the default width (9). If fewer widths are specified than the number of columns in the table, the last width specified is used for the remaining columns.

separator(#) specifies how often separator lines should be drawn between rows of the table. The default is separator(0), meaning that no separator lines should be displayed.

saving(*filename*[, replace]) creates a Stata data file (.dta file) containing the table values with variable names corresponding to the displayed *colnames*. replace indicates that *filename* be overwritten, if it exists. saving() is only appropriate with tabular output.

The following options are available with stpower exponential but are not shown in the dialog box:

noheader prevents the table header from displaying. This option is useful when the command is issued repeatedly, such as within a loop. noheader implies notitle.

continue draws a continuation border at the bottom of the table. This option is useful when the command is issued repeatedly within a loop.

Remarks

Remarks are presented under the following headings:

> Introduction
> Other ways of specifying the effect size
> Sample-size determination by using different approximations
> Sample-size determination in the presence of censoring
> Nonuniform accrual and exponential losses to follow-up
> The conditional versus unconditional approaches
> Link to the sample-size and power computation for the log-rank test
> Power determination

Introduction

Let $S_1(t)$ and $S_2(t)$ be the exponential survivor functions with hazard rates λ_1 and λ_2 in the control and experimental groups, respectively. Define ψ to be the treatment effect that can be expressed as a difference, $\delta = \lambda_2 - \lambda_1$, between hazard rates or as the log of the hazard ratio (a difference between log hazard-rates), $\ln(\Delta) = \ln(\lambda_2/\lambda_1) = \ln(\lambda_2) - \ln(\lambda_1)$. Negative values of the treatment effect ψ imply the superiority of the experimental treatment over the standard (control) treatment. Denote R and T to be the length of the accrual period and the total duration of the study, respectively. Then, the follow-up period f is $f = T - R$.

Consider a study designed to compare the exponential survivor functions, $S_1(t) = e^{-\lambda_1 t}$ and $S_2(t) = e^{-\lambda_2 t}$, of the two treatment groups. The disparity in survivor functions may be tested using the hazards λ_1 and λ_2 for the exponential model. Depending on the definition of the treatment effect ψ, two test statistics based on the difference and on the log ratio of the hazards may be used to conduct tests of the difference between survivor functions using respective null hypotheses, $H_0: \delta = 0$ and $H_0: \ln(\Delta) = 0$.

The basic formula for the sample-size and power calculations for the test of H_0: $\delta = 0$ is proposed by Lachin (1981). He also derives the equation relating the sample size and power allowing for uniform accrual of subjects into the study over the period from 0 to R. Lachin and Foulkes (1986) extend this formula to truncated exponential accrual over the interval 0 to R and exponential losses to follow-up over the interval 0 to T.

The simplest method for the sample-size and power calculations for the test of H_0: $\ln(\Delta) = 0$ is presented by George and Desu (1974). Rubinstein, Gail, and Santner (1981) extend their method to account for uniform accrual and exponential losses to follow-up and apply it to planning the duration of a survival study. The formula that relates the sample size and power for this test and takes into account the uniform accrual and exponential losses to follow-up is formulated by Lakatos and Lan (1992), based on the derivations of Rubinstein, Gail, and Santner (1981).

By default, stpower exponential computes the sample size required to achieve a specified power to detect a difference between hazard rates, $\delta_a = \lambda_{2a} - \lambda_{1a}$, using the method of Lachin (1981). If loghazard is specified, the sample size required to detect a log of the hazard ratio $\ln(\Delta_a) = \ln(\lambda_{2a}/\lambda_{1a})$ with specified power is reported using the formula derived by George and Desu (1974). In the presence of an accrual period, the methods of Lachin and Foulkes (1986) or for uniform accrual only, Rubinstein, Gail, and Santner (1981) (if loghazard and unconditional are specified), are used.

If power determination is desired, sample size n() must be specified. Otherwise, sample-size determination is assumed with power(0.8) (or, equivalently, beta(0.2)). The default setting for power or the probability of a *type II error*, a failure to reject the null hypothesis when the alternative hypothesis is true, may be changed by using power() or beta(), respectively.

The default probability of a *type I error*, a rejection of the null hypothesis when the null hypothesis is true, of a test is 0.05 but may be changed by using the alpha() option. One-sided tests may be requested by using the onesided option. The default equal-group allocation may be changed by specifying either p1() or nratio().

By default, the estimates of sample sizes or power for the test of the difference between hazards are reported. This may be changed to the test versus the difference between log hazards by using the loghazard option. The default conditional approach may be replaced with the unconditional approach by using unconditional; see *The conditional versus unconditional approaches*.

If neither the length of a follow-up period, f, nor the length of an accrual period, R, is specified in fperiod() or aperiod(), respectively, the study is assumed to continue until all subjects experience an event (failure), regardless of how much time is required. If either of the two options is supplied, a study of length $T = R + f$ is assumed.

If an accrual period of length R is specified in the aperiod() option, uniform accrual over the period $[0, R]$ is assumed. The accrual distribution may be changed to truncated exponential when the shape parameter is specified in ashape(). The combination of the aprob() and aptime() (or atime()) options may be used in place of the ashape() option to request the desired shape of the truncated exponential accrual. To take into account exponential losses to follow-up, the losshaz() or lossprob() and losstime() options may be used. For examples, see *Nonuniform accrual and exponential losses to follow-up*.

Optionally, results may be displayed in a table by using table or columns() as demonstrated in *Sample-size determination in the presence of censoring* below and in *Creating output tables* of [ST] **stpower**. Refer to example 7 in *Power and effect-size determination* of [ST] **stpower logrank** and [ST] **stpower** to see how to obtain a graph of a power curve.

Other ways of specifying the effect size

Here we demonstrate how to provide the information about the disparity in the two exponential survivor functions, also known as the effect size, to stpower exponential. We separately consider the two tests, based on the hazard difference and based on the log of the hazard ratio (or on the log-hazard difference).

▷ Example 1: Effect size for the test on the difference between hazards

Consider a fictional study for which the goal is to compare the two exponential survivor functions of the control and the experimental groups, using the test based on the difference between hazard rates. The yearly hazard rate in the control group is known to be 0.4, corresponding to roughly 45% survival by 2 years. The effect size, expressed as the hazard ratio, of 0.5 corresponding to the reduction in hazard of the experimental group from 0.4 to 0.2 (or increase in survival to 67% by 2 years), is of interest. The investigators need an estimate of the required sample size to detect this effect size with 80% power with a two-sided, 0.05-level test, the default settings of stpower exponential. For simplicity, we also assume that there are enough resources to monitor the subjects until all of them fail (a type I study, as defined in the next section).

The hazard rate in the control group and the effect size are required in the computation. The value of the hazard rate in the control group may be supplied directly as argument h_1 (the first syntax) or computed using the supplied value of the survival probability, s_1, at the reference time specified in the t() option (the second syntax). The effect size may be supplied by using the hratio() option, or by directly supplying the hazard rate in the experimental group as argument h_2 or, if the t() option is specified, by supplying the survival probability in the experimental group as argument s_2 at reference time t().

We demonstrate all the above by using our fictional example. To obtain the required sample size, we supply the hazard rate in the control group, 0.4, as argument h_1 and the hazard ratio of 0.5 in hratio().

```
. stpower exponential 0.4, hratio(0.5)
Note: input parameters are hazard rates.
Estimated sample sizes for two-sample comparison of survivor functions
Exponential test, hazard difference, conditional
Ho: h2-h1 = 0

Input parameters:

         alpha =    0.0500   (two sided)
            h1 =    0.4000
            h2 =    0.2000
         h2-h1 =   -0.2000
         power =    0.8000
            p1 =    0.5000

Estimated sample sizes:
             N =      74
            N1 =      37
            N2 =      37
```

Instead of the hazard rate, we specify the survival probability of 0.45 in the control group as argument s_2 and a reference survival time 2 in option t().

```
. stpower exponential 0.45, t(2) hratio(0.5)
Note: input parameters are survival probabilities.

Estimated sample sizes for two-sample comparison of survivor functions
Exponential test, hazard difference, conditional
Ho: h2-h1 = 0

Input parameters:

        alpha =    0.0500  (two sided)
           s1 =    0.4500
           s2 =    0.6708
            t =    2.0000
        h2-h1 =   -0.1996
        power =    0.8000
           p1 =    0.5000

Estimated sample sizes:
            N =        74
           N1 =        37
           N2 =        37
```

Because the hazard ratio of 0.5 is the default of hratio(), we could have omitted this option from the above syntaxes.

Now rather than specifying the effect size by using the hratio(0.5) option above, we supply the hazard rate in the experimental group, 0.2, as argument h_2,

```
. stpower exponential 0.4 0.2
```
(output omitted)

or the survival probability in the experimental group, 0.67, as argument s_2 and the reference survival time of 2 years in t(),

```
. stpower exponential 0.45 0.67, t(2)
```
(output omitted)

and obtain the same estimates as in the above output: 74 of the total sample and 37 of the group sample sizes.

◁

▷ Example 2: Effect size for the test on the log of the hazard ratio

Continuing with the above example, suppose that the two survivor functions are to be compared using the test based on the difference between the log hazards (or test based on the log of the hazard ratio). To request this test, the loghazard option must be specified.

Unlike the hazard-difference test, the log hazard-ratio test does not require the hazard rate in the control group for the computation when the study is continued until all subjects fail. The only required information is the size of the effect desired to be detected by this test.

To obtain the estimate of the sample size for the test based on the log of the hazard ratio, we specify the loghazard option and use the default value of the hazard ratio of 0.5:

(Continued on next page)

```
. stpower exponential, loghazard
Estimated sample sizes for two-sample comparison of survivor functions
Exponential test, log-hazard difference, conditional
Ho: ln(h2/h1) = 0

Input parameters:

        alpha =    0.0500  (two sided)
    ln(h2/h1) =   -0.6931
        power =    0.8000
           p1 =    0.5000

Estimated sample sizes:
            N =        66
           N1 =        33
           N2 =        33
```

Similarly to the examples for the hazard-difference test, the effect size may also be supplied via hazard rates,

```
. stpower exponential 0.4 0.2, loghazard
```
(output omitted)

or survival probabilities and reference survival time,

```
. stpower exponential 0.45 0.67, t(2) loghazard
```
(output omitted)

For a fixed-duration study, when either fperiod() or aperiod() is specified (see *Sample-size determination in the presence of censoring*), the sample-size computations for the log hazard-ratio test do require the value of the hazard rate in the control group.

◁

Sample-size determination by using different approximations

Consider the following two types of survival studies: the first type, a *type I study*, is when investigators have enough resources to monitor the subjects until all of them experience an event (failure) and the second type, a *type II study*, is when the study terminates after a fixed period of time, regardless of whether all subjects experienced an event by that time.

Here we explore sample-size estimates using different approximations for a type I study. Examples of sample-size determination for a type II study are presented in the next section.

In survival studies, the requirement for the sample size is based on the requirement to observe a certain number of events (failures) to ensure a prespecified power of a test to detect a difference in survivor functions. For a type I study, the number of subjects required for the study is the same as the number of events required to be observed in the study because all subjects experience an event by the end of the study.

▷ Example 3: Sample size using the Lachin method

Consider an example from Lachin (1981, 107). A clinical trial is to be conducted to compare the survivor functions in the control and the experimental groups with a one-sided exponential test, based on the difference between hazards, of the superiority of a new treatment (H_a: $\delta < 0$) for a disease with moderate levels of mortality. Subjects in the control group get a standard treatment and subjects in the experimental group receive a new treatment. From previous studies the yearly hazard rate for the standard treatment was found to be $\lambda_1 = 0.3$, corresponding to 50% survival after 2.3 years. The investigators would like to know how many subjects are required to detect a reduction in hazard to $\lambda_2 = 0.2$ (H_a: $\delta = -0.1$), which corresponds to an increase in survival to 63% at 2.3 years, with 90% power, equal-sized groups, and a significance level, α, of 0.05.

To obtain the estimate of the sample size for the above study, we supply hazard rates 0.3 and 0.2 as arguments and specify the power(0.9) option for 90% power and the onesided option for a one-sided test.

```
. stpower exponential 0.3 0.2, power(0.9) onesided
Note: input parameters are hazard rates.

Estimated sample sizes for two-sample comparison of survivor functions
Exponential test, hazard difference, conditional
Ho: h2-h1 = 0

Input parameters:

          alpha =    0.0500   (one sided)
             h1 =    0.3000
             h2 =    0.2000
          h2-h1 =   -0.1000
          power =    0.9000
             p1 =    0.5000

Estimated sample sizes:

              N =       218
             N1 =       109
             N2 =       109
```

From the output, a total of 218 events (subjects) must be observed (recruited) in a study to ensure a power of 90% of a one-sided exponential test to detect a 13% increase in survival probability of subjects in the experimental group with $\alpha = 0.05$. Our estimate of 218 of the total number of subjects (109 per group) required for the study is the same as the one reported in Lachin (1981, 107).

◁

▷ Example 4: Sample size using the George–Desu method

The syntax above reports the sample size obtained using the approximation of Lachin (1981) for the test based on the hazard difference. To obtain the sample size using the approximation of George and Desu (1974), for the equivalent alternative H_a: $\ln(\Delta) = -0.4055$ (a test based on the log of the hazard ratio), we need to specify the loghazard option.

```
. stpower exponential 0.3 0.2, power(0.9) onesided loghazard
Note: input parameters are hazard rates.

Estimated sample sizes for two-sample comparison of survivor functions
Exponential test, log-hazard difference, conditional
Ho: ln(h2/h1) = 0

Input parameters:

          alpha =    0.0500   (one sided)
             h1 =    0.3000
             h2 =    0.2000
       ln(h2/h1) =  -0.4055
          power =    0.9000
             p1 =    0.5000

Estimated sample sizes:

              N =       210
             N1 =       105
             N2 =       105
```

The George–Desu method yields a slightly smaller estimate, 210, of the total number of events (subjects). George and Desu (1974) studied the accuracy of the two approximations based on δ and $\ln(\Delta)$ and concluded that the former is slightly conservative; that is, it gives slightly larger sample-size estimates. The latter was found to be accurate to one or two units of the exact solution for equal-sized groups.

We could have also obtained the same results by using the hratio() option to specify the hazard ratio $0.2/0.3 = 0.66667$ instead of supplying arguments 0.3 and 0.2:

```
. stpower exponential, hratio(0.66667) power(0.9) onesided loghazard
```
(output omitted)

❑ Technical note

The above approach may also be used to obtain an approximation to the sample size or power for the exact F test of equality of two exponential mean analysis (life) times (using the relation between a mean and a hazard rate of the exponential distribution, $\mu = 1/\lambda$).

For example, the sample size of 210 obtained above may be used as an approximation to the number of subjects required in a study of which the goal is to detect an increase in a mean analysis (life) time of the experimental group from $3.33 = 1/0.3$ to $5 = 1/0.2$ by using the exact, one-sided, 0.05-level F test with 90% power.

The test statistic of the F test is a ratio of two sample means from two exponential distributions that has an exact F distribution. The George–Desu method is based on the normal approximation of the distribution of the log of this test statistic. George and Desu (1974) studied this approximation for equal-sized groups and some common values of significance levels, powers, and hazard ratios and found it to be accurate to one or two units of the exact solution.

❑ Technical note

We can obtain the same results from stpower exponential as in the examples above by using the respective survival probabilities 0.5 and 0.63 at $t = 2.3$ as arguments in place of hazard rates 0.3 and 0.2. For example,

```
. stpower exponential 0.5 0.63, t(2.3) power(0.9) onesided
Note: input parameters are survival probabilities.
```
(output omitted)

Other alternative syntaxes, using the hratio() option, are

```
. stpower exponential 0.3, hratio(0.66667) power(0.9) onesided
Note: input parameters are hazard rates.
```
(output omitted)

where 0.3 is interpreted as a hazard rate because t() is not specified, and

```
. stpower exponential 0.5, t(2.3) hratio(0.6667) power(0.9) onesided
Note: input parameters are survival probabilities.
```
(output omitted)

Sample-size determination in the presence of censoring

Often in practice, investigators may not have enough resources to continue a study until all subjects experience an event and therefore plan to terminate the study after a fixed period, T. Some subjects may not experience an event by the end of the study, in which case the (administrative) censoring of subjects occurs. In the presence of censoring, the number of subjects required in a study will be larger than the number of events required to be observed in the study.

We investigate how terminating the study after some fixed period, T, before all subjects experience an event, affects the requirements for the sample size. The duration of a study is divided into two phases: an accrual phase of a length R, during which subjects are recruited to the study, and a follow-up phase of a length f, during which subjects are followed up until the end of the study and no new subjects enter the study. The duration of a study, T, is the sum of the lengths of the two phases.

Consider the following study designs. In the first study design, A, each subject is followed up for a length of time T. Here the minimum follow-up time f is equal to T, and, consequently, $R = 0$. In practice, however, subjects will often enter the study at random times and will be followed up until the end of a study at time T, in which case the subjects who come under observation later in time will have a shorter follow-up than subjects who entered the study at the beginning. Therefore, the minimum follow-up time f will be less than T, and R will be equal to $T - f$. Here the length of the accrual period, R, must be taken into account in the computations. In the presence of an accrual period, subjects may be recruited continuously during a period of length T ($R = T$, $f = 0$), the second study design, B. Or subjects may be recruited for a fixed period, R, and then followed up for a period of time, f, during which no new subjects enter the trial, so that the total duration of study is $T = R + f$, the third design, C.

▷ Example 5: Sample size in the presence of accrual and follow-up periods

Continuing example 3, assume that the investigators have resources to continue the study for only 5 years, $T = 5$. We tabulate sample-size values for different combinations of an accrual period, R, and a minimum follow-up period, f, such that $T = f + R = 5$. Because the `aperiod()` and `fperiod()` options allow specifying only one number, we can obtain the table by issuing `stpower exponential` repeatedly within a `forvalues` loop.

```
. local cont continue
. local columns columns(power n aperiod fperiod h1 h2 alpha)
. local T = 5
. forvalues R = 0/`T' {
  2.         if `R' == `T' {
  3.                 local cont
  4.         }
  5.         local f = `T' - `R'
  6.         stpower exponential 0.3 0.2, power(0.9) onesided \\\
>                 aperiod(`R') fperiod(`f') `header' `cont' `columns'
  7.         local header noheader
  8. }
Note: input parameters are hazard rates.
```

(Continued on next page)

```
Estimated sample sizes for two-sample comparison of survivor functions
Exponential test, hazard difference, conditional
Ho: h2-h1 = 0

     Power         N        AP+        FP         H1         H2       Alpha*

      .9          304        0          5         .3         .2        .05
      .9          322        1          4         .3         .2        .05
      .9          344        2          3         .3         .2        .05
      .9          378        3          2         .3         .2        .05
      .9          426        4          1         .3         .2        .05
      .9          502        5          0         .3         .2        .05

 * one sided
 + uniform accrual; 50.00% accrued by 50.00% of AP
```

For each iteration of the loop, the value of a follow-up period, f, stored in macro f (see [P] **macro**), is computed using the current value of an accrual period, R, and the fixed-time study T, defined by macros R and T, respectively. The values 'R' and 'f' are then passed to the aperiod() and fperiod() options, respectively. We use the columns() option to obtain a customized table with columns power, n, aperiod, fperiod, h1, h2, and alpha. We store this option in local columns simply to avoid specifying a long expression in stpower exponential. We specify noheader in the syntax of stpower exponential each time it is called in the loop except for the first time, and we specify continue each time except for the final one.

The first and the last entries of the above table correspond to the extreme cases of no accrual (design A) and no follow-up (design B), respectively.

For a design A, the estimate of the sample size, 304, is larger than the earlier estimate of 218 from example 3. That is, if the study in example 3 terminates after 5 years, the requirement for the sample size increases by 39% to ensure that the same number of 218 events is observed.

By trying different values of the follow-up period, we may find that a 30-year follow-up is required if the investigators can recruit no more than 218 subjects: 30 years is required to observe all subjects to fail in this study.

```
. stpower exponential 0.3 0.2, power(0.9) onesided fperiod(30)
Note: input parameters are hazard rates.

Estimated sample sizes for two-sample comparison of survivor functions
Exponential test, hazard difference, conditional
Ho: h2-h1 = 0

Input parameters:

          alpha =    0.0500   (one sided)
             h1 =    0.3000
             h2 =    0.2000
          h2-h1 =   -0.1000
          power =    0.9000
             p1 =    0.5000

Accrual and follow-up information:
       duration =   30.0000
      follow-up =   30.0000

Estimated sample sizes:
              N =      218
             N1 =      109
             N2 =      109
```

For design B, instead of being monitored for 5 years, subjects are being continuously recruited during all of those 5 years; the total sample size increases from 304 to 502. The reason for such an increase is that the average analysis time (the time when a subject is at risk of a failure) decreases from 5 to 2.5 and therefore reduces the probability of a subject failing by the end of the study.

In general, the estimates of the total sample size steadily increase as the length of the follow-up decreases. That is, the presence of a follow-up period reduces the requirement for the number of subjects in the study. For example, a clinical trial with a 3-year uniform accrual and a 2-year follow-up needs a total of 378 subjects (189 per group) compared with the total of 502 subjects required for a study with no follow-up and a 5-year accrual.

Nonuniform accrual and exponential losses to follow-up

In the presence of an accrual period, stpower exponential performs computations assuming uniform accrual over the period of time R, specified in aperiod(). The assumption of uniform accrual may be relaxed by requesting a truncated exponential accrual over the interval 0 to R with shape γ as specified in ashape(#). If the estimate of the shape parameter, γ, is unavailable, an alternative is to specify the proportion of subjects, $G(t^*)$, expected to be recruited by a fixed time, t^* (or by a proportion of the accrual period R, t^*/R), in the aprob() and atime() (or aptime()) options, respectively. This information is used to find the corresponding γ by using

$$G(t^*) = \{1 - \exp(-\gamma t^*)\}/\{1 - \exp(-\gamma R)\}$$

Also see *Methods and formulas*.

▷ Example 6: Truncated exponential entry distribution

Continuing example 5, we investigate the influence of nonuniform accrual on the estimate of the sample size for a study with a 3-year accrual and a 2-year follow-up. Suppose that the recruitment of subjects to the study is slow for most of the accrual period and increases rapidly toward the end of the recruitment. Consider an extreme case of such an accrual corresponding to shape parameter -6. The graph of uniform and truncated exponential with shape -6 entry distributions over $[0, 3]$ is given below.

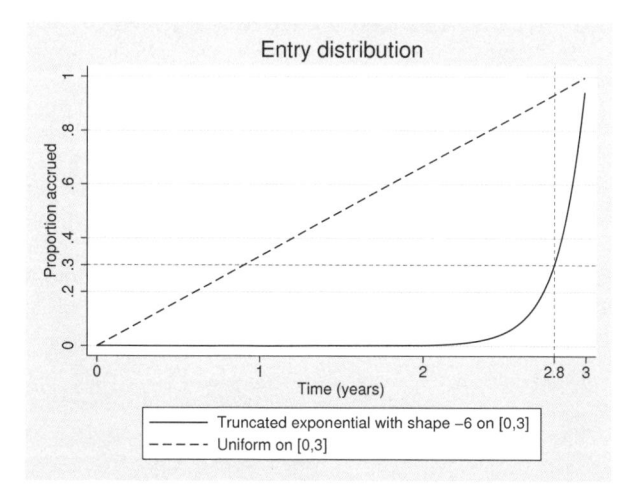

From the above graph, the accrual of subjects is extremely slow during most of the recruitment period, with 70% of subjects being recruited within the last few months of a 3-year accrual period. More precisely, according to the graph, 30% of subjects are expected to be recruited after 2.8 years.

To obtain the estimate of the sample size for this study, we type

```
. stpower exponential 0.3 0.2, power(0.9) onesided aperiod(3) fperiod(2) ashape(-6)
Note: input parameters are hazard rates.

Estimated sample sizes for two-sample comparison of survivor functions
Exponential test, hazard difference, conditional
Ho: h2-h1 = 0

Input parameters:
         alpha =     0.0500   (one sided)
            h1 =     0.3000
            h2 =     0.2000
         h2-h1 =    -0.1000
         power =     0.9000
            p1 =     0.5000

Accrual and follow-up information:
      duration =     5.0000
     follow-up =     2.0000
       accrual =     3.0000   (exponential)
    accrued(%) =      50.00   (by time t*)
            t* =     2.8845   (96.15% of accrual)

Estimated sample sizes:
             N =        516
            N1 =        258
            N2 =        258
```

and conclude that 516 subjects have to be recruited to this study. This sample size ensures 90% power of a one-sided, 0.05-level test to detect a reduction in hazard from 0.3 to 0.2 when the accrual of subjects follows the considered truncated exponential distribution. For this extreme case of a negative truncated exponential entry distribution (the concave entry distribution), the estimate of the sample 516 increases significantly compared with an estimate of 378 for a uniform entry distribution. On the other hand, a truncated exponential distribution with positive values of the shape parameter (convex entry distribution) will reduce the requirement for the sample size when compared with uniform accrual.

Suppose that we do not know (or do not wish to guess) the value of the shape parameter. The only information available to us from the above graph is that 30% of subjects are expected to be recruited after 2.8 years (or, equivalently, 50% of subjects are expected to be recruited by 96.15% of the accrual period from the previous output). We submit this information in the aprob() and atime() or aprob() and aptime() options, as shown below, and obtain the same estimate of the sample size 516.

```
. stpower exponential 0.3 0.2, power(0.9) onesided aperiod(3) fperiod(2)
> aprob(0.3) atime(2.8)
Note: input parameters are hazard rates.

Estimated sample sizes for two-sample comparison of survivor functions
Exponential test, hazard difference, conditional
Ho: h2-h1 = 0
```

Input parameters:

```
        alpha =    0.0500  (one sided)
           h1 =    0.3000
           h2 =    0.2000
        h2-h1 =   -0.1000
        power =    0.9000
           p1 =    0.5000
```

Accrual and follow-up information:

```
     duration =    5.0000
    follow-up =    2.0000
      accrual =    3.0000  (exponential)
   accrued(%) =      30.00 (by time t*)
           t* =    2.8000  (93.33% of accrual)
```

Estimated sample sizes:

```
            N =       516
           N1 =       258
           N2 =       258
```

or, equivalently,

```
. stpower exponential 0.3 0.2, power(0.9) onesided aperiod(3) fperiod(2)
> aptime(0.9615)
  (output omitted)
```

Because 50% is the default value of `aprob()`, we omit this option in the above.

◁

Apart from administrative censoring, subjects may not experience an event by the end of the study because of being lost to follow-up for various reasons. See *Theory and terminology* in [ST] **stpower** and [ST] **Glossary** for a more detailed description. Rubinstein, Gail, and Santner (1981) and Lachin and Foulkes (1986) extend sample-size and power computations to take into account exponentially distributed losses to follow-up. In addition to being exponentially distributed, losses to follow-up are assumed to be independent of the survival times.

▷ Example 7: Exponential losses to follow-up

Suppose that in example 5, in the study with a 3-year uniform accrual and a 2-year follow-up, yearly loss hazards in the control and the experimental groups are 0.2. Groups may have different loss hazards, which can be specified in `losshaz()`.

(*Continued on next page*)

```
. stpower exponential 0.3 0.2, power(0.9) onesided aperiod(3) fperiod(2)
> losshaz(0.2 0.2) detail
Note: input parameters are hazard rates.

Estimated sample sizes for two-sample comparison of survivor functions
Exponential test, hazard difference, conditional
Ho: h2-h1 = 0

Input parameters:
          alpha =    0.0500   (one sided)
             h1 =    0.3000
             h2 =    0.2000
          h2-h1 =   -0.1000
          power =    0.9000
             p1 =    0.5000

Accrual and follow-up information:
       duration =    5.0000
      follow-up =    2.0000
        accrual =    3.0000   (uniform)
            lh1 =    0.2000
            lh2 =    0.2000

Estimated sample sizes:
              N =       500
             N1 =       250
             N2 =       250

Estimated expected number of events:
          E|Ha =       213       E|Ho =       216
         E1|Ha =       121      E1|Ho =       108
         E2|Ha =        92      E2|Ho =       108

Estimated expected number of losses to follow-up:
          L|Ha =       173       L|Ho =       172
         L1|Ha =        81      L1|Ho =        86
         L2|Ha =        92      L2|Ho =        86
```

The sample size required for a one-sided, 0.05-level test to detect a reduction in hazard from 0.3 to 0.2 with 90% power increases from 378 (see example 5) to 500. We observe that for the extreme case of losses to follow-up, sample size increases significantly. A conservative adjustment commonly applied in practice is $N(1+p_L)$, where p_L is the expected proportion of losses to follow-up in both groups combined. For this example, p_L may be computed as $0.5(0.369+0.324) \approx 0.35$ from table 2 of Lachin and Foulkes (1986). Then the conservative estimate of the sample size is $378(1+0.35) = 510$, which is slightly greater than 500, the actual required sample size.

We also requested that additional information about the expected number of events and losses to follow-up under the null and under the alternative hypothesis be displayed by using the `detail` option. From the above output, a total of 173 subjects (81 from the control and 92 from the experimental groups) are expected to be lost in the study with exponentially distributed losses with yearly rates of 0.2 in each group under the alternative hypothesis.

If the proportion of subjects lost to follow-up by a fixed period in each group is available rather than loss to follow-up rates, it can also be supplied by using the `lossprob()` and `losstime()` options. For example, in the above study approximately 33%, $1 - \exp(-0.2 \times 2) \approx 0.33$, of subjects in each group are lost at time 2 (years). We can obtain the same estimates of sample sizes with

```
. stpower exponential 0.3 0.2, power(0.9) onesided aperiod(3) fperiod(2)
> lossprob(0.33 0.33) losstime(2)
```
  (output omitted)

The conditional versus unconditional approaches

Denote ψ to be the effect size, and $\widehat{\lambda}_1$ and $\widehat{\lambda}_2$ to be the maximum likelihood estimates of the respective hazard-rate parameters. Consider the two effect-size estimators based on the difference between the hazard rates, $\widehat{\lambda}_2 - \widehat{\lambda}_1$, and based on the log of the hazard ratio, $\ln(\widehat{\lambda}_2/\widehat{\lambda}_1)$. Both estimators are asymptotically normal under the null and under the alternative hypothesis.

We adopt Chow, Shao, and Wang (2003, 171) terminology when referring to the conditional and unconditional tests. The *conditional test* is the test that uses the constraint $\lambda_2 = \lambda_1$ (conditional on H_0) when computing the variance of the effect-size estimator under the null. The *unconditional test* is the test that does not use the above constraint when computing the variance of the effect-size estimator under the null. The score and the Wald tests are each one of the examples of conditional and unconditional tests, respectively. Chow, Shao, and Wang (2003) note that neither of the two tests (conditional or unconditional) is always more powerful than the other under the alternative hypothesis. Therefore, there is no definite recommendation of which one is preferable to be used in practice.

The *conditional approach* relies on the following relationship between sample size and power (given in Lachin 1981) to compute estimates of required sample size or power,

$$|\psi| = z_{1-\alpha}\left\{\mathrm{Var}(\psi, H_0)\right\}^{1/2} + z_{1-\beta}\left\{\mathrm{Var}(\psi, H_a)\right\}^{1/2}$$

where $z_{1-\alpha}$ and $z_{1-\beta}$ are the $(1-\alpha)$th and the $(1-\beta)$th quantiles of the standard normal distribution, and $\mathrm{Var}(\psi, H_0)$ and $\mathrm{Var}(\psi, H_a)$ are the asymptotic variances under the null and under the alternative, respectively, of the effect-size estimator, $\widehat{\psi}$. This approach uses the variance of the estimator conditional on the hypothesis type.

The *unconditional approach* replaces $\mathrm{Var}(\psi, H_0)$ with $\mathrm{Var}(\psi, H_a)$ in the above and uses the variance under the alternative to compute the estimates of sample size and power:

$$|\psi| = (z_{1-\alpha} + z_{1-\beta})\left\{\mathrm{Var}(\psi, H_a)\right\}^{1/2}$$

Therefore, the resulting formulas based on the two approaches are different.

Lakatos and Lan (1992) formulate the sample-size formula for the test of the log of the hazard ratio based on the method of Rubinstein, Gail, and Santner (1981). This formula is based on the unconditional approach. Lachin and Foulkes (1986) provide the sample-size formula for the test of the log of the hazard ratio that uses the conditional approach. They also present both conditional and unconditional versions of formulas for the test based on the difference between hazards. As noted by Lachin and Foulkes (1986), sample sizes estimated based on the unconditional approach will be larger than the estimates based on the conditional approach for equal-sized groups.

Both approaches are available with `stpower exponential`; the conditional is the default and the unconditional may be requested by specifying the `unconditional` option. Refer to *Methods and formulas* for the formulas underlying these approaches.

◁

▷ Example 6: Sample size using the Rubinstein–Gail–Santner method

Consider the following scenario in Lakatos and Lan (1992, table I). A 10-year survival study with a 1-year accrual period and a 9-year follow-up is conducted to compare the survivor functions of the two groups by using a two-sided, 0.05 exponential test based on the log of the hazard ratio. The probability of surviving to the end of a study for subjects in the control group is 0.8 ($S_1(t) = 0.8$, $t = 10$).

Subjects are recruited uniformly over the interval $[0, 1]$. Lakatos and Lan (1992) report an estimate of 664 for the sample size required to detect a change in the hazard of the experimental group corresponding to the hazard ratio $\Delta = 0.5$ with 90% power by using the Rubinstein–Gail–Santner method. To obtain the estimates according to this method, we need to specify both `loghazard` and `unconditional`.

```
. stpower exponential 0.8, t(10) power(0.9) aperiod(1) fperiod(9) loghazard
> unconditional
Note: input parameters are survival probabilities.

Estimated sample sizes for two-sample comparison of survivor functions
Exponential test, log-hazard difference, unconditional
Ho: ln(h2/h1) = 0

Input parameters:
           alpha =    0.0500  (two sided)
              s1 =    0.8000
              s2 =    0.8944
               t =   10.0000
        ln(h2/h1) =   -0.6931
           power =    0.9000
              p1 =    0.5000

Accrual and follow-up information:
        duration =   10.0000
       follow-up =    9.0000
         accrual =    1.0000  (uniform)

Estimated sample sizes:
               N =       664
              N1 =       332
              N2 =       332
```

◁

Because the default value of the hazard ratio is 0.5, we omit the `hratio(0.5)` option in the above. From the output, we obtain the same estimate of 664 of the sample size as reported in Lakatos and Lan (1992).

❏ Technical note

In the absence of censoring, the estimates of the sample size or power based on the test of log of the hazard ratio are the same for the conditional and the unconditional approaches. For example, both

```
. stpower exponential 0.8, t(10) power(0.9) loghazard
```
(output omitted)

and

```
. stpower exponential 0.8, t(10) power(0.9) loghazard unconditional
```
(output omitted)

produce the same estimate of the sample size (88). The asymptotic variance of maximum likelihood estimates of the log of the hazard ratio does not depend on hazard rates when there is no censoring and therefore does not depend on the type of hypothesis, $\text{Var}(\widehat{\psi}, H_0) = \text{Var}(\widehat{\psi}, H_a) = 2/N$.

❏

Although all the above examples demonstrated sample-size computation for equal-sized groups, unequal allocation may be taken into account by using the `p1()` or `nratio()` option.

Link to the sample-size and power computation for the log-rank test

▷ Example 7: Sample size using the Freedman and the Schoenfeld methods

Continuing examples 3 and 4, Lachin (1981, 106) gives another approximation to obtain the estimate of the sample size under the equal-group allocation. This approximation coincides with the formula derived by Freedman (1982) for the number of events in the context of the log-rank test. We can obtain such an estimate by using `stpower logrank` and by specifying the hazard ratio computed earlier of 0.66667:

```
. stpower logrank, hratio(0.66667) power(0.9) onesided
Estimated sample sizes for two-sample comparison of survivor functions
Log-rank test, Freedman method
Ho: S1(t) = S2(t)

Input parameters:
        alpha =    0.0500  (one sided)
       hratio =    0.6667
        power =    0.9000
           p1 =    0.5000

Estimated number of events and sample sizes:
            E =       216
            N =       216
           N1 =       108
           N2 =       108
```

The estimate, 216, of the sample size is the same as given in Lachin (1981, 107) and is slightly smaller than the estimate, 218, obtained in example 3, and larger than 210, obtained using the George–Desu method in example 4.

The approximation due to George and Desu (1974) is the same as the approximation to the number of events derived by Schoenfeld (1981) in application to the log-rank test. We can confirm that by using

```
. stpower logrank, hratio(0.66667) power(0.9) onesided schoenfeld
```
(output omitted)

and obtain the same estimate of 210 as using `stpower exponential` with the `loghazard` option in example 4.

◁

Power determination

Power determination may be requested by specifying the `n()` option. We verify the power computation for the study from example 6. We expect the power estimate to be close to 0.9.

▷ Example 8: Power determination

The only thing we change in the syntax of `stpower exponential` from example 6 is replacing the `power(0.9)` option with the `n(664)` option:

```
. stpower exponential 0.8, t(10) n(664) aperiod(1) fperiod(9) loghazard unconditional
Note: input parameters are survival probabilities.
Estimated power for two-sample comparison of survivor functions
Exponential test, log-hazard difference, unconditional
Ho: ln(h2/h1) = 0
Input parameters:
            alpha =    0.0500   (two sided)
               s1 =    0.8000
               s2 =    0.8944
                t =   10.0000
         ln(h2/h1) =   -0.6931
                N =        664
               p1 =    0.5000
   Accrual and follow-up information:
         duration =   10.0000
        follow-up =    9.0000
          accrual =    1.0000   (uniform)
Estimated power:
            power =    0.9000
```

and obtain the estimate of power 0.9.

◁

For examples of how such information may be used to construct power curves, see [ST] **stpower** and example 7 in *Power and effect-size determination* of [ST] **stpower logrank**.

▷ Example 9: Using streg to perform the log hazard-ratio test

In this example, we demonstrate the importance of sample-size computations to ensure a high power of a test to detect a difference between exponential survivor functions. We consider an asymptotic Wald (or normal z) test to test the equality of the log of the hazard ratio to zero.

Continuing example 8, suppose that the investigators have only 100 subjects available for the study. As we see below, the power to detect a 50% risk reduction (the hazard ratio of 0.5) of the experimental group reduces from 90% to 24%:

```
. stpower exponential 0.8, t(10) n(100) aperiod(1) fperiod(9) loghazard unconditional
Note: input parameters are survival probabilities.
Estimated power for two-sample comparison of survivor functions
Exponential test, log-hazard difference, unconditional
Ho: ln(h2/h1) = 0
Input parameters:
            alpha =    0.0500   (two sided)
               s1 =    0.8000
               s2 =    0.8944
                t =   10.0000
         ln(h2/h1) =   -0.6931
                N =        100
               p1 =    0.5000
   Accrual and follow-up information:
         duration =   10.0000
        follow-up =    9.0000
          accrual =    1.0000   (uniform)
Estimated power:
            power =    0.2414
```

To demonstrate the implication of this reduction, consider the following example. We generate the data according to the study from example 8 with the following code.

```
program simdata
        args n h1 h2 R
        set obs 'n'
        generate double entry = 'R'*runiform()
        generate double u = runiform()
        /* random allocation to two groups of equal sizes */
        generate double u1 = runiform()
        generate double u2 = runiform()
        sort u1 u2, stable
        generate byte drug = (_n<='n'/2)
        /* exponential failure times with rates h1 and h2 */
        generate double failtime = entry - ln(1-u)/'h1' if drug==0
        replace failtime = entry - ln(1-u)/'h2' if drug==1
end

. clear
. set seed 234
. quietly simdata 100 0.0223 0.0112 1
```

The entry times of subjects are generated from a uniform $[0, 1)$ distribution and stored in variable entry. The subjects are randomized to two groups of equal size of 50 subjects each. The survival times are generated from exponential distribution with the hazard rate of $-\ln(0.8)/10 = 0.0223$ in the control group, drug $= 0$, and the hazard rate of $0.5 \times 0.0223 = 0.0112$ in the experimental group, drug $= 1$, conditional on subjects' entry times in entry.

Before analyzing these survival data, we need to set it up properly using stset. The failure-time variable is failtime. The study terminates at $t = 10$, so we use exit(time 10) with stset to specify that all failure times past 10 are to be treated as censored. Because subjects enter the study at random times (entry) and become at risk of a failure upon entering the study, we also specify the origin(entry) option to ensure that the analysis time is adjusted for the entry times. For more details, see [ST] stset.

```
. stset failtime, exit(time 10) origin(entry)
        failure event:  (assumed to fail at time=failtime)
    obs. time interval: (origin, failtime]
    exit on or before:  time 10
       t for analysis:  (time-origin)
               origin:  time entry

      100  total obs.
        0  exclusions

      100  obs. remaining, representing
       15  failures in single record/single failure data
  891.5574 total analysis time at risk, at risk from t =         0
                              earliest observed entry t =         0
                                   last observed exit t =   9.99908
```

To perform the log hazard-ratio test, we fit an exponential regression model on drug by using streg (see [ST] streg). We can express the log of the hazard ratio in terms of regression coefficients as follows: $\ln(\Delta) = \ln(\lambda_2/\lambda_1) = \ln\{\exp(\beta_0 + \beta_1)/\exp(\beta_0)\} = \beta_1$, where β_0 and β_1 are the estimated coefficients for the constant and drug in the regression model. Then the test of H_0: $\ln(\Delta) = 0$ may be rewritten in terms of a coefficient on drug as H_0: $\beta_1 = 0$. This test is part of the standard output after streg.

```
. streg drug, distribution(exponential) nohr
         failure _d:  1 (meaning all fail)
   analysis time _t:  (failtime-origin)
             origin:  time entry
exit on or before:    time 10
Iteration 0:   log likelihood = -54.114272
Iteration 1:   log likelihood = -53.090859
Iteration 2:   log likelihood = -53.053454
Iteration 3:   log likelihood = -53.053406
Iteration 4:   log likelihood = -53.053406
Exponential regression -- log relative-hazard form
No. of subjects =         100              Number of obs   =        100
No. of failures =          15
Time at risk    = 891.5574257
                                            LR chi2(1)      =       2.12
Log likelihood  =  -53.053406               Prob > chi2     =     0.1452
```

_t	Coef.	Std. Err.	z	P>|z|	[95% Conf. Interval]
drug	-.7729235	.5477226	-1.41	0.158	-1.84644 .300593
_cons	-3.756554	.3162278	-11.88	0.000	-4.376349 -3.136759

From the output table above, the p-value for a two-sided test of the coefficient for drug, 0.158, is greater than 0.05. On that basis, we do not have evidence to reject the null hypothesis of no difference between the two exponential survivor functions and therefore make an incorrect decision because we simulated the data with different group hazard rates. If we were to repeat this, say, 100 times, using different datasets simulated according to the alternative H_a: $\ln(\Delta) = \ln(0.5) = -0.6931$ (see [R] **simulate**), for roughly 76 of them we would have failed to reject the null hypothesis of no difference (a type II error). Therefore, more subjects are required to be able to detect the log of the hazard ratio of -0.4055 in this study.

◁

▷ Example 10: Using results from streg to perform the Wald test of hazard difference

We obtain the power of the test based on the difference between hazards for the study in the above example (omit the loghazard option from the previous syntax).

```
. stpower exponential 0.8, t(10) n(100) aperiod(1) fperiod(9) unconditional
Note: input parameters are survival probabilities.
Estimated power for two-sample comparison of survivor functions
Exponential test, hazard difference, unconditional
Ho: h2-h1 = 0
Input parameters:
        alpha =    0.0500   (two sided)
           s1 =    0.8000
           s2 =    0.8944
            t =   10.0000
        h2-h1 =   -0.0112
            N =      100
           p1 =    0.5000
Accrual and follow-up information:
     duration =   10.0000
    follow-up =    9.0000
      accrual =    1.0000   (uniform)
Estimated power:
        power =    0.2458
```

We obtain a power estimate of 0.2458, which is close to 0.2414 from example 9.

To test the difference between hazard rates by using the Wald test, we express this difference in terms of coefficients, $\lambda_2 - \lambda_1 = \exp(\beta_0)\{\exp(\beta_1) - 1\}$, and use `testnl` ([R] **testnl**) after `streg` to perform the nonlinear hypothesis test of H_0: $\exp(\beta_0)\{\exp(\beta_1) - 1\} = 0$.

```
. testnl exp(_b[_cons])*(exp(_b[drug])-1) = 0
 (1)   exp(_b[_cons])*(exp(_b[drug])-1) = 0
              chi2(1) =        2.03
          Prob > chi2 =      0.1540
```

We obtain the same conclusions from the Wald test based on the difference between hazards as in example 9. That is, based on the p-value of 0.1540 we fail to reject the null hypothesis of no difference between hazards of two groups (or miss the alternative H_a: $\delta = -0.0112$ corresponding to reduction in hazard from roughly 0.02 to 0.01) for the data from example 9.

◁

❑ Technical note

Often in practice, to test the disparity in two exponential survivor functions, the log-rank test is used instead of the hazard-difference test. Also the Wald (or the score) test from the Cox model is used instead of the exponential log hazard-ratio test. Refer to [ST] **sts test** and [ST] **stcox** for examples on how to perform these tests (also see [ST] **stpower logrank** and [ST] **stpower cox**).

❑

❑ Technical note

Sometimes the estimates of sample size and power obtained under the assumption of the exponential model are used as an approximation to the results used in a more general context of the log-rank test or the Cox proportional hazards model. Refer to Lachin (2000, 409–410) for the rationale behind this. Also see Lakatos and Lan (1992) for a discussion of the circumstances under which sample-size estimates obtained assuming the exponential model may be inaccurate when used with more general proportional hazards models.

❑

Saved results

`stpower exponential` saves the following in `r()`:

Scalars

`r(power)`	power of test
`r(alpha)`	significance level of test
`r(hratio)`	hazard ratio
`r(onesided)`	1 if one-sided test, 0 otherwise
`r(h1)`	hazard in the control group (if specified)
`r(h2)`	hazard in the experimental group
`r(t)`	reference survival time (if `t()` is specified)
`r(p1)`	proportion of subjects in the control group
`r(fperiod)`	length of the follow-up period (if specified)
`r(aperiod)`	length of the accrual period (if specified)
`r(ashape)`	shape parameter (if `aperiod()` is specified)
`r(lh1)`	loss hazard in the control group (if specified)
`r(lh2)`	loss hazard in the experimental group (if specified)
`r(lt)`	reference loss to follow-up time (if `losstime()` is specified)

Macros
    r(method)   type of method (hazard difference or log-hazard difference)
    r(accrual)   type of entry distribution (uniform or exponential) (if requested)
    r(type)   type of approach (conditional or unconditional)

Matrices
    r(N)   1×3 matrix of required sample sizes
    r(Pr)   1×4 matrix of probabilities of an event (when computed)
    r(Ea)   1×3 matrix of expected number of events under the alternative (when computed)
    r(Eo)   1×3 matrix of expected number of events under the null (when computed)
    r(La)   1×3 matrix of expected number of losses under the alternative (when computed)
    r(Lo)   1×3 matrix of expected number of losses under the null (when computed)

Methods and formulas

stpower exponential is implemented as an ado-file.

In addition to the notation given in *Introduction*, denote N, N_1, and N_2 to be the total number of subjects required for the study, the number of subjects in the control group, and the number of subjects in the experimental group, respectively. Let $p_1 = N_1/N$ and $p_2 = N_2/N = 1 - p_1$ be the proportions of subjects allocated to the control and the experimental groups; γ be the shape parameter of the truncated exponential distribution with p.d.f. $g(z) = \gamma \exp(-\gamma z)/\{1 - \exp(-\gamma R)\}$, $0 \leq z \leq R$, $\gamma \neq 0$; η_1 and η_2 be the loss hazards in the control and the experimental groups; and $z_{(1-\alpha/k)}$ and $z_{(1-\beta)}$ be the $(1 - \alpha/k)$th and the $(1 - \beta)$th quantiles of the standard normal distribution, with $k = 1$ for the one-sided test and $k = 2$ for the two-sided test. Denote $\overline{\lambda} = p_1\lambda_1 + p_2\lambda_2$. Recall that the difference between hazards is denoted by $\delta = \lambda_2 - \lambda_1$ and the hazard ratio is denoted by $\Delta = \lambda_2/\lambda_1$.

If survival probabilities $S_1(t)$ and $S_2(t)$ at a fixed time t are specified rather than hazard rates, the hazard rates are computed as $\lambda_i = -\ln\{S_i(t)\}/t$, $i = 1, 2$. If loss to follow-up probabilities $L_1(t)$ and $L_2(t)$ at a fixed time t are given instead of loss to follow-up hazard rates, the loss hazard rates are computed as $\eta_i = -\ln\{1 - L_i(t)\}/t$, $i = 1, 2$.

All formulas below are derived under the assumption of exponential survival distributions with hazard rates in the control and the experimental groups λ_1 and λ_2, respectively, and rely on large-sample properties of the maximum likelihood estimates of λ_1 and λ_2.

Denote $\xi_o = \zeta(\overline{\lambda}, \gamma, \eta_1)p_1^{-1} + \zeta(\overline{\lambda}, \gamma, \eta_2)p_2^{-1}$ and $\xi_a = \zeta(\lambda_1, \gamma, \eta_1)p_1^{-1} + \zeta(\lambda_2, \gamma, \eta_2)p_2^{-1}$.

The formula for the sample-size calculation using the conditional approach is

$$N = \frac{\left(z_{1-\alpha/k}\xi_o^{1/2} + z_{1-\beta}\xi_a^{1/2}\right)^2}{\psi^2}$$

and using the unconditional approach is

$$N = \frac{(z_{1-\alpha/k} + z_{1-\beta})^2 \xi_a}{\psi^2}$$

where $\zeta(\lambda, \gamma, \eta) = \lambda^2/p_{\rm E}$ if $\psi = \delta$, $\zeta(\lambda, \gamma, \eta) = 1/p_{\rm E}$ if $\psi = \ln(\Delta)$, and $p_{\rm E}$ is to be defined later. λ and η denote a failure hazard rate and a loss to follow-up hazard rate.

In the absence of censoring, the overall probability of an event (failure), p_E, is set to 1. Here the resulting formula for the sample size for the log hazard-ratio test depends only on the ratio of hazards and not on the individual group hazard rates. The resulting sample size formula for the test of the difference may also be rewritten as a function of the ratio of hazards only. Therefore, under no censoring, for a fixed value of the hazard ratio $\Delta = \lambda_2/\lambda_1$, the estimates of the sample size (or power) will be constant with respect to varying hazard rates λ_1 and λ_2.

In the presence of censoring, when each subject is followed up for a fixed period $f = T$,

$$p_E = p_E(\lambda, \eta) = \frac{\lambda}{\lambda + \eta}[1 - \exp\{-(\lambda + \eta)T\}]$$

In the presence of an accrual period, the probability of an event is defined as

$$p_E = p_E(\lambda, \eta) = \frac{\lambda}{(\lambda + \eta)}\left[1 - \frac{\exp\{-(\lambda + \eta)(T - R)\} - \exp\{-(\lambda + \eta)T\}}{(\lambda + \eta)R}\right]$$

or

$$p_E = p_E(\lambda, \gamma, \eta) = \frac{\lambda}{(\lambda + \eta)}\left(1 + \frac{\gamma \exp\{-(\lambda + \eta)T\}[1 - \exp\{(\lambda + \eta - \gamma)R\}]}{(\lambda + \eta - \gamma)\{1 - \exp(-\gamma R)\}}\right)$$

under uniform or truncated exponential accrual with shape γ over $[0, R]$, respectively. Uniform accrual is assumed for $|\gamma| < 10^{-6}$.

The formulas are obtained from Lachin (1981), Lachin and Foulkes (1986), and Lakatos and Lan (1992). To avoid division by 0 in the case $\lambda + \eta = \gamma$, the probability of an event is taken to be the limit of the above expression, $p_E = \lim_{\lambda+\eta->\gamma} p_E(\lambda, \gamma, \eta)$.

The final estimate of the sample size is rounded up to the nearest even integer under the equal-group allocation and rounded up to the nearest integer otherwise. The number of subjects required to be recruited in each group is obtained as $N_1 = \pi_1 N$ and $N_2 = N - N_1$, where N_1 is rounded down to the nearest integer.

The expected number of events and losses to follow-up are computed as suggested by Lachin and Foulkes (1986). Under the null hypothesis,

$$E_{H_0} = N_1 p_E(\overline{\lambda}, \gamma, \eta_1) + N_2 p_E(\overline{\lambda}, \gamma, \eta_2)$$
$$L_{H_0} = N_1(\eta_1/\overline{\lambda}) p_E(\overline{\lambda}, \gamma, \eta_1) + N_2(\eta_2/\overline{\lambda}) p_E(\overline{\lambda}, \gamma, \eta_2)$$

and under the alternative hypothesis,

$$E_{H_a} = N_1 p_E(\lambda_1, \gamma, \eta_1) + N_2 p_E(\lambda_2, \gamma, \eta_2)$$
$$L_{H_a} = N_1(\eta_1/\lambda_1) p_E(\lambda_1, \gamma, \eta_1) + N_2(\eta_2/\lambda_2) p_E(\lambda_2, \gamma, \eta_2)$$

For unconditional tests, the expected number of events and losses to follow-up under the null is computed by setting $\overline{\lambda} = \lambda_1$. The estimates of the expected number of events and losses to follow-up in each group are rounded to the nearest integer.

To obtain the estimate of the power, $1 - \beta$, the formulas for the sample size are solved for $z_{(1-\beta)}$ and the normal cumulative distribution function is used to obtain the corresponding probability $1 - \beta$.

To obtain the unknown shape parameter, γ, of a truncated exponential entry distribution, an iterative procedure is used to solve the equation

$$\pi = G(t^*) = \frac{1 - \exp(-\gamma t^*)}{1 - \exp(-\gamma R)}$$

for a given proportion of subjects π recruited at a given time, t^*, for $t^* \in [0, R]$.

References

Chow, S.-C., J. Shao, and H. Wang. 2003. *Sample Size Calculations in Clinical Research*. New York: Marcel Dekker.

Freedman, L. S. 1982. Tables of the number of patients required in clinical trials using the logrank test. *Statistics in Medicine* 1: 121–129.

George, S. L., and M. M. Desu. 1974. Planning the size and duration of a clinical trial studying the time to some critical event. *Journal of Chronic Diseases* 27: 15–24.

Lachin, J. M. 1981. Introduction to sample size determination and power analysis for clinical trials. *Controlled Clinical Trials* 2: 93–113.

——. 2000. *Biostatistical Methods: The Assessment of Relative Risks*. New York: Wiley.

Lachin, J. M., and M. A. Foulkes. 1986. Evaluation of sample size and power for analyses of survival with allowance for nonuniform patient entry, losses to follow-up, noncompliance, and stratification. *Biometrics* 42: 507–519.

Lakatos, E., and K. K. G. Lan. 1992. A comparison of sample size methods for the logrank statistic. *Statistics in Medicine* 11: 179–191.

Rubinstein, L. V., M. H. Gail, and T. J. Santner. 1981. Planning the duration of a comparative clinical trial with loss to follow-up and a period of continued observation. *Journal of Chronic Diseases* 34: 469–479.

Schoenfeld, D. A. 1981. The asymptotic properties of nonparametric tests for comparing survival distributions. *Biometrika* 68: 316–319.

Also see [ST] **stpower** for more references.

Also see

[ST] **stpower** — Sample-size, power, and effect-size determination for survival analysis

[ST] **stpower logrank** — Sample size, power, and effect size for the log-rank test

[ST] **stpower cox** — Sample size, power, and effect size for the Cox proportional hazards model

[R] **sampsi** — Sample size and power determination

[ST] **streg** — Parametric survival models

[R] **test** — Test linear hypotheses after estimation

[ST] **Glossary**

Title

stpower logrank — Sample size, power, and effect size for the log-rank test

Syntax

Sample-size determination

> **stpower logrank** [*surv$_1$* [*surv$_2$*]] [, *options*]

Power determination

> **stpower logrank** [*surv$_1$* [*surv$_2$*]] , **n**(*numlist*) [*options*]

Effect-size determination

> **stpower logrank** [*surv$_1$*] , **n**(*numlist*) { **power**(*numlist*) | **beta**(*numlist*) } [*options*]

where

surv$_1$ is the survival probability in the control group at the end of the study t^*;

surv$_2$ is the survival probability in the experimental group at the end of the study t^*.

surv$_1$ and *surv$_2$* may each be specified either as one number or as a list of values (see [U] **11.1.8 numlist**) enclosed in parentheses.

options	description
Main	
*alpha(*numlist*)	significance level; default is alpha(0.05)
*power(*numlist*)	power; default is power(0.8)
*beta(*numlist*)	probability of type II error; default is beta(0.2)
*n(*numlist*)	sample size; required to compute power or effect size
*hratio(*numlist*)	hazard ratio (effect size) of the experimental to the control group; default is hratio(0.5)
onesided	one-sided test; default is two sided
*p1(*numlist*)	proportion of subjects in the control group; default is p1(0.5), meaning equal group sizes
*nratio(*numlist*)	ratio of sample sizes, N_2/N_1; default is nratio(1), meaning equal group sizes
schoenfeld	use the formula based on the log hazard-ratio in calculations; default is to use the formula based on the hazard ratio
parallel	treat number lists in starred options as parallel (do not enumerate all possible combinations of values) when multiple values per option are specified

Censoring

simpson(# # # | matname)	survival probabilities in the control group at three specific time points to compute the probability of an event (failure), using Simpson's rule under uniform accrual
st1(varname$_s$ varname$_t$)	variables varname$_s$, containing survival probabilities in the control group, and varname$_t$, containing respective time points, to compute the probability of an event (failure), using numerical integration under uniform accrual
wdprob(#)	the proportion of subjects anticipated to withdraw from the study; default is wdprob(0)

Reporting

table	display results in a table with default columns
columns(colnames)	display results in a table with specified colnames columns
notitle	suppress table title
nolegend	suppress table legend
colwidth(# [# ...])	column widths; default is colwidth(9)
separator(#)	draw a horizontal separator line every # lines; default is separator(0), meaning no separator lines
saving(filename [, replace])	save the table data to filename; use replace to overwrite existing filename
†noheader	suppress table header; seldom used
†continue	draw a continuation border in the table output; seldom used

*Starred options may be specified either as one number or as a list of values (see [U] **11.1.8 numlist**).

†noheader and continue are not shown in the dialog box.

colnames	description
alpha	significance level
power	power
beta	type II error probability
n	total number of subjects
n1	number of subjects in the control group
n2	number of subjects in the experimental group
e	total number of events (failures)
hr	hazard ratio
loghr	log of the hazard ratio
s1	survival probability in the control group
s2	survival probability in the experimental group
p1	proportion of subjects in the control group
nratio	ratio of sample sizes, experimental to control
w	proportion of withdrawals

stpower logrank — Sample size, power, and effect size for the log-rank test

By default, the following *colnames* are displayed:

power, n, n1, n2, e, and alpha are always displayed;

hr is displayed, unless the schoenfeld option is specified, in which case loghr is displayed;

s1 and s2 is displayed if survival probabilities are specified; and

w is displayed if withdrawal proportion (wdprob() option) is specified.

Menu

Statistics > Survival analysis > Power and sample size > Log-rank test

Description

stpower logrank estimates required sample size, power, and effect size for survival analysis comparing survivor functions in two groups by using the log-rank test. It also reports the number of events (failures) required to be observed in a study. This command supports two methods to obtain the estimates, those according to Freedman (1982) and Schoenfeld (1981). The command provides options to take into account unequal allocation of subjects between the two groups and possible withdrawal of subjects from the study (loss to follow-up). Optionally, the estimates can be adjusted for uniform accrual of subjects into the study. Also the minimal effect size (minimal detectable value of the hazard ratio or the log hazard-ratio) may be obtained for given power and sample size.

You can use stpower logrank to

- calculate required number of events and sample size when you know power and effect size (expressed as a hazard ratio) for uncensored and censored survival data,

- calculate power when you know sample size (number of events) and effect size (expressed as a hazard ratio) for uncensored and censored survival data, and

- calculate effect size (hazard ratio or log hazard-ratio if the schoenfeld option is specified) when you know sample size (number of events) and power for uncensored and censored survival data.

stpower logrank's input parameters, $surv_1$ and $surv_2$, are the values of survival probabilities in the control group (or the less favorable of the two groups), s_1, and in the experimental group, s_2, at the end of the study t^*.

Options

    Main

alpha(*numlist*) sets the significance level of the test. The default is alpha(0.05).

power(*numlist*) sets the power of the test. The default is power(0.8). If beta() is specified, this value is set to be 1−beta(). Only one of power() or beta() may be specified.

beta(*numlist*) sets the probability of a type II error of the test. The default is beta(0.2). If power() is specified, this value is set to be 1−power(). Only one of beta() or power() may be specified.

n(*numlist*) specifies the number of subjects in the study to be used to compute the power of the test or the minimal effect size (minimal detectable value of the hazard ratio or log hazard-ratio) if power() or beta() is also specified.

hratio(*numlist*) specifies the hazard ratio (effect size) of the experimental group to the control group. The default is hratio(0.5). This value defines the clinically significant improvement of the experimental procedure over the control desired to be detected by the log-rank test, with a certain power specified in power(). If both arguments $surv_1$ and $surv_2$ are specified, hratio() is not allowed and the hazard ratio is instead computed as $\ln(surv_2)/\ln(surv_1)$.

onesided indicates a one-sided test. The default is two sided.

p1(*numlist*) specifies the proportion of subjects in the control group. The default is p1(0.5), meaning equal allocation of subjects to the control and the experimental groups. Only one of p1() or nratio() may be specified.

nratio(*numlist*) specifies the sample-size ratio of the experimental group relative to the control group, N_2/N_1. The default is nratio(1), meaning equal allocation between the two groups. Only one of nratio() or p1() may be specified.

schoenfeld requests calculations using the formula based on the log hazard-ratio, according to Schoenfeld (1981). The default is to use the formula based on the hazard ratio, according to Freedman (1982).

parallel reports results sequentially (in parallel) over the list of numbers supplied to options allowing *numlist*. By default, results are computed over all combinations of the number lists in the following order of nesting: alpha(); p1() or nratio(); list of arguments $surv_1$ and $surv_2$; hratio(); power() or beta(); and n(). This option requires that options with multiple values each contain the same number of elements.

Censoring

simpson(# # # | *matname*) specifies survival probabilities in the control group at three specific time points, to compute the probability of an event (failure) using Simpson's rule, under the assumption of uniform accrual. Either the actual values or a 1×3 matrix, *matname*, containing these values can be specified. By default, the probability of an event is approximated as an average of the failure probabilities $1-s_1$ and $1-s_2$; see *Methods and formulas*. simpson() may not be combined with st1() and may not be used if arguments $surv_1$ or $surv_2$ are specified.

st1(*varname$_s$ varname$_t$*) specifies variables *varname$_s$*, containing survival probabilities in the control group, and *varname$_t$*, containing respective time points, to compute the probability of an event (failure) using numerical integration, under the assumption of uniform accrual; see [R] **dydx**. The minimum and the maximum values of *varname$_t$* must be the length of the follow-up period and the duration of the study, respectively. By default, the probability of an event is approximated as an average of the failure probabilities $1-s_1$ and $1-s_2$; see *Methods and formulas*. st1() may not be combined with simpson() and may not be used if arguments $surv_1$ or $surv_2$ are specified.

wdprob(#) specifies the proportion of subjects anticipated to withdraw from the study. The default is wdprob(0). wdprob() may not be combined with n().

Reporting

table displays results in a tabular format and is implied if any number list contains more than one element. This option is useful if you are producing results one case at a time and wish to construct your own custom table using a forvalues loop.

columns(*colnames*) specifies results in a table with specified *colnames* columns. The order of the columns in the output table is the same as the order of *colnames* specified in columns(). Column names in columns() must be space-separated.

notitle prevents the table title from displaying.

nolegend prevents the table legend from displaying and column headers from being marked.

colwidth(# [# ...]) specifies column widths. The default is 9 for all columns. The number of specified values may not exceed the number of columns in the table. A missing value (.) may be specified for any column to indicate the default width (9). If fewer widths are specified than the number of columns in the table, the last width specified is used for the remaining columns.

separator(#) specifies how often separator lines should be drawn between rows of the table. The default is separator(0), meaning that no separator lines should be displayed.

saving(*filename* [, replace]) creates a Stata data file (.dta file) containing the table values with variable names corresponding to the displayed *colnames*. replace indicates that *filename* be overwritten, if it exists. saving() is only appropriate with tabular output.

The following options are available with stpower logrank but are not shown in the dialog box:

noheader prevents the table header from displaying. This option is useful when the command is issued repeatedly, such as within a loop. noheader implies notitle.

continue draws a continuation border at the bottom of the table. This option is useful when the command is issued repeatedly within a loop.

Remarks

Remarks are presented under the following headings:

> Introduction
> Computing sample size in the absence of censoring
> Computing sample size in the presence of censoring
> Withdrawal of subjects from the study
> Including information about subject accrual
> Power and effect-size determination
> Performing the analysis using the log-rank test

Introduction

Consider a survival study comparing the survivor functions in two groups using the log-rank test. Let $S_1(t)$ and $S_2(t)$ denote the survivor functions of the control and the experimental groups, respectively. The key assumption of the log-rank test is that the hazard functions are proportional. That is, $h_2(t) = \Delta h_1(t)$ for any t or, equivalently, $S_2(t) = \{S_1(t)\}^\Delta$, where Δ is the hazard ratio. If $\Delta < 1$, the survival in the experimental group is higher relative to the survival in the control group; the new treatment is superior to the standard treatment. If $\Delta > 1$, then the standard treatment is superior to the new treatment. Under the proportional-hazards assumption, the test of the equality of the two survivor functions H_0: $S_1(t) = S_2(t)$ versus H_a: $S_1(t) \neq S_2(t)$ is equivalent to the test H_0: $\Delta = 1$ versus H_a: $\Delta \neq 1$ or H_0: $\ln(\Delta) = 0$ versus H_a: $\ln(\Delta) \neq 0$.

The methods implemented in stpower logrank for sample-size or power determination relate the power of the log-rank test directly to the number of events observed in the study. Depending on whether censoring occurs in a study, the required number of subjects is either equal to the number of events or is computed using the estimates of the number of events and the combined probability of an event (failure). Thus, in the presence of censoring, in addition to the number of events, the probability of a subject not being censored (failing) needs to be estimated to obtain the final estimate of the required number of subjects in the study.

To determine the required number of events, the investigator must specify the size or significance level, α, and the clinically significant difference between the two treatments (effect size) to be detected by the log-rank test, H_a: $\Delta = \Delta_a$, with prespecified power $1 - \beta$. The significance level, α, represents the probability of a *type I error*, a rejection of the null hypothesis when it is true. β represents the probability of a *type II error*, a failure to reject the null hypothesis when the alternative hypothesis is true. The significance level is often set to 0.05, and values for power() usually vary from 0.8 to 0.95. By default, stpower logrank uses power(0.8) (or, equivalently, beta(0.2)) and alpha(0.05). The effect size, a difference between the two treatments, is usually expressed as a hazard ratio, Δ_a, using the hratio() option. Under an unequal allocation of subjects between the two groups, the proportion of subjects in the control group may be specified in p1(), or the ratio of sample sizes may be supplied to nratio(). Optionally, results for the one-sided log-rank test may be requested by using onesided.

When all subjects fail by the end of the study (no censoring), a *type I study*, the information above is sufficient to obtain the number of subjects required in the study. Often, in practice, not all subjects fail by the end of the study, in which case censoring of subjects occurs (a *type II study*). Here the estimates of the survival probabilities in the control and experimental groups are necessary to estimate the probability of an event and, then, the required sample size.

By default, stpower logrank performs computations for the uncensored data (a type I study). It uses the hazard ratio specified in hratio() or the default hazard ratio of 0.5 to obtain required sample size or power. For censored data (a type II study), under administrative censoring, the value of the survival probability in the control group (supplied as argument $surv_1$ or, in the presence of an accrual period, in the simpson() or st1() option must be specified. If the value of the survival probability in the experimental group, $surv_2$, is omitted, $surv_1$ and the value of the hazard ratio in hratio() are used to compute the survival probability in the experimental group, s_2. If both arguments $surv_1$ and $surv_2$ are specified, the hazard ratio, Δ_a, is computed using these values and the hratio() option is not allowed.

If power determination is desired, sample size n() must be specified. If both n() and power() (or beta()) are specified, the minimal effect size (minimal value of the hazard ratio or log hazard-ratio) that can be detected by the log-rank test with requested power and fixed sample size is computed.

stpower logrank supports two methods, those of Freedman (1982) and Schoenfeld (1981), to obtain the estimates of the number of events or power (see also Marubini and Valsecchi [1997, 127, 134] and Collett [2003b, 301, 306]). The latter is used if option schoenfeld is specified. The final estimates of the sample size are based on the approximation of the probability of an event due to Freedman (1982), the default, or, for uniform accrual, due to Schoenfeld (1983) (see also Collett 2003b) if option simpson() is specified.

Optionally, the results may be displayed in a table by using table or columns(), as demonstrated in [ST] **stpower**. Refer to [ST] **stpower** and to example 7 in *Power and effect-size determination* to see how to obtain a graph of a power curve.

Computing sample size in the absence of censoring

We demonstrate several examples of how to use stpower logrank to obtain the estimates of sample size and number of events using Freedman (1982) and Schoenfeld (1981) methods for uncensored data, a type I study (when no censoring of subjects occurs).

▷ Example 1: Number of events (failures)

Consider a survival study to be conducted to compare the survivor function of subjects receiving a treatment (the experimental group) to the survivor function of those receiving a placebo or no treatment (the control group) using the log-rank test. Suppose that the study continues until all subjects fail (no censoring). The investigator wants to know how many events need to be observed in the study to achieve a power of 80% of a two-sided log-rank test with $\alpha = 0.05$, to detect a 50% reduction in the hazard of the experimental group ($\Delta_a = 0.5$). Because the default settings of `stpower logrank` are `power(0.8)`, `alpha(0.05)`, and `hratio(0.5)`, to obtain the estimate of the required number of events for the above study using the Freedman method (the default), we simply type

```
. stpower logrank

Estimated sample sizes for two-sample comparison of survivor functions
Log-rank test, Freedman method
Ho: S1(t) = S2(t)

Input parameters:
        alpha =    0.0500  (two sided)
       hratio =    0.5000
        power =    0.8000
           p1 =    0.5000

Estimated number of events and sample sizes:
            E =        72
            N =        72
           N1 =        36
           N2 =        36
```

From the output, a total of 72 events (failures) must be observed to achieve the required power of 80%. Because all subjects experience an event by the end of the study, the number of subjects required to be recruited to the study is equal to the number of events. That is, the investigator needs to recruit a total of 72 subjects (36 per group) to the study.

We can request the Schoenfeld method by specifying the `schoenfeld` option:

```
. stpower logrank, schoenfeld

Estimated sample sizes for two-sample comparison of survivor functions
Log-rank test, Schoenfeld method
Ho: S1(t) = S2(t)

Input parameters:
        alpha =    0.0500  (two sided)
    ln(hratio) =   -0.6931
        power =    0.8000
           p1 =    0.5000

Estimated number of events and sample sizes:
            E =        66
            N =        66
           N1 =        33
           N2 =        33
```

We obtain a slightly smaller estimate, 66, of the total number of events.

◁

❏ Technical note

Freedman (1982) and Schoenfeld (1981) derive the formulas for the number of events based on the asymptotic distribution of the log-rank test statistic. Freedman (1982) uses the asymptotic mean and variance of the log-rank test statistic expressed as a function of the true hazard ratio, Δ, whereas Schoenfeld (1981) (see also Collett [2003b, 302]) bases the derivation on the asymptotic mean of the log-rank test statistic as a function of the true log hazard-ratio, $\ln(\Delta)$. We label the corresponding approaches as "Freedman method" and "Schoenfeld method" in the output.

For values of the hazard ratio close to one, the two methods tend to give similar results. In general, the Freedman method gives higher estimates than the Schoenfeld method. The performance of the Freedman method was studied by Lakatos and Lan (1992) and was found to slightly overestimate the sample size under the assumption of proportional hazards. Hsieh (1992) investigates the performance of the two methods under the unequal allocation and concludes that Freedman's formula predicts the highest power for the log-rank test when the sample-size ratio of the two groups equals the reciprocal of the hazard ratio. Schoenfeld's formula predicts highest powers when sample sizes in the two groups are equal.

❑

Computing sample size in the presence of censoring

Because of limited costs and time, it is often infeasible to continue the study until all subjects experience an event. Instead, the study terminates at some prespecified point in time. As a result, some subjects may not experience an event by the end of the study; that is, administrative censoring of subjects occurs. This increases the requirement on the number of subjects in the study to ensure that a certain number of events is observed.

In the presence of censoring (for a type II study), Freedman (1982) assumes the following. The analysis occurs at a fixed time, t^*, after the last patient was accrued, and all information about subject follow-up beyond time, t^*, is excluded. To minimize an overestimation of the sample size because of neglecting this information, the author suggests choosing t^* as the minimum follow-up time, f, beyond which the frequency of occurrence of events is low (the time at which, say, 85% of the total events expected are observed). Under this assumption, the number of required subjects does not depend on the rates of accrual and occurrence of events but only on the proportions of patients in the two treatment groups, s_1 and s_2, surviving after the minimum follow-up time, f. See *Including information about subject accrual* about how to compute the sample size in the presence of a long accrual.

If censoring of subjects occurs, the probability of a subject not being censored needs to be estimated to obtain an accurate estimate of the required sample size. The assumption above justifies a simple procedure, suggested by Freedman (1982) and used by default by stpower logrank, to compute this probability using the estimates of survival probabilities at the end of the study in the control and the experimental groups. Therefore, for a type II study (under administrative censoring), these probabilities must be supplied to stpower logrank.

There are three ways of providing the information about survival of subjects in two groups. The first way is to supply both survival probabilities as arguments $surv_1$ and $surv_2$. The second way is to specify the survival probability in the control group as $surv_1$ and a hazard ratio in hratio(). Finally, the third way is to supply survival in the control group $surv_1$ only and rely on the default hratio(0.5). Below we demonstrate the first way.

▷ Example 2: Sample size in the presence of censoring

Consider an example from Machin et al. (2009, 91) of a study of patients with resectable colon cancer. The goal of the study was to compare the efficacy of the drug levamisole against a placebo with respect to relapse-free survival, using a one-sided log-rank test with a significance level of 5%. The investigators anticipated a 10% increase (from 50% to 60%, with a respective hazard ratio of 0.737) in the survival of the experimental group with respect to the survival of the control (placebo) group at the end of the study. They wanted to detect this increase with a power of 80%. To obtain the required sample size, we enter the survival probabilities 0.5 and 0.6 as arguments and specify the onesided option to request a one-sided test.

```
. stpower logrank 0.5 0.6, onesided

Estimated sample sizes for two-sample comparison of survivor functions
Log-rank test, Freedman method
Ho: S1(t) = S2(t)

Input parameters:

        alpha =    0.0500   (one sided)
           s1 =    0.5000
           s2 =    0.6000
       hratio =    0.7370
        power =    0.8000
           p1 =    0.5000

Estimated number of events and sample sizes:
            E =      270
            N =      600
           N1 =      300
           N2 =      300
```

From the above output, the investigators would have to observe a total of 270 events (relapses) to detect a 26% decrease in the hazard ($\Delta_a = 0.737$) of the experimental group relative to the hazard of the control group with a power of 80% using a one-sided log-rank test with $\alpha = 0.05$. They would have to recruit a total of 600 patients (300 per group) to observe that many events. In the absence of censoring, only 270 subjects would have been required to detect a decrease in hazard corresponding to $\Delta_a = 0.737$:

```
. stpower logrank, hratio(0.737) onesided

Estimated sample sizes for two-sample comparison of survivor functions
Log-rank test, Freedman method
Ho: S1(t) = S2(t)

Input parameters:

        alpha =    0.0500   (one sided)
       hratio =    0.7370
        power =    0.8000
           p1 =    0.5000

Estimated number of events and sample sizes:
            E =      270
            N =      270
           N1 =      135
           N2 =      135
```

Similarly, using the Schoenfeld method,

```
. stpower logrank 0.5 0.6, onesided schoenfeld
  (output omitted)
```

we find that 590 subjects are required in the study to observe a total of 266 events to ensure a power of a test of 80%.

◁

Although all examples demonstrated above assume equal group sizes, the information about the unequal allocation of subjects between the two groups may be provided by using p1() or nratio().

(Continued on next page)

Withdrawal of subjects from the study

Under administrative censoring, the subject is known to have experienced either of the two outcomes by the end of the study: survival or failure. Often, in practice, subjects may withdraw from the study before it terminates and therefore may not experience an event by the end of the study (or be censored), but for reasons other than administrative. Withdrawal of subjects from a study may greatly affect the estimate of the sample size and must be accounted for in the computations. Refer to [ST] **stpower** and [ST] **Glossary** for a formal definition of withdrawal.

Freedman (1982) suggests a conservative adjustment for the estimate of the sample size in the presence of withdrawal. Withdrawal is assumed to be independent of failure (event) times and administrative censoring.

The proportion of subjects anticipated to withdraw from a study may be specified by using wdprob().

▷ Example 3: Withdrawal of subjects from the study

Continuing example 2, suppose that a withdrawal rate of 10% is expected in the study of colon cancer patients. To account for this, we also specify wdprob(0.1):

```
. stpower logrank 0.5 0.6, onesided wdprob(0.1)
Estimated sample sizes for two-sample comparison of survivor functions
Log-rank test, Freedman method
Ho: S1(t) = S2(t)

Input parameters:

          alpha =     0.0500   (one sided)
             s1 =     0.5000
             s2 =     0.6000
         hratio =     0.7370
          power =     0.8000
             p1 =     0.5000
     withdrawal =    10.00%

Estimated number of events and sample sizes:

              E =        270
              N =        666
             N1 =        333
             N2 =        333
```

The estimate of the total sample size using the Freedman method increases from 600 to 666 when the withdrawal rate is assumed to be 10%. The adjustment of the estimate of the sample size for the withdrawal of subjects is conservative. It assumes equal withdrawals from each group; that is, 10% of subjects are lost by the end of the study in each group. This adjustment affects only the estimates of the sample sizes but not the number of events. The reasons for this are the following: withdrawal is assumed to be independent of event times, and the ratio of subjects surviving until the end of the study in the two groups does not change under equal withdrawals.

◁

We could use the alternative syntax and specify the survival probability in the control group, 0.5, with the value of the hazard ratio 0.737 in hratio() instead of supplying the two survival probabilities:

```
. stpower logrank 0.5, hratio(0.737) onesided wdprob(0.1)
```
  (*output omitted*)

Including information about subject accrual

Many clinical studies have an accrual period of R, during which the subjects are recruited to the study, and a follow-up period of $f = T - R$, during which the subjects are followed up until the end of the study, T, and no new subjects enter the study. The information about the duration of an accrual and a follow-up period affects the probability of a subject failing during the study.

Freedman (1982) suggests approximating the combined event-free probability as an average of the survival probabilities in the control and the experimental groups at the minimum follow-up time, $t^* = f$ (the default approach used in stpower logrank). However, for a long accrual of subjects, this approach may overestimate the required number of subjects, often seriously, because it does not take into account the information about subject follow-up beyond time f. Here Freedman (1982) proposes to use the survival probabilities at the *average follow-up time*, defined as $t^* = (f+T)/2 = f + 0.5R$, instead of the minimum follow-up time, f.

Alternatively, Schoenfeld (1983) (see also Collett [2003b, 306]) presents a formula for the required number of subjects allowing for uniform entry (accrual, recruitment) over $[0, R]$ and a follow-up period, f. This information is incorporated into the formula for the probability of a failure. The formula involves the integrals of the survivor functions of the control and the experimental groups. Schoenfeld (1983) suggests approximating the integral by using Simpson's rule, which requires the estimates of the survivor function at three specific time points, f, $0.5R + f$, and $T = R + f$. It is sufficient to provide the estimates of these three survival probabilities, $S_1(f)$, $S_1(0.5R+f)$, and $S_1(T)$, for the control group only. The corresponding survival probabilities of the experimental group are automatically computed using the value of the hazard ratio in hratio() and the proportional-hazards assumption.

The three estimates of the survival probabilities of the control group may be supplied by using the simpson() option to adjust the estimates of the sample size or power for uniform entry and a follow-up period. If the estimate of the survivor function over an array of values in the range $[f, T]$ is available from a previous study, it can be supplied using the st1() option to form a more accurate approximation of the probability of an event using numerical integration (see [R] **dydx**). Here the value of the length of the accrual period is needed for the computation. It is computed as the difference between the maximum and the minimum values of the time variable *varname$_t$*, supplied using st1(), that is, $R = T - f = \max(varname_t) - \min(varname_t)$.

▷ Example 4: Sample size in the presence of accrual and follow-up periods

Consider an example described in Collett (2003b, 309) of a survival study of chronic active hepatitis. A new treatment is to be compared with a standard treatment with respect to the survival times of the patients with this disease. The investigators desire to detect a change in a hazard ratio of 0.57 with 90% power and a 5% two-sided significance level. Also subjects are to be entered into the study uniformly over a period of 18 months and then followed up for 24 months. From the Kaplan–Meier estimate of the survivor function available for the control group, the survival probabilities at $f = 24$, $0.5R + f = 33$, and $T = 42$ months are 0.70, 0.57, and 0.45, respectively.

(Continued on next page)

```
. stpower logrank, hratio(0.57) power(0.9) schoenfeld simpson(0.7 0.57 0.45)
Note: probability of an event is computed using Simpson's rule with
      S1(f) = 0.70, S1(f+R/2) = 0.57, S1(T) = 0.45
      S2(f) = 0.82, S2(f+R/2) = 0.73, S2(T) = 0.63
Estimated sample sizes for two-sample comparison of survivor functions
Log-rank test, Schoenfeld method
Ho: S1(t) = S2(t)

Input parameters:

        alpha =   0.0500  (two sided)
   ln(hratio) =  -0.5621
        power =   0.9000
           p1 =   0.5000

Estimated number of events and sample sizes:
            E =      134
            N =      380
           N1 =      190
           N2 =      190
```

Collett (2003b, 305) reports the required number of events to be 133, which, apart from roundoff errors, agrees with our estimate of 134. In a later example, Collett (2003, 309) uses the number of events, rounded to 140, to compute the required sample size as $140/0.35 = 400$, where 0.35 is the estimate of the combined probability of an event. By hand, without rounding the number of events, we compute the required sample size as $133/0.35 = 380$ and obtain the same estimate of the total sample size as in the output.

Using the average follow-up time suggested by Freedman (1982), we obtain the following:

```
. stpower logrank 0.57, hratio(0.57) power(0.9) schoenfeld
Estimated sample sizes for two-sample comparison of survivor functions
Log-rank test, Schoenfeld method
Ho: S1(t) = S2(t)

Input parameters:

        alpha =   0.0500  (two sided)
           s1 =   0.5700
           s2 =   0.7259
   ln(hratio) =  -0.5621
        power =   0.9000
           p1 =   0.5000

Estimated number of events and sample sizes:
            E =      134
            N =      378
           N1 =      189
           N2 =      189
```

We specify the survival probability in the control group at $t^* = 0.5R + f = 0.5 \times 18 + 24 = 33$ as $S_1(33) = 0.57$ and the hazard ratio of 0.57 (coincidentally). The respective survival probability in the experimental group is $S_2(33) = S_1(33)^\Delta = 0.57^{0.57} = 0.726$. Here we obtain the estimate, 378, of the sample size, which is close to the estimate of 380 computed using the more complicated approximation. In this example, the two approximations produce similar results, but this may not always be the case.

◁

Technical note

The approximation suggested by Schoenfeld (1983) and Collett (2003b) is considered to be more accurate because it takes into account information about the patient survival beyond the average follow-up time. In general, the Freedman and Schoenfeld approximations will tend to give similar results when $\{\widetilde{S}(f) + \widetilde{S}(T)\}/2 \approx \widetilde{S}(0.5R + f)$; see *Methods and formulas* for a formal definition of $\widetilde{S}(\cdot)$.

❑

If we use the survival probability in the control group, $S_1(24) = 0.7$, at a follow-up time $t^* = f = 24$ instead of the average follow-up time $t^* = 33$ in the presence of an accrual period,

```
. stpower logrank 0.7, hratio(0.57) power(0.9) schoenfeld
(output omitted)
```

we obtain the estimate of the total sample size of 550, which is significantly greater than the previously estimated sample sizes of 380 and 378.

Power and effect-size determination

Sometimes the number of subjects available for the enrollment into the study is limited. In such cases, the researchers may want to investigate with what power they can detect a desired treatment effect for a given sample size.

▷ Example 5: Using stpower logrank to compute power

Recall the colon cancer study described in example 2. Suppose that only 100 subjects are available to be recruited to the study. We find out how this affects the power to detect a hazard ratio of 0.737.

```
. stpower logrank 0.5, hratio(0.737) onesided n(100)
Estimated power for two-sample comparison of survivor functions
Log-rank test, Freedman method
Ho: S1(t) = S2(t)

Input parameters:
        alpha =    0.0500  (one sided)
           s1 =    0.5000
           s2 =    0.6000
       hratio =    0.7370
            N =       100
           p1 =    0.5000

Estimated number of events and power:
            E =        46
        power =    0.2646
```

The power to detect an alternative H_a: $\Delta = 0.737$ decreased from 0.8 to 0.2646 when the sample size decreased from 600 to 100 (the number of events decreased from 270 to 46).

◁

▷ Example 6: Using stpower logrank to compute effect size

Continuing the above example, we can find that the value of the hazard ratio that can be detected for a fixed sample size of 100 with 80% power is approximately 0.42, corresponding to an increase in survival probability from 0.5 to roughly 0.75.

```
. stpower logrank 0.5, n(100) power(0.8) onesided
Estimated hazard ratio for two-sample comparison of survivor functions
Log-rank test, Freedman method
Ho: S1(t) = S2(t)

Input parameters:
        alpha =    0.0500  (one sided)
           s1 =    0.5000
           s2 =    0.7455
            N =       100
        power =    0.8000
           p1 =    0.5000

Estimated number of events and hazard ratio:
            E =        38
       hratio =    0.4237
```

◁

▷ Example 7: Plotting power curves

Here we demonstrate how to produce a graph of power curves over a range of hazard-ratio values. Continuing example 5, we visualize the effect of reducing the sample size from 600 to 100 on a power of the log-rank test to detect a hazard ratio of 0.737 by plotting two power curves for the sample sizes $N = 100$ and $N = 600$.

First, we generate a dataset named mypower containing the table values by using the saving() option. We request to compute the power for each of the two sample sizes over 100 values of the hazard ratio from 0.01 to 0.99 with 0.01 step size by supplying number lists 100, 600, and 0.01(0.01)0.99 to the n() and hratio() options, respectively. The values of hazard ratios, sample sizes, and powers are saved in variables hr, n, and power, respectively.

```
. quietly stpower logrank 0.5, hratio(0.01(0.01)0.99) n(100 600) onesided
> saving(mypower)
```

Next we generate the graph:

```
. use mypower
. twoway (line power hr if n==100) (line power hr if n==600),
    yline( .8, lstyle(foreground) lwidth(vvthin))
    xline(.42, lstyle(foreground) lwidth(vvthin))
    yline(.26, lstyle(foreground) lwidth(vvthin))
    xline(.74, lstyle(foreground) lwidth(vvthin))
    legend(label(1 "N = 100") label(2 "N = 600"))
    text(.85 .5 "(.42, .8)" .3 .81 "(.74, .26)" .85 .81 "(.74, .8)")
    title("Power curves") note("s1 = .5, alpha = .05 (one sided)")
    xtitle("Hazard ratio") ytitle("Power")
```

Power curves (s1 = .5, alpha = .05 (one sided); N = 100 solid, N = 600 dashed)

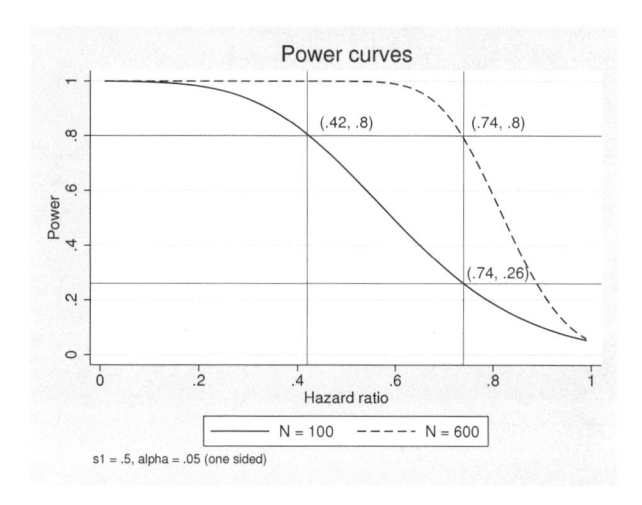

❑ Technical note

The decrease in sample size reduces the number of events observed in the study and therefore changes the estimates of the power. If the number of events were fixed, power would have been independent of the sample size, provided that all other parameters were held constant, because the formulas relate power directly to the number of events and not the number of subjects.

❑

Examples 5 and 7 demonstrate that a significant reduction in a sample size (a number of events) greatly reduces the power of the log-rank test to detect a desired change in survival of the two groups. Indeed, we examine this further in the next section.

Performing the analysis using the log-rank test

▷ Example 8: Using the log-rank test to detect a change in survival for a fixed sample size

Continuing example 5, consider the generated dataset drug.dta, consisting of variables drug, a drug type, and failtime, a time to failure.

```
. use http://www.stata-press.com/data/r11/drug
(Patient Survival in Drug Trial)
. tabulate drug
```

Treatment type	Freq.	Percent	Cum.
Placebo	50	33.33	33.33
Drug A	50	33.33	66.67
Drug B	50	33.33	100.00
Total	150	100.00	

```
. by drug, sort: summarize failtime
```

-> drug = Placebo

Variable	Obs	Mean	Std. Dev.	Min	Max
failtime	50	1.03876	.5535538	.1687701	2.382302

-> drug = Drug A

Variable	Obs	Mean	Std. Dev.	Min	Max
failtime	50	1.191802	.5927507	.2366922	2.277536

-> drug = Drug B

Variable	Obs	Mean	Std. Dev.	Min	Max
failtime	50	1.717314	.8350659	.5511715	3.796102

Failure times of the control group (Placebo) were generated from the Weibull distribution with $\lambda_w = 0.693$ and $p = 2$ (see [ST] **streg**); failure times of the two experimental groups, Drug A and Drug B, were generated from Weibull distributions with hazard functions proportional to the hazard of the control group in ratios 0.737 and 0.42, respectively. The Weibull family of survival distributions is chosen arbitrarily, and the Weibull parameter, λ_w, is chosen such that the survival at 1 year, $t = 1$, is roughly equal to 0.5. Subjects are randomly allocated to one of the three groups in equal proportions. Subjects with failure times greater than $t = 1$ will be censored at $t = 1$.

Before analyzing these survival data, we need to set it up using stset. After that, we can use sts test, logrank to test the survivor functions separately for Drug A against Placebo and Drug B against Placebo by using the log-rank test. See [ST] **stset** and [ST] **sts test** for more information about these two commands.

```
. stset failtime, exit(time 1)

     failure event:  (assumed to fail at time=failtime)
obs. time interval:  (0, failtime]
 exit on or before:  time 1

      150  total obs.
        0  exclusions

      150  obs. remaining, representing
       59  failures in single record/single failure data
 128.9845  total analysis time at risk, at risk from t =         0
                               earliest observed entry t =         0
                                    last observed exit t =         1

. sts test drug if drug!=2, logrank

         failure _d:  1 (meaning all fail)
   analysis time _t:  failtime
  exit on or before:  time 1
```

Log-rank test for equality of survivor functions

drug	Events observed	Events expected
Placebo	25	22.17
Drug A	21	23.83
Total	46	46.00

```
             chi2(1) =     0.70
             Pr>chi2 =     0.4028
```

```
. sts test drug if drug!=1, logrank

        failure _d:  1 (meaning all fail)
  analysis time _t:  failtime
  exit on or before:  time 1
```

Log-rank test for equality of survivor functions

drug	Events observed	Events expected
Placebo	25	16.61
Drug B	13	21.39
Total	38	38.00
	chi2(1) =	7.55
	Pr>chi2 =	0.0060

From the results from sts test for the Drug A group, we fail to reject the null hypothesis of no difference between the survivor functions in the two groups; the test made a type II error. On the other hand, for the Drug B group the one-sided p-value of 0.003, computed as $0.006/2 = 0.003$, suggests that the null hypothesis of nonsuperiority of the experimental treatment be rejected at the 0.005 significance level. We correctly conclude that the data provide the evidence that Drug B is superior to the Placebo.

Results from sts test, logrank for the two experimental groups agree with findings from examples 5 and 7. For the sample size of 100, the power of the log-rank test to detect the hazard ratio of 0.737 (10% increase in survival) is low (26%), whereas this sample size is sufficient for the test to detect a change in a hazard of 0.42 (25% increase in survival) with approximately 80% power.

Here we simulated our data from the alternative hypothesis and therefore can determine whether the correct decision or a type II error was made by the test. In practice, however, there is no way of determining the accuracy of the decision from the test. All we know is that in a long series of trials, there is a 5% chance that a particular test will incorrectly reject the null hypothesis and a 74% and a 20% chance that the test will miss the alternatives $H_a: \Delta = 0.737$ and $H_a: \Delta = 0.42$, respectively.

◁

Saved results

stpower logrank saves the following in r():

Scalars
    r(E)   total number of events (failures)
    r(power)   power of test
    r(alpha)   significance level of test
    r(hratio)   hazard ratio
    r(onesided)   type of test (0 if two-sided test, 1 if one-sided test)
    r(s1)   survival probability in the control group (if specified)
    r(s2)   survival probability in the experimental group (if specified)
    r(p1)   proportion of subjects in the control group
    r(w)   proportion of withdrawals (if specified)
    r(Pr_E)   probability of an event (failure) (when computed)

Macros
    r(method)   type of method (Freedman or Schoenfeld)

Matrices
    r(N)   1×3 matrix of required sample sizes

Methods and formulas

`stpower logrank` is implemented as an ado-file.

Let $S_1(t)$ and $S_2(t)$ denote the survivor functions of the control and the experimental groups and $\Delta(t) = \ln\{S_2(t)\}/\ln\{S_1(t)\}$ denote the hazard ratio at time t of the experimental to the control groups. Thus, for a given constant hazard ratio Δ, the survivor function of the experimental group at any time $t > 0$ may be computed as $S_2(t) = \{S_1(t)\}^\Delta$ under the assumption of proportional hazards. Define E and N to be the total number of events and the total number of subjects required for the study; w to be the proportion of subjects withdrawn from the study (lost to follow-up); $z_{(1-\alpha/k)}$ and $z_{(1-\beta)}$ to be the $(1-\alpha/k)$th and the $(1-\beta)$th quantiles of the standard normal distribution, with $k = 1$ for the one-sided test and $k = 2$ for the two-sided test. Let λ be the allocation ratio to the experimental group with respect to the control group, i.e., $N_2 = \lambda N_1$. If π_1 is the proportion of subjects allocated to the control group, then $\lambda = (1 - \pi_1)/\pi_1$.

The total number of events required to be observed in a study to ensure a power of $1 - \beta$ of the log-rank test to detect the hazard ratio Δ with significance level α, according to Freedman (1982), is

$$E = \frac{1}{\lambda}(z_{1-\alpha/k} + z_{1-\beta})^2 \left(\frac{\lambda\Delta + 1}{\Delta - 1}\right)^2$$

and, according to Schoenfeld (1983) and Collett (2003a, 301), is

$$E = \frac{(z_{1-\alpha/k} + z_{1-\beta})^2}{\pi_1(1-\pi_1)\ln^2(\Delta)} = \frac{1}{\lambda}(z_{1-\alpha/k} + z_{1-\beta})^2 \left\{\frac{1+\lambda}{\ln(\Delta)}\right\}^2$$

Both formulas are approximations and rely on a set of assumptions such as distinct failure times, all subjects completing the course of the study (no withdrawal), and a constant ratio, λ, of subjects at risk in two groups at each failure time.

The total sample size required to observe the total number of events, E, is given by

$$N = \frac{E}{p_E}$$

The estimate of the sample size is rounded up to the nearest even integer, for an equal allocation, or rounded up to the nearest integer otherwise. The number of subjects required to be recruited in each group is obtained as $N_1 = \pi_1 N$ and $N_2 = N - N_1$, where N_1 is rounded down to the nearest integer.

By default, the probability of an event (failure), p_E, is approximated as suggested by Freedman (1982):

$$p_E = 1 - \frac{S_1(t^*) + \lambda S_2(t^*)}{1 + \lambda}$$

where t^* is the minimum follow-up time, f, or, in the presence of an accrual period, the average follow-up time, $(f + T)/2 = f + 0.5R$.

If `simpson()` is specified, the probability of an event is approximated using Simpson's rule as suggested by Schoenfeld (1983):

$$p_E = 1 - \frac{1}{6}\left\{\widetilde{S}(f) + 4\widetilde{S}(0.5R + f) + \widetilde{S}(T)\right\}$$

where $\widetilde{S}(t) = \{S_1(t) + \lambda S_2(t)\}/(1+\lambda)$ and f, R, and $T = f + R$ are the follow-up period, the accrual period, and the total duration of the study, respectively.

The methods do not incorporate time explicitly but rather use it to determine values of the survival probabilities $S_1(t)$ and $S_2(t)$ used in the computations.

If st1() is used, the integral in the expression for the probability of an event

$$p_E = 1 - \frac{1}{R}\int_f^T \widetilde{S}(t)dt$$

is computed numerically using cubic splines ([R] **dydx**). The value of R is computed as the difference between the maximum and the minimum values of *varname$_t$* in st1(), $R = T - f = \max(varname_t) - \min(varname_t)$.

To account for the proportion of subjects, w, withdrawn from the study (lost to follow-up), a conservative adjustment to the total sample size is applied as follows:

$$N_w = \frac{N}{1-w}$$

Equal withdrawal rates are assumed in the adjustment of the group sample sizes for the withdrawal of subjects. Equal withdrawals do not affect the estimates of the number of events, provided that withdrawal is independent of event times and the ratio of subjects at risk in two groups remains constant at each failure time.

The power for each method is estimated using the formula

$$1 - \beta = \Phi\{|\psi|^{-1}(\lambda N p_E)^{1/2} - z_{1-\alpha/k}\}$$

where $\Phi(\cdot)$ is the standard normal cumulative distribution function; $\psi = (\lambda\Delta + 1)/(\Delta - 1)$ or $\psi = (1+\lambda)/\ln(\Delta)$ if the schoenfeld option is specified.

The estimate of the hazard ratio (or log hazard-ratio) for fixed power and sample size is computed (iteratively for censoring) using the formulas for the sample size given above. The value of the hazard ratio (log hazard-ratio) corresponding to the reduction in a hazard of the experimental group relative to the control group is reported.

References

Collett, D. 2003a. *Modelling Binary Data.* 2nd ed. London: Chapman & Hall/CRC.

———. 2003b. *Modelling Survival Data in Medical Research.* 2nd ed. London: Chapman & Hall/CRC.

Freedman, L. S. 1982. Tables of the number of patients required in clinical trials using the logrank test. *Statistics in Medicine* 1: 121–129.

Hsieh, F. Y. 1992. Comparing sample size formulae for trials with unbalanced allocation using the logrank test. *Statistics in Medicine* 11: 1091–1098.

Lakatos, E., and K. K. G. Lan. 1992. A comparison of sample size methods for the logrank statistic. *Statistics in Medicine* 11: 179–191.

Machin, D., M. J. Campbell, S. B. Tan, and S. H. Tan. 2009. *Sample Size Tables for Clinical Studies.* 3rd ed. Chichester, UK: Wiley–Blackwell.

Marubini, E., and M. G. Valsecchi. 1997. *Analysing Survival Data from Clinical Trials and Observational Studies.* Chichester, UK: Wiley.

Schoenfeld, D. A. 1981. The asymptotic properties of nonparametric tests for comparing survival distributions. *Biometrika* 68: 316–319.

——. 1983. Sample-size formula for the proportional-hazards regression model. *Biometrics* 39: 499–503.

Also see [ST] **stpower** for more references.

Also see

[ST] **stpower** — Sample-size, power, and effect-size determination for survival analysis

[ST] **stpower cox** — Sample size, power, and effect size for the Cox proportional hazards model

[ST] **stpower exponential** — Sample size and power for the exponential test

[R] **sampsi** — Sample size and power determination

[ST] **sts test** — Test equality of survivor functions

[ST] **stcox** — Cox proportional hazards model

[R] **test** — Test linear hypotheses after estimation

[ST] **Glossary**

Title

> **stptime** — Calculate person-time, incidence rates, and SMR

Syntax

> stptime [*if*] [, *options*]

options	description
Main	
at(*numlist*)	compute person-time at specified intervals; default is to compute overall person-time and incidence rates
trim	exclude observations ≤ minimum or > maximum of at()
by(*varname*)	compute incidence rates or SMRs by *varname*
Options	
per(*#*)	units to be used in reported rates
dd(*#*)	number of decimal digits to be displayed
smr(*groupvar ratevar*)	use *groupvar* and *ratevar* in using() dataset to calculate SMRs
using(*filename*)	specify filename to merge that contains smr() variables
level(*#*)	set confidence level; default is level(95)
noshow	do not show st setting information
Advanced	
jackknife	jackknife confidence intervals
title(*string*)	label output table with *string*
output(*filename*[, replace])	save summary dataset as *filename*; use replace to overwrite existing *filename*

You must stset your data before using stptime; see [ST] **stset**.
by is allowed; see [D] **by**.
fweights, iweights, and pweights may be specified using stset; see [ST] **stset**.

Menu

Statistics > Survival analysis > Summary statistics, tests, and tables > Person-time, incidence rates, and SMR

Description

stptime calculates person-time and incidence rates. stptime computes standardized mortality/morbidity ratios (SMRs) after merging the data with a suitable file of standard rates specified with the using() option.

Options

◻ Main ◻

at(*numlist*) specifies intervals at which person-time is to be computed. The intervals are specified in analysis time t units. If at() is not specified, overall person-time and incidence rates are computed.

If, for example, you specify at(5(5)20) and the trim option is not specified, person-time is reported for the intervals $t = (0 - 5]$, $t = (5 - 10]$, $t = (10 - 15]$, and $t = (15 - 20]$.

trim specifies that observations less than or equal to the minimum or greater than the maximum value listed in at() be excluded from the computations.

by(*varname*) specifies a categorical variable by which incidence rates or SMRs are to be computed.

◻ Options ◻

per(*#*) specifies the units to be used in reported rates. For example, if the analysis time is in years, specifying per(1000) results in rates per 1,000 person-years.

dd(*#*) specifies the maximum number of decimal digits to be reported for rates, ratios, and confidence intervals. This option affects only how values are displayed, not how they are calculated.

smr(*groupvar ratevar*) specifies two variables in the using() dataset. The *groupvar* identifies the age-group or calendar-period variable used to match the data in memory and the using() dataset. The *ratevar* variable contains the appropriate reference rates. stptime then calculates SMRs rather than incidence rates.

using(*filename*) specifies the filename that contains a file of standard rates that is to be merged with the data so that SMRs can be calculated.

level(*#*) specifies the confidence level, as a percentage, for confidence intervals. The default is level(95) or as set by set level; see [U] **20.7 Specifying the width of confidence intervals**.

noshow prevents stptime from showing the key st variables. This option is seldom used because most people type stset, show or stset, noshow to set whether they want to see these variables mentioned at the top of the output of every st command; see [ST] **stset**.

◻ Advanced ◻

jackknife specifies that jackknife confidence intervals be produced. This is the default if pweights or iweights were specified when the dataset was stset.

title(*string*) replaces the default "person-time" label on the output table with *string*.

output(*filename* [, replace]) saves a summary dataset in *filename*. The file contains counts of failures and person-time, incidence rates (or SMRs), confidence limits, and categorical variables identifying the time intervals. This dataset could be used for further calculations or simply as input to the table command.

replace indicates that *filename* be overwritten, if it exists. This option is not shown in the dialog box.

Remarks

stptime computes and tabulates the person-time and incidence rate (formed from the number of failures divided by the person-time). If you use the by() option, this will be calculated by different levels of one or more categorical explanatory variables specified by *varname*. Confidence intervals for the rate are also given. By default, the confidence intervals are calculated using the quadratic approximation to the Poisson log likelihood for the log-rate parameter. However, whenever the Poisson assumption is questionable, such as when `pweights` or `iweights` are used, jackknife confidence intervals can also be calculated.

`stptime` can also calculate and report SMRs if the data have been merged with a suitable file of reference rates.

If `pweights` or `iweights` were specified when the dataset was `stset`, `stptime` calculates jackknife confidence intervals by default.

The summary dataset can be saved to a file specified with the `output()` option for further analysis or a more elaborate graphical display.

▷ Example 1

We begin with a simple fictitious example from Clayton and Hills (1993, 42). Thirty subjects were monitored until the development of a particular disease. Here are the data for the first five subjects:

```
. use http://www.stata-press.com/data/r11/stptime
. list in 1/5
```

	id	year	fail
1.	1	19.6	1
2.	2	10.8	1
3.	3	14.1	1
4.	4	3.5	1
5.	5	4.8	1

The `id` variable identifies the subject, `year` records the time to failure in years, and `fail` is the failure indicator, which is one for all 30 subjects in the data. To use `stptime`, we must first `stset` the data.

```
. stset year, fail(fail) id(id)
                id:  id
     failure event:  fail != 0 & fail < .
obs. time interval:  (year[_n-1], year]
 exit on or before:  failure

       30  total obs.
        0  exclusions

       30  obs. remaining, representing
       30  subjects
       30  failures in single failure-per-subject data
    261.9  total analysis time at risk, at risk from t =         0
                                  earliest observed entry t =    0
                                       last observed exit t =   36.5
```

We can use `stptime` to obtain the overall person-time of observation and disease incidence rate.

```
. stptime, title(person-years)
         failure _d:  fail
   analysis time _t:  year
                 id:  id
```

Cohort	person-years	failures	rate	[95% Conf. Interval]
total	261.9	30	.11454754	.08009 .1638299

The total 261.9 person-years reported by stptime matches what stset reported as total analysis time at risk. stptime computed an incidence rate of 0.11454754 per person-year. In epidemiology, incidence rates are often presented per 1,000 person-years. We can do this by specifying per(1000).

```
. stptime, title(person-years) per(1000)
        failure _d:  fail
  analysis time _t:  year
               id:   id
```

Cohort	person-years	failures	rate	[95% Conf. Interval]
total	261.9	30	114.54754	80.09001 163.8299

More interesting would be to compare incidence rates at 10-year intervals. We will specify dd(4) to display rates to four decimal places.

```
. stptime, per(1000) at(0(10)40) dd(4)
        failure _d:  fail
  analysis time _t:  year
               id:   id
```

Cohort	person-time	failures	rate	[95% Conf. Interval]
(0 - 10]	188.8000	18	95.3390	60.0676 151.3215
(10 - 20]	55.1000	10	181.4882	97.6506 337.3044
(20 - 30]	11.5000	1	86.9565	12.2490 617.3106
> 30	6.5000	1	153.8462	21.6713 1092.1648
total	261.9000	30	114.5475	80.0900 163.8299

◁

▷ Example 2

Using the diet data (Clayton and Hills 1993) described in example 1 of [ST] **stsplit**, we will use stptime to tabulate age-specific person-years and coronary heart disease (CHD) incidence rates. In this dataset, CHD has been coded as fail = 1, 3, or 13.

We first stset the data: failure codes for CHD are specified; origin is set to date of birth, making age the analysis time; and the scale is set to 365.25, so analysis time is measured in years.

```
. use http://www.stata-press.com/data/r11/diet
(Diet data with dates)
. stset dox, origin(time dob) enter(time doe) id(id) scale(365.25)
> fail(fail==1 3 13)
               id:  id
    failure event:  fail == 1 3 13
obs. time interval:  (dox[_n-1], dox]
 enter on or after:  time doe
  exit on or before:  failure
    t for analysis:  (time-origin)/365.25
            origin:  time dob

      337  total obs.
        0  exclusions

      337  obs. remaining, representing
      337  subjects
       46  failures in single failure-per-subject data
 4603.669  total analysis time at risk, at risk from t =         0
                                earliest observed entry t =  30.07529
                                   last observed exit t =  69.99863
```

The incidence of CHD per 1,000 person-years can be tabulated in 10-year intervals.

```
. stptime, per(1000) at(40(10)70) trim
          failure _d:  fail == 1 3 13
    analysis time _t:  (dox-origin)/365.25
              origin:  time dob
   enter on or after:  time doe
                  id:  id
                note:  _group<=40 trimmed
```

Cohort	person-time	failures	rate	[95% Conf. Interval]	
(40 - 50]	907.00616	6	6.6151701	2.971936	14.72457
(50 - 60]	2107.0418	18	8.5427828	5.382317	13.55906
(60 - 70]	1493.2923	22	14.732548	9.700656	22.37457
total	4507.3402	46	10.205575	7.644246	13.62512

The SMR for a cohort is the ratio of the total number of observed deaths to the number expected from age-specific reference rates. This expected number can be found by multiplying the person-time in each cohort by the reference rate for that cohort. Using the smr option to define the cohort variable and reference rate variable in the using() dataset, stptime calculates SMRs and confidence intervals. You must specify the per() option. For example, if the reference rates were per 100,000, you would specify per(100000).

▷ Example 3

In smrchd.dta, we have age-specific CHD rates per 1,000 person-years for a reference population. We can merge these data with our current data and use stptime to obtain SMRs and confidence intervals.

```
. stptime, smr(ageband rate) using(http://www.stata-press.com/data/r11/smrchd)
> per(1000) at(40(10)70) trim
          failure _d:  fail == 1 3 13
    analysis time _t:  (dox-origin)/365.25
              origin:  time dob
   enter on or after:  time doe
                  id:  id
                note:  _group<=40 trimmed
```

Cohort	person-time	observed failures	expected failures	SMR	[95% Conf. Interval]	
(40 - 50]	907.00616	6	5.62344	1.067	.4793445	2.374931
(50 - 60]	2107.0418	18	18.7527	.95986	.6047547	1.52349
(60 - 70]	1493.2923	22	22.8474	.96291	.6340298	1.46239
total	4507.3402	46	47.2235	.97409	.7296205	1.300477

The stptime command can also calculate person-time and incidence rates or SMRs by categories of the explanatory variable. In our diet data, the variable hienergy is coded 1 if the total energy consumption is more than 2.75 Mcal and 0 otherwise. We want to compute the person-years and incidence rates for these two levels of hienergy.

```
. stptime, by(hienergy) per(1000)
         failure _d:  fail == 1 3 13
     analysis time _t:  (dox-origin)/365.25
              origin:  time dob
    enter on or after:  time doe
                 id:  id
```

hienergy	person-time	failures	rate	[95% Conf. Interval]	
0	2059.4305	28	13.595992	9.387478	19.69123
1	2544.2382	18	7.0748093	4.457431	11.2291
total	4603.6687	46	9.9920309	7.484296	13.34002

We can also compute the incidence rate for the two levels of hienergy and the three previously defined age cohorts:

```
. stptime, by(hienergy) per(1000) at(40(10)70) trim
         failure _d:  fail == 1 3 13
     analysis time _t:  (dox-origin)/365.25
              origin:  time dob
    enter on or after:  time doe
                 id:  id
```

hienergy	person-time	failures	rate	[95% Conf. Interval]	
0					
(40 - 50]	346.87474	2	5.76577	1.442006	23.05407
(50 - 60]	979.34018	12	12.253148	6.958681	21.57587
> 60	699.13758	14	20.024671	11.85966	33.81104
1					
(40 - 50]	560.13142	4	7.1411813	2.680213	19.02702
(50 - 60]	1127.7016	6	5.3205566	2.390317	11.84292
> 60	794.15469	8	10.073604	5.037786	20.14327
total	4507.3402	46	10.205575	7.644246	13.62512

Or we can compute the corresponding SMR:

```
. stptime, smr(ageband rate) using(http://www.stata-press.com/data/r11/smrchd)
> by(hienergy) per(1000) at(40(10)70) trim
         failure _d:  fail == 1 3 13
     analysis time _t:  (dox-origin)/365.25
              origin:  time dob
    enter on or after:  time doe
                 id:  id
```

hienergy	person-time	observed failures	expected failures	SMR	[95% Conf. Interval]	
0						
(40 - 50]	346.87474	2	2.15062	.9299629	.2325815	3.718399
(50 - 60]	979.34018	12	8.71613	1.376758	.7818743	2.424256
> 60	699.13758	14	10.6968	1.308802	.7751411	2.209872
1						
(40 - 50]	560.13142	4	3.47281	1.151803	.4322924	3.068875
(50 - 60]	1127.7016	6	10.0365	.5978154	.2685749	1.330665
> 60	794.15469	8	12.1506	.6584055	.329267	1.316554
total	4507.3402	46	47.2235	.9740917	.7296205	1.300477

Saved results

stptime saves the following in r():

Scalars
    r(ptime)   person-time
    r(failures)   observed failures
    r(expected)   expected number of failures
    r(smr)   standardized mortality ratio
    r(lb)   lower bound for SMR
    r(ub)   upper bound for SMR

Methods and formulas

stptime is implemented as an ado-file.

Reference

Clayton, D. G., and M. Hills. 1993. *Statistical Models in Epidemiology*. Oxford: Oxford University Press.

Also see

[ST] **strate** — Tabulate failure rates and rate ratios

[ST] **stci** — Confidence intervals for means and percentiles of survival time

[ST] **stir** — Report incidence-rate comparison

[ST] **stset** — Declare data to be survival-time data

[ST] **stsplit** — Split and join time-span records

[ST] **epitab** — Tables for epidemiologists

Title

strate — Tabulate failure rates and rate ratios

Syntax

Tabulate failure rates

strate [*varlist*] [*if*] [*in*] [, *strate_options*]

Calculate rate ratios with the Mantel–Haenszel method

stmh *varname* [*varlist*] [*if*] [*in*] [, *options*]

Calculate rate ratios with the Mantel–Cox method

stmc *varname* [*varlist*] [*if*] [*in*] [, *options*]

strate_options	description
Main	
per(*#*)	units to be used in reported rates
smr(*varname*)	use *varname* as reference-rate variable to calculate SMRs
cluster(*varname*)	cluster variable to be used by the jackknife
jackknife	report jackknife confidence intervals
missing	include missing values as extra categories
level(*#*)	set confidence level; default is level(95)
output(*filename*[, replace])	save summary dataset as *filename*; use replace to overwrite existing *filename*
nolist	suppress listed output
graph	graph rates against exposure category
nowhisker	omit confidence intervals from the graph
Plot	
marker_options	change look of markers (color, size, etc.)
marker_label_options	add marker labels; change look or position
cline_options	affect rendition of the plotted points
CI plot	
ciopts(*rspike_options*)	affect rendition of the confidence intervals (whiskers)
Add plots	
addplot(*plot*)	add other plots to the generated graph
Y axis, X axis, Titles, Legend, Overall	
twoway_options	any options other than by() documented in [G] ***twoway_options***

strate — Tabulate failure rates and rate ratios

options	description
Main	
by(*varlist*)	tabulate rate ratio on *varlist*
compare(*num1*,*den2*)	compare categories of exposure variable
missing	include missing values as extra categories
level(*#*)	set confidence level; default is level(95)

You must stset your data before using strate, stmh, and stmc; see [ST] **stset**.
by is allowed with stmh and stmc; see [D] **by**.
fweights, iweights, and pweights may be specified using stset; see [ST] **stset**.

Menu

strate

Statistics > Survival analysis > Summary statistics, tests, and tables > Tabulate failure rates and rate ratios

stmh

Statistics > Survival analysis > Summary statistics, tests, and tables > Tabulate Mantel-Haenszel rate ratios

stmc

Statistics > Survival analysis > Summary statistics, tests, and tables > Tabulate Mantel-Cox rate ratios

Description

strate tabulates rates by one or more categorical variables declared in *varlist*. You can also save an optional summary dataset, which includes event counts and rate denominators, for further analysis or display. The combination of the commands stsplit and strate implements most of, if not all, the functions of the special-purpose person-years programs in widespread use in epidemiology. See Clayton and Hills (1993) and [ST] **stsplit**. If your interest is solely in calculating person-years, see [ST] **stptime**.

stmh calculates stratified rate ratios and significance tests by using a Mantel–Haenszel-type method.

stmc calculates rate ratios that are stratified finely by time by using the Mantel–Cox method. The corresponding significance test (the log-rank test) is also calculated.

Both stmh and stmc can estimate the failure-rate ratio for two categories of the explanatory variable specified by the first argument of *varlist*. You can define categories to be compared by specifying them with the compare() option. The remaining variables in *varlist* before the comma are categorical variables, which are to be "controlled for" using stratification. Strata are defined by cross-classification of these variables.

You can also use stmh and stmc to carry out trend tests for a metric explanatory variable. Here a one-step Newton approximation to the log-linear Poisson regression coefficient is computed.

Options for strate

Main

per(*#*) specifies the units to be used in reported rates. For example, if the analysis time is in years, specifying per(1000) results in rates per 1,000 person-years.

smr(*varname*) specifies a reference-rate variable. strate then calculates SMRs rather than rates. This option will usually follow stsplit to separate the follow-up records by age bands and possibly calendar periods.

cluster(*varname*) defines a categorical variable that indicates clusters of data to be used by the jackknife. If the jackknife option is selected and this option is not specified, the cluster variable is taken as the id variable defined in the st data. Specifying cluster() implies jackknife.

jackknife specifies that jackknife confidence intervals be produced. This is the default if weights were specified when the dataset was stset.

missing specifies that missing values of the explanatory variables be treated as extra categories. The default is to exclude such observations.

level(*#*) specifies the confidence level, as a percentage, for confidence intervals. The default is level(95) or as set by set level; see [U] **20.7 Specifying the width of confidence intervals**.

output(*filename*[, replace]) saves a summary dataset in *filename*. The file contains counts of failures and person-time, rates (or SMRs), confidence limits, and all the categorical variables in the *varlist*. This dataset could be used for further calculations or simply as input to the table command.

> replace indicates that *filename* be overwritten, if it exists. This option is not shown in the dialog box.

nolist suppresses the output. This is used only when saving results to a file specified by output().

graph produces a graph of the rate against the numerical code used for the categories of *varname*.

nowhisker omits the confidence intervals from the graph.

┌─ Plot ───

marker_options affect the rendition of markers drawn at the plotted points, including their shape, size, color, and outline; see [G] ***marker_options***.

marker_label_options specify if and how the markers are to be labeled; see [G] ***marker_label_options***.

cline_options affect whether lines connect the plotted points and the rendition of those lines; see [G] ***cline_options***.

┌─ CI plot ──

ciopts(*rspike_options*) affects the rendition of the confidence intervals (whiskers); see [G] ***rspike_options***.

┌─ Add plots ──

addplot(*plot*) provides a way to add other plots to the generated graph; see [G] ***addplot_option***.

┌─ Y axis, X axis, Titles, Legend, Overall ──────────────────────────────────

twoway_options are any of the options documented in [G] ***twoway_options***, excluding by(). These include options for titling the graph (see [G] ***title_options***) and for saving the graph to disk (see [G] ***saving_option***).

Options for stmh and stmc

⌐ Main ⌐

by(*varlist*) specifies categorical variables by which the rate ratio is to be tabulated.

A separate rate ratio is produced for each category or combination of categories of *varlist*, and a test for unequal rate ratios (effect modification) is displayed.

compare(*num1*,*den2*) specifies the categories of the exposure variable to be compared. The first code defines the numerator categories, and the second code defines the denominator categories.

When compare is absent and there are only two categories, the larger is compared with the smaller; when there are more than two categories, compare analyzes log-linear trend.

missing specifies that missing values of the explanatory variables be treated as extra categories. The default is to exclude such observations.

level(*#*) specifies the confidence level, as a percentage, for confidence intervals. The default is level(95) or as set by set level; see [U] **20.7 Specifying the width of confidence intervals**.

Remarks

Remarks are presented under the following headings:

> *Tabulation of rates by using strate*
> *Stratified rate ratios using stmh*
> *Log-linear trend test for metric explanatory variables using stmh*
> *Controlling for age with fine strata by using stmc*

Tabulation of rates by using strate

strate tabulates the rate, formed from the number of failures divided by the person-time, by different levels of one or more categorical explanatory variables specified by *varlist*. Confidence intervals for the rate are also given. By default, the confidence intervals are calculated using the quadratic approximation to the Poisson log likelihood for the log-rate parameter. However, whenever the Poisson assumption is questionable, jackknife confidence intervals can also be calculated. The jackknife option also allows for multiple records for the same cluster (usually subject).

strate can also calculate and report SMRs if the data have been merged with a suitable file of reference rates.

You can save the summary dataset saved to a file specified with the output() option for further analysis or more elaborate graphical display.

If weights were specified when the dataset was stset, strate calculates jackknife confidence intervals by default.

▷ Example 1

Using the diet data (Clayton and Hills 1993) described in example 1 of [ST] **stsplit**, we will use strate to tabulate age-specific coronary heart disease (CHD). In this dataset, CHD has been coded as fail = 1, 3, or 13.

We first stset the data: failure codes for CHD are specified; origin is set to date of birth, making age analysis time; and the scale is set to 365.25, so analysis time is measured in years.

```
. use http://www.stata-press.com/data/r11/diet
(Diet data with dates)

. stset dox, origin(time doe) id(id) scale(365.25) fail(fail==1 3 13)
                id:  id
     failure event:  fail == 1 3 13
obs. time interval:  (dox[_n-1], dox]
 exit on or before:  failure
     t for analysis:  (time-origin)/365.25
             origin:  time doe

        337  total obs.
          0  exclusions

        337  obs. remaining, representing
        337  subjects
         46  failures in single failure-per-subject data
   4603.669  total analysis time at risk, at risk from t =         0
                                   earliest observed entry t =         0
                                      last observed exit t =  20.04107
```

Now we stsplit the data into 10-year age bands.

```
. stsplit ageband, at(40(10)70) after(time=dob) trim
(26 + 0 obs. trimmed due to lower and upper bounds)
(418 observations (episodes) created)
```

stsplit added 418 observations to the dataset in memory and generated a new variable, ageband, which identifies each observation's age group.

The CHD rate per 1,000 person-years can now be tabulated for categories of ageband:

```
. strate ageband, per(1000) graph
        failure _d:  fail == 1 3 13
   analysis time _t:  (dox-origin)/365.25
            origin:  time doe
                id:  id
              note:  ageband<=40 trimmed

Estimated rates (per 1000) and lower/upper bounds of 95% confidence intervals
(729 records included in the analysis)
```

ageband	D	Y	Rate	Lower	Upper
40	6	0.9070	6.6152	2.9719	14.7246
50	18	2.1070	8.5428	5.3823	13.5591
60	22	1.4933	14.7325	9.7007	22.3746

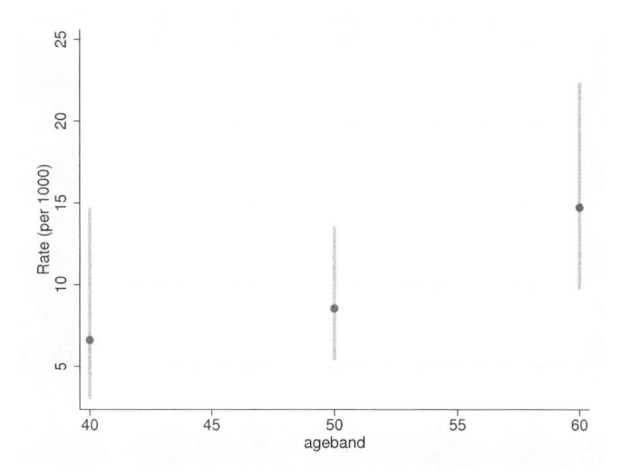

Because we specified the graph option, strate also generated a plot of the estimated rates and confidence intervals.

◁

The SMR for a cohort is the ratio of the total number of observed deaths to the number expected from age-specific reference rates. This expected number can be found by first expanding on age, using stsplit, and then multiplying the person-years in each age band by the reference rate for that band. merge (see [D] **merge**) can be used to add the reference rates to the dataset. Using the smr option to define the variable containing the reference rates, strate calculates SMRs and confidence intervals. You must specify the per() option. For example, if the reference rates were per 100,000, you would specify per(100000). When reference rates are available by age and calendar period, you must call stsplit twice to expand on both time scales before merging the data with the reference-rate file.

▷ Example 2

In smrchd.dta, we have age-specific CHD rates per 1,000 person-years for a reference population. We can merge these data with our current data and use strate to obtain SMRs and confidence intervals.

(*Continued on next page*)

```
. sort ageband
. merge m:1 ageband using http://www.stata-press.com/data/r11/smrchd
ageband was byte now float

    Result                          # of obs.

    not matched                           26
        from master                       26  (_merge==1)
        from using                         0  (_merge==2)
    matched                              729  (_merge==3)

. strate ageband, per(1000) smr(rate)
         failure _d:  fail == 1 3 13
   analysis time _t:  (dox-origin)/365.25
             origin:  time doe
                 id:  id
               note:  ageband<=40 trimmed
Estimated SMRs and lower/upper bounds of 95% confidence intervals
(729 records included in the analysis)

    ageband    D      E       SMR     Lower    Upper

        40     6     5.62   1.0670   0.4793   2.3749
        50    18    18.75   0.9599   0.6048   1.5235
        60    22    22.85   0.9629   0.6340   1.4624
```

◁

Stratified rate ratios using stmh

The stmh command is used for estimating rate ratios, controlled for confounding, using stratification. You can use it to estimate the ratio of the rates of failure for two categories of the explanatory variable. Categories to be compared may be defined by specifying the codes of the levels with compare().

The first variable listed on the command line after stmh is the explanatory variable used in comparing rates, and any remaining variables, if any, are categorical variables, which are to be controlled for by using stratification.

▷ Example 3

To illustrate this command, let's return to the diet data. The variable hienergy is coded 1 if the total energy consumption is more than 2.75 Mcal and 0 otherwise. We want to compare the rate for hienergy level 1 with the rate for level 0, controlled for ageband.

To do this, we first stset and stsplit the data into age bands as before, and then we use stmh:

```
. use http://www.stata-press.com/data/r11/diet, clear
(Diet data with dates)

. stset dox, origin(time dob) enter(time doe) id(id) scale(365.25)
> fail(fail==1 3 13)
  (output omitted)

. stsplit ageband, at(40(10)70) after(time=dob) trim
(26 + 0 obs. trimmed due to lower and upper bounds)
(418 observations (episodes) created)
```

```
. stmh hienergy, by(ageband)
        failure _d:  fail == 1 3 13
   analysis time _t: (dox-origin)/365.25
             origin: time dob
   enter on or after: time doe
                 id: id
               note: ageband<=40 trimmed
```
Maximum likelihood estimate of the rate ratio
  comparing hienergy==1 vs. hienergy==0
  by ageband
RR estimate, and lower and upper 95% confidence limits

ageband	RR	Lower	Upper
40	1.24	0.23	6.76
50	0.43	0.16	1.16
60	0.50	0.21	1.20

Overall estimate controlling for ageband

RR	chi2	P>chi2	[95% Conf. Interval]	
0.534	4.36	0.0369	0.293	0.972

Approx test for unequal RRs (effect modification): chi2(2) =   1.19
                                                   Pr>chi2 =   0.5514

Because the RR estimates are approximate, the test for unequal rate ratios is also approximate.

We can also compare the effect of hienergy between jobs, controlling for ageband.

```
. stmh hienergy ageband, by(job)
        failure _d:  fail == 1 3 13
   analysis time _t: (dox-origin)/365.25
             origin: time dob
   enter on or after: time doe
                 id: id
               note: ageband<=40 trimmed
```
Mantel-Haenszel estimate of the rate ratio
  comparing hienergy==1 vs. hienergy==0
  controlling for ageband
  by job
RR estimate, and lower and upper 95% confidence limits

job	RR	Lower	Upper
0	0.42	0.13	1.33
1	0.64	0.22	1.87
2	0.51	0.21	1.26

Overall estimate controlling for ageband job

RR	chi2	P>chi2	[95% Conf. Interval]	
0.521	4.88	0.0271	0.289	0.939

Approx test for unequal RRs (effect modification): chi2(2) =   0.28
                                                   Pr>chi2 =   0.8695

Log-linear trend test for metric explanatory variables using stmh

stmh may also be used to carry out trend tests for a metric explanatory variable. A one-step Newton approximation to the log-linear Poisson regression coefficient is also computed.

The diet dataset contains the height for each patient recorded in the variable height. We can test for a trend of heart disease rates with height controlling for ageband by typing

```
. stmh height ageband
           failure _d:  fail == 1 3 13
    analysis time _t:  (dox-origin)/365.25
              origin:  time dob
      enter on or after:  time doe
                  id:  id
                note:  ageband<=40 trimmed
Score test for trend of rates with height
  with an approximate estimate of the
  rate ratio for a one unit increase in height
  controlling for ageband
RR estimate, and lower and upper 95% confidence limits
```

RR	chi2	P>chi2	[95% Conf. Interval]	
0.906	18.60	0.0000	0.866	0.948

stmh tested for trend of heart disease rates with height within age bands and provided a rough estimate of the rate ratio for a 1-cm increase in height—this estimate is a one-step Newton approximation to the maximum likelihood estimate. It is not consistent, but it does provide a useful indication of the size of the effect.

The rate ratio is significantly less than 1, so there is clear evidence for a decreasing rate with increasing height (about 9% decrease in rate per centimeter increase in height).

Controlling for age with fine strata by using stmc

The stmc (Mantel–Cox) command is used to control for variation of rates on a time scale by breaking up time into short intervals, or *clicks*.

Usually this approach is used only to calculate significance tests, but the rate ratio estimated remains just as useful as in the coarsely stratified analysis from stmh. The method may be viewed as an approximate form of Cox regression.

The rate ratio produced by stmc is controlled for analysis time separately for each level of the variables specified with by() and then combined to give a rate ratio controlled for both time and the by() variables.

▷ Example 4

For example, to obtain the effect of high energy controlled for age by stratifying finely, we first stset the data specifying the date of birth, dob, as the origin (so analysis time is age), and then we use stmc:

```
. stset dox, origin(time dob) enter(time doe) id(id) scale(365.25)
> fail(fail==1 3 13)
  (output omitted)
```

```
. stmc hienergy
            failure _d:  fail == 1 3 13
      analysis time _t:  (dox-origin)/365.25
                origin:  time dob
     enter on or after:  time doe
                    id:  id
Mantel-Cox comparisons

Mantel-Haenszel estimates of the rate ratio
   comparing hienergy==1 vs. hienergy==0
   controlling for time (by clicks)

Overall Mantel-Haenszel estimate, controlling for time from dob
```

RR	chi2	P>chi2	[95% Conf. Interval]	
0.537	4.20	0.0403	0.293	0.982

The rate ratio of 0.537 is close to that obtained with stmh when controlling for age by using 10-year age bands.

◁

Saved results

stmh and stmc save the following in r():

Scalars
r(RR)   overall rate ratio

Methods and formulas

strate, stmh, and stmc are implemented as ado-files.

Nathan Mantel (1919–2002) was an American biostatistician who grew up in New York. He worked at the National Cancer Institute from 1947 to 1974 on a wide range of medical problems and was also later affiliated with George Washington University and the American University in Washington.

William M. Haenszel (1910–1998) was an American biostatistician and epidemiologist who graduated from the University of Buffalo. He also worked at the National Cancer Institute and later at the University of Illinois.

Acknowledgments

The original versions of strate, stmh, and stmc were written by David Clayton, Cambridge Institute for Medical Research, and Michael Hills, London School of Hygiene and Tropical Medicine (retired).

References

Clayton, D. G., and M. Hills. 1993. *Statistical Models in Epidemiology*. Oxford: Oxford University Press.

——. 1995. ssa7: Analysis of follow-up studies. *Stata Technical Bulletin* 27: 19–26. Reprinted in *Stata Technical Bulletin Reprints*, vol. 5, pp. 219–227. College Station, TX: Stata Press.

——. 1997. ssa10: Analysis of follow-up studies with Stata 5.0. *Stata Technical Bulletin* 40: 27–39. Reprinted in *Stata Technical Bulletin Reprints*, vol. 7, pp. 253–268. College Station, TX: Stata Press.

Gail, M. H. 1997. A conversation with Nathan Mantel. *Statistical Science* 12: 88–97.

Hankey, B. 1997. A conversation with William M. Haenszel. *Statistical Science* 12: 108–112.

Also see

[ST] **stci** — Confidence intervals for means and percentiles of survival time

[ST] **stir** — Report incidence-rate comparison

[ST] **stptime** — Calculate person-time, incidence rates, and SMR

[ST] **stset** — Declare data to be survival-time data

Title

streg — Parametric survival models

Syntax

streg [*varlist*] [*if*] [*in*] [, *options*]

options	description
Model	
<u>nocon</u>stant	suppress constant term
<u>d</u>istribution(<u>e</u>xponential)	exponential survival distribution
<u>d</u>istribution(<u>go</u>mpertz)	Gompertz survival distribution
<u>d</u>istribution(<u>logl</u>ogistic)	loglogistic survival distribution
<u>d</u>istribution(<u>ll</u>ogistic)	synonym for distribution(loglogistic)
<u>d</u>istribution(<u>w</u>eibull)	Weibull survival distribution
<u>d</u>istribution(<u>logn</u>ormal)	lognormal survival distribution
<u>d</u>istribution(<u>ln</u>ormal)	synonym for distribution(lognormal)
<u>d</u>istribution(<u>ga</u>mma)	generalized gamma survival distribution
<u>fr</u>ailty(<u>g</u>amma)	gamma frailty distribution
<u>fr</u>ailty(<u>inv</u>gaussian)	inverse-Gaussian distribution
<u>ti</u>me	use accelerated failure-time metric
Model 2	
<u>str</u>ata(*varname*)	strata ID variable
<u>off</u>set(*varname*)	include *varname* in model with coefficient constrained to 1
<u>sh</u>ared(*varname*)	shared frailty ID variable
<u>anc</u>illary(*varlist*)	use *varlist* to model the first ancillary parameter
anc2(*varlist*)	use *varlist* to model the second ancillary parameter
<u>con</u>straints(*constraints*)	apply specified linear constraints
<u>col</u>linear	keep collinear variables
SE/Robust	
vce(*vcetype*)	*vcetype* may be oim, <u>r</u>obust, <u>cl</u>uster *clustvar*, opg, <u>boot</u>strap, or <u>jack</u>knife
Reporting	
<u>l</u>evel(*#*)	set confidence level; default is level(95)
nohr	do not report hazard ratios
tr	report time ratios
<u>nos</u>how	do not show st setting information
<u>nohe</u>ader	suppress header from coefficient table
<u>nolr</u>test	do not perform likelihood-ratio test
<u>nocns</u>report	do not display constraints
display_options	control spacing and display of omitted variables and base and empty cells

Maximization

maximize_options	control the maximization process; seldom used
† <u>coef</u>legend	display coefficients' legend instead of coefficient table

† coeflegend does not appear in the dialog box.

You must stset your data before using streg; see [ST] **stset**.

varlist may contain factor variables; see [U] **11.4.3 Factor variables**.

bootstrap, by, fracpoly, jackknife, mfp, mi estimate, nestreg, statsby, stepwise, and svy are allowed; see [U] **11.1.10 Prefix commands**.

vce(bootstrap) and vce(jackknife) are not allowed with the mi estimate prefix.

shared(), vce(), and noheader are not allowed with the svy prefix.

fweights, iweights, and pweights may be specified using stset; see [ST] **stset**. However, weights may not be specified if you are using the bootstrap prefix with the streg command.

See [U] **20 Estimation and postestimation commands** for more capabilities of estimation commands.

Menu

Statistics > Survival analysis > Regression models > Parametric survival models

Description

streg performs maximum likelihood estimation for parametric regression survival-time models. streg can be used with single- or multiple-record or single- or multiple-failure st data. Survival models currently supported are exponential, Weibull, Gompertz, lognormal, loglogistic, and generalized gamma. Parametric frailty models and shared-frailty models are also fit using streg.

Also see [ST] **stcox** for proportional hazards models.

Options

Model

noconstant; see [R] **estimation options**.

distribution(*distname*) specifies the survival model to be fit. A specified distribution() is remembered from one estimation to the next when distribution() is not specified.

For instance, typing streg x1 x2, distribution(weibull) fits a Weibull model. Subsequently, you do not need to specify distribution(weibull) to fit other Weibull regression models.

The currently supported distributions are listed in the options table, but new ones may have been added since this manual was printed. Type help streg for an up-to-date list.

All Stata estimation commands, including streg, redisplay results when you type the command name without arguments. To fit a model with no explanatory variables, type streg, distribution(*distname*)....

frailty(gamma | invgaussian) specifies the assumed distribution of the frailty, or heterogeneity. The estimation results, in addition to the standard parameter estimates, will contain an estimate of the variance of the frailties and a likelihood-ratio test of the null hypothesis that this variance is zero. When this null hypothesis is true, the model reduces to the model with frailty(*distname*) not specified.

A specified frailty() is remembered from one estimation to the next when distribution() is not specified. When you specify distribution(), the previously remembered specification of frailty() is forgotten.

time specifies that the model be fit in the accelerated failure-time metric rather than in the log relative-hazard metric. This option is valid only for the exponential and Weibull models because these are the only models that have both a proportional hazards and an accelerated failure-time parameterization. Regardless of metric, the likelihood function is the same, and models are equally appropriate viewed in either metric; it is just a matter of changing the interpretation.

time must be specified at estimation.

Model 2

strata(*varname*) specifies the stratification ID variable. Observations with equal values of the variable are assumed to be in the same stratum. Stratified estimates (with equal coefficients across strata but intercepts and ancillary parameters unique to each stratum) are then obtained. This option is not available if frailty(*distname*) is specified.

offset(*varname*); see [R] **estimation options**.

shared(*varname*) is valid with frailty() and specifies a variable defining those groups over which the frailty is shared, analogous to a random-effects model for panel data where *varname* defines the panels. frailty() specified without shared() treats the frailties as occurring at the observation level.

A specified shared() is remembered from one estimation to the next when distribution() is not specified. When you specify distribution(), the previously remembered specification of shared() is forgotten.

shared() may not be used with distribution(gamma), vce(robust), vce(cluster *clustvar*), vce(opg), or the svy prefix.

If shared() is specified without frailty() and there is no remembered frailty() from the previous estimation, frailty(gamma) is assumed to provide behavior analogous to stcox; see [ST] **stcox**.

ancillary(*varlist*) specifies that the ancillary parameter for the Weibull, lognormal, Gompertz, and loglogistic distributions and that the first ancillary parameter (sigma) of the generalized log-gamma distribution be estimated as a linear combination of *varlist*. This option may not be used with frailty(*distname*).

When an ancillary parameter is constrained to be strictly positive, the logarithm of the ancillary parameter is modeled as a linear combination of *varlist*.

anc2(*varlist*) specifies that the second ancillary parameter (kappa) for the generalized log-gamma distribution be estimated as a linear combination of *varlist*. This option may not be used with frailty(*distname*).

constraints(*constraints*), collinear; see [R] **estimation options**.

SE/Robust

vce(*vcetype*) specifies the type of standard error reported, which includes types that are derived from asymptotic theory, that are robust to some kinds of misspecification, that allow for intragroup correlation, and that use bootstrap or jackknife methods; see [R] ***vce_option***.

⌐ Reporting ⌐

level(#) specifies the confidence level, as a percentage, for confidence intervals. The default is level(95) or as set by set level; see [U] **20.7 Specifying the width of confidence intervals**.

nohr, which may be specified at estimation or upon redisplaying results, specifies that coefficients rather than exponentiated coefficients be displayed, i.e., that coefficients rather than hazard ratios be displayed. This option affects only how coefficients are displayed, not how they are estimated.

This option is valid only for models with a natural proportional-hazards parameterization: exponential, Weibull, and Gompertz. These three models, by default, report hazard ratios (exponentiated coefficients).

tr specifies that exponentiated coefficients, which are interpreted as time ratios, be displayed. tr is appropriate only for the loglogistic, lognormal, and gamma models, or for the exponential and Weibull models when fit in the accelerated failure-time metric.

tr may be specified at estimation or upon replay.

noshow prevents streg from showing the key st variables. This option is rarely used because most people type stset, show or stset, noshow to set once and for all whether they want to see these variables mentioned at the top of the output of every st command; see [ST] **stset**.

noheader suppresses the output header, either at estimation or upon replay.

nolrtest is valid only with frailty models, in which case it suppresses the likelihood-ratio test for significant frailty.

nocnsreport; see [R] **estimation options**.

display_options: <u>noomit</u>ted, <u>vsquish</u>, <u>noempty</u>cells, <u>base</u>levels, <u>allbase</u>levels; see [R] **estimation options**.

⌐ Maximization ⌐

maximize_options: <u>diff</u>icult, <u>tech</u>nique(*algorithm_spec*), <u>iter</u>ate(#), [<u>no</u>]<u>log</u>, <u>tra</u>ce, gradient, showstep, <u>hess</u>ian, <u>showtol</u>erance, <u>tol</u>erance(#), <u>ltol</u>erance(#), <u>nrtol</u>erance(#), nonrtolerance, from(*init_specs*); see [R] **maximize**. These options are seldom used.

Setting the optimization type to technique(bhhh) resets the default *vcetype* to vce(opg).

The following option is available with streg but is not shown in the dialog box:

coeflegend; see [R] **estimation options**.

Remarks

Remarks are presented under the following headings:

> *Introduction*
> *Distributions*
> > *Weibull and exponential models*
> > *Gompertz model*
> > *Lognormal and loglogistic models*
> > *Generalized gamma model*
>
> *Examples*
> *Parameterization of ancillary parameters*
> *Stratified estimation*
> *(Unshared-) frailty models*
> *Shared-frailty models*

Introduction

What follows is a brief summary of what you can do with streg. For a complete tutorial, see Cleves et al. (2008), which devotes four chapters to this topic.

Two often-used models for adjusting survivor functions for the effects of covariates are the accelerated failure-time (AFT) model and the multiplicative or proportional hazards (PH) model. In the AFT model, the natural logarithm of the survival time, $\log t$, is expressed as a linear function of the covariates, yielding the linear model

$$\log t_j = \mathbf{x}_j \boldsymbol{\beta} + z_j$$

where $\mathbf{x}_j$ is a vector of covariates, $\boldsymbol{\beta}$ is a vector of regression coefficients, and z_j is the error with density $f()$. The distributional form of the error term determines the regression model. If we let $f()$ be the normal density, the lognormal regression model is obtained. Similarly, by letting $f()$ be the logistic density, the loglogistic regression is obtained. Setting $f()$ equal to the extreme-value density yields the exponential and the Weibull regression models.

The effect of the AFT model is to change the time scale by a factor of $\exp(-\mathbf{x}_j\boldsymbol{\beta})$. Depending on whether this factor is greater or less than 1, time is either accelerated or decelerated (degraded). That is, if a subject at baseline experiences a probability of survival past time t equal to $S(t)$, then a subject with covariates $\mathbf{x}_j$ would have probability of survival past time t equal to $S()$ evaluated at the point $\exp(-\mathbf{x}_j\boldsymbol{\beta})t$, instead. Thus accelerated failure time does not imply a positive acceleration of time with the increase of a covariate but instead implies a deceleration of time or, equivalently, an increase in the expected waiting time for failure.

In the PH model, the concomitant covariates have a multiplicative effect on the hazard function

$$h(t_j) = h_0(t)g(\mathbf{x}_j)$$

for some $h_0(t)$, and for $g(\mathbf{x}_j)$, a nonnegative function of the covariates. A popular choice, and the one adopted here, is to let $g(\mathbf{x}_j) = \exp(\mathbf{x}_j\boldsymbol{\beta})$. The function $h_0(t)$ may either be left unspecified, yielding the Cox proportional hazards model (see [ST] **stcox**), or take a specific parametric form. For the streg command, $h_0(t)$ is assumed to be parametric. Three regression models are currently implemented as PH models: the exponential, Weibull, and Gompertz models. The exponential and Weibull models are implemented as both AFT and PH models, and the Gompertz model is implemented only in the PH metric.

The above model allows for the presence of an intercept term, β_0, within $\mathbf{x}_j\boldsymbol{\beta}$. Thus what is commonly referred to as the *baseline hazard function*—the hazard when all covariates are zero—is actually equal to $h_0(t)\exp(\beta_0)$. That is, the intercept term serves to scale the baseline hazard. Of course, specifying noconstant suppresses the intercept or equivalently constrains β_0 to equal zero.

streg is suitable only for data that have been stset. By stsetting your data, you define the variables _t0, _t, and _d, which serve as the trivariate response variable (t_0, t, d). Each response corresponds to a period under observation, $(t_0, t]$, resulting in either failure ($d = 1$) or right-censoring ($d = 0$) at time t. As a result, streg is appropriate for data exhibiting delayed entry, gaps, time-varying covariates, and even multiple-failure data.

Distributions

Six parametric survival distributions are currently supported by streg. The parameterization and ancillary parameters for each distribution are summarized in table 1:

Table 1. Parametric survival distributions supported by streg

Distribution	Metric	Survivor function	Parameterization	Ancillary parameters
Exponential	PH	$\exp(-\lambda_j t_j)$	$\lambda_j = \exp(\mathbf{x}_j \boldsymbol{\beta})$	
Exponential	AFT	$\exp(-\lambda_j t_j)$	$\lambda_j = \exp(-\mathbf{x}_j \boldsymbol{\beta})$	
Weibull	PH	$\exp(-\lambda_j t_j^p)$	$\lambda_j = \exp(\mathbf{x}_j \boldsymbol{\beta})$	p
Weibull	AFT	$\exp(-\lambda_j t_j^p)$	$\lambda_j = \exp(-p\mathbf{x}_j \boldsymbol{\beta})$	p
Gompertz	PH	$\exp\{-\lambda_j \gamma^{-1}(e^{\gamma t_j} - 1)\}$	$\lambda_j = \exp(\mathbf{x}_j \boldsymbol{\beta})$	γ
Lognormal	AFT	$1 - \Phi\left\{\frac{\log(t_j) - \mu_j}{\sigma}\right\}$	$\mu_j = \mathbf{x}_j \boldsymbol{\beta}$	σ
Loglogistic	AFT	$\{1 + (\lambda_j t_j)^{1/\gamma}\}^{-1}$	$\lambda_j = \exp(-\mathbf{x}_j \boldsymbol{\beta})$	γ
Generalized gamma				
if $\kappa > 0$	AFT	$1 - I(\gamma, u)$	$\mu_j = \mathbf{x}_j \boldsymbol{\beta}$	σ, κ
if $\kappa = 0$	AFT	$1 - \Phi(z)$	$\mu_j = \mathbf{x}_j \boldsymbol{\beta}$	σ, κ
if $\kappa < 0$	AFT	$I(\gamma, u)$	$\mu_j = \mathbf{x}_j \boldsymbol{\beta}$	σ, κ

where PH = proportional hazards, AFT = accelerated failure time, and $\Phi(z)$ is the standard normal cumulative distribution. For the generalized gamma, $\gamma = |\kappa|^{-2}$, $u = \gamma \exp(|\kappa|z)$, $I(a, x)$ is the incomplete gamma function, and $z = \text{sign}(\kappa)\{\log(t_j) - \mu_j\}/\sigma$.

Plotted in figure 1 are example hazard functions for five of the six distributions. The exponential hazard (not separately plotted) is a special case of the Weibull hazard when the Weibull ancillary parameter $p = 1$. The generalized gamma (not plotted) is extremely flexible and therefore can take many shapes.

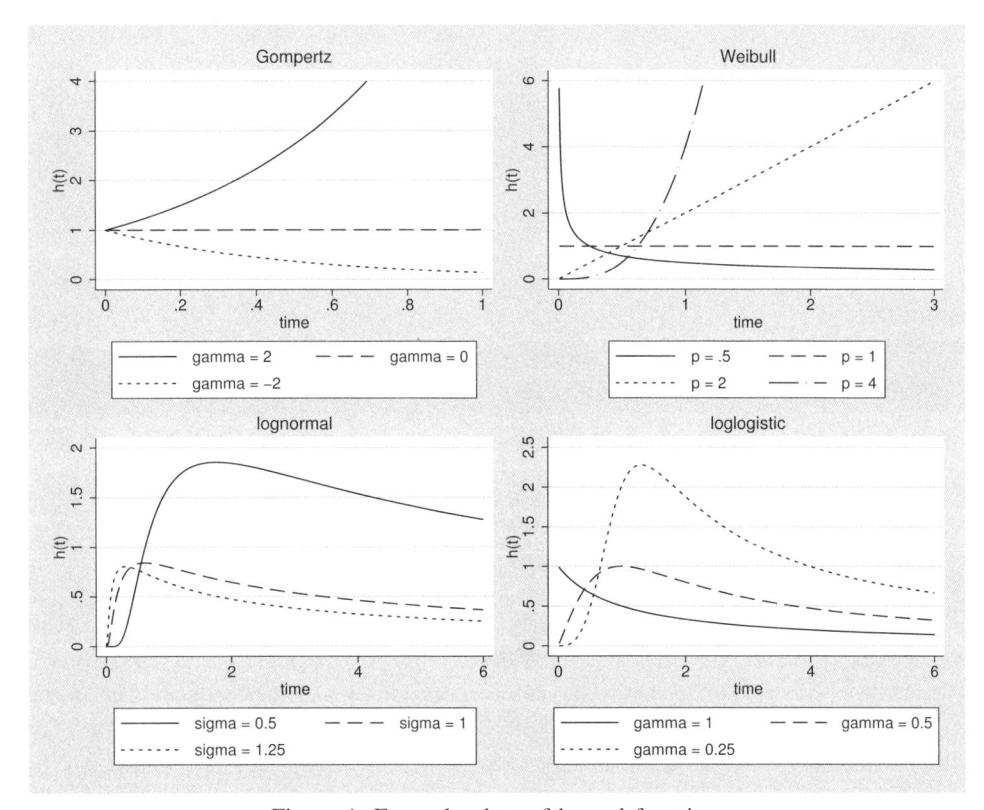

Figure 1. Example plots of hazard functions

Weibull and exponential models

The Weibull and exponential models are parameterized as both PH and AFT models. The Weibull distribution is suitable for modeling data with monotone hazard rates that either increase or decrease exponentially with time, whereas the exponential distribution is suitable for modeling data with constant hazard (see figure 1).

For the PH model, $h_0(t) = 1$ for exponential regression, and $h_0(t) = p\, t^{p-1}$ for Weibull regression, where p is the shape parameter to be estimated from the data. Some authors refer not to p but to $\sigma = 1/p$.

The AFT model is written as

$$\log(t_j) = \mathbf{x}_j \boldsymbol{\beta}^* + z_j$$

where z_j has an extreme-value distribution scaled by σ. Let $\boldsymbol{\beta}$ be the vector of regression coefficients derived from the PH model so that $\boldsymbol{\beta}^* = -\sigma \boldsymbol{\beta}$. This relationship holds only if the ancillary parameter, p, is a constant; it does not hold when the ancillary parameter is parameterized in terms of covariates.

streg uses, by default, for the exponential and Weibull models, the proportional-hazards metric simply because it eases comparison with those results produced by stcox. You can, however, specify the time option to choose the accelerated failure-time parameterization.

The Weibull hazard and survivor functions are

$$h(t) = p\lambda t^{p-1}$$

$$S(t) = \exp(-\lambda t^p)$$

where λ is parameterized as described in table 1. If $p = 1$, these functions reduce to those of the exponential.

Gompertz model

The Gompertz regression is parameterized only as a PH model. First described in 1825, this model has been extensively used by medical researchers and biologists modeling mortality data. The Gompertz distribution implemented is the two-parameter function as described in Lee and Wang (2003), with the following hazard and survivor functions:

$$h(t) = \lambda \exp(\gamma t)$$

$$S(t) = \exp\{-\lambda \gamma^{-1}(e^{\gamma t} - 1)\}$$

The model is implemented by parameterizing $\lambda_j = \exp(\mathbf{x}_j \boldsymbol{\beta})$, implying that $h_0(t) = \exp(\gamma t)$, where γ is an ancillary parameter to be estimated from the data.

This distribution is suitable for modeling data with monotone hazard rates that either increase or decrease exponentially with time (see figure 1).

When γ is positive, the hazard function increases with time; when γ is negative, the hazard function decreases with time; and when γ is zero, the hazard function is equal to λ for all t, so the model reduces to an exponential.

Some recent survival analysis texts, such as Klein and Moeschberger (2003), restrict γ to be strictly positive. If $\gamma < 0$, then as t goes to infinity, the survivor function, $S(t)$, exponentially decreases to a nonzero constant, implying that there is a nonzero probability of never failing (living forever). That is, there is always a nonzero hazard rate, yet it decreases exponentially. By restricting γ to be positive, we know that the survivor function always goes to zero as t tends to infinity.

Although the above argument may be desirable from a mathematical perspective, in Stata's implementation, we took the more traditional approach of not restricting γ. We did this because, in survival studies, subjects are not monitored forever—there is a date when the study ends, and in many investigations, specifically in medical research, an exponentially decreasing hazard rate is clinically appealing.

Lognormal and loglogistic models

The lognormal and loglogistic models are implemented only in the AFT form. These two distributions are similar and tend to produce comparable results. For the lognormal distribution, the natural logarithm of time follows a normal distribution; for the loglogistic distribution, the natural logarithm of time follows a logistic distribution.

The lognormal survivor and density functions are

$$S(t) = 1 - \Phi\left\{\frac{\log(t) - \mu}{\sigma}\right\}$$

$$f(t) = \frac{1}{t\sigma\sqrt{2\pi}} \exp\left[\frac{-1}{2\sigma^2}\left\{\log(t) - \mu\right\}^2\right]$$

where $\Phi(z)$ is the standard normal cumulative distribution function.

The lognormal regression is implemented by setting $\mu_j = \mathbf{x}_j\boldsymbol{\beta}$ and treating the standard deviation, σ, as an ancillary parameter to be estimated from the data.

The loglogistic regression is obtained if z_j has a logistic density. The loglogistic survivor and density functions are

$$S(t) = \left\{1 + (\lambda t)^{1/\gamma}\right\}^{-1}$$

$$f(t) = \frac{\lambda^{1/\gamma} t^{1/\gamma - 1}}{\gamma\left\{1 + (\lambda t)^{1/\gamma}\right\}^2}$$

This model is implemented by parameterizing $\lambda_j = \exp(-\mathbf{x}_j\boldsymbol{\beta})$ and treating the scale parameter γ as an ancillary parameter to be estimated from the data.

Unlike the exponential, Weibull, and Gompertz distributions, the lognormal and the loglogistic distributions are indicated for data exhibiting nonmonotonic hazard rates, specifically initially increasing and then decreasing rates (figure 1).

Generalized gamma model

The generalized gamma model is implemented only in the AFT form. The three-parameter generalized gamma survivor and density functions are

$$S(t) = \begin{cases} 1 - I(\gamma, u), & \text{if } \kappa > 0 \\ 1 - \Phi(z), & \text{if } \kappa = 0 \\ I(\gamma, u), & \text{if } \kappa < 0 \end{cases}$$

$$f(t) = \begin{cases} \frac{\gamma^\gamma}{\sigma t \sqrt{\gamma}\Gamma(\gamma)} \exp(z\sqrt{\gamma} - u), & \text{if } \kappa \neq 0 \\ \frac{1}{\sigma t \sqrt{2\pi}} \exp(-z^2/2), & \text{if } \kappa = 0 \end{cases}$$

where $\gamma = |\kappa|^{-2}$, $z = \text{sign}(\kappa)\{\log(t) - \mu\}/\sigma$, $u = \gamma \exp(|\kappa|z)$, $\Phi(z)$ is the standard normal cumulative distribution function, and $I(a, x)$ is the incomplete gamma function. See the gammap(a,x) entry in [D] **functions** to see how the incomplete gamma function is implemented in Stata.

This model is implemented by parameterizing $\mu_j = \mathbf{x}_j\boldsymbol{\beta}$ and treating the parameters κ and σ as ancillary parameters to be estimated from the data.

The hazard function of the generalized gamma distribution is extremely flexible, allowing for many possible shapes, including as special cases the Weibull distribution when $\kappa = 1$, the exponential when $\kappa = 1$ and $\sigma = 1$, and the lognormal distribution when $\kappa = 0$. The generalized gamma model is, therefore, commonly used for evaluating and selecting an appropriate parametric model for the data. The Wald or likelihood-ratio test can be used to test the hypotheses that $\kappa = 1$ or that $\kappa = 0$.

Examples

▷ Example 1

The Weibull distribution provides a good illustration of `streg` because this distribution is parameterized as both AFT and PH and serves to compare and contrast the two approaches.

We wish to analyze an experiment testing the ability of emergency generators with new-style bearings to withstand overloads. This dataset is described in [ST] **stcox**. This time, we wish to fit a Weibull model:

```
. use http://www.stata-press.com/data/r11/kva
(Generator experiment)

. streg load bearings, distribution(weibull)

        failure _d:  1 (meaning all fail)
   analysis time _t:  failtime

Fitting constant-only model:

Iteration 0:   log likelihood = -13.666193
Iteration 1:   log likelihood = -9.7427276
Iteration 2:   log likelihood = -9.4421169
Iteration 3:   log likelihood = -9.4408287
Iteration 4:   log likelihood = -9.4408286

Fitting full model:

Iteration 0:   log likelihood = -9.4408286
Iteration 1:   log likelihood =  -2.078323
Iteration 2:   log likelihood =  5.2226016
Iteration 3:   log likelihood =  5.6745808
Iteration 4:   log likelihood =  5.6934031
Iteration 5:   log likelihood =  5.6934189
Iteration 6:   log likelihood =  5.6934189

Weibull regression -- log relative-hazard form

No. of subjects =       12                       Number of obs   =        12
No. of failures =       12
Time at risk    =      896
                                                 LR chi2(2)      =     30.27
Log likelihood  =   5.6934189                    Prob > chi2     =    0.0000
```

_t	Haz. Ratio	Std. Err.	z	P>|z|	[95% Conf. Interval]
load	1.599315	.1883807	3.99	0.000	1.269616	2.014631
bearings	.1887995	.1312109	-2.40	0.016	.0483546	.7371644
/ln_p	2.051552	.2317074	8.85	0.000	1.597414	2.505691
p	7.779969	1.802677			4.940241	12.25202
1/p	.1285352	.0297826			.0816192	.2024193

Because we did not specify otherwise, the estimation took place in the hazard metric, which is the default for `distribution(weibull)`. The estimates are directly comparable to those produced by `stcox`: `stcox` estimated a hazard ratio of 1.526 for `load` and 0.0636 for `bearings`.

However, we estimated the baseline hazard function as well, assuming that it is Weibull. The estimates are the full maximum-likelihood estimates. The shape parameter is fit as $\ln p$, but `streg` then reports p and $1/p = \sigma$ so that you can think about the parameter however you wish.

We find that p is greater than 1, which means that the hazard of failure increases with time and, here, increases dramatically. After 100 hours, the bearings are more than 1 million times more likely to fail per second than after 10 hours (or, to be precise, $(100/10)^{7.78-1}$). From our knowledge of generators, we would expect this; it is the accumulation of heat due to friction that causes bearings to expand and seize.

◁

❑ Technical note

Regression results are often presented in a metric other than the natural regression coefficients, i.e., as hazard ratios, relative risk ratios, odds ratios, etc. In those cases, standard errors are calculated using the delta method.

However, the Z test and p-values given are calculated from the natural regression coefficients and standard errors. Although a test based on, say, a hazard ratio and its standard error would be asymptotically equivalent to that based on a regression coefficient, in real samples a hazard ratio will tend to have a more skewed distribution because it is an exponentiated regression coefficient. Also, it is more natural to think of these tests as testing whether a regression coefficient is nonzero, rather than testing whether a transformed regression coefficient is unequal to some nonzero value (one for a hazard ratio).

Finally, the confidence intervals given are obtained by transforming the endpoints of the corresponding confidence interval for the untransformed regression coefficient. This ensures that, say, strictly positive quantities such as hazard ratios have confidence intervals that do not overlap zero.

❑

▷ Example 2

The previous estimation took place in the PH metric, and exponentiated coefficients—hazard ratios—were reported. If we want to see the unexponentiated coefficients, we could redisplay results and specify the nohr option:

```
. streg, nohr
Weibull regression -- log relative-hazard form
No. of subjects =           12                  Number of obs    =         12
No. of failures =           12
Time at risk    =          896
                                                LR chi2(2)       =      30.27
Log likelihood  =    5.6934189                  Prob > chi2      =     0.0000
```

_t	Coef.	Std. Err.	z	P>|z|	[95% Conf. Interval]	
load	.4695753	.1177884	3.99	0.000	.2387143	.7004363
bearings	-1.667069	.6949745	-2.40	0.016	-3.029194	-.3049443
_cons	-45.13191	10.60663	-4.26	0.000	-65.92053	-24.34329
/ln_p	2.051552	.2317074	8.85	0.000	1.597414	2.505691
p	7.779969	1.802677			4.940241	12.25202
1/p	.1285352	.0297826			.0816192	.2024193

◁

▷ Example 3

We could just as well have fit this model in the AFT metric:

```
. streg load bearings, d(weibull) time nolog
        failure _d:  1 (meaning all fail)
   analysis time _t: failtime

Weibull regression -- accelerated failure-time form
No. of subjects =        12                  Number of obs   =        12
No. of failures =        12
Time at risk    =       896
                                             LR chi2(2)      =     30.27
Log likelihood  =  5.6934189                 Prob > chi2     =    0.0000
```

_t	Coef.	Std. Err.	z	P>|z|	[95% Conf. Interval]	
load	-.060357	.0062214	-9.70	0.000	-.0725507	-.0481632
bearings	.2142771	.0746451	2.87	0.004	.0679753	.3605789
_cons	5.80104	.1752301	33.11	0.000	5.457595	6.144485
/ln_p	2.051552	.2317074	8.85	0.000	1.597414	2.505691
p	7.779969	1.802677			4.940241	12.25202
1/p	.1285352	.0297826			.0816192	.2024193

This is the same model we previously fit, but it is presented in a different metric. Calling the previous coefficients b, these coefficients are $-\sigma b = -b/p$. For instance, in the previous example, the coefficient on load was reported as 0.4695753, and $-0.4695753/7.779969 = -0.06035696$. d() is a convenient shorthand for distribution().

◁

▷ Example 4

streg may also be applied to more complicated data. Below we have multiple records per subject on a failure that can occur repeatedly:

```
. use http://www.stata-press.com/data/r11/mfail3
. stdescribe
```

			per subject		
Category	total	mean	min	median	max
no. of subjects	926				
no. of records	1734	1.87257	1	2	4
(first) entry time		0	0	0	0
(final) exit time		470.6857	1	477	960
subjects with gap	6				
time on gap if gap	411	68.5	16	57.5	133
time at risk	435444	470.2419	1	477	960
failures	808	.8725702	0	1	3

In this dataset, subjects have up to four records (most have two) and have up to three failures (most have one) and, although you cannot tell from the above output, the data have time-varying covariates, as well. There are even six subjects with gaps in their histories, meaning that, for a while, they went unobserved. Although we could estimate in the AFT metric, it is easier to interpret results in the PH metric (or the log relative-hazard metric, as it is also known):

```
. streg x1 x2, d(weibull) vce(robust)

Fitting constant-only model:

Iteration 0:   log pseudolikelihood = -1398.2504
Iteration 1:   log pseudolikelihood = -1382.8224
Iteration 2:   log pseudolikelihood = -1382.7457
Iteration 3:   log pseudolikelihood = -1382.7457

Fitting full model:

Iteration 0:   log pseudolikelihood = -1382.7457
Iteration 1:   log pseudolikelihood = -1328.4186
Iteration 2:   log pseudolikelihood = -1326.4483
Iteration 3:   log pseudolikelihood = -1326.4449
Iteration 4:   log pseudolikelihood = -1326.4449

Weibull regression -- log relative-hazard form

No. of subjects    =        926           Number of obs   =       1734
No. of failures    =        808
Time at risk       =     435444
                                           Wald chi2(2)    =     154.45
Log pseudolikelihood =  -1326.4449         Prob > chi2     =     0.0000
```

(Std. Err. adjusted for 926 clusters in id)

	Haz. Ratio	Robust Std. Err.	z	P>|z|	[95% Conf. Interval]	
x1	2.240069	.1812848	9.97	0.000	1.911504	2.625111
x2	.3206515	.0504626	-7.23	0.000	.2355458	.436507
/ln_p	.1771265	.0310111	5.71	0.000	.1163458	.2379071
p	1.193782	.0370205			1.123384	1.268591
1/p	.8376738	.0259772			.7882759	.8901674

A one-unit change in x1 approximately doubles the hazard of failure, whereas a one-unit change in x2 cuts the hazard to one-third its previous value. We also see that these data are close to being exponentially distributed; p is nearly 1.

Above we mentioned that interpreting results in the PH metric is easier, though regression coefficients are not difficult to interpret in the AFT metric. A positive coefficient means that time is decelerated by a unit increase in the covariate in question. This may seem awkward, but think of this instead as a unit increase in the covariate causing a delay in failure and thus *increasing* the expected time until failure.

The difficulty that arises with the AFT metric is merely that it places an emphasis on log(time-to-failure) rather than risk (hazard) of failure. With this emphasis usually comes a desire to predict the time to failure, and therein lies the difficulty with complex survival data. Predicting the log(time to failure) with predict assumes that the subject is at risk from time 0 until failure and has a fixed covariate pattern over this period. With these data, such assumptions produce predictions having little to do with the test subjects, who exhibit not only time-varying covariates but also multiple failures.

Predicting time to failure with complex survival data is difficult regardless of the metric under which estimation took place. Those who estimate in the PH metric are probably used to dealing with results from Cox regression, of which predicted time to failure is typically not the focus.

◁

▷ Example 5

The multiple-failure data above are close enough to exponentially distributed that we will reestimate using exponential regression:

```
. streg x1 x2, d(exp) vce(robust)
Iteration 0:    log pseudolikelihood = -1398.2504
Iteration 1:    log pseudolikelihood = -1343.6083
Iteration 2:    log pseudolikelihood = -1341.5932
Iteration 3:    log pseudolikelihood = -1341.5893
Iteration 4:    log pseudolikelihood = -1341.5893
Exponential regression -- log relative-hazard form
No. of subjects    =         926           Number of obs    =        1734
No. of failures    =         808
Time at risk       =      435444
                                            Wald chi2(2)     =      166.92
Log pseudolikelihood =   -1341.5893         Prob > chi2      =      0.0000
                         (Std. Err. adjusted for 926 clusters in id)
```

		Robust			
_t	Haz. Ratio	Std. Err.	z	P>|z|	[95% Conf. Interval]
x1	2.19065	.1684399	10.20	0.000	1.884186 2.54696
x2	.3037259	.0462489	-7.83	0.000	.2253552 .4093511

◁

❑ Technical note

For our "complex" survival data, we specified vce(robust) when fitting the Weibull and exponential models. This was because these data were stset with an id() variable, and given the time-varying covariates and multiple failures, it is important not to assume that the observations within each subject are independent. When we specified vce(robust), it was implicit that we were "clustering" on the groups defined by the id() variable.

You might sometimes have multiple observations per subject, which exist merely as a result of the data-organization mechanism and are not used to record gaps, time-varying covariates, or multiple failures. Such data could be collapsed into single-observation-per-subject data with no loss of information. In these cases, we refer to splitting the observations to form multiple observations per subject as *noninformative*. When the episode-splitting is noninformative, the model-based (nonrobust) standard errors produced will be the same as those produced when the data are collapsed into single records per subject. Thus, for these type of data, clustering of these multiple observations that results from specifying vce(robust) is not critical.

❑

▷ Example 6

A reasonable question to ask is, "Given that we have several possible parametric models, how can we select one?" When parametric models are nested, the likelihood-ratio or Wald test can be used to discriminate between them. This can certainly be done for Weibull versus exponential or gamma versus Weibull or lognormal. When models are not nested, however, these tests are inappropriate, and the task of discriminating between models becomes more difficult. A common approach to this problem is to use the Akaike information criterion (AIC). Akaike (1974) proposed penalizing each log likelihood to reflect the number of parameters being estimated in a particular model and then comparing them. Here the AIC can be defined as

$$\text{AIC} = -2(\log \text{ likelihood}) + 2(c + p + 1)$$

where c is the number of model covariates and p is the number of model-specific ancillary parameters listed in table 1. Although the best-fitting model is the one with the largest log likelihood, the preferred model is the one with the smallest AIC value. The AIC value may be obtained by using the estat ic postestimation command.

Using cancer.dta distributed with Stata, let's first fit a generalized gamma model and test the hypothesis that $\kappa = 0$ (test for the appropriateness of the lognormal) and then test the hypothesis that $\kappa = 1$ (test for the appropriateness of the Weibull).

```
. use http://www.stata-press.com/data/r11/cancer
(Patient Survival in Drug Trial)
. stset studytime, failure(died)
  (output omitted)
. replace drug = drug==2 | drug==3       // 0, placebo : 1, nonplacebo
(48 real changes made)
. streg drug age, d(gamma) nolog

         failure _d:  died
   analysis time _t:  studytime

Gamma regression -- accelerated failure-time form

No. of subjects =           48                 Number of obs   =         48
No. of failures =           31
Time at risk    =          744
                                               LR chi2(2)      =      36.07
Log likelihood  =   -42.452006                 Prob > chi2     =     0.0000
```

_t	Coef.	Std. Err.	z	P>|z|	[95% Conf. Interval]	
drug	1.394658	.2557198	5.45	0.000	.893456	1.895859
age	-.0780416	.0227978	-3.42	0.001	-.1227245	-.0333587
_cons	6.456091	1.238457	5.21	0.000	4.02876	8.883421
/ln_sig	-.3793632	.183707	-2.07	0.039	-.7394222	-.0193041
/kappa	.4669252	.5419478	0.86	0.389	-.595273	1.529123
sigma	.684297	.1257101			.4773897	.980881

The Wald test of the hypothesis that $\kappa = 0$ (test for the appropriateness of the lognormal) is reported in the output above. The p-value is 0.389, suggesting that lognormal might be an adequate model for these data.

The Wald test for $\kappa = 1$ is

```
. test [kappa]_cons = 1
 ( 1)  [kappa]_cons = 1
           chi2(  1) =    0.97
         Prob > chi2 =  0.3253
```

providing some support against rejecting the Weibull model.

We now fit the exponential, Weibull, loglogistic, and lognormal models separately. To directly compare coefficients, we will ask Stata to report the exponential and Weibull models in AFT form by specifying the time option. The output from fitting these models and the results from the generalized gamma model are summarized in table 2.

Table 2. Results obtained from streg, using cancer.dta with drug as an indicator variable

Parameter	Exponential	Weibull	Lognormal	Loglogistic	Generalized gamma
Age	−.0886715	−.0714323	−.0833996	−.0803289	−.078042
Drug	1.682625	1.305563	1.445838	1.420237	1.394658
Constant	7.146218	6.289679	6.580887	6.446711	6.456091
Ancillary		1.682751	0.751136	0.429276	0.684297
Kappa					0.466925
Log-likelihood	−48.397094	−42.931335	−42.800864	−43.21698	−42.452006
AIC	102.7942	93.86267	93.60173	94.43396	94.90401

The largest log likelihood was obtained for the generalized gamma model; however, the lognormal model is preferred by the AIC.

◁

Parameterization of ancillary parameters

By default, all ancillary parameters are estimated as being constant. For example, the ancillary parameter, p, of the Weibull distribution is assumed to be a constant that is not dependent on any covariates. streg's ancillary() and anc2() options allow for complete parameterization of parametric survival models. By specifying, for example,

```
. streg x1 x2, d(weibull) ancillary(x2 z1 z2)
```

both λ and the ancillary parameter, p, are parameterized in terms of covariates.

Ancillary parameters are usually restricted to be strictly positive, in which case the logarithm of the ancillary parameter is modeled using a linear predictor, which can assume any value on the real line.

▷ Example 7

Consider a dataset in which we model the time until hip fracture as Weibull for patients on the basis of age, sex, and whether the patient wears a hip-protective device (variable protect). We believe that the hazard is scaled according to sex and the presence of the device but believe the hazards for both sexes to be of different *shapes*.

```
. use http://www.stata-press.com/data/r11/hip3, clear
(hip fracture study)
. streg protect age, d(weibull) ancillary(male) nolog
         failure _d:  fracture
   analysis time _t:  time1
                id:   id

Weibull regression -- log relative-hazard form
No. of subjects =         148                 Number of obs    =        206
No. of failures =          37
Time at risk    =        1703
                                              LR chi2(2)       =      39.80
Log likelihood  =   -69.323532                Prob > chi2      =     0.0000
```

_t	Coef.	Std. Err.	z	P>|z|	[95% Conf. Interval]	
_t						
protect	-2.130058	.3567005	-5.97	0.000	-2.829178	-1.430938
age	.0939131	.0341107	2.75	0.006	.0270573	.1607689
_cons	-10.17575	2.551821	-3.99	0.000	-15.17722	-5.174269
ln_p						
male	-.4887189	.185608	-2.63	0.008	-.8525039	-.1249339
_cons	.4540139	.1157915	3.92	0.000	.2270667	.6809611

From our estimation results, we see that $\widehat{\ln(p)} = 0.454$ for females and $\widehat{\ln(p)} = 0.454 - 0.489 = -0.035$ for males. Thus $\widehat{p} = 1.57$ for females and $\widehat{p} = 0.97$ for males. When we combine this with the main equation in the model, the estimated hazards are then

$$\widehat{h}(t_j|\mathbf{x}_j) = \begin{cases} \exp\left(-10.18 - 2.13\text{protect}_j + 0.09\text{age}_j\right) 1.57 t_j^{0.57} & \text{if female} \\ \exp\left(-10.18 - 2.13\text{protect}_j + 0.09\text{age}_j\right) 0.97 t_j^{-0.03} & \text{if male} \end{cases}$$

If we believe this model, we would say that the hazard for males given age and protect is almost constant over time.

Contrast this with what we obtain if we type

```
. streg protect age if male, d(weibull)
. streg protect age if !male, d(weibull)
```

which is completely general, because not only will the shape parameter, p, differ over both sexes, but the regression coefficients will as well.

◁

The anc2() option is for use only with the gamma regression model, because it contains two ancillary parameters—anc2() is used to parametrize κ.

Stratified estimation

When we type

```
. streg xvars, d(distname) strata(varname)
```

we are asking that a completely stratified model be fit. By *completely stratified*, we mean that both the model's intercept and any ancillary parameters are allowed to vary for each level of the strata variable. That is, we are constraining the coefficients on the covariates to be the same across strata but allowing the intercept and ancillary parameters to vary.

▷ Example 8

We demonstrate this by fitting a stratified lognormal model to the cancer data, with the drug variable left in its original state: drug==1 refers to the placebo, and drug==2 and drug==3 correspond to two alternative treatments.

```
. use http://www.stata-press.com/data/r11/cancer
(Patient Survival in Drug Trial)
. stset studytime died
 (output omitted )
. streg age, d(weibull) strata(drug) nolog
        failure _d:  died
  analysis time _t:  studytime

Weibull regression -- log relative-hazard form

No. of subjects =          48                   Number of obs   =         48
No. of failures =          31
Time at risk    =         744
                                                LR chi2(3)      =      16.58
Log likelihood  =   -41.113074                  Prob > chi2     =     0.0009
```

_t	Coef.	Std. Err.	z	P>|z|	[95% Conf. Interval]
_t					
age	.1212332	.0367538	3.30	0.001	.049197 .1932694
_Sdrug_2	-4.561178	2.339448	-1.95	0.051	-9.146411 .0240556
_Sdrug_3	-3.715737	2.595986	-1.43	0.152	-8.803776 1.372302
_cons	-10.36921	2.341022	-4.43	0.000	-14.95753 -5.780896
ln_p					
_Sdrug_2	.4872195	.332019	1.47	0.142	-.1635257 1.137965
_Sdrug_3	.2194213	.4079989	0.54	0.591	-.5802418 1.019084
_cons	.4541282	.1715663	2.65	0.008	.1178645 .7903919

◁

Completely stratified models are fit by first generating stratum-specific indicator variables (dummy variables) and then adding these as independent variables in the model and as covariates in the ancillary parameter. The strata() option is thus merely a shorthand method for generating the indicator variables from the drug categories and then placing these indicators in *both* the main equation and the ancillary equation(s).

We associate the term *stratification* with this process by noting that the intercept term of the main equation is a component of the baseline hazard (or baseline survivor) function. By allowing this intercept, as well as the ancillary shape parameter, to vary with respect to the strata, we allow the baseline functions to completely vary over the strata, analogous to a stratified Cox model.

▷ Example 9

We can produce a less-stratified model by specifying a factor variable in the ancillary() option.

```
. streg age, d(weibull) ancillary(i.drug) nolog
         failure _d:  died
   analysis time _t:  studytime
Weibull regression -- log relative-hazard form
No. of subjects =          48                   Number of obs   =         48
No. of failures =          31
Time at risk    =         744
                                                LR chi2(1)      =       9.61
Log likelihood  =  -44.596379                   Prob > chi2     =     0.0019
```

	_t	Coef.	Std. Err.	z	P>|z|	[95% Conf.	Interval]
_t							
	age	.1126419	.0362786	3.10	0.002	.0415373	.1837466
	_cons	-10.95772	2.308489	-4.75	0.000	-15.48227	-6.433162
ln_p							
	drug						
	2	-.3279568	.11238	-2.92	0.004	-.5482176	-.107696
	3	-.4775351	.1091141	-4.38	0.000	-.6913948	-.2636755
	_cons	.6684086	.1327284	5.04	0.000	.4082657	.9285514

By doing this, we are restricting not only the coefficients on the covariates to be the same across "strata" but also the intercept, while allowing only the ancillary parameter to differ.

◁

By using ancillary() or strata(), we may thus consider a wide variety of models, depending on what we believe about the effect of the covariate(s) in question. For example, when fitting a Weibull PH model to the cancer data, we may choose from many models, depending on what we want to assume is the effect of the categorical variable drug. For all models considered below, we assume implicitly that the effect of age is proportional on the hazard function.

1. drug has no effect:

    . streg age, d(weibull)

2. The effect of drug is proportional on the hazard (scale), and the effect of age is the same for each level of drug:

    . streg age i.drug, d(weibull)

3. drug affects the shape of the hazard, and the effect of age is the same for each level of drug:

    . streg age, d(weibull) ancillary(i.drug)

4. drug affects both the scale and shape of the hazard, and the effect of age is the same for each level of drug:

    . streg age, d(weibull) strata(drug)

5. drug affects both the scale and shape of the hazard, and the effect of age is different for each level of drug:

    . streg drug##c.age, d(weibull) strata(drug)

These models may be compared using Wald or likelihood-ratio tests when the models in question are nested (such as 3 nested within 4) or by using the AIC for nonnested models.

Everything we said regarding the modeling of ancillary parameters and stratification applies to AFT models as well, for which interpretations may be stated in terms of the baseline survivor function, i.e., the unaccelerated probability of survival past time t.

❏ Technical note

When fitting PH models as we have done in the previous three examples, streg will, by default, display the exponentiated regression coefficients, labeled as hazard ratios. However, in our previous examples using `ancillary()` and `strata()`, the regression outputs displayed the untransformed coefficients instead. This change in behavior has to do with the modeling of the ancillary parameter. When we use one or more covariates from the main equation to model an ancillary parameter, hazard ratios (and time ratios for AFT models) lose their interpretation. streg, as a precaution, disallows the display of hazard/time ratios when `ancillary()`, `anc2()`, or `strata()` is specified.

Keep this in mind when comparing results across various model specifications. For example, when comparing a stratified Weibull PH model to a standard Weibull PH model, be sure that the latter is displayed using the `nohr` option.

❏

(Unshared-) frailty models

A frailty model is a survival model with unobservable heterogeneity, or *frailty*. At the observation level, frailty is introduced as an unobservable multiplicative effect, α, on the hazard function, such that

$$h(t|\alpha) = \alpha h(t)$$

where $h(t)$ is a nonfrailty hazard function, say, the hazard function of any of the six parametric models supported by streg described earlier in this entry. The frailty, α, is a random positive quantity and, for model identifiability, is assumed to have mean 1 and variance θ.

Exploiting the relationship between the cumulative hazard function and survivor function yields the expression for the survivor function, given the frailty

$$S(t|\alpha) = \exp\left\{-\int_0^t h(u|\alpha)du\right\} = \exp\left\{-\alpha \int_0^t \frac{f(u)}{S(u)}du\right\} = \{S(t)\}^\alpha$$

where $S(t)$ is the survivor function that corresponds to $h(t)$.

Because α is unobservable, it must be integrated out of $S(t|\alpha)$ to obtain the unconditional survivor function. Let $g(\alpha)$ be the probability density function of α, in which case an estimable form of our frailty model is achieved as

$$S_\theta(t) = \int_0^\infty S(t|\alpha)g(\alpha)d\alpha = \int_0^\infty \{S(t)\}^\alpha g(\alpha)d\alpha$$

Given the unconditional survivor function, we can obtain the unconditional hazard and density in the usual way:

$$f_\theta(t) = -\frac{d}{dt}S_\theta(t) \qquad h_\theta(t) = \frac{f_\theta(t)}{S_\theta(t)}$$

Hence, an unshared-frailty model is merely a typical parametric survival model, with the additional estimation of an overdispersion parameter, θ. In a standard survival regression, the likelihood calculations are based on $S(t)$, $h(t)$, and $f(t)$. In an unshared-frailty model, the likelihood is based analogously on $S_\theta(t)$, $h_\theta(t)$, and $f_\theta(t)$.

At this stage, the only missing piece is the choice of frailty distribution, $g(\alpha)$. In theory, any continuous distribution supported on the positive numbers that has expectation 1 and finite variance θ is allowed here. For mathematical tractability, however, we limit the choice to either the gamma$(1/\theta, \theta)$ distribution or the inverse-Gaussian distribution with parameters 1 and $1/\theta$, denoted as IG$(1, 1/\theta)$. The gamma(a, b) distribution has probability density function

$$g(x) = \frac{x^{a-1} e^{-x/b}}{\Gamma(a) b^a}$$

and the IG(a, b) distribution has density

$$g(x) = \left(\frac{b}{2\pi x^3}\right)^{1/2} \exp\left\{-\frac{b}{2a}\left(\frac{x}{a} - 2 + \frac{a}{x}\right)\right\}$$

Therefore, performing the integrations described above will show that specifying frailty(gamma) will result in the frailty survival model (in terms of the nonfrailty survivor function, $S(t)$)

$$S_\theta(t) = [1 - \theta \log\{S(t)\}]^{-1/\theta}$$

Specifying frailty(invgaussian) will give

$$S_\theta(t) = \exp\left\{\frac{1}{\theta}\left(1 - [1 - 2\theta \log\{S(t)\}]^{1/2}\right)\right\}$$

Regardless of the choice of frailty distribution, $\lim_{\theta \to 0} S_\theta(t) = S(t)$, and thus the frailty model reduces to $S(t)$ when there is no heterogeneity present.

When using frailty models, distinguish between the hazard faced by the individual (subject), $\alpha h(t)$, and the "average" hazard for the population, $h_\theta(t)$. Similarly, an individual will have probability of survival past time t equal to $\{S(t)\}^\alpha$, whereas $S_\theta(t)$ will measure the proportion of the population surviving past time t. You specify $S(t)$ as before with distribution(*distname*), and the list of possible parametric forms for $S(t)$ is given in table 1. Thus when you specify distribution() you are specifying a model for an individual with frailty equal to one. Specifying frailty(*distname*) determines which of the two above forms for $S_\theta(t)$ is used.

The output of the estimation remains unchanged from the nonfrailty version, except for the additional estimation of θ and a likelihood-ratio test of $H_0: \theta = 0$. For more information on frailty models, Hougaard (1986) offers an excellent introduction. For a Stata-specific overview, see Gutierrez (2002).

▷ Example 10

Consider as an example a survival analysis of data on women with breast cancer. Our hypothetical dataset consists of analysis times on 80 women with covariates age, smoking, and dietfat, which measures the average weekly calories from fat ($\times 10^3$) in the patient's diet over the course of the study.

```
. use http://www.stata-press.com/data/r11/bc
. list in 1/12
```

	age	smoking	dietfat	t	dead
1.	30	1	4.919	14.2	0
2.	50	0	4.437	8.21	1
3.	47	0	5.85	5.64	1
4.	49	1	5.149	4.42	1
5.	52	1	4.363	2.81	1
6.	29	0	6.153	35	0
7.	49	1	3.82	4.57	1
8.	27	1	5.294	35	0
9.	47	0	6.102	3.74	1
10.	59	0	4.446	2.29	1
11.	35	0	6.203	15.3	0
12.	26	0	4.515	35	0

The data are well fit by a Weibull model for the distribution of survival time conditional on age, smoking, and dietary fat. By omitting the dietfat variable from the model, we hope to introduce unobserved heterogeneity.

```
. stset t, fail(dead)
 (output omitted )
. streg age smoking, d(weibull) frailty(gamma)
        failure _d: dead
  analysis time _t: t

Fitting Weibull model:

Fitting constant-only model:

Iteration 0:    log likelihood = -137.15363
Iteration 1:    log likelihood =  -136.3927
Iteration 2:    log likelihood = -136.01557
Iteration 3:    log likelihood = -136.01202
Iteration 4:    log likelihood = -136.01201

Fitting full model:

Iteration 0:    log likelihood = -85.933969
Iteration 1:    log likelihood =  -73.61173
Iteration 2:    log likelihood = -68.999447
Iteration 3:    log likelihood = -68.340858
Iteration 4:    log likelihood = -68.136187
Iteration 5:    log likelihood = -68.135804
Iteration 6:    log likelihood = -68.135804
```

```
Weibull regression -- log relative-hazard form
                 Gamma frailty
No. of subjects =           80                 Number of obs   =         80
No. of failures =           58
Time at risk    =      1257.07
                                               LR chi2(2)      =     135.75
Log likelihood  =   -68.135804                 Prob > chi2     =     0.0000
```

_t	Haz. Ratio	Std. Err.	z	P>|z|	[95% Conf. Interval]	
age	1.475948	.1379987	4.16	0.000	1.228811	1.772788
smoking	2.788548	1.457031	1.96	0.050	1.00143	7.764894
/ln_p	1.087761	.222261	4.89	0.000	.6521376	1.523385
/ln_the	.3307466	.5250758	0.63	0.529	-.698383	1.359876
p	2.967622	.6595867			1.91964	4.587727
1/p	.3369701	.0748953			.2179729	.520931
theta	1.392007	.7309092			.4973889	3.895711

```
Likelihood-ratio test of theta=0: chibar2(01) =     22.57 Prob>=chibar2 = 0.000
```

We could also use an inverse-Gaussian distribution to model the heterogeneity.

```
. streg age smoking, dist(weibull) frailty(invgauss) nolog
        failure _d: dead
   analysis time _t: t

Weibull regression -- log relative-hazard form
               Inverse-Gaussian frailty
No. of subjects =           80                 Number of obs   =         80
No. of failures =           58
Time at risk    =      1257.07
                                               LR chi2(2)      =     125.44
Log likelihood  =   -73.838578                 Prob > chi2     =     0.0000
```

_t	Haz. Ratio	Std. Err.	z	P>|z|	[95% Conf. Interval]	
age	1.284133	.0463256	6.93	0.000	1.196473	1.378217
smoking	2.905409	1.252785	2.47	0.013	1.247892	6.764528
/ln_p	.7173904	.1434382	5.00	0.000	.4362567	.9985241
/ln_the	.2374778	.8568064	0.28	0.782	-1.441832	1.916788
p	2.049079	.2939162			1.546906	2.714273
1/p	.4880241	.0700013			.3684228	.6464518
theta	1.268047	1.086471			.2364941	6.799082

```
Likelihood-ratio test of theta=0: chibar2(01) =     11.16 Prob>=chibar2 = 0.000
```

The results are similar with respect to the choice of frailty distribution, with the gamma frailty model producing a slightly higher likelihood. Both models show a statistically significant level of unobservable heterogeneity because the p-value for the likelihood-ratio (LR) test of H_0: $\theta = 0$ is virtually zero in both cases.

❑ Technical note

With gamma-distributed or inverse-Gaussian–distributed frailty, hazard ratios decay over time in favor of the *frailty effect*, and thus the displayed "Haz. Ratio" in the above output is actually the hazard ratio only for $t = 0$. The degree of decay depends on θ. Should the estimated θ be close to zero, the hazard ratios regain their usual interpretation. The rate of decay and the limiting hazard ratio differ between the gamma and inverse-Gaussian models; see Gutierrez (2002) for details.

For this reason, many researchers prefer fitting frailty models in the AFT metric because the interpretation of regression coefficients is unchanged by the frailty—the factors in question serve to either accelerate or decelerate the survival experience. The only difference is that with frailty models, the unconditional probability of survival is described by $S_\theta(t)$ rather than $S(t)$.

❑

❑ Technical note

The LR test of $\theta = 0$ is a boundary test and thus requires careful consideration concerning the calculation of its p-value. In particular, the null distribution of the LR test statistic is not the usual χ_1^2 but rather is a 50:50 mixture of a χ_0^2 (point mass at zero) and a χ_1^2, denoted as $\overline{\chi}_{01}^2$. See Gutierrez, Carter, and Drukker (2001) for more details.

❑

To verify that the significant heterogeneity is caused by the omission of dietfat, we now refit the Weibull/inverse-Gaussian frailty model with dietfat included.

```
. streg age smoking dietfat, d(weibull) frailty(invgauss) nolog
        failure _d:  dead
   analysis time _t:  t

Weibull regression -- log relative-hazard form
                     Inverse-Gaussian frailty
No. of subjects =          80                Number of obs    =         80
No. of failures =          58
Time at risk    =     1257.07
                                             LR chi2(3)       =     246.41
Log likelihood  =  -13.352142                Prob > chi2      =     0.0000
```

_t	Haz. Ratio	Std. Err.	z	P>|z|	[95% Conf. Interval]	
age	1.74928	.0985246	9.93	0.000	1.566453	1.953447
smoking	5.203552	1.704943	5.03	0.000	2.737814	9.889992
dietfat	9.229842	2.219331	9.24	0.000	5.761312	14.78656
/ln_p	1.431742	.0978847	14.63	0.000	1.239892	1.623593
/ln_the	-14.29293	2673.364	-0.01	0.996	-5253.995	5225.399
p	4.185987	.4097439			3.455224	5.071278
1/p	.2388923	.0233839			.197189	.2894155
theta	6.17e-07	.0016502			0	.

```
Likelihood-ratio test of theta=0: chibar2(01) =      0.00 Prob>=chibar2 = 1.000
```

The estimate of the frailty variance component θ is near zero, and the p-value of the test of $H_0: \theta = 0$ equals one, indicating negligible heterogeneity. A regular Weibull model could be fit to these data (with dietfat included), producing almost identical estimates of the hazard ratios and ancillary parameter, p, so such an analysis is omitted here.

Also hazard ratios now regain their original interpretation. Thus an increase in weekly calories from fat of 1,000 would increase the risk of death by more than ninefold.

◁

Shared-frailty models

A generalization of the frailty models considered in the previous section is the *shared-frailty* model, where the frailty is assumed to be group specific; this is analogous to a panel-data regression model. For observation j from the ith group, the hazard is

$$h_{ij}(t|\alpha_i) = \alpha_i h_{ij}(t)$$

for $i = 1, \ldots, n$ and $j = 1, \ldots, n_i$, where by $h_{ij}(t)$ we mean $h(t|\mathbf{x}_{ij})$, which is the individual hazard given covariates $\mathbf{x}_{ij}$.

Shared-frailty models are appropriate when you wish to model the frailties as being specific to groups of subjects, such as subjects within families. Here a shared-frailty model may be used to model the degree of correlation within groups; i.e., the subjects within a group are correlated because they share the same common frailty.

▷ Example 11

Consider the data from a study of 38 kidney dialysis patients, as described in McGilchrist and Aisbett (1991). The study is concerned with the prevalence of infection at the catheter-insertion point. Two recurrence times (in days) are measured for each patient, and each recorded time is the time from initial insertion (onset of risk) to infection or censoring.

```
. use http://www.stata-press.com/data/r11/catheter
(Kidney data, McGilchrist and Aisbett, Biometrics, 1991)

. list patient time infect age female in 1/10
```

	patient	time	infect	age	female
1.	1	16	1	28	0
2.	1	8	1	28	0
3.	2	13	0	48	1
4.	2	23	1	48	1
5.	3	22	1	32	0
6.	3	28	1	32	0
7.	4	318	1	31.5	1
8.	4	447	1	31.5	1
9.	5	30	1	10	0
10.	5	12	1	10	0

Each patient (patient) has two recurrence times (time) recorded, with each catheter insertion resulting in either infection (infect==1) or right-censoring (infect==0). Among the covariates measured are age and sex (female==1 if female, female==0 if male).

One subtlety to note concerns the use of the generic term *subjects*. In this example, the subjects are the individual catheter insertions, not the patients themselves. This is a function of how the data were recorded—the onset of risk occurs at catheter insertion (of which there are two for each patient) not, say, at the time of admission of the patient into the study. Thus we have two subjects (insertions) within each group (patient).

It is reasonable to assume independence of patients but unreasonable to assume that recurrence times within each patient are independent. One solution would be to fit a standard survival model, adjusting the standard errors of the parameter estimates to account for the possible correlation by specifying vce(cluster patient).

We could also model the correlation by assuming that the correlation is the result of a latent patient-level effect, or frailty. That is, rather than fitting a standard model and specifying vce(cluster patient), we fit a frailty model and specify shared(patient). Assuming that the time to infection, given age and female, follows a Weibull distribution, and inverse-Gaussian distributed frailties, we get

```
. stset time, fail(infect)
(output omitted)
. streg age female, d(weibull) frailty(invgauss) shared(patient) nolog

         failure _d:  infect
   analysis time _t:  time

Weibull regression --
        log relative-hazard form              Number of obs    =       76
        Inverse-Gaussian shared frailty       Number of groups =       38
Group variable: patient
No. of subjects =         76                  Obs per group: min =       2
No. of failures =         58                                 avg =       2
Time at risk    =       7424                                 max =       2
                                              LR chi2(2)       =     9.84
Log likelihood  =   -99.093527                Prob > chi2      =   0.0073
```

_t	Haz. Ratio	Std. Err.	z	P>|z|	[95% Conf. Interval]	
age	1.006918	.013574	0.51	0.609	.9806623	1.033878
female	.2331376	.1046382	-3.24	0.001	.0967322	.5618928
/ln_p	.1900625	.1315342	1.44	0.148	-.0677398	.4478649
/ln_the	.0357272	.7745362	0.05	0.963	-1.482336	1.55379
p	1.209325	.1590676			.9345036	1.564967
1/p	.8269074	.1087666			.638991	1.070087
theta	1.036373	.8027085			.2271066	4.729362

Likelihood-ratio test of theta=0: chibar2(01) =   8.70 Prob>=chibar2 = 0.002

Contrast this with what we obtain by assuming a subject-level lognormal model:

```
. streg age female, d(lnormal) frailty(invgauss) shared(patient) nolog
        failure _d: infect
   analysis time _t: time
Lognormal regression --
        accelerated failure-time form          Number of obs     =         76
        Inverse-Gaussian shared frailty        Number of groups  =         38
Group variable: patient
No. of subjects =            76                Obs per group: min =          2
No. of failures =            58                             avg =          2
Time at risk    =          7424                             max =          2
                                               LR chi2(2)       =      16.34
Log likelihood  =    -97.614583                Prob > chi2      =     0.0003
```

_t	Coef.	Std. Err.	z	P>|z|	[95% Conf. Interval]	
age	-.0066762	.0099457	-0.67	0.502	-.0261694	.0128171
female	1.401719	.3334931	4.20	0.000	.7480844	2.055354
_cons	3.336709	.4972641	6.71	0.000	2.362089	4.311329
/ln_sig	.0625872	.1256185	0.50	0.618	-.1836205	.3087949
/ln_the	-1.606248	1.190775	-1.35	0.177	-3.940125	.7276282
sigma	1.064587	.1337318			.8322516	1.361783
theta	.2006389	.2389159			.0194458	2.070165

```
Likelihood-ratio test of theta=0: chibar2(01) =     1.53 Prob>=chibar2 = 0.108
```

The frailty effect is insignificant at the 10% level in the latter model yet highly significant in the former. We thus have two possible stories to tell concerning these data: If we believe the first model, we believe that the individual hazard of infection continually rises over time (Weibull), but there is a significant frailty effect causing the population hazard to begin falling after some time. If we believe the second model, we believe that the individual hazard first rises and then declines (lognormal), meaning that if a given insertion does not become infected initially, the chances that it will become infected begin to decrease after a certain point. Because the frailty effect is insignificant, the population hazard mirrors the individual hazard in the second model.

As a result, both models view the population hazard as rising initially and then falling past a certain point. The second version of our story corresponds to higher log likelihood, yet perhaps not significantly higher given the limited data. More investigation is required. One idea is to fit a more distribution-agnostic form of a frailty model, such as a piecewise exponential (Cleves et al. 2008, 320–323) or a Cox model with frailty; see [ST] **stcox**.

◁

Shared-frailty models are also appropriate when the frailties are subject specific yet there exist multiple records per subject. Here you would share frailties across the same id() variable previously stset. When you have subject-specific frailties and uninformative episode splitting, it makes no difference whether you fit a shared or an unshared frailty model. The estimation results will be the same. However, there is a distinction between unshared- and shared-frailty models for single-record-per-subject data with delayed entry; see Gutierrez (2002) for details.

Saved results

streg saves the following in e():

Scalars
e(N)	number of observations
e(N_sub)	number of subjects
e(N_fail)	number of failures
e(N_g)	number of groups
e(k)	number of parameters
e(k_eq)	number of equations
e(k_eq_model)	number of equations in model Wald test
e(k_aux)	number of auxiliary parameters
e(k_dv)	number of dependent variables
e(k_autoCns)	number of base, empty, and omitted constraints
e(df_m)	model degrees of freedom
e(ll)	log likelihood
e(ll_0)	log likelihood, constant-only model
e(ll_c)	log likelihood, comparison model
e(N_clust)	number of clusters
e(chi2)	χ^2
e(chi2_c)	χ^2, comparison model
e(risk)	total time at risk
e(g_min)	smallest group size
e(g_avg)	average group size
e(g_max)	largest group size
e(theta)	frailty parameter
e(aux_p)	ancillary parameter (weibull)
e(gamma)	ancillary parameter (gompertz, loglogistic)
e(sigma)	ancillary parameter (gamma, lnormal)
e(kappa)	ancillary parameter (gamma)
e(p)	significance
e(p_c)	significance, comparison model
e(rank)	rank of e(V)
e(rank0)	rank of e(V), constant-only model
e(ic)	number of iterations
e(rc)	return code
e(converged)	1 if converged, 0 otherwise

Macros
    e(cmd)   model or regression name
    e(cmd2)   streg
    e(cmdline)   command as typed
    e(dead)   _d
    e(depvar)   _t
    e(title)   title in estimation output
    e(title2)   secondary title in estimation output
    e(clustvar)   name of cluster variable
    e(shared)   frailty grouping variable
    e(fr_title)   title in output identifying frailty
    e(wtype)   weight type
    e(wexp)   weight expression
    e(t0)   _t0
    e(vce)   *vcetype* specified in vce()
    e(vcetype)   title used to label Std. Err.
    e(frm2)   hazard or time
    e(chi2type)   Wald or LR; type of model χ^2 test
    e(offset1)   offset for main equation
    e(stcurve)   stcurve
    e(diparm#)   display transformed parameter #
    e(opt)   type of optimization
    e(which)   max or min; whether optimizer is to perform maximization or minimization
    e(ml_method)   type of ml method
    e(user)   name of likelihood-evaluator program
    e(technique)   maximization technique
    e(singularHmethod) m-marquardt or hybrid; method used when Hessian is singular
    e(crittype)   optimization criterion
    e(properties)   b V
    e(predict)   program used to implement predict
    e(predict_sub)   predict subprogram
    e(footnote)   program used to implement the footnote display
    e(asbalanced)   factor variables fvset as asbalanced
    e(asobserved)   factor variables fvset as asobserved

Matrices
    e(b)   coefficient vector
    e(Cns)   constraints matrix
    e(ilog)   iteration log (up to 20 iterations)
    e(gradient)   gradient vector
    e(V)   variance–covariance matrix of the estimators
    e(V_modelbased)   model-based variance

Functions
    e(sample)   marks estimation sample

Methods and formulas

`streg` is implemented as an ado-file.

For an introduction to survival models, see Cleves et al. (2008). For an introduction to survival analysis directed at social scientists, see Box-Steffensmeier and Jones (2004).

Consider for $j = 1, \ldots, n$ observations the trivariate response, (t_{0j}, t_j, d_j), representing a period of observation, $(t_{0j}, t_j]$, ending in either failure ($d_j = 1$) or right-censoring ($d_j = 0$). This structure allows analysis of a wide variety of models and may be used to account for delayed entry, gaps, time-varying covariates, and multiple failures per subject. Regardless of the structure of the data, once they are `stset`, the data may be treated in a common manner by `streg`: the `stset`-created variable `_t0` holds the t_{0j}, `_t` holds the t_j, and `_d` holds the d_j.

For a given survivor function, $S(t)$, the density function is obtained as

$$f(t) = -\frac{d}{dt}S(t)$$

and the hazard function (the instantaneous rate of failure) is obtained as $h(t) = f(t)/S(t)$. Available forms for $S(t)$ are listed in table 1. For a set of covariates from the jth observation, $\mathbf{x}_j$, define $S_j(t) = S(t|\mathbf{x} = \mathbf{x}_j)$, and similarly define $h_j(t)$ and $f_j(t)$. For example, in a Weibull PH model, $S_j(t) = \exp\{-\exp(\mathbf{x}_j\boldsymbol{\beta})t^p\}$.

Parameter estimation

In this command, $\boldsymbol{\beta}$ and the ancillary parameters are estimated via maximum likelihood. A subject known to fail at time t_j contributes to the likelihood function the value of the density at time t_j conditional on the entry time t_{0j}, $f_j(t_j)/S_j(t_{0j})$. A censored observation, known to survive only up to time t_j, contributes $S_j(t_j)/S_j(t_{0j})$, which is the probability of surviving beyond time t_j conditional on the entry time, t_{0j}. The log likelihood is thus given by

$$\log L = \sum_{j=1}^{n} \{d_j \log f_j(t_j) + (1-d_j) \log S_j(t_j) - \log S_j(t_{0j})\}$$

Implicit in the above log-likelihood expression are the regression parameters, $\boldsymbol{\beta}$, and the ancillary parameters because both are components of the chosen $S_j(t)$ and its corresponding $f_j(t)$; see table 1. `streg` reports maximum likelihood estimates of $\boldsymbol{\beta}$ and of the ancillary parameters (if any for the chosen model). The reported log-likelihood value is $\log L_r = \log L + T$, where $T = \sum \log(t_j)$ is summed over uncensored observations. The adjustment removes the time units from $\log L$. Whether the adjustment is made makes no difference to any test or result since such tests and results depend on differences in log-likelihood functions or their second derivatives, or both.

Specifying `ancillary()`, `anc2()`, or `strata()` will parameterize the ancillary parameter(s) by using the linear predictor, $\mathbf{z}_j\boldsymbol{\alpha}_z$, where the covariates, $\mathbf{z}_j$, need not be distinct from $\mathbf{x}_j$. Here `streg` will report estimates of $\boldsymbol{\alpha}_z$ in addition to estimates of $\boldsymbol{\beta}$. The log likelihood here is simply the log likelihood given above, with $\mathbf{z}_j\boldsymbol{\alpha}_z$ substituted for the ancillary parameter. If the ancillary parameter is constrained to be strictly positive, its logarithm is parameterized instead; i.e., we substitute the linear predictor for the logarithm of the ancillary parameter in the above log likelihood. The gamma model has two ancillary parameters, σ and κ; we parameterize σ by using `ancillary()` and κ by using `anc2()`, and the linear predictors used for each may be distinct. Specifying `strata()` creates indicator variables for the strata, places these indicators in the main equation, and uses the indicators to parameterize any ancillary parameters that exist for the chosen model.

Unshared-frailty models have a log likelihood of the above form, with $S_\theta(t)$ and $f_\theta(t)$ substituted for $S(t)$ and $f(t)$, respectively. Equivalently, for gamma-distributed frailties,

$$\log L = \sum_{j=1}^{n} \left[\theta^{-1} \log\{1 - \theta \log S_j(t_{0j})\} - (\theta^{-1} + d_j) \log\{1 - \theta \log S_j(t_j)\} + d_j \log h_j(t_j) \right]$$

and for inverse-Gaussian–distributed frailties,

$$\log L = \sum_{j=1}^{n} \left[\theta^{-1} \{1 - 2\theta \log S_j(t_{0j})\}^{1/2} - \theta^{-1} \{1 - 2\theta \log S_j(t_j)\}^{1/2} + \right.$$

$$\left. d_j \log h_j(t_j) - \frac{1}{2} d_j \log\{1 - 2\theta \log S_j(t_j)\} \right]$$

In a shared-frailty model, the frailty is common to a group of observations. Thus, to form an unconditional likelihood, the frailties must be integrated out at the group level. The data are organized as $i = 1, \ldots, n$ groups with the ith group comprising $j = 1, \ldots, n_i$ observations. The log likelihood is the sum of the log-likelihood contributions for each group. Define $D_i = \sum_j d_{ij}$ as the number of failures in the ith group. For gamma frailties, the log-likelihood contribution for the ith group is

$$\log L_i = \sum_{j=1}^{n_i} d_{ij} \log h_{ij}(t_{ij}) - (1/\theta + D_i) \log\left\{1 - \theta \sum_{j=1}^{n_i} \log \frac{S_{ij}(t_{ij})}{S_{ij}(t_{0ij})}\right\} +$$

$$D_i \log \theta + \log \Gamma(1/\theta + D_i) - \log \Gamma(1/\theta)$$

For inverse-Gaussian frailties, define

$$C_i = \left\{1 - 2\theta \sum_{j=1}^{n_i} \log \frac{S_{ij}(t_{ij})}{S_{ij}(t_{0ij})}\right\}^{-1/2}$$

The log-likelihood contribution for the ith group then becomes

$$\log L_i = \theta^{-1}(1 - C_i^{-1}) + B(\theta C_i, D_i) + \sum_{j=1}^{n_i} d_{ij} \{\log h_{ij}(t_{ij}) + \log C_i\}$$

The function $B(a, b)$ is related to the modified Bessel function of the third kind, commonly known as the BesselK function; see Wolfram (1996, 746). In particular,

$$B(a, b) = a^{-1} + \frac{1}{2} \left\{\log\left(\frac{2}{\pi}\right) - \log a\right\} + \log \text{BesselK}\left(\frac{1}{2} - b, a^{-1}\right)$$

For both unshared- and shared-frailty models, estimation of θ takes place jointly with the estimation of β and the ancillary parameters.

This command supports the Huber/White/sandwich estimator of the variance and its clustered version using vce(robust) and vce(cluster *clustvar*), respectively. See [P] _robust, in particular, in *Maximum likelihood estimators* and *Methods and formulas*. If observations in the dataset represent repeated observations on the same subjects (that is, there are time-varying covariates), the assumption of independence of the observations is highly questionable, meaning that the conventional estimate of variance is not appropriate. We strongly advise that you use the vce(robust) and vce(cluster *clustvar*) options here. (streg knows to specify vce(cluster *clustvar*) if you specify vce(robust).) vce(robust) and vce(cluster *clustvar*) do not apply in shared-frailty models, where the correlation within groups is instead modeled directly.

streg also supports estimation with survey data. For details on VCEs with survey data, see [SVY] **variance estimation**.

> Benjamin Gompertz (1779–1865) came from a Jewish family who left Holland and settled in England. He was self-educated in mathematics and in 1824 was appointed as actuary and head clerk of the Alliance Assurance Company. He carried out pioneer work on the application of differential calculus to actuarial questions, particularly the dependence of mortality on age, and contributed to astronomy and the study of scientific instruments.
>
> Ernst Hjalmar Waloddi Weibull (1887–1979) was a Swedish applied physicist most famous for his work on the statistics of material properties. He worked in Germany and Sweden as an inventor and a consulting engineer, publishing his first paper on the propagation of explosive waves in 1914, thereafter becoming a full professor at the Royal Institute of Technology in 1924. Weibull's ideas about the statistical distributions of material strength came to the attention of engineers in the late 1930s with the publication of two important papers: "Investigations into strength properties of brittle materials" and "The phenomenon of rupture in solids".

References

Akaike, H. 1974. A new look at the statistical model identification. *IEEE Transactions on Automatic Control* 19: 716–723.

Box-Steffensmeier, J. M., and B. S. Jones. 2004. *Event History Modeling: A Guide for Social Scientists*. Cambridge: Cambridge University Press.

Cleves, M. A., W. W. Gould, R. G. Gutierrez, and Y. Marchenko. 2008. *An Introduction to Survival Analysis Using Stata*. 2nd ed. College Station, TX: Stata Press.

Cox, D. R., and D. Oakes. 1984. *Analysis of Survival Data*. London: Chapman & Hall/CRC.

Crowder, M. J., A. C. Kimber, R. L. Smith, and T. J. Sweeting. 1991. *Statistical Analysis of Reliability Data*. London: Chapman & Hall/CRC.

Cui, J. 2005. Buckley–James method for analyzing censored data, with an application to a cardiovascular disease and an HIV/AIDS study. *Stata Journal* 5: 517–526.

Fisher, R. A., and L. H. C. Tippett. 1928. Limiting forms of the frequency distribution of the largest or smallest member of a sample. *Mathematical Proceedings of the Cambridge Philosophical Society* 24: 180–190.

Gutierrez, R. G. 2002. Parametric frailty and shared frailty survival models. *Stata Journal* 2: 22–44.

Gutierrez, R. G., S. Carter, and D. M. Drukker. 2001. sg160: On boundary-value likelihood-ratio tests. *Stata Technical Bulletin* 60: 15–18. Reprinted in *Stata Technical Bulletin Reprints*, vol. 10, pp. 269–273. College Station, TX: Stata Press.

Hooker, P. F. 1965. Benjamin Gompertz. *Journal of the Institute of Actuaries* 91: 203–212.

Hosmer Jr., D. W., S. Lemeshow, and S. May. 2008. *Applied Survival Analysis: Regression Modeling of Time to Event Data*. 2nd ed. New York: Wiley.

Hougaard, P. 1986. Survival models for heterogeneous populations derived from stable distributions. *Biometrika* 73: 387–396.

Kalbfleisch, J. D., and R. L. Prentice. 2002. *The Statistical Analysis of Failure Time Data*. 2nd ed. New York: Wiley.

Klein, J. P., and M. L. Moeschberger. 2003. *Survival Analysis: Techniques for Censored and Truncated Data*. 2nd ed. New York: Springer.

Lee, E. T., and J. W. Wang. 2003. *Statistical Methods for Survival Data Analysis*. 3rd ed. New York: Wiley.

McGilchrist, C. A., and C. W. Aisbett. 1991. Regression with frailty in survival analysis. *Biometrics* 47: 461–466.

Olshansky, S. J., and B. A. Carnes. 1997. Ever since Gompertz. *Demography* 34: 1–15.

Peto, R., and P. Lee. 1973. Weibull distributions for continuous-carcinogenesis experiments. *Biometrics* 29: 457–470.

Pike, M. C. 1966. A method of analysis of a certain class of experiments in carcinogenesis. *Biometrics* 22: 142–161.

Royston, J. P. 2006. Explained variation for survival models. *Stata Journal* 6: 83–96.

Schoenfeld, D. A. 1982. Partial residuals for the proportional hazards regression model. *Biometrika* 69: 239–241.

Scotto, M. G., and A. Tobías. 1998. sg83: Parameter estimation for the Gumbel distribution. *Stata Technical Bulletin* 43: 32–35. Reprinted in *Stata Technical Bulletin Reprints*, vol. 8, pp. 133–137. College Station, TX: Stata Press.

———. 2000. sg146: Parameter estimation for the generalized extreme value distribution. *Stata Technical Bulletin* 56: 40–43. Reprinted in *Stata Technical Bulletin Reprints*, vol. 10, pp. 205–210. College Station, TX: Stata Press.

Weibull, W. 1939. A statistical theory of the strength of materials. In *Ingeniörs Vetenskaps Akademien Handlingar*, vol. 151. Stockholm: Generalstabens Litografiska Anstalts Förlag.

Wolfram, S. 1996. *The Mathematica Book*. 3rd ed. Champaign, IL: Wolfram Media.

Also see

[ST] **streg postestimation** — Postestimation tools for streg

[ST] **stcurve** — Plot survivor, hazard, cumulative hazard, or cumulative incidence function

[ST] **stcrreg** — Competing-risks regression

[ST] **sts** — Generate, graph, list, and test the survivor and cumulative hazard functions

[ST] **stset** — Declare data to be survival-time data

[ST] **stcox** — Cox proportional hazards model

[U] **20 Estimation and postestimation commands**

Title

streg postestimation — Postestimation tools for streg

Description

The following postestimation command is of special interest after `streg`:

command	description
stcurve	plot the survivor, hazard, and cumulative hazard functions

For information on `stcurve`, see [ST] **stcurve**.

The following standard postestimation commands are also available:

command	description
estat	AIC, BIC, VCE, and estimation sample summary
estat (svy)	postestimation statistics for survey data
estimates	cataloging estimation results
lincom	point estimates, standard errors, testing, and inference for linear combinations of coefficients
linktest	link test for model specification
lrtest[1]	likelihood-ratio test
margins	marginal means, predictive margins, marginal effects, and average marginal effects
nlcom	point estimates, standard errors, testing, and inference for nonlinear combinations of coefficients
predict	predictions, residuals, influence statistics, and other diagnostic measures
predictnl	point estimates, standard errors, testing, and inference for generalized predictions
suest	seemingly unrelated estimation
test	Wald tests of simple and composite linear hypotheses
testnl	Wald tests of nonlinear hypotheses

[1] lrtest is not appropriate with svy estimation results.

See the corresponding entries in the *Base Reference Manual* for details, but see [SVY] **estat** for details about `estat (svy)`.

Special-interest postestimation command

`stcurve` is used after `streg` (or `stcox`) to plot the cumulative hazard, survivor, and hazard functions at the mean value of the covariates or at values specified by the `at()` options.

Syntax for predict

> predict [*type*] *newvar* [*if*] [*in*] [, *statistic* oos no<u>off</u>set <u>alpha</u>1 <u>uncond</u>itional <u>part</u>ial]

> predict [*type*] { *stub** | *newvarlist* } [*if*] [*in*] , <u>sc</u>ores

statistic	description
Main	
<u>medi</u>an time	median survival time; the default
<u>medi</u>an <u>lnt</u>ime	median ln(survival time)
mean time	mean survival time
mean <u>lnt</u>ime	mean ln(survival time)
<u>hazard</u>	hazard
hr	hazard ratio, also known as the relative hazard
xb	linear prediction $\mathbf{x}_j\boldsymbol{\beta}$
stdp	standard error of the linear prediction; $\text{SE}(\mathbf{x}_j\boldsymbol{\beta})$
<u>s</u>urv	$S(t|t_0)$
* <u>cs</u>urv	$S(t\,|\,\text{earliest } t_0 \text{ for subject})$
* <u>csn</u>ell	Cox–Snell residuals
* <u>mg</u>ale	martingale-like residuals
* <u>dev</u>iance	deviance residuals

Unstarred statistics are available both in and out of sample; type predict ... if e(sample) ... if wanted only for the estimation sample. Starred statistics are calculated for the estimation sample by default, but the oos option makes them available both in and out of sample.

When no option is specified, the predicted median survival time is calculated for all models. The predicted hazard ratio, option hr, is available only for the exponential, Weibull, and Gompertz models. The mean time and mean lntime options are not available for the Gompertz model. Unconditional estimates of mean time and mean lntime are not available if frailty() was specified with streg.

csnell, mgale, and deviance are not allowed with svy estimation results.

Menu

Statistics > Postestimation > Predictions, residuals, etc.

Options for predict

> Main

median time calculates the predicted median survival time in analysis-time units. This is the prediction from time 0 conditional on constant covariates. When no options are specified with predict, the predicted median survival time is calculated for all models.

median lntime calculates the natural logarithm of what median time produces.

mean time calculates the predicted mean survival time in analysis-time units. This is the prediction from time 0 conditional on constant covariates. This option is not available for Gompertz regressions and is available for frailty models only if alpha1 is specified, in which case what you obtain is an estimate of the mean survival time conditional on a frailty effect of one.

mean lntime predicts the mean of the natural logarithm of time. This option is not available for Gompertz regression and is available for frailty models only if alpha1 is specified, in which case what you obtain is an estimate of the mean log-survival time conditional on a frailty effect of one.

hazard calculates the predicted hazard.

hr calculates the hazard ratio. This option is valid only for models having a proportional-hazards parameterization.

xb calculates the linear prediction from the fitted model. That is, you fit the model by estimating a set of parameters, $\beta_0, \beta_1, \beta_2, \ldots, \beta_k$, and the linear prediction is $\widehat{y}_j = \widehat{\beta}_0 + \widehat{\beta}_1 x_{1j} + \widehat{\beta}_2 x_{2j} + \cdots + \widehat{\beta}_k x_{kj}$, often written in matrix notation as $\widehat{y}_j = \mathbf{x}_j \widehat{\beta}$.

The $x_{1j}, x_{2j}, \ldots, x_{kj}$ used in the calculation are obtained from the data currently in memory and need not correspond to the data on the independent variables used in estimating β.

stdp calculates the standard error of the prediction, that is, the standard error of $\widehat{y}_j$.

surv calculates each observation's predicted survivor probability, $S(t|t_0)$, where t_0 is _t0, the analysis time at which each record became at risk. For multiple-record data, see the csurv option below.

csurv calculates the predicted $S(t|\text{earliest } t_0)$ for each subject in multiple-record data by calculating the conditional survivor values, $S(t|t_0)$ (see the surv option above), and then multiplying them together.

What you obtain from surv will differ from what you obtain from csurv only if you have multiple records for that subject.

csnell calculates the Cox–Snell generalized residuals. For multiple-record-per-subject data, by default only one value per subject is calculated and it is placed on the last record for the subject.

Adding the partial option will produce partial Cox–Snell residuals, one for each record within subject; see partial below. Partial Cox–Snell residuals are the additive contributions to a subject's overall Cox–Snell residual. In single-record-per-subject data, the partial Cox–Snell residuals are the Cox–Snell residuals.

mgale calculates the martingale-like residuals. For multiple-record data, by default only one value per subject is calculated and it is placed on the last record for the subject.

Adding the partial option will produce partial martingale residuals, one for each record within subject; see partial below. Partial martingale residuals are the additive contributions to a subject's overall martingale residual. In single-record data, the partial martingale residuals are the martingale residuals.

deviance calculates the deviance residuals. Deviance residuals are martingale residuals that have been transformed to be more symmetric about zero. For multiple-record data, by default only one value per subject is calculated and it is placed on the last record for the subject.

Adding the partial option will produce partial deviance residuals, one for each record within subject; see partial below. Partial deviance residuals are transformed partial martingale residuals. In single-record data, the partial deviance residuals are the deviance residuals.

oos makes csurv, csnell, mgale, and deviance available both in and out of sample. oos also dictates that summations and other accumulations take place over the sample as defined by if and in. By default, the summations are taken over the estimation sample, with if and in merely determining which values of *newvar* are to be filled in once the calculation is finished.

nooffset is relevant only if you specified offset(*varname*) with streg. It modifies the calculations made by predict so that they ignore the offset variable; the linear prediction is treated as $\mathbf{x}\beta$ rather than $\mathbf{x}\beta + \text{offset}$.

alpha1, when used after fitting a frailty model, specifies that *statistic* be predicted conditional on a frailty value equal to one. This is the default for shared-frailty models.

unconditional, when used after fitting a frailty model, specifies that *statistic* be predicted unconditional on the frailty. That is, the prediction is averaged over the frailty distribution. This is the default for unshared-frailty models.

partial is relevant only for multiple-record data and is valid with csnell, mgale, and deviance. Specifying partial will produce "partial" versions of these statistics, where one value is calculated for each record instead of one for each subject. The subjects are determined by the id() option of stset.

Specify partial if you wish to perform diagnostics on individual records rather than on individual subjects. For example, a partial deviance can be used to diagnose the fitted characteristics of an individual record rather than those of the set of records for a given subject.

scores calculates equation-level score variables. The number of score variables created depends upon the chosen distribution.

The first new variable will always contain $\partial \ln L / \partial (\mathbf{x}_j \beta)$.

The subsequent new variables will contain the partial derivative of the log likelihood with respect to the ancillary parameters.

Remarks

predict after streg is used to generate a variable containing predicted values or residuals.

For a more detailed discussion on residuals, read *Residuals and diagnostic measures* in the [ST] **stcox postestimation** entry. Many of the concepts and ideas presented there also apply to streg models.

Regardless of the metric used, predict can generate predicted median survival times and median log-survival times for all models, and predicted mean times and mean log-survival times where available. Predicted survival, hazard, and residuals are also available for all models. The predicted hazard ratio can be calculated only for models with a proportional-hazards parameterization, i.e., the Weibull, exponential, and Gompertz models. However, the estimation need not take place in the log-hazard metric. You can perform, for example, a Weibull regression specifying the time option and then ask that hazard ratios be predicted.

After fitting a frailty model, you can use predict with the alpha1 option to generate predicted values based on $S(t)$ or use the unconditional option to generate predictions based on $S_\theta(t)$; see [ST] **streg**.

▷ Example 1

Let's return to example 1 of [ST] **streg** concerning the ability of emergency generators with new-style bearings to withstand overloads. Assume that, as before, we fit a proportional hazards Weibull model:

(Continued on next page)

```
. use http://www.stata-press.com/data/r11/kva
(Generator experiment)

. streg load bearings, d(weibull) nolog
        failure _d:  1 (meaning all fail)
   analysis time _t:  failtime
```

Weibull regression -- log relative-hazard form

No. of subjects =   12   Number of obs   =   12
No. of failures =   12
Time at risk   =   896
                                                LR chi2(2)   =   30.27
Log likelihood =   5.6934189   Prob > chi2   =   0.0000

_t	Haz. Ratio	Std. Err.	z	P>|z|	[95% Conf. Interval]	
load	1.599315	.1883807	3.99	0.000	1.269616	2.014631
bearings	.1887995	.1312109	-2.40	0.016	.0483546	.7371644
/ln_p	2.051552	.2317074	8.85	0.000	1.597414	2.505691
p	7.779969	1.802677			4.940241	12.25202
1/p	.1285352	.0297826			.0816192	.2024193

Now we can predict both the median survival time and the log-median survival time for each observation:

```
. predict time, time
(option median time assumed; predicted median time)

. predict lntime, lntime
(option log median time assumed; predicted median log time)

. format time lntime %9.4f

. list failtime load bearings time lntime
```

	failtime	load	bearings	time	lntime
1.	100	15	0	127.5586	4.8486
2.	140	15	1	158.0407	5.0629
3.	97	20	0	94.3292	4.5468
4.	122	20	1	116.8707	4.7611
5.	84	25	0	69.7562	4.2450
6.	100	25	1	86.4255	4.4593
7.	54	30	0	51.5845	3.9432
8.	52	30	1	63.9114	4.1575
9.	40	35	0	38.1466	3.6414
10.	55	35	1	47.2623	3.8557
11.	22	40	0	28.2093	3.3397
12.	30	40	1	34.9504	3.5539

◁

▷ Example 2

Using the cancer data of example 6 in [ST] **streg**, again with drug remapped into a drug-treatment indicator, we can examine the various residuals that Stata produces. For a more detailed discussion on residuals, read *Residuals and diagnostic measures* in the [ST] **stcox postestimation** entry. Many of the concepts and ideas presented there also apply to streg models. For a more technical presentation of these residuals, see *Methods and formulas*.

We will begin by requesting the generalized Cox–Snell residuals with the command predict cs, csnell. The csnell option causes predict to create a new variable, cs, containing the Cox–Snell residuals. If the model fits the data, these residuals should have a standard exponential distribution with $\lambda = 1$. One way to verify the fit is to calculate an empirical estimate of the cumulative hazard function—based, for example, on the Kaplan–Meier survival estimates or the Aalen–Nelson estimator, taking the Cox–Snell residuals as the time variable and the censoring variable as before—and plotting it against cs. If the model fits the data, the plot should be a straight line with a slope of 1.

To do this after fitting the model, we first stset the data, specifying cs as our new failure-time variable and died as the failure indicator. We then use the sts generate command to generate the variable km containing the Kaplan–Meier survival estimates. Finally, we generate a new variable, H (cumulative hazard), and plot it against cs. The commands are

```
. use http://www.stata-press.com/data/r11/cancer, clear
(Patient Survival in Drug Trial)
. replace drug = drug==2 | drug==3         // 0, placebo : 1, nonplacebo
(48 real changes made)
. stset studytime, failure(died)
 (output omitted)
. qui streg age drug, d(exp)
. predict double cs, csnell
. qui stset cs, failure(died)
. qui sts generate km=s
. qui generate double H=-ln(km)
. line H cs cs, sort
```

We specified cs twice in the graph command so that a reference $45°$ line was plotted. We did this separately for each of four distributions. Results are plotted in figure 1:

(*Continued on next page*)

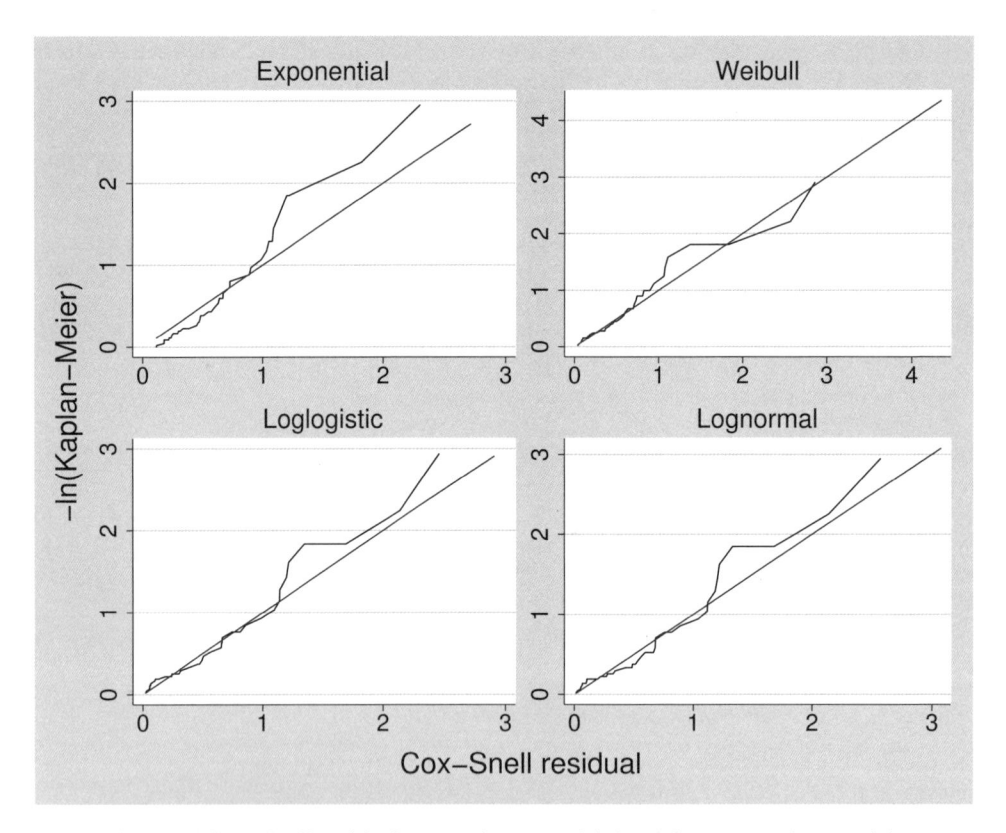

Figure 1. Cox–Snell residuals to evaluate model fit of four regression models

The plots indicate that the Weibull and lognormal models fit the data best and that the exponential model fits poorly. These results are consistent with our previous results (in [ST] **streg**) based on Akaike's information criterion.

◁

▷ Example 3

Let's now look at the martingale-like and deviance residuals. We use the term "martingale-like" because, although these residuals do not arise naturally from martingale theory for parametric survival models as they do for the Cox proportional hazards model, they do share similar form. We can generate these residuals by using predict's mgale option. Martingale residuals take values between $-\infty$ and 1 and therefore are difficult to interpret. The deviance residuals are a rescaling of the martingale-like residuals so that they are symmetric about zero and thus more like residuals obtained from linear regression. Plots of either deviance residuals against the linear predictor (i.e., the log relative hazard in PH models) or of deviance residuals against individual predictors can be useful in identifying aberrant observations and in assessing model fit. Continuing with our modified cancer data, we plot the deviance residual obtained after fitting a lognormal model:

```
. qui streg age drug, d(lnormal)
. predict dev, deviance
```

```
. scatter dev studytime, yline(0) m(o)
```

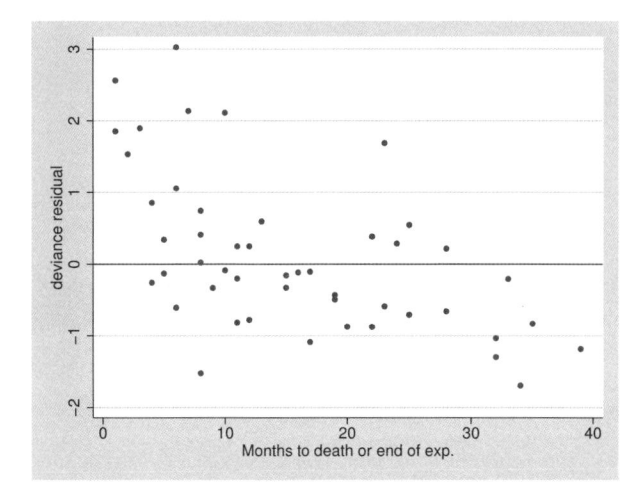

Figure 2. Deviance residuals to evaluate model fit of lognormal model

Figure 2 shows the deviance residuals to be relatively well behaved, with a few minor early exceptions.

◁

Methods and formulas

All postestimation commands listed above are implemented as ado-files.

predict *newvar*, *options* may be used after streg to predict various quantities, according to the following *options*:

median time:
$$newvar_j = \{t : \widehat{S}_j(t) = 1/2\}$$

where $\widehat{S}_j(t)$ is $S_j(t)$ with the parameter estimates "plugged in".

median lntime:
$$newvar_j = \left\{y : \widehat{S}_j(e^y) = 1/2\right\}$$

mean time:
$$newvar_j = \int_0^\infty \widehat{S}_j(t)dt$$

mean lntime:
$$newvar_j = \int_{-\infty}^\infty y e^y \widehat{f}_j(e^y)dy$$

where $\widehat{f}_j(t)$ is $f_j(t)$ with the parameter estimates plugged in.

hazard:
$$newvar_j = \widehat{f}_j(t_j)/\widehat{S}_j(t_j)$$

hr (PH models only):
$$newvar_j = \exp(\mathbf{x}_j\widehat{\boldsymbol{\beta}})$$

xb:
$$newvar_j = \mathbf{x}_j\widehat{\boldsymbol{\beta}}$$

stdp:
$$newvar_j = \widehat{\mathrm{se}}(\mathbf{x}_j\widehat{\boldsymbol{\beta}})$$

surv and csurv:
$$newvar_j = \widehat{S}_j(t_j)/\widehat{S}_j(t_{0j})$$

The above represents the probability of survival past time t_j given survival up until t_{0j} and represents what you obtain when you specify surv. If csurv is specified, these probabilities are multiplied (in time order) over a subject's multiple observations. What is obtained is then equal to the probability of survival past time t_j, given survival through the earliest observed t_{0j}, and given the subject's (possibly time-varying) covariate history. In single-record-per-subject data, surv and csurv are identical.

csnell:
$$newvar_j = -\log \widehat{S}_j(t_j)$$

The Cox–Snell (1968) residual, CS_j, for observation j at time t_j is defined as $\widehat{H}_j(t_j) = -\log \widehat{S}_j(t_j)$, which is the estimated cumulative hazard function obtained from the fitted model (Collett 2003, 111–112). Cox and Snell argued that if the correct model has been fit to the data, these residuals are n observations from an exponential distribution with unit mean. Thus a plot of the cumulative hazard rate of the residuals against the residuals themselves should result in a straight line of slope 1. Cox–Snell residuals can never be negative and therefore are not symmetric about zero. The options csnell and partial store in each observation that observation's contribution to the subject's Cox–Snell residual, which we refer to as a partial Cox–Snell residual. If only csnell is specified, partial residuals are summed within each subject to obtain one overall Cox–Snell residual for that subject. If there is only 1 observation per subject, partial has no effect.

mgale:
$$newvar_j = d_j - CS_j$$

Martingale residuals follow naturally from martingale theory for Cox proportional hazards, but their development does not carry over for parametric survival models. However, martingale-like residuals similar to those obtained for Cox can be derived from the Cox–Snell residuals: $M_j = d_j - CS_j$, where CS_j are the Cox–Snell residuals, as previously described.

Because martingale-like residuals are calculated from the Cox–Snell residuals, they also could be partial or not. Partial martingale residuals are generated with the mgale and partial options, and overall martingale residuals are generated with the mgale option.

Martingale residuals can be interpreted as the difference over time between the number of deaths in the data and the expected number from the fitted model. These residuals take values between $-\infty$ and 1 and have an expected value of zero, although, like the Cox–Snell residuals, they are not symmetric about zero, making them difficult to interpret.

deviance:
$$newvar_j = \text{sign}(M_j)\left[-2\left\{M_j + d_j \log(d_j - M_j)\right\}\right]^{1/2}$$

Deviance residuals are a scaling of the martingale-like residuals in an attempt to make them symmetric about zero. When the model fits the data, these residuals are symmetric about zero and thus can be more readily used to examine the data for outliers. If you also specify the partial option, you obtain partial deviance residuals, one for each observation.

predict also allows two options for use after fitting frailty models: alpha1 and unconditional. If unconditional is specified, the above predictions are modified to be based on $S_\theta(t)$ and $f_\theta(t)$, rather than $S(t)$ and $f(t)$; see [ST] streg. If alpha1 is specified, the predictions are as described above.

References

Collett, D. 2003. *Modelling Binary Data*. 2nd ed. London: Chapman & Hall/CRC.

Cox, D. R., and E. J. Snell. 1968. A general definition of residuals (with discussion). *Journal of the Royal Statistical Society, Series B* 30: 248–275.

Also see

[ST] **streg** — Parametric survival models

[ST] **stcurve** — Plot survivor, hazard, cumulative hazard, or cumulative incidence function

[U] **20 Estimation and postestimation commands**

Title

sts — Generate, graph, list, and test the survivor and cumulative hazard functions

Syntax

sts [graph] [*if*] [*in*] [, ...]

sts list [*if*] [*in*] [, ...]

sts test *varlist* [*if*] [*in*] [, ...]

sts generate *newvar* = ... [*if*] [*in*] [, ...] .

You must stset your data before using sts; see [ST] **stset**.
fweights, iweights, and pweights may be specified using stset; see [ST] **stset**.

See [ST] **sts graph**, [ST] **sts list**, [ST] **sts test**, and [ST] **sts generate** for details of syntax.

Description

sts reports and creates variables containing the estimated survivor and related functions, such as the Nelson–Aalen cumulative hazard function. For the survivor function, sts tests and produces Kaplan–Meier estimates or, via Cox regression, adjusted estimates.

sts graph is equivalent to typing sts by itself—it graphs the survivor function.

sts list lists the estimated survivor and related functions.

sts test tests the equality of the survivor function across groups.

sts generate creates new variables containing the estimated survivor function, the Nelson–Aalen cumulative hazard function, or related functions.

sts can be used with single- or multiple-record or single- or multiple-failure st data.

▷ Example 1

. use http://www.stata-press.com/data/r11/drugtr

Graph the Kaplan–Meier survivor function	. sts graph
	. sts graph, by(drug)
Graph the Nelson–Aalen cumulative hazard function	. sts graph, cumhaz
	. sts graph, cumhaz by(drug)
Graph the estimated hazard function	. sts graph, hazard
	. sts graph, hazard by(drug)
List the Kaplan–Meier survivor function	. sts list
	. sts list, by(drug) compare
List the Nelson–Aalen cumulative hazard function	. sts list, cumhaz
	. sts list, cumhaz by(drug) compare

Generate variable containing the Kaplan–Meier survivor function	. sts gen surv = s . sts gen surv_by_drug = s, by(drug)
Generate variable containing the Nelson–Aalen cumulative hazard function	. sts gen haz = na . sts gen haz_by_drug = na, by(drug)
Test equality of survivor functions	. sts test drug . gen agecat = autocode(age,4,47,67) . sts test drug, strata(agecat)

◁

Remarks

Remarks are presented under the following headings:

>   Listing, graphing, and generating variables
>   Comparing survivor or cumulative hazard functions
>   Testing equality of survivor functions
>   Adjusted estimates
>   Counting the number lost due to censoring

sts concerns the survivor function, $S(t)$; the probability of surviving to t or beyond; the cumulative hazard function, $H(t)$; and the hazard function, $h(t)$. Its subcommands can list and generate variables containing $\widehat{S}(t)$ and $\widehat{H}(t)$ and test the equality of $S(t)$ over groups. Also:

- All subcommands share a common syntax.

- All subcommands deal with either the Kaplan–Meier product-limit or the Nelson–Aalen estimates unless you request adjusted survival estimates.

- If you request an adjustment, all subcommands perform the adjustment in the same way, which is described below.

The full details of each subcommand are found in the entries following this one, but each subcommand provides so many options to control exactly how the listing looks, how the graph appears, the form of the test to be performed, or what exactly is to be generated that the simplicity of sts can be easily overlooked.

So, without getting burdened by the details of syntax, let us demonstrate the sts commands by using the Stanford heart transplant data introduced in [ST] **stset**.

Listing, graphing, and generating variables

You can list the overall survivor function by typing sts list, and you can graph it by typing sts graph or sts. sts assumes that you mean graph when you do not type a subcommand.

Or, you can list the Nelson–Aalen cumulative hazard function by typing sts list, cumhaz, and you can graph it by typing sts graph, cumhaz.

When you type sts list, you are shown all the details:

```
. use http://www.stata-press.com/data/r11/stan3
(Heart transplant data)
. stset, noshow
```

```
. sts list
```

Time	Beg. Total	Fail	Net Lost	Survivor Function	Std. Error	[95% Conf. Int.]	
1	103	1	0	0.9903	0.0097	0.9331	0.9986
2	102	3	0	0.9612	0.0190	0.8998	0.9852
3	99	3	0	0.9320	0.0248	0.8627	0.9670
5	96	1	0	0.9223	0.0264	0.8507	0.9604
(output omitted)							
1586	2	0	1	0.1519	0.0493	0.0713	0.2606
1799	1	0	1	0.1519	0.0493	0.0713	0.2606

When you type sts graph or just sts, you are shown a graph of the same result detailed by list:

```
. sts graph
```

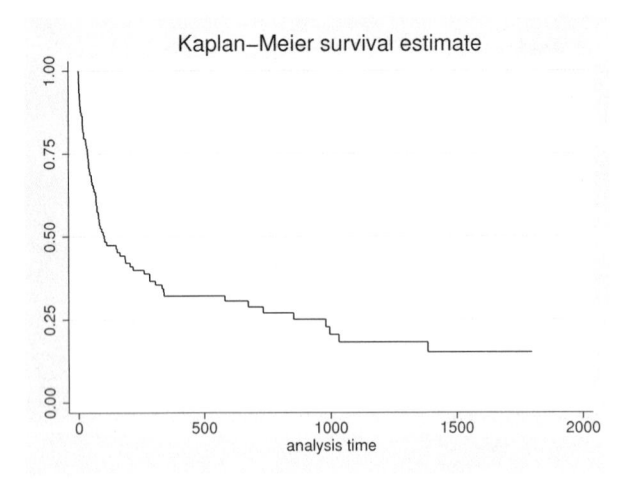

Kaplan–Meier survival estimate

sts generate is a rarely used command. Typing sts generate survf = s creates a new variable, survf, containing the same survivor function that list just listed and graph just graphed:

```
. sts gen survf = s
. sort t1
. list t1 survf in 1/10
```

	t1	survf
1.	1	.99029126
2.	1	.99029126
3.	1	.99029126
4.	1	.99029126
5.	2	.96116505
6.	2	.96116505
7.	2	.96116505
8.	2	.96116505
9.	2	.96116505
10.	2	.96116505

sts generate is provided if you want to make a calculation, listing, or graph that sts cannot already do for you.

Comparing survivor or cumulative hazard functions

sts allows you to compare survivor or cumulative hazard functions. sts graph and sts graph, cumhaz are probably most successful at this. For example, survivor functions can be plotted using

. sts graph, by(posttran)

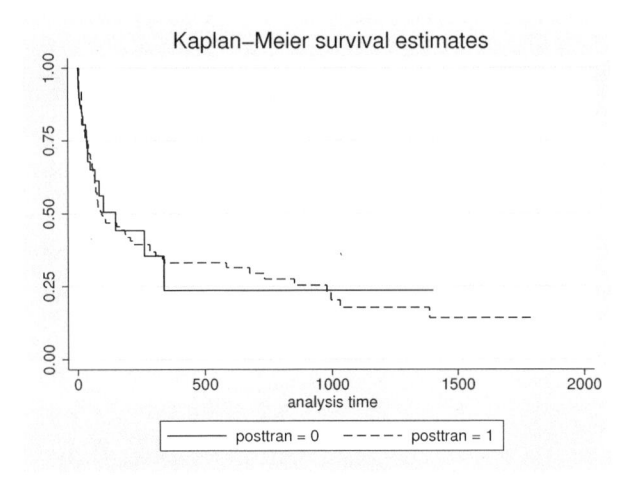

and Nelson–Aalen cumulative hazard functions can be plotted using

. sts graph, cumhaz by(posttran)

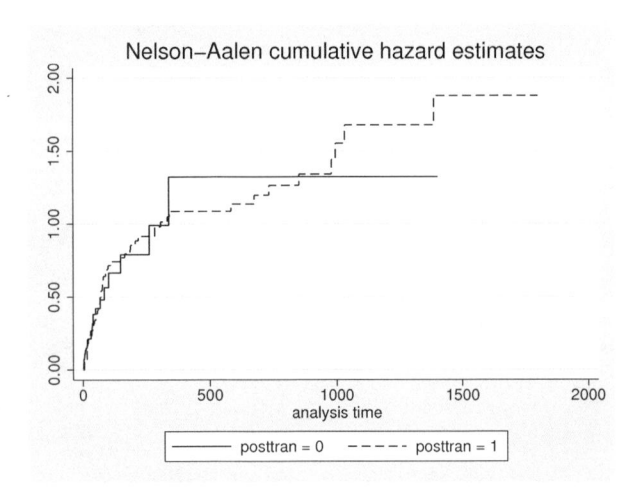

To compare survivor functions, we typed sts graph, just as before, and then we added by(posttran) to see the survivor functions for the groups designated by posttran. Here there are two groups, but as far as the sts command is concerned, there could have been more. cumhaz was also added to compare cumulative hazard functions.

You can also plot and compare estimated hazard functions by using sts graph, hazard. The hazard is estimated as a kernel smooth of the increments that sum to form the estimated cumulative hazard. The increments themselves do not estimate the hazard, but the smooth is weighted so that it estimates the hazard; see [ST] **sts graph**.

Just as you can compare survivor functions graphically by typing sts graph, by(posttran) and cumulative hazard functions by typing sts graph, cumhaz by(posttran), you can obtain detailed listings by typing sts list, by(posttran) and sts list, cumhaz by(posttran), respectively. Below we list the survivor function and specify enter, which adds a number-who-enter column:

```
. sts list, by(posttran) enter
```

Time	Beg. Total	Fail	Lost	Enter	Survivor Function	Std. Error	[95% Conf. Int.]	
posttran=0								
0	0	0	0	103	1.0000	.	.	.
1	103	1	3	0	0.9903	0.0097	0.9331	0.9986
2	99	3	3	0	0.9603	0.0195	0.8976	0.9849
(output omitted)								
427	2	0	1	0	0.2359	0.1217	0.0545	0.4882
1400	1	0	1	0	0.2359	0.1217	0.0545	0.4882
posttran=1								
1	0	0	0	3	1.0000	.	.	.
2	3	0	0	3	1.0000	.	.	.
3	6	0	0	3	1.0000	.	.	.
4	9	0	0	2	1.0000	.	.	.
5	11	0	0	3	1.0000	.	.	.
5.1	14	1	0	0	0.9286	0.0688	0.5908	0.9896
6	13	0	0	1	0.9286	0.0688	0.5908	0.9896
8	14	0	0	2	0.9286	0.0688	0.5908	0.9896
10	16	0	0	2	0.9286	0.0688	0.5908	0.9896
(output omitted)								
1586	2	0	1	0	0.1420	0.0546	0.0566	0.2653
1799	1	0	1	0	0.1420	0.0546	0.0566	0.2653

sts list's compare option allows you to compare survivor or cumulative hazard functions by listing the groups side by side.

```
. sts list, by(posttran) compare
```

		Survivor Function	
posttran		0	1
time	1	0.9903	1.0000
	225	0.4422	0.3934
	449	0.2359	0.3304
	673	0.2359	0.3139
	897	0.2359	0.2535
	1121	0.2359	0.1774
	1345	0.2359	0.1774
	1569	.	0.1420
	1793	.	0.1420
	2017	.	.

If we include the cumhaz option, the cumulative hazard functions are listed:

```
. sts list, cumhaz by(posttran) compare
                Nelson-Aalen Cum. Haz.
posttran              0          1

time       1      0.0097     0.0000
         225      0.7896     0.9145
         449      1.3229     1.0850
         673      1.3229     1.1350
         897      1.3229     1.3411
        1121      1.3229     1.6772
        1345      1.3229     1.6772
        1569         .       1.8772
        1793         .       1.8772
        2017         .          .
```

When you specify compare, the same detailed survivor or cumulative hazard function is calculated, but it is then evaluated at 10 or so given times, and those evaluations are listed. Above we left it to sts list to choose the comparison times, but we can specify them ourselves with the at() option:

```
. sts list, by(posttran) compare at(0 100 to 1700)
                Survivor Function
posttran              0          1

time       0      1.0000     1.0000
         100      0.5616     0.4814
         200      0.4422     0.4184
         300      0.3538     0.3680
         400      0.2359     0.3304
         500      0.2359     0.3304
         600      0.2359     0.3139
         700      0.2359     0.2942
         800      0.2359     0.2746
         900      0.2359     0.2535
        1000      0.2359     0.2028
        1100      0.2359     0.1774
        1200      0.2359     0.1774
        1300      0.2359     0.1774
        1400      0.2359     0.1420
        1500         .       0.1420
        1600         .       0.1420
        1700         .       0.1420
```

Testing equality of survivor functions

sts test tests equality of survivor functions:

```
. sts test posttran
Log-rank test for equality of survivor functions
              Events      Events
posttran     observed    expected

0               30         31.20
1               45         43.80

Total           75         75.00
              chi2(1) =      0.13
              Pr>chi2 =    0.7225
```

When you do not specify otherwise, `sts test` performs the log-rank test, but it can also perform the Wilcoxon test:

```
. sts test posttran, wilcoxon
```

Wilcoxon (Breslow) test for equality of survivor functions

posttran	Events observed	Events expected	Sum of ranks
0	30	31.20	-85
1	45	43.80	85
Total	75	75.00	0

$$\text{chi2(1)} = 0.14$$
$$\text{Pr>chi2} = 0.7083$$

`sts test` also performs stratified tests; see [ST] **sts test**.

Adjusted estimates

All the estimates of the survivor function we have seen so far are the Kaplan–Meier product-limit estimates. `sts` can make adjusted estimates to the survivor function. We want to illustrate this and explain how it is done.

The heart transplant dataset is not the best for demonstrating this feature because we are starting with survivor functions that are similar already, so let's switch to data on a fictional drug trial:

```
. use http://www.stata-press.com/data/r11/drug2, clear
(Patient Survival in Drug Trial)
. stset, noshow
. stdescribe
```

			per subject		
Category	total	mean	min	median	max
no. of subjects	48				
no. of records	48	1	1	1	1
(first) entry time		0	0	0	0
(final) exit time		15.5	1	12.5	39
subjects with gap	0				
time on gap if gap	0				
time at risk	744	15.5	1	12.5	39
failures	31	.6458333	0	1	1

This dataset contains 48 subjects, all observed from time 0. The `st` command shows us how the dataset is currently declared:

```
. st
-> stset studytime, failure(died) noshow
       failure event:  died != 0 & died < .
  obs. time interval:  (0, studytime]
   exit on or before:  failure
```

The dataset contains variables age and drug:

```
. summarize age drug
```

Variable	Obs	Mean	Std. Dev.	Min	Max
age	48	47.125	9.492718	32	67
drug	48	.5833333	.4982238	0	1

We are comparing the outcomes of drug = 1 with that of the placebo, drug = 0. Here are the survivor curves for the two groups:

```
. sts graph, by(drug)
```

Kaplan–Meier survival estimates

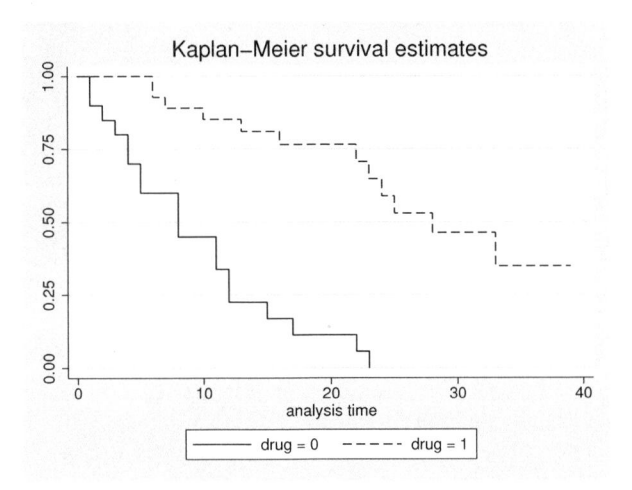

Here are the survivor curves adjusted for age (and scaled to age 50):

```
. generate age50 = age-50
. sts graph, by(drug) adjustfor(age50)
```

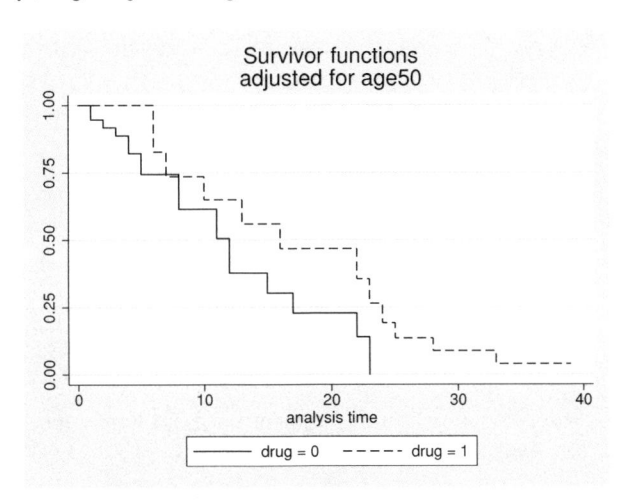

Survivor functions adjusted for age50

The age difference between the two samples accounts for much of the difference between the survivor functions.

When you type by(*group*) adjustfor(*vars*), sts fits a separate Cox proportional hazards model on *vars* (estimation via stcox) and retrieves the separately estimated baseline survivor functions. sts graph graphs the baseline survivor functions, sts list lists them, and sts generate saves them.

Thus sts list can list what sts graph plots:

```
. sts list, by(drug) adjustfor(age50) compare
            Adjusted Survivor Function
drug                        0         1

time        1          0.9463    1.0000
            5          0.7439    1.0000
            9          0.6135    0.7358
           13          0.3770    0.5588
           17          0.2282    0.4668
           21          0.2282    0.4668
           25               .    0.1342
           29               .    0.0872
           33               .    0.0388
           37               .    0.0388
           41               .         .

Survivor function adjusted for age50
```

In both the graph and the listing, we must adjust for variable age50 = age − 50 and not just age. Adjusted survivor functions are adjusted to the adjustfor() variables and scaled to correspond to the adjustfor() variables set to 0. Here is the result of adjusting for age, which is 0 at birth:

```
. sts list, by(drug) adjustfor(age) compare
            Adjusted Survivor Function
drug                        0         1

time        1          0.9994    1.0000
            5          0.9970    1.0000
            9          0.9951    0.9995
           13          0.9903    0.9990
           17          0.9853    0.9987
           21          0.9853    0.9987
           25               .    0.9965
           29               .    0.9958
           33               .    0.9944
           37               .    0.9944
           41               .         .

Survivor function adjusted for age
```

These are equivalent to what we obtained previously but not nearly so informative because of the scaling of the survivor function. The adjustfor(age) option scales the survivor function to correspond to age = 0. age is calendar age, so the survivor function is scaled to correspond to a newborn.

There is another way that sts will adjust the survivor function. Rather than specifying by(*group*) adjustfor(*vars*), we can specify strata(*group*) adjustfor(*vars*):

```
. sts list, strata(drug) adjustfor(age50) compare
         Adjusted Survivor Function
drug                     0           1

time         1      0.9526      1.0000
             5      0.7668      1.0000
             9      0.6417      0.7626
            13      0.4080      0.5995
            17      0.2541      0.5139
            21      0.2541      0.5139
            25           .      0.1800
            29           .      0.1247
            33           .      0.0614
            37           .      0.0614
            41           .           .

Survivor function adjusted for age50
```

When we specify strata() instead of by(), instead of fitting separate Cox models for each stratum, sts list fits one stratified Cox model and retrieves the stratified baseline survivor function. That is, strata() rather than by() constrains the effect of the adjustfor() variables to be the same across strata.

Counting the number lost due to censoring

sts list, in the detailed output, shows the number lost in the fourth column:

```
. sts list
              Beg.              Net        Survivor    Std.
    Time     Total    Fail     Lost        Function    Error      [95% Conf. Int.]

        1       48       2        0          0.9583    0.0288    0.8435    0.9894
        2       46       1        0          0.9375    0.0349    0.8186    0.9794
        3       45       1        0          0.9167    0.0399    0.7930    0.9679
   (output omitted)
        8       36       3        1          0.7061    0.0661    0.5546    0.8143
        9       32       0        1          0.7061    0.0661    0.5546    0.8143
       10       31       1        1          0.6833    0.0678    0.5302    0.7957
   (output omitted)
       39        1       0        1          0.1918    0.0791    0.0676    0.3634
```

sts graph, if you specify the lost option, will show that number, too:

(Continued on next page)

```
. sts graph, lost
```

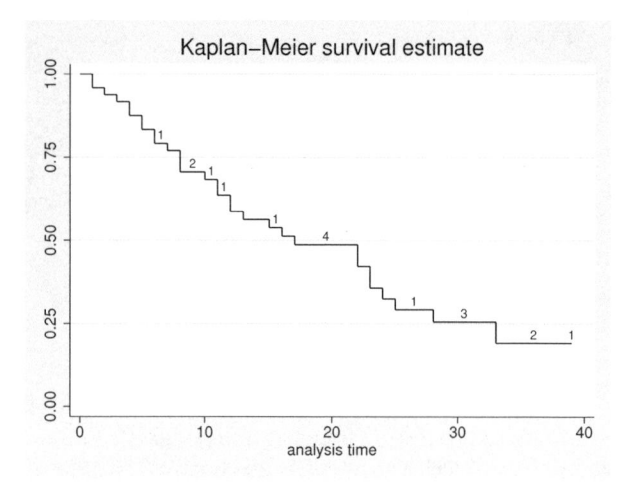

Kaplan–Meier survival estimate

The number on the listing and on the graph is the number of net lost, defined as the number of censored minus the number who enter. With simple survival data—with 1 observation per subject—net lost corresponds to lost.

With more complicated survival data—meaning delayed entry or multiple records per subject—the number of net lost may surprise you. With complicated data, the vague term *lost* can mean many things. Sometimes subjects are lost, but mostly there are many censorings followed by reentries—a subject is censored at time 5 immediately to reenter the data with different covariates. This is called thrashing.

There are other possibilities: a subject can be lost, but only for a while, and so reenter the data with a gap; a subject can be censored out of one stratum to enter another. There are too many possibilities to dedicate a column in a table or a plotting symbol in a graph to each one. sts's solution is to define *lost* as *net lost*, meaning censored minus entered, and show that number. How we define *lost* does not affect the calculation of the survivor function; it merely affects a number that researchers often report.

Defining lost as censored minus entered results in exactly what is desired for simple survival data. Because everybody enters at time 0, calculating censored minus entered amounts to calculating censored − 0. The number of net lost is the number of censored.

In more complicated data, calculating censored minus entered results in the number really lost if there are no gaps and no delayed entry. Then the subtraction smooths the thrashing. In an interval, five might be censored and three reenter, so $5 - 3 = 2$ were lost.

In even more complicated data, calculating censored minus entered results in something reasonable once you understand how to interpret negative numbers and are cautious in interpreting positive ones. Five might be censored and three might enter (from the five? who can say?), resulting in two net lost; or three might be censored and five enter, resulting in -2 being lost.

sts, by default, reports the net lost but will, if you specify the enter option, report the pure number censored and the pure number who enter. Sometimes you will want to do that. Earlier in this entry, we used sts list to display the survivor functions in the Stanford heart transplant data for subjects pre- and posttransplantation, and we slipped in an enter option:

```
. use http://www.stata-press.com/data/r11/stan3, clear
(Heart transplant data)
```

sts — Generate, graph, list, and test the survivor and cumulative hazard functions

```
. stset, noshow
. sts list, by(posttran) enter
```

Time	Beg. Total	Fail	Lost	Enter	Survivor Function	Std. Error	[95% Conf. Int.]	
posttran=0								
0	0	0	0	103	1.0000	.	.	.
1	103	1	3	0	0.9903	0.0097	0.9331	0.9986
2	99	3	3	0	0.9603	0.0195	0.8976	0.9849
3	93	3	3	0	0.9293	0.0258	0.8574	0.9657
(output omitted)								
427	2	0	1	0	0.2359	0.1217	0.0545	0.4882
1400	1	0	1	0	0.2359	0.1217	0.0545	0.4882
posttran=1								
1	0	0	0	3	1.0000	.	.	.
2	3	0	0	3	1.0000	.	.	.
3	6	0	0	3	1.0000	.	.	.
4	9	0	0	2	1.0000	.	.	.
5	11	0	0	3	1.0000	.	.	.
5.1	14	1	0	0	0.9286	0.0688	0.5908	0.9896
6	13	0	0	1	0.9286	0.0688	0.5908	0.9896
8	14	0	0	2	0.9286	0.0688	0.5908	0.9896
(output omitted)								
1586	2	0	1	0	0.1420	0.0546	0.0566	0.2653
1799	1	0	1	0	0.1420	0.0546	0.0566	0.2653

We did that to keep you from being shocked at negative numbers for the net lost. In this complicated dataset, the value of posttran changes over time. All patients start with posttran = 0, and later some change to posttran = 1.

Thus, at time 1 in the posttran = 0 group, three are lost—to the group but not to the experiment. Simultaneously, in the posttran = 1 group, we see that three enter. Had we not specified the enter option, you would not have seen that three enter, and you would have seen that −3 were, in net, lost:

```
. sts list, by(posttran)
```

Time	Beg. Total	Fail	Net Lost	Survivor Function	Std. Error	[95% Conf. Int.]	
posttran=0							
1	103	1	3	0.9903	0.0097	0.9331	0.9986
2	99	3	3	0.9603	0.0195	0.8976	0.9849
3	93	3	3	0.9293	0.0258	0.8574	0.9657
(output omitted)							
427	2	0	1	0.2359	0.1217	0.0545	0.4882
1400	1	0	1	0.2359	0.1217	0.0545	0.4882
posttran=1							
1	0	0	−3	1.0000	.	.	.
2	3	0	−3	1.0000	.	.	.
3	6	0	−3	1.0000	.	.	.
4	9	0	−2	1.0000	.	.	.
5	11	0	−3	1.0000	.	.	.
5.1	14	1	0	0.9286	0.0688	0.5908	0.9896
6	13	0	−1	0.9286	0.0688	0.5908	0.9896
8	14	0	−2	0.9286	0.0688	0.5908	0.9896
(output omitted)							
1586	2	0	1	0.1420	0.0546	0.0566	0.2653
1799	1	0	1	0.1420	0.0546	0.0566	0.2653

Here specifying enter makes the table easier to explain, but do not jump to the conclusion that specifying enter is always a good idea. In this same dataset, let's look at the overall survivor function, first with the enter option:

```
. sts list, enter
```

Time	Beg. Total	Fail	Lost	Enter	Survivor Function	Std. Error	[95% Conf. Int.]	
0	0	0	0	103	1.0000	.	.	.
1	103	1	3	3	0.9903	0.0097	0.9331	0.9986
2	102	3	3	3	0.9612	0.0190	0.8998	0.9852
3	99	3	3	3	0.9320	0.0248	0.8627	0.9670
(output omitted)								
1571	3	0	1	0	0.1519	0.0493	0.0713	0.2606
1586	2	0	1	0	0.1519	0.0493	0.0713	0.2606
1799	1	0	1	0	0.1519	0.0493	0.0713	0.2606

At time 1, three are lost and three enter. There is no delayed entry in this dataset, and there are no gaps; so, it is the same three that were lost and reentered, and no one was really lost. At time 1571, on the other hand, a patient really was lost. This is all more clearly revealed when we do not specify the enter option:

```
. sts list
```

Time	Beg. Total	Fail	Net Lost	Survivor Function	Std. Error	[95% Conf. Int.]	
1	103	1	0	0.9903	0.0097	0.9331	0.9986
2	102	3	0	0.9612	0.0190	0.8998	0.9852
3	99	3	0	0.9320	0.0248	0.8627	0.9670
(output omitted)							
1571	3	0	1	0.1519	0.0493	0.0713	0.2606
1586	2	0	1	0.1519	0.0493	0.0713	0.2606
1799	1	0	1	0.1519	0.0493	0.0713	0.2606

Thus, to summarize:

- The sts list and graph commands will show the number lost or censored. sts list, by default, shows this number on the detailed output. sts graph shows the number when you specify the lost option.

- By default, the number lost is the net number lost, defined as censored minus entered.

- Both commands allow you to specify the enter option and then show the number who actually entered, and the number lost becomes the actual number censored, not censored minus entered.

❑ Technical note

There is one other issue about the Kaplan–Meier estimator regarding delayed entry. When the earliest entry into the study occurs after $t = 0$, one may still calculate the Kaplan–Meier estimation, but the interpretation changes. Rather than estimating $S(t)$, you are now estimating $S(t|t_{\min})$, the probability of surviving past time t given survival to time $t_{\min}$, where $t_{\min}$ is the earliest entry time.

❑

Saved results

sts test saves the following in r():

Scalars
    r(df)   degrees of freedom
    r(chi2) χ^2

Methods and formulas

sts is implemented as an ado-file.

Unless adjusted estimates are requested, sts estimates the survivor function by using the Kaplan–Meier product-limit method.

When the cumhaz option is specified, sts estimates the cumulative hazard function by using the Nelson–Aalen estimator.

For an introduction to the Kaplan–Meier product-limit method and the log-rank test, see Pagano and Gauvreau (2000, 495–499); for a detailed discussion, see Cox and Oakes (1984), Kalbfleisch and Prentice (2002), or Klein and Moeschberger (2003). For an introduction to survival analysis with examples using the sts commands, see Dupont (2009).

Let t_j, $j = 1, \ldots,$ denote the times at which failure occurs. Let n_j be the number at risk of failure just before time t_j and d_j be the number of failures at time t_j. Then the nonparametric maximum likelihood estimate of the survivor function (Kaplan and Meier 1958) is

$$\widehat{S}(t) = \prod_{j|t_j \leq t} \left(\frac{n_j - d_j}{n_j} \right)$$

(Kalbfleisch and Prentice 2002, 15).

The failure function, $\widehat{F}(t)$, is defined as $1 - \widehat{S}(t)$.

The standard error reported is given by Greenwood's formula (Greenwood 1926):

$$\widehat{\mathrm{Var}}\{\widehat{S}(t)\} = \widehat{S}^2(t) \sum_{j|t_j \leq t} \frac{d_j}{n_j(n_j - d_j)}$$

(Kalbfleisch and Prentice 2002, 17–18). These standard errors, however, are not used for confidence intervals. Instead, the asymptotic variance of $\ln[-\ln \widehat{S}(t)]$,

$$\widehat{\sigma}^2(t) = \frac{\sum \frac{d_j}{n_j(n_j - d_j)}}{\left\{ \sum \ln\left(\frac{n_j - d_j}{n_j}\right) \right\}^2}$$

is used, where sums are calculated over $j|t_j \leq t$ (Kalbfleisch and Prentice 2002, 18). The confidence bounds are then $\widehat{S}(t)^{\exp(\pm z_{\alpha/2} \widehat{\sigma}(t))}$, where $z_{\alpha/2}$ is the $(1 - \alpha/2)$ quantile of the normal distribution. sts suppresses reporting the standard error and confidence bounds if the data are pweighted because these formulas are no longer appropriate.

When the `adjustfor()` option is specified, the survivor function estimate, $\widehat{S}(t)$, is the baseline survivor function estimate $\widehat{S}_0(t)$ of stcox; see [ST] **stcox**. If, `by()`, is specified, $\widehat{S}(t)$ is obtained from fitting separate Cox models on `adjustfor()` for each of the `by()` groups. If instead `strata()` is specified, one Cox model on `adjustfor()`, stratified by `strata()`, is fit.

The Nelson–Aalen estimator of the cumulative hazard rate function is derived from Nelson (1972) and Aalen (1978) and is defined up to the largest observed time as

$$\widehat{H}(t) = \sum_{j|t_j \leq t} \frac{d_j}{n_j}$$

Its variance (Aalen 1978) may be estimated by

$$\widehat{\text{Var}}\{\widehat{H}(t)\} = \sum_{j|t_j \leq t} \frac{d_j}{n_j{}^2}$$

Pointwise confidence intervals are calculated using the asymptotic variance of $\ln \widehat{H}(t)$,

$$\widehat{\phi}^2(t) = \frac{\widehat{\text{Var}}\{\widehat{H}(t)\}}{\{\widehat{H}(t)\}^2}$$

The confidence bounds are then $\widehat{H}(t)\exp\{\pm z_{\alpha/2}\widehat{\phi}(t)\}$. If the data are `pweighted`, these formulas are not appropriate, and then confidence intervals are not reported.

Major Greenwood (1880–1949) was born in London to a medical family. His given name, "Major", was also that of his father and grandfather. Greenwood trained as a doctor but followed a career in medical research, learning statistics from Karl Pearson. He worked as a medical statistician and epidemiologist at the Lister Institute, the Ministry of Health, and the London School of Hygiene and Tropical Medicine. With interests ranging from clinical trials to historical subjects, Greenwood played a major role in developing biostatistics in the first half of the twentieth century.

Edward Lynn Kaplan (1920–2006) was working at Bell Telephone Laboratories on the lifetimes of vacuum tubes when, through John W. Tukey, he became aware of the work of Paul Meier on essentially the same statistical problem. They had both previously been graduate students at Princeton. Their two separate papers were merged and the result was published after some years. Kaplan became a professor of mathematics at Oregon State University, where he wrote a book on mathematical programming and games.

Paul Meier (1924–) took degrees at Oberlin and Princeton and then served on the faculty at Johns Hopkins, Chicago, and Columbia. In addition to his key contribution with Kaplan, the most cited paper in statistical science, he worked as a biostatistician, making many theoretical and applied contributions in the area of clinical trials.

Wayne B. Nelson (1936–) was born in Chicago and received degrees in physics and statistics from Caltech and the University of Illinois. A longtime employee of General Electric, he now works as a consultant, specializing in reliability analysis and accelerated testing.

Odd Olai Aalen (1947–) was born in Oslo, Norway, and studied there and at Berkeley, where he was awarded a PhD in 1975 for a thesis on counting processes. He is a professor of statistics at the University of Oslo and works on survival and event history analysis. Aalen was one of the prime movers in introducing martingale ideas to this branch of statistics.

Nelson and Aalen met for the first time at a conference at the University of South Carolina in 2003.

References

Aalen, O. O. 1978. Nonparametric inference for a family of counting processes. *Annals of Statistics* 6: 701–726.

Breslow, N. E. 1992. Kaplan and Meier (1958) "Nonparametric estimation from incomplete observations". In *Breakthroughs in Statistics, Vol. II: Methodology and Distribution*, ed. S. Kotz and N. L. Johnson, 311–338. New York: Springer.

Cleves, M. A. 1999. stata53: censored option added to sts graph command. *Stata Technical Bulletin* 50: 34–36. Reprinted in *Stata Technical Bulletin Reprints*, vol. 9, pp. 4–7. College Station, TX: Stata Press.

Cox, D. R., and D. Oakes. 1984. *Analysis of Survival Data*. London: Chapman & Hall/CRC.

Dupont, W. D. 2009. *Statistical Modeling for Biomedical Researchers: A Simple Introduction to the Analysis of Complex Data*. 2nd ed. Cambridge: Cambridge University Press.

Greenwood, M. 1926. The natural duration of cancer. *Reports on Public Health and Medical Subjects* 33: 1–26.

Hogben, L. 1950. Major Greenwood, 1880–1949. *Obituary Notices of Fellows of the Royal Society* 7: 139–154.

Kalbfleisch, J. D., and R. L. Prentice. 2002. *The Statistical Analysis of Failure Time Data*. 2nd ed. New York: Wiley.

Kaplan, E. L., and P. Meier. 1958. Nonparametric estimation from incomplete observations. *Journal of the American Statistical Association* 53: 457–481.

Klein, J. P., and M. L. Moeschberger. 2003. *Survival Analysis: Techniques for Censored and Truncated Data*. 2nd ed. New York: Springer.

Linhart, J. M., J. S. Pitblado, and J. Hassell. 2004. From the help desk: Kaplan–Meier plots with stsatrisk. *Stata Journal* 4: 56–65.

Nelson, W. 1972. Theory and applications of hazard plotting for censored failure data. *Technometrics* 14: 945–966.

Newman, S. C. 2001. *Biostatistical Methods in Epidemiology*. New York: Wiley.

Pagano, M., and K. Gauvreau. 2000. *Principles of Biostatistics*. 2nd ed. Belmont, CA: Duxbury.

Smythe, B. 2006. Obituary: Edward Kaplan 1920–2006. *IMS Bulletin* 35: 7.

Wilkinson, L. 1998. Greenwood, Major. In Vol. 2 of *Encyclopedia of Biostatistics*, ed. P. Armitage and T. Colton, 1778–1780. Chichester, UK: Wiley.

Also see

[ST] **stci** — Confidence intervals for means and percentiles of survival time

[ST] **stcox** — Cox proportional hazards model

[ST] **sts generate** — Create variables containing survivor and related functions

[ST] **sts graph** — Graph the survivor and cumulative hazard functions

[ST] **sts list** — List the survivor or cumulative hazard function

[ST] **sts test** — Test equality of survivor functions

[ST] **stset** — Declare data to be survival-time data

[ST] **st** — Survival-time data

[ST] **survival analysis** — Introduction to survival analysis & epidemiological tables commands

Title

> **sts generate** — Create variables containing survivor and related functions

Syntax

> sts generate *newvar* =
>
> $\{$ s $|$ se(s) $|$ h $|$ se(lls) $|$ lb(s) $|$ ub(s) $|$ na $|$ se(na) $|$ lb(na) $|$ ub(na) $|$ n $|$ d $\}$
>
> [*newvar* = {...} ...] [*if*] [*in*] [, *options*]

options	description
Options	
by(*varlist*)	calculate separately on different groups of *varlist*
adjustfor(*varlist*)	adjust the estimates to zero values of *varlist*
strata(*varlist*)	stratify on different groups of *varlist*
level(*#*)	set confidence level; default is level(95)

You must stset your data before using sts generate; see [ST] **stset**.

Menu

Statistics > Survival analysis > Summary statistics, tests, and tables > Create survivor, hazard, and other variables

Description

sts generate creates new variables containing the estimated survivor (failure) function, the Nelson–Aalen cumulative hazard (integrated hazard) function, and related functions. See [ST] **sts** for an introduction to this command.

sts generate can be used with single- or multiple-record or single- or multiple-failure st data.

Functions

Main

s produces the Kaplan–Meier product-limit estimate of the survivor function, $\widehat{S}(t)$, or, if adjustfor() is specified, the baseline survivor function from a Cox regression model on the adjustfor() variables.

se(s) produces the Greenwood, pointwise standard error, $\widehat{se}\{\widehat{S}(t)\}$. This option may not be used with adjustfor().

h produces the estimated hazard component, $\Delta H_j = H(t_j) - H(t_{j-1})$, where t_j is the current failure time and t_{j-1} is the previous one. This is mainly a utility function used to calculate the estimated cumulative hazard, $H(t_j)$, yet you can estimate the hazard via a kernel smooth of the ΔH_j; see [ST] **sts graph**. It is recorded at all the points at which a failure occurs and is computed as d_j/n_j, where d_j is the number of failures occurring at time t_j and n_j is the number at risk at t_j before the occurrence of the failures.

se(lls) produces $\widehat{\sigma}(t)$, the standard error of $\ln\{-\ln \widehat{S}(t)\}$. This option may not be used with adjustfor().

lb(s) produces the lower bound of the confidence interval for $\widehat{S}(t)$ based on $\ln\{-\ln \widehat{S}(t)\}$: $\widehat{S}(t)^{\exp(-z_{\alpha/2}\widehat{\sigma}(t))}$, where $z_{\alpha/2}$ is the $(1-\alpha/2)$ quantile of the standard normal distribution. This option may not be used with adjustfor().

ub(s) produces the upper bound of the confidence interval for $\widehat{S}(t)$ based on $\ln\{-\ln \widehat{S}(t)\}$: $\widehat{S}(t)^{\exp(z_{\alpha/2}\widehat{\sigma}(t))}$, where $z_{\alpha/2}$ is the $(1-\alpha/2)$ quantile of the standard normal distribution. This option may not be used with adjustfor().

na produces the Nelson–Aalen estimate of the cumulative hazard function. This option may not be used with adjustfor().

se(na) produces pointwise standard error for the Nelson–Aalen estimate of the cumulative hazard function, $\widehat{H}(t)$. This option may not be used with adjustfor().

lb(na) produces the lower bound of the confidence interval for $\widehat{H}(t)$ based on the log-transformed cumulative hazard function. This option may not be used with adjustfor().

ub(na) produces the corresponding upper bound. This option may not be used with adjustfor().

n produces n_j, the number at risk just before time t_j. This option may not be used with adjustfor().

d produces d_j, the number failing at time t_j. This option may not be used with adjustfor().

Options

by(*varlist*) produces separate survivor or cumulative hazard functions by making separate calculations for each group identified by equal values of the variables in *varlist*. by() may not be combined with strata().

adjustfor(*varlist*) adjusts the estimate of the survivor (failure) or hazard function to that for 0 values of *varlist*. This option is available only with functions s or h. See [ST] **sts graph** for an example of how to adjust for values different from 0.

If you specify adjustfor() with by(), sts fits separate Cox regression models for each group, using the adjustfor() variables as covariates. The separately calculated baseline survivor functions are then retrieved.

If you specify adjustfor() with strata(), sts fits a stratified-on-group Cox regression model using the adjustfor() variables as covariates. The stratified, baseline survivor function is then retrieved.

strata(*varlist*) requests estimates of the survivor (failure) or hazard functions stratified on variables in *varlist*. It requires specifying adjustfor() and may not be combined with by().

level(*#*) specifies the confidence level, as a percentage, for the lb(s), ub(s), lb(na), and ub(na) functions. The default is level(95) or as set by set level; see [U] **20.7 Specifying the width of confidence intervals**.

Remarks

sts generate is a seldom-used command that gives you access to the calculations listed by sts list and graphed by sts graph.

Use of this command is demonstrated in [ST] **sts**.

Methods and formulas

See [ST] **sts**.

References

See [ST] **sts** for references.

Also see

[ST] **sts** — Generate, graph, list, and test the survivor and cumulative hazard functions

[ST] **sts graph** — Graph the survivor and cumulative hazard functions

[ST] **sts list** — List the survivor or cumulative hazard function

[ST] **sts test** — Test equality of survivor functions

[ST] **stset** — Declare data to be survival-time data

Title

> **sts graph** — Graph the survivor and cumulative hazard functions

Syntax

sts graph [*if*] [*in*] [, *options*]

options	description
Main	
survival	graph Kaplan–Meier survivor function; the default
failure	graph Kaplan–Meier failure function
cumhaz	graph Nelson–Aalen cumulative hazard function
hazard	graph smoothed hazard estimate
by(*varlist*)	calculate separately on different groups of *varlist*
adjustfor(*varlist*)	adjust the estimates to zero values of *varlist*
strata(*varlist*)	stratify on different groups of *varlist*
separate	show curves on separate graphs; default is to show curves one on top of another
ci	show pointwise confidence bands
At-risk table	
risktable	show table of number at risk beneath graph
risktable(*risk_spec*)	show customized table of number at risk beneath graph
Options	
level(*#*)	set confidence level; default is level(95)
per(*#*)	units to be used in reported rates
noshow	do not show st setting information
tmax(*#*)	show graph for $t \leq \#$
tmin(*#*)	show graph for $t \geq \#$
noorigin	begin survival (failure) curve at first exit time; default is to begin at $t = 0$
width(*#* [*#*...])	override default bandwidth(s)
kernel(*kernel*)	kernel function; use with hazard
noboundary	no boundary correction; use with hazard
lost	show number lost
enter	show number entered and number lost
atrisk	show numbers at risk at beginning of each interval
censored(single)	show one hash mark at each censoring time, no matter what number is censored
censored(number)	show one hash mark at each censoring time and number censored above hash mark
censored(multiple)	show multiple hash marks for multiple censoring at the same time
censopts(*hash_options*)	affect rendition of hash marks
lostopts(*marker_label_options*)	affect rendition of numbers lost
atriskopts(*marker_label_options*)	affect rendition of numbers at risk

Plot	
plotopts(cline_options)	affect rendition of the plotted lines
plot#opts(cline_options)	affect rendition of the #th plotted line; may not be combined with separate
CI plot	
ciopts(area_options)	affect rendition of the confidence bands
ci#opts(area_options)	affect rendition of the #th confidence band; may not be combined with separate
Add plots	
addplot(plot)	add other plots to the generated graph
Y axis, X axis, Titles, Legend, Overall	
twoway_options	any options documented in [G] *twoway_options*
byopts(byopts)	how subgraphs are combined, labeled, etc.

See [R] **kdensity** for information on *kernel*, and see [G] *marker_label_options*, [G] *cline_options*, and [G] *area_options*.

where *risk_spec* is

　　　[*numlist*] [, *table_options* group(*group*)]

numlist specifies the points at which the number at risk is to be evaluated, *table_options* customizes the table of number at risk, and group(*group*) specifies a specific group/row for *table_options* to be applied.

table_options	description
Main	
axis_label_options	control table by using axis labeling options; seldom used
order(order_spec)	select which rows appear and their order
righttitles	place titles on right side of the table
failevents	show number failed in the at-risk table
text_options	affect rendition of table elements and titles
Row titles	
rowtitle([text] [, rtext_options])	change title for a row
Title	
title([text] [, ttext_options])	change overall table title

See [G] *axis_label_options*.

where *order_spec* is

　　　# ["*text*" ["*text*" ...]] [...]

sts graph — Graph the survivor and cumulative hazard functions

text_options	description
size(*textsizestyle*)	size of text
color(*colorstyle*)	color of text
justification(*justificationstyle*)	text left-justified, centered, or right-justified
format(%*fmt*)	format values per %*fmt*
topgap(*relativesize*)	margin above rows
bottomgap(*relativesize*)	margin beneath rows
† style(*textstyle*)	overall style of text

† style() does not appear in the dialog box.
See [G] *textsizestyle*, [G] *colorstyle*, [G] *justificationstyle*, [G] *relativesize*, and [G] *textstyle*.

rtext_options	description
size(*textsizestyle*)	size of text
color(*colorstyle*)	color of text
justification(*justificationstyle*)	text left-justified, centered, or right-justified
at(*#*)	override x position of titles
topgap(*relativesize*)	margin above rows
† style(*textstyle*)	overall style of text

† style() does not appear in the dialog box.
See [G] *textsizestyle*, [G] *colorstyle*, [G] *justificationstyle*, [G] *relativesize*, and [G] *textstyle*.

ttext_options	description
size(*textsizestyle*)	size of text
color(*colorstyle*)	color of text
justification(*justificationstyle*)	text left-justified, centered, or right-justified
at(*#*)	override x position of titles
topgap(*relativesize*)	margin above rows
bottomgap(*relativesize*)	margin beneath rows
† style(*textstyle*)	overall style of text

† style() does not appear in the dialog box.
See [G] *textsizestyle*, [G] *colorstyle*, [G] *justificationstyle*, [G] *relativesize*, and [G] *textstyle*.

group	description
#*rownum*	specify group by row number in table
value	specify group by value of group
label	specify group by text of value label associated with group

hash_options	description
line_options	change look of dropped lines
marker_label_options	add marker labels

See [G] *line_options* and [G] *marker_label_options*, except for mlabel().

risktable() may be repeated and is *merged-explicit*; see [G] **concept: repeated options**.

You must stset your data before using sts graph; see [ST] **stset**.

fweights, iweights, and pweights may be specified using stset; see [ST] **stset**.

Menu

Statistics > Survival analysis > Graphs > Survivor and cumulative hazard functions

Description

sts graph graphs the estimated survivor (failure) function, the Nelson–Aalen estimated cumulative (integrated) hazard function, or the estimated hazard function. See [ST] **sts** for an introduction to this command.

sts graph can be used with single- or multiple-record or single- or multiple-failure st data.

Options

Main

survival, failure, cumhaz, and hazard specify the function to graph.

> survival specifies that the Kaplan–Meier survivor function be plotted. This option is the default if a function is not specified.
>
> failure specifies that the Kaplan–Meier failure function, $1 - S(t+0)$, be plotted.
>
> cumhaz specifies that the Nelson–Aalen estimate of the cumulative hazard function be plotted.
>
> hazard specifies that an estimate of the hazard function be plotted. This estimate is calculated as a weighted kernel-density estimate using the estimated hazard contributions, $\Delta \widehat{H}(t_j) = \widehat{H}(t_j) - \widehat{H}(t_{j-1})$. These hazard contributions are the same as those obtained by sts generate *newvar* = h.

by(*varlist*) produces separate survivor, cumulative hazard, or hazard functions by making separate calculations for each group identified by equal values of the variables in *varlist* and plots all the groups on one graph. by() may not be combined with strata().

> If you have more than one by() variable but need only one, use egen to create it; see [D] **egen**.

adjustfor(*varlist*) adjusts the estimate of the survivor or hazard functions to that for 0 values of *varlist*. If you want to adjust the function to values different from 0, you need to center the variables around those values before issuing the command. Say that you want to plot the survivor function adjusted to age of patients and the ages in your sample are 40–60 years. Then

> . sts graph, adjustfor(age)

will graph the survivor function adjusted to age 0. If you want to adjust the function to age 40, type

> . gen age40 = age - 40
> . sts graph, adjustfor(age40)

adjustfor() is not available with cumhaz or ci.

If you specify adjustfor() with by(), sts fits separate Cox regression models for each group, using the adjustfor() variables as covariates. The separately calculated baseline survivor functions are then retrieved.

If you specify adjustfor() with strata(), sts fits a stratified-on-group Cox regression model using the adjustfor() variables as covariates. The stratified, baseline survivor function is then retrieved.

strata(*varlist*) produces estimates of the survivor (failure) or hazard functions stratified on variables in *varlist* and plots all the groups on one graph. It requires specifying adjustfor() and may not be combined with by().

If you have more than one strata() variable but need only one, use egen to create it; see [D] **egen**.

separate is meaningful only with by() or strata(); it requests that each group be placed on its own graph rather than one on top of the other. Sometimes curves have to be placed on separate graphs—such as when you specify ci—because otherwise it would be too confusing.

ci includes pointwise confidence bands. The default is not to produce these bands. ci is not allowed with adjustfor() or pweights.

At-risk table

risktable[([*numlist*] [, *table_options*])] displays a table showing the number at risk beneath the plot. risktable may not be used with separate or adjustfor().

risktable displays the table in the default format with number at risk shown for each time reported on the x axis.

risktable([*numlist*] [, *table_options*]) specifies that the number at risk be evaluated at the points specified in *numlist* or that the rendition of the table be changed by *table_options*.

There are two ways to change the points at which the numbers at risk are evaluated.

1. The x axis of the graph may be altered. For example:

    . sts graph, xlabel(0(5)40) risktable

2. A *numlist* can be specified directly in the risktable() option, which affects only the at-risk table. For example:

    . sts graph, risktable(0(5)40)

The two examples produce the same at-risk table, but the first also changes the time labels on the graph's x axis.

table_options affect the rendition of the at-risk table and may be any of the following:

group(#*rownum* | *value* | *label*) specifies that all the suboptions specified in the risktable() apply only to the specified group. Because the risktable() option may be repeated, this option allows different rows of the at-risk table to be displayed with different colors, font sizes, etc.

When both a value and a value label are matched, the value label takes precedence.

risktable() may be specified with or without the group() suboption. When specified without group(), each suboption is applied to all available groups or rows. risktable() specified without group() is considered to be global and is itself merged-explicit. See [G] **concept: repeated options** for more information on how repeated options are merged.

Consider the following example:

```
. sts graph, by(drug) risktable(, color(red) size(small))
> risktable(, color(navy))
```

The example above would produce a table where all rows are colored navy with small text.

Combining global risktable() options with group-specific risktable() options can be useful. When global options are combined with group-specific options, group-specific options always take precedence.

Consider the following example:

```
. sts graph, by(drug) risktable(, color(navy))
> risktable(, color(red) group(#1))
```

The example above would produce a table with the first row colored red and all remaining rows colored navy.

Main

axis_label_options control the table by using axis labeling options. These options are seldom used. See [G] *axis_label_options*.

order() specifies which and in what order rows are to appear in the at-risk table. Optionally, order() can be used to override the default text.

order(# # # ...) is the syntax used for identifying which rows to display and their order. order(1 2 3) would specify that row 1 is to appear first in the table, followed by row 2, followed by row 3. order(1 2 3) is the default if there are three groups. If there were four groups, order(1 2 3 4) would be the default, and so on. If there were four groups and you specified order(1 2 3), the fourth row would not appear in the at-risk table. If you specified order(2 1 3), row 2 would appear first, followed by row 1, followed by row 3.

order(# "*text*" # "*text*" ...) is the syntax used for specifying the row order and alternate row titles.

Consider the following at-risk table:

```
drug = 1    20    8     2
drug = 2    14    10    4    1
drug = 3    14    13    10   5
```

Specifying order(1 "Placebo" 3 2) would produce

```
Placebo     20    8     2
drug = 3    14    13    10   5
drug = 2    14    10    4    1
```

and specifying order(1 "Placebo" 3 "Drug 2" 2 "Drug 1") would produce

```
Placebo     20    8     2
Drug 2      14    13    10   5
Drug 1      14    10    4    1
```

righttitles specifies that row titles be placed to the right of the at-risk values. The default is to place row titles to the left of the at-risk values.

failevents specifies that the number of failure events be shown in parentheses, after the time in which the risk values were calculated.

text_options affect the rendition of both row titles and number at risk and may be any of the following:

size(*textsizestyle*) specifies the size of text.

color(*colorstyle*) specifies the color of text.

justification(*justificationstyle*) specifies how text elements are to be justified.

format(%*fmt*) specifies how numeric values are to be formatted.

topgap(*relativesize*) specifies how much space is to be placed above each row.

bottomgap(*relativesize*) specifies how much space is to be placed beneath each row.

style(*textstyle*) specifies the style of text. This option does not appear in the dialog box.

Row titles

rowtitle([*text*][, *rtext_options*]) changes the default text or rendition of row titles. Specifying rowtitle(, color(navy)) would change the color of all row titles to navy.

rowtitle() is often combined with group() to change the text or rendition of a title. Specifying rowtitle(Placebo) group(#2) would change the title of the second row to Placebo. Specifying rowtitle(, color(red)) group(#3) would change the color of the row title for the third row to red.

Row titles may include more than one line. Lines are specified one after the other, each enclosed in double quotes. Specifying rowtitle("Experimental drug") group(#1) would produce a one-line row title, and specifying rowtitle("Experimental" "Drug") group(#1) would produce a multiple-line row title.

rtext_options affect the rendition of both row titles and number at risk and may be any of the following:

size(*textsizestyle*) specifies the size of text.

color(*colorstyle*) specifies the color of text.

justification(*justificationstyle*) specifies how text elements are to be justified.

at(*#*) allows you to reposition row titles or the overall table title to align with a specific location on the x axis.

topgap(*relativesize*) specifies how much space is to be placed above each row.

style(*textstyle*) specifies the style of text. This option does not appear on the dialog box.

Title

title([*title*][, *ttext_options*]) may be used to override the default title for the at-risk table and affect the rendition of its text.

Titles may include one line of text or multiple lines. title("At-risk table") will produce a one-line title, and title("At-risk" "table") will produce a multiple-line title.

ttext_options affect the rendition of both row titles and number at risk and may be any of the following:

size(*textsizestyle*) specifies the size of text.

color(*colorstyle*) specifies the color of text.

justification(*justificationstyle*) specifies how text elements are to be justified.

at(*#*) allows you to reposition row titles or the overall table title to align with a specific location on the x axis.

at(rowtitles) places the overall table title at the default position calculated for the row titles. This option is sometimes useful for alignment when the default justification has not been used.

topgap(*relativesize*) specifies how much space is to be placed above each row.

bottomgap(*relativesize*) specifies how much space is to be placed beneath each row.

style(*textstyle*) specifies the style of text. This option does not appear on the dialog box.

Options

level(*#*) specifies the confidence level, as a percentage, for the pointwise confidence interval around the survivor, failure, or cumulative hazard function; see [U] **20.7 Specifying the width of confidence intervals**.

per(*#*) specifies the units used to report the survival or failure rates. For example, if the analysis time is in years, specifying per(100) results in rates per 100 person-years.

noshow prevents sts graph from showing the key st variables. This option is seldom used because most people type stset, show or stset, noshow to set whether they want to see these variables mentioned at the top of the output of every st command; see [ST] **stset**.

tmax(*#*) specifies that the plotted curve be graphed only for $t \leq \#$. This option does not affect the calculation of the function, rather the portion that is displayed.

tmin(*#*) specifies that the plotted curve be graphed only for $t \geq \#$. This option does not affect the calculation of the function, rather the portion that is displayed.

noorigin requests that the plot of the survival (failure) curve begin at the first exit time instead of beginning at $t = 0$ (the default). This option is ignored when cumhaz or hazard is specified.

width(*#* [*#*...]) is for use with hazard and specifies the bandwidth to be used in the kernel smooth used to plot the estimated hazard function. If width() is not specified, a default bandwidth is used as described in [R] **kdensity**. If it is used with by(), multiple bandwidths may be specified, one for each group. If there are more groups than the k bandwidths specified, the default bandwidth is used for the $k + 1, \ldots$ remaining groups. If any bandwidth is specified as . (dot), the default bandwidth is used for that group.

kernel(*kernel*) is for use with hazard and specifies the kernel function to be used in calculating the weighted kernel-density estimate required to produce a smoothed hazard-function estimator. The default kernel is Epanechnikov, yet *kernel* may be any of the kernels supported by kdensity; see [R] **kdensity**.

noboundary is for use with hazard. It specifies that no boundary-bias adjustments are to be made when calculating the smoothed hazard-function estimator. By default, the smoothed hazards are adjusted near the boundaries. If the epan2, biweight, or rectangular kernel is used, the bias correction near the boundary is performed using boundary kernels. For other kernels, the plotted range of the smoothed hazard function is restricted to be within one bandwidth of each endpoint. For these other kernels, specifying noboundary merely removes this range restriction.

lost specifies that the numbers lost be shown on the plot. These numbers are shown as small numbers over the flat parts of the function.

If `enter` is not specified, the numbers displayed are the number censored minus the number who enter. If you do specify `enter`, the numbers displayed are the pure number censored. The underlying logic is described in [ST] **sts**.

`lost` may not be used with `hazard`.

`enter` specifies that the number who enter be shown on the graph, as well as the number lost. The number who enter are shown as small numbers beneath the flat parts of the plotted function.

`enter` may not be used with `hazard`.

`atrisk` specifies that the numbers at risk at the beginning of each interval be shown on the plot. The numbers at risk are shown as small numbers beneath the flat parts of the plotted function.

`atrisk` may not be used with `hazard`.

`censored(single | number | multiple)` specifies that hash marks be placed on the graph to indicate censored observations.

> `censored(single)` places one hash mark at each censoring time, regardless of the number of censorings at that time.
>
> `censored(number)` places one hash mark at each censoring time and displays the number of censorings about the hash mark.
>
> `censored(multiple)` places multiple hash marks for multiple censorings at the same time. For instance, if 3 observations are censored at time 5, three hash marks are placed at time 5. `censored(multiple)` is intended for use when there are few censored observations; if there are too many censored observations, the graph can look bad. In such cases, we recommend that `censored(number)` be used.
>
> `censored()` may not be used with `hazard`.

`censopts(`*hash_options*`)` specifies options that affect how the hash marks for censored observations are rendered; see [G] *line_options*. When combined with `censored(number)`, `censopts()` also specifies how the count of censoring is rendered; see [G] *marker_label_options*, except `mlabel()` is not allowed.

`lostopts(`*marker_label_options*`)` specifies options that affect how the numbers lost are rendered; see [G] *marker_label_options*. This option implies the `lost` option.

`atriskopts(`*marker_label_options*`)` specifies options that affect how the numbers at risk are rendered; see [G] *marker_label_options*. This option implies the `atrisk` option.

Plot

`plotopts(`*cline_options*`)` affects the rendition of the plotted lines; see [G] *cline_options*. This option may not be combined with `by(`*varlist*`)` or `strata(`*varlist*`)`, unless `separate` is also specified.

`plot#opts(`*cline_options*`)` affects the rendition of the #th plotted line; see [G] *cline_options*. This option may not be combined with `separate`.

CI plot

`ciopts(`*area_options*`)` affects the rendition of the confidence bands; see [G] *area_options*. This option may not be combined with `by(`*varlist*`)` or `strata(`*varlist*`)`, unless `separate` is also specified.

`ci#opts(`*area_options*`)` affects the rendition of the #th confidence band; see [G] *area_options*. This option may not be combined with `separate`.

⎡ Add plots ⎤

addplot(*plot*) provides a way to add other plots to the generated graph; see [G] *addplot_option*.

⎡ Y axis, X axis, Titles, Legend, Overall ⎤

twoway_options are any of the options documented in [G] *twoway_options*. These include options for titling the graph (see [G] *title_options*) and for saving the graph to disk (see [G] *saving_option*).

byopts(*byopts*) affects the appearance of the combined graph when by() or adjustfor() is specified, including the overall graph title and the organization of subgraphs. byopts() may not be specified with separate. See [G] *by_option*.

Remarks

Remarks are presented under the following headings:

>   Including the number lost on the graph
>   Graphing the Nelson–Aalen cumulative hazard function
>   Graphing the hazard function
>   Adding an at-risk table
>   On boundary bias for smoothed hazards

If you have not read [ST] **sts**, please do so.

By default, sts graph displays the Kaplan–Meier product-limit estimate of the survivor (failure) function. Only one of sts graph's options, adjustfor(), modifies the calculation. All the other options merely determine how the results of the calculation are graphed.

We demonstrate many of sts graph's features in [ST] **sts**. This discussion picks up where that entry leaves off.

Including the number lost on the graph

In *Adjusted estimates* in [ST] **sts**, we introduced a simple drug-trial dataset with 1 observation per subject. Here is a graph of the survivor functions, by drug, including the number lost because of censoring:

```
. use http://www.stata-press.com/data/r11/drug2
(Patient Survival in Drug Trial)
. sts graph, by(drug) lost
        failure _d:  died
  analysis time _t:  studytime
```

Kaplan–Meier survival estimates

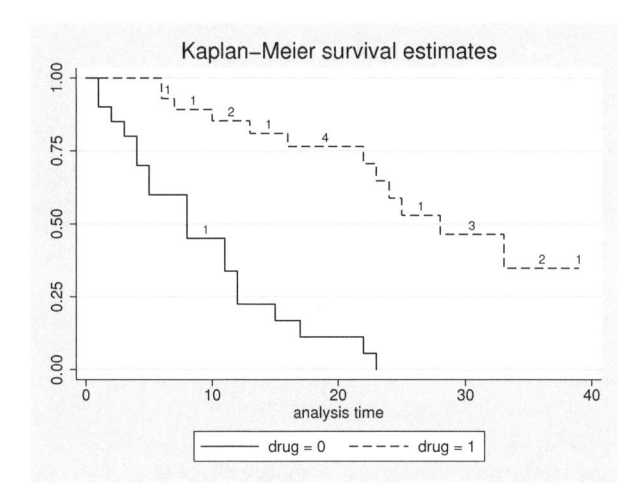

There is no late entry in these data, so we modify the data so that a few subjects entered late. Here is the same graph on the modified data:

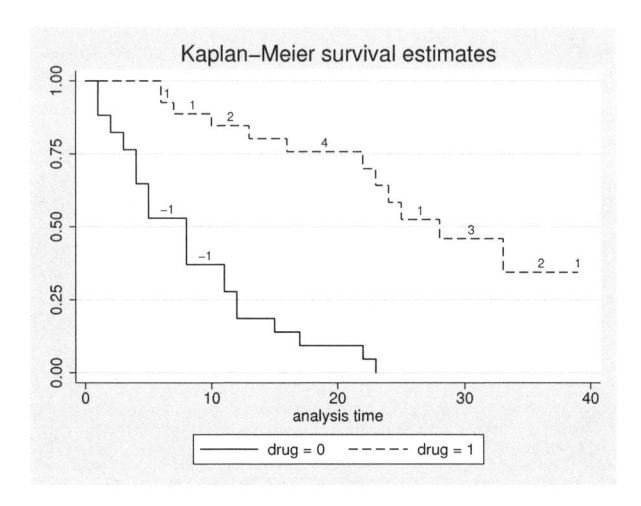

```
. use http://www.stata-press.com/data/r11/drug2b
(Patient Survival in Drug Trial)
. sts graph, by(drug) lost
        failure _d:  died
  analysis time _t:  studytime
```

Kaplan–Meier survival estimates

Note the negative numbers. These occur because, by default, `lost` means censored minus entered. Here −1 means that 1 entered, or 2 entered and 1 was lost, etc. If we specify the `enter` option, we will see the censored and entered separately:

```
. sts graph, by(drug) lost enter
        failure _d:  died
   analysis time _t: studytime
```

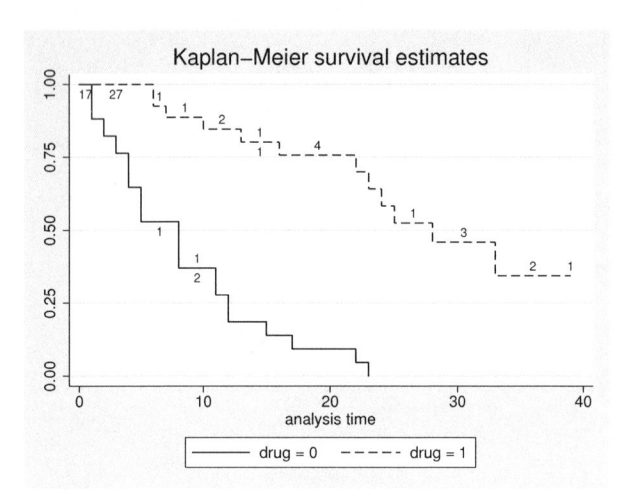

Kaplan–Meier survival estimates

Although it might appear that specifying enter with lost is a good idea, that is not always true.

We have yet another version of the data—the correct data not adjusted to have late entry—but in this version we have multiple records per subject. The data are the same, but where there was one record in the first dataset, sometimes there are now two because we have a covariate that is changing over time. From this dataset, here is the graph with the number lost shown:

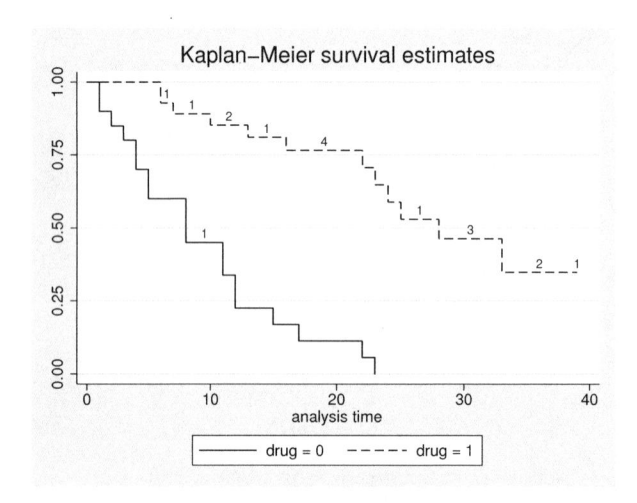

```
. use http://www.stata-press.com/data/r11/drug2c
(Patient Survival in Drug Trial)
. sts graph, by(drug) lost
        failure _d:  died
   analysis time _t: studytime
                id:  id
```

Kaplan–Meier survival estimates

This looks just like the first graph we presented, as indeed it should. Again we emphasize that the data are logically, if not physically, equivalent. If, however, we graph the number lost and entered, we get a graph showing a lot of activity:

```
. sts graph, by(drug) lost enter
       failure _d:  died
  analysis time _t:  studytime
              id:  id
```

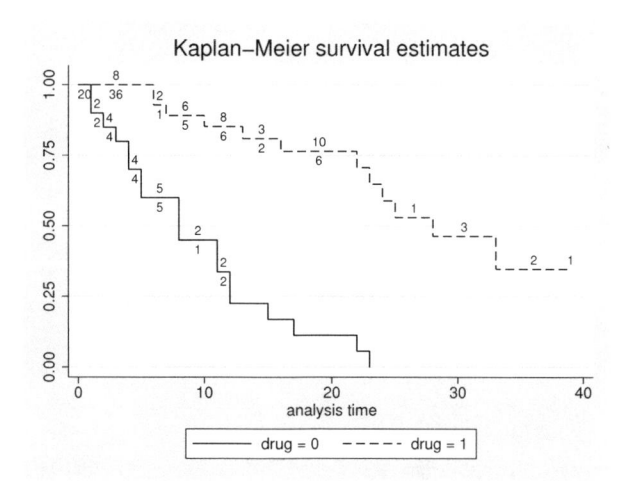

Kaplan–Meier survival estimates

All that activity goes by the name *thrashing*—subjects are being censored to enter the data again, but with different covariates. This graph was better when we did not specify enter because the censored-minus-entered calculation smoothed out the thrashing.

Graphing the Nelson–Aalen cumulative hazard function

We can plot the Nelson–Aalen estimate of the cumulative (integrated) hazard function by specifying the cumhaz option. For example, from the 1-observation-per-subject drug-trial dataset, here is a graph of the cumulative hazard functions by drug:

(*Continued on next page*)

```
. use http://www.stata-press.com/data/r11/drug2
(Patient Survival in Drug Trial)
. stset, noshow
. sts graph, cumhaz by(drug)
```

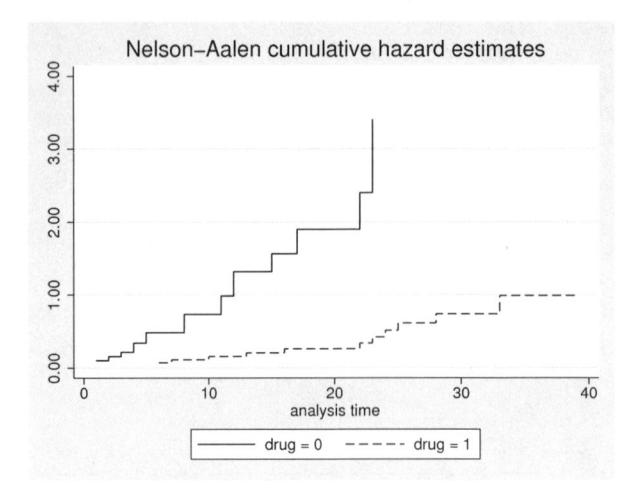

And here is a plot including the number lost because of censoring:

```
. sts graph, cumhaz by(drug) lost
```

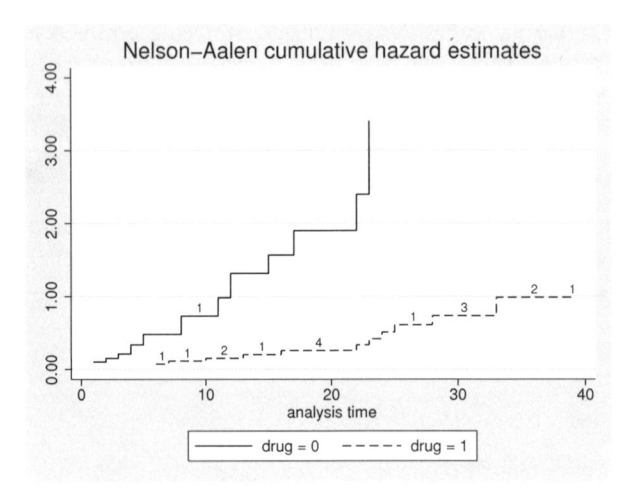

Graphing the hazard function

sts graph may also be used to plot an estimate of the hazard function. This graph is based on a weighted kernel smooth of the estimated hazard contributions, $\Delta \widehat{H}(t_j) = \widehat{H}(t_j) - \widehat{H}(t_{j-1})$, obtained by sts generate *newvar* = h. There are thus issues associated with selecting a kernel function and a bandwidth, although sts graph will use defaults if we do not want to worry about this.

```
. sts graph, hazard by(drug)
```

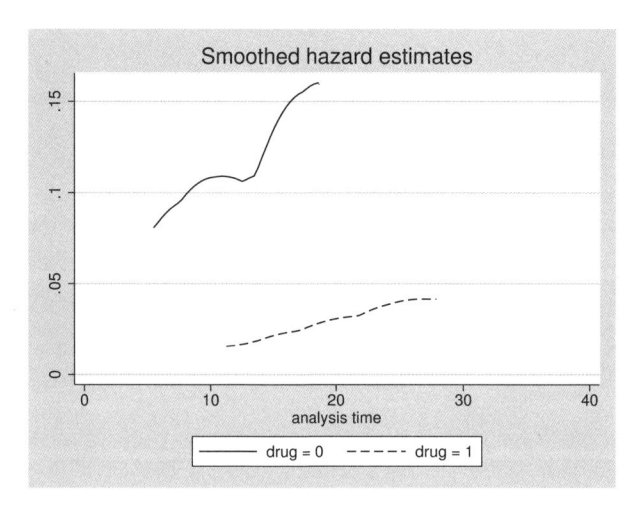

[Graph: Smoothed hazard estimates, comparing drug = 0 and drug = 1 over analysis time]

We can also adjust and customize the kernel smooth.

```
. sts graph, hazard by(drug) kernel(gauss) width(5 7)
> title(Comparison of hazard functions)
```

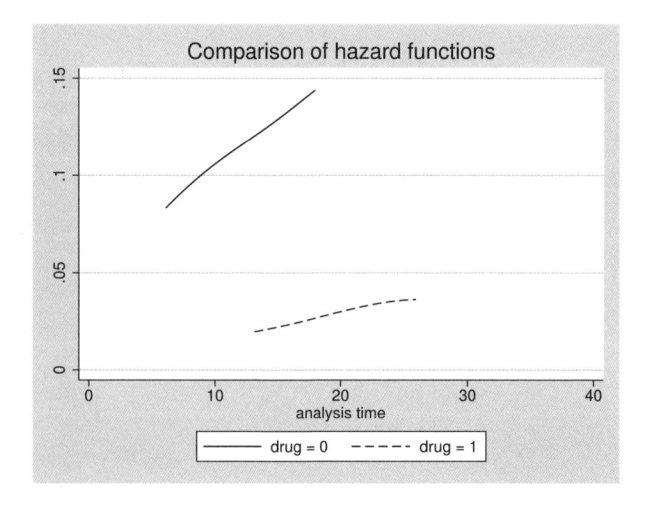

[Graph: Comparison of hazard functions, comparing drug = 0 and drug = 1 over analysis time]

Adding an at-risk table

A table showing the number at risk may be added beneath a survivor, failure, or Nelson–Aalen cumulative hazard plot.

(*Continued on next page*)

```
. sts graph, by(drug) risktable
```

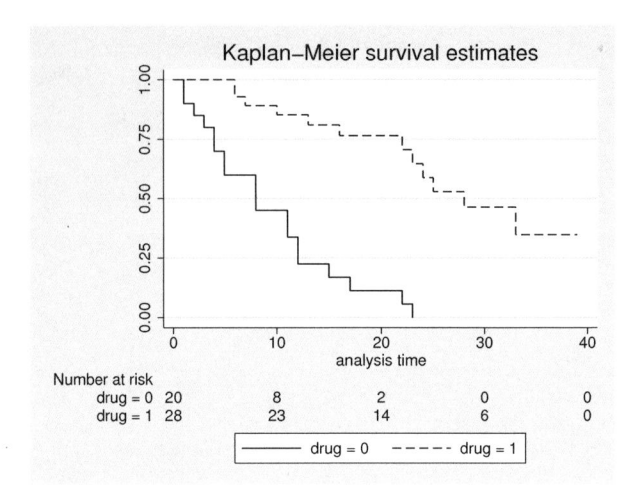

By default, both the legend and the at-risk table share space at the bottom of the graph. Placing the legend in an empty area inside the plot may often be desirable.

```
. sts graph, by(drug) risktable legend(ring(0) position(2) rows(2))
```

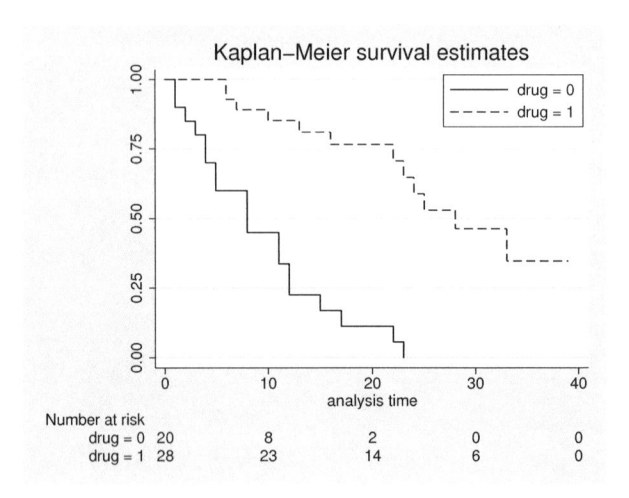

By default, row titles are placed on the left of the at-risk table and are right-justified. We can illustrate this by changing the text of the row titles to have an unequal length.

sts graph — Graph the survivor and cumulative hazard functions

```
. sts graph, by(drug) risktable(, order(1 "Placebo" 2 "Test drug"))
```

Kaplan–Meier survival estimates graph with risk table showing Placebo (20, 8, 2, 0, 0) and Test drug (28, 23, 14, 6, 0) at times 0, 10, 20, 30, 40.

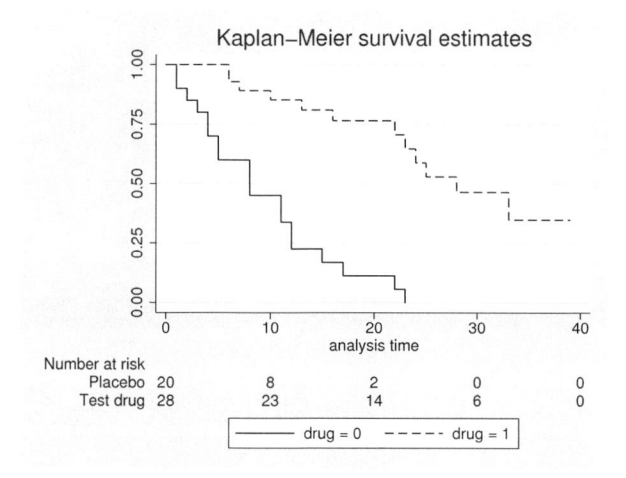

If desired, the text of row titles can be left-justified.

```
. sts graph, by(drug) risktable(, order(1 "Placebo" 2 "Test drug")
> rowtitle(, justification(left)))
```

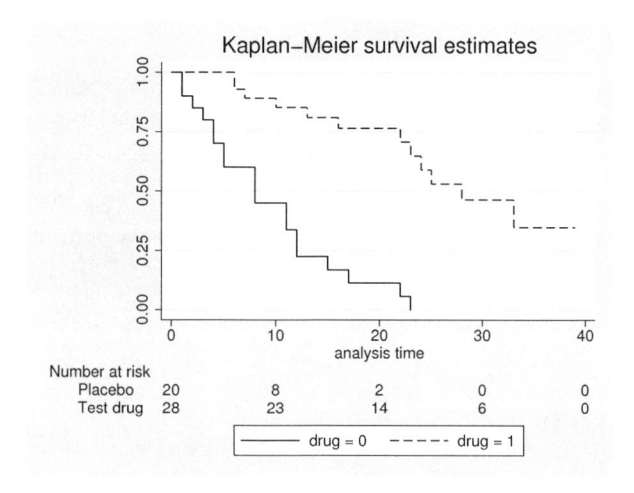

Kaplan–Meier survival estimates graph with left-justified row titles in risk table: Placebo (20, 8, 2, 0, 0) and Test drug (28, 23, 14, 6, 0) at times 0, 10, 20, 30, 40.

(Continued on next page)

In addition to left justification, the table title can be aligned with the row titles.

```
. sts graph, by(drug) risktable(, order(1 "Placebo" 2 "Test drug")
> rowtitle(, justification(left)) title(, at(rowtitle)))
```

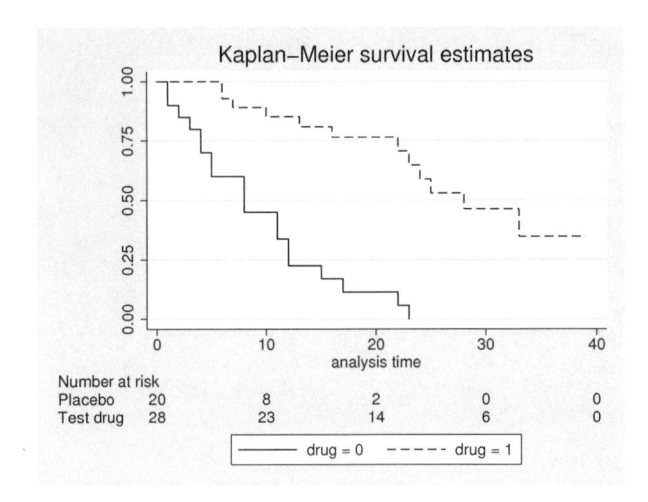

On boundary bias for smoothed hazards

sts graph uses the usual smoothing kernel technique to estimate the hazard function. Kernel estimators commonly encounter bias when estimating near the boundaries of the data range, and therefore estimates of the hazard function in the boundary regions are generally less reliable. To alleviate this problem, estimates that use the epan2, biweight, and rectangular kernels are adjusted at the boundaries with what are known as *boundary kernels* (e.g., Müller and Wang [1994]; Hess, Serachitopol, and Brown [1999]). For estimates using other kernels, no boundary adjustment is made. Instead, the default graphing range is constrained to be the range $[L+b, R-b]$, where L and R are the respective minimum and maximum analysis times at which failure occurred and b is the bandwidth.

Methods and formulas

See [ST] **sts**.

The estimated hazard is calculated as a kernel smooth of the estimated hazard contributions, $\Delta \widehat{H}(t_j) = \widehat{H}(t_j) - \widehat{H}(t_{j-1})$, using

$$\widehat{h}(t) = b^{-1} \sum_{j=1}^{D} K_t \left(\frac{t - t_j}{b} \right) \Delta \widehat{H}(t_j)$$

where $K_t()$ is the kernel (Müller and Wang 1994) function, b is the bandwidth, and the summation is over the D times at which failure occurs (Klein and Moeschberger 2003, 167). If adjustfor() is specified, the $\Delta \widehat{H}(t_j)$ are instead obtained from stcox as the estimated baseline contributions from a Cox model; see [ST] **stcox** for details on how the $\Delta \widehat{H}(t_j)$ are calculated in this case.

Pointwise confidence bands for smoothed hazard functions are calculated using the method based on a log transformation,

$$\widehat{h}(t)\exp\left[\pm\frac{Z_{1-\alpha/2}\sigma\{\widehat{h}(t)\}}{\widehat{h}(t)}\right]$$

See Klein and Moeschberger (2003, 168) for details.

References

Hess, K. R., D. M. Serachitopol, and B. W. Brown. 1999. Hazard function estimators: A simulation study. *Statistics in Medicine* 18: 3075–3088.

Klein, J. P., and M. L. Moeschberger. 2003. *Survival Analysis: Techniques for Censored and Truncated Data*. 2nd ed. New York: Springer.

Müller, H.-G., and J.-L. Wang. 1994. Hazard rate estimation under random censoring with varying kernels and bandwidths. *Biometrics* 50: 61–76.

Also see [ST] **sts** for more references.

Also see

[ST] **sts** — Generate, graph, list, and test the survivor and cumulative hazard functions

[ST] **sts generate** — Create variables containing survivor and related functions

[ST] **sts list** — List the survivor or cumulative hazard function

[ST] **sts test** — Test equality of survivor functions

[ST] **stset** — Declare data to be survival-time data

[R] **kdensity** — Univariate kernel density estimation

Title

sts list — List the survivor or cumulative hazard function

Syntax

sts list [*if*] [*in*] [, *options*]

options	description
Main	
survival	report Kaplan–Meier survivor function; the default
failure	report Kaplan–Meier failure function
cumhaz	report Nelson–Aalen cumulative hazard function
by(*varlist*)	calculate separately on different groups of *varlist*
adjustfor(*varlist*)	adjust the estimates to zero values of *varlist*
strata(*varlist*)	stratify on different groups of *varlist*
Options	
level(#)	set confidence level; default is level(95)
at(# | *numlist*)	report estimated survivor/cumulative hazard function at specified times; default is to report at all unique time values
enter	report number lost as pure censored instead of censored minus lost
noshow	do not show st setting information
compare	report groups of survivor/cumulative hazard functions side by side
saving(*filename*[, replace])	save results to *filename*; use replace to overwrite existing *filename*

You must stset your data before using sts list; see [ST] **stset**.
fweights, iweights, and pweights may be specified using stset; see [ST] **stset**.

Menu

Statistics > Survival analysis > Summary statistics, tests, and tables > List survivor and cumulative hazard functions

Description

sts list lists the estimated survivor (failure) or the Nelson–Aalen estimated cumulative (integrated) hazard function. See [ST] **sts** for an introduction to this command.

sts list can be used with single- or multiple-record or single- or multiple-failure st data.

Options

☐ Main ☐

survival, failure, and cumhaz specify the function to report.

survival specifies that the Kaplan–Meier survivor function be listed. This option is the default if a function is not specified.

failure specifies that the Kaplan–Meier failure function $1 - S(t+0)$ be listed.

cumhaz specifies that the Nelson–Aalen estimate of the cumulative hazard function be listed.

by(*varlist*) produces separate survivor or cumulative hazard functions by making separate calculations for each group identified by equal values of the variables in *varlist*. by() may not be combined with strata().

adjustfor(*varlist*) adjusts the estimate of the survivor (failure) function to that for 0 values of *varlist*. This option is not available with the Nelson–Aalen function. See [ST] **sts graph** for an example of how to adjust for values different from 0.

If you specify adjustfor() with by(), sts fits separate Cox regression models for each group, using the adjustfor() variables as covariates. The separately calculated baseline survivor functions are then retrieved.

If you specify adjustfor() with strata(), sts fits a stratified-on-group Cox regression model, using the adjustfor() variables as covariates. The stratified, baseline survivor function is then retrieved.

strata(*varlist*) requests estimates of the survivor (failure) function stratified on variables in *varlist*. It requires specifying adjustfor() and may not be combined with by().

Options

level(*#*) specifies the confidence level, as a percentage, for the Greenwood pointwise confidence interval of the survivor (failure) or for the pointwise confidence interval of the Nelson–Aalen cumulative hazard function; see [U] **20.7 Specifying the width of confidence intervals**.

at(*#* | *numlist*) specifies the time values at which the estimated survivor (failure) or cumulative hazard function is to be listed.

The default is to list the function at all the unique time values in the data, or if functions are being compared, at about 10 times chosen over the observed interval. In any case, you can control the points chosen.

at(5 10 20 30 50 90) would display the function at the designated times.

at(10 20 to 100) would display the function at times 10, 20, 30, 40, ..., 100.

at(0 5 10 to 100 200) would display the function at times 0, 5, 10, 15, ..., 100, and 200.

at(20) would display the curve at (roughly) 20 equally spaced times over the interval observed in the data. We say roughly because Stata may choose to increase or decrease your number slightly if that would result in rounder values of the chosen times.

enter specifies that the table contain the number who enter and, correspondingly, that the number lost be displayed as the pure number censored rather than censored minus entered. The logic underlying this is explained in [ST] **sts**.

noshow prevents sts list from showing the key st variables. This option is seldom used because most people type stset, show or stset, noshow to set whether they want to see these variables mentioned at the top of the output of every st command; see [ST] **stset**.

compare is specified only with by() or strata(). It compares the survivor (failure) or cumulative hazard functions and lists them side by side rather than first one and then the next.

saving(*filename*[, replace]) saves the results in a Stata data file (.dta file).

Remarks

Only one of `sts list`'s options—`adjustfor()`—modifies the calculation. All the other options merely determine how the results of the calculation are displayed.

If you do not specify `adjustfor()` or `cumhaz`, `sts list` displays the Kaplan–Meier product-limit estimate of the survivor (failure) function. Specify `by()` to perform the calculation separately on the different groups.

Specify `adjustfor()` to calculate an adjusted survival curve. Now if you specify `by()` or `strata()`, this further modifies how the adjustment is made.

`sts list, cumhaz` displays the Nelson–Aalen estimate of the cumulative hazard function.

We demonstrate many of `sts list`'s features in [ST] **sts**. This discussion picks up where that entry leaves off.

By default, `sts list` will bury you in output. With the Stanford heart transplant data introduced in [ST] **stset**, the following commands produce 154 lines of output.

```
. use http://www.stata-press.com/data/r11/stan3
(Heart transplant data)
. stset, noshow
. sts list, by(posttran)
```

Time	Beg. Total	Fail	Net Lost	Survivor Function	Std. Error	[95% Conf. Int.]	
posttran=0							
1	103	1	3	0.9903	0.0097	0.9331	0.9986
2	99	3	3	0.9603	0.0195	0.8976	0.9849
3	93	3	3	0.9293	0.0258	0.8574	0.9657
(output omitted)							
1400	1	0	1	0.2359	0.1217	0.0545	0.4882
posttran=1							
1	0	0	-3	1.0000	.	.	.
2	3	0	-3	1.0000	.	.	.
(output omitted)							
5.1	14	1	0	0.9286	0.0688	0.5908	0.9896
6	13	0	-1	0.9286	0.0688	0.5908	0.9896
(output omitted)							
1799	1	0	1	0.1420	0.0546	0.0566	0.2653

`at()` and `compare` are the solutions. Here is another detailed, but more useful, view of the heart transplant data:

```
. sts list, at(10 40 to 170) by(posttran)
                    Beg.                 Survivor      Std.
         Time      Total      Fail       Function      Error       [95% Conf. Int.]

posttran=0
           10         74        12         0.8724      0.0346      0.7858    0.9256
           40         31        11         0.6781      0.0601      0.5446    0.7801
           70         17         2         0.6126      0.0704      0.4603    0.7339
          100         11         1         0.5616      0.0810      0.3900    0.7022
          130         10         1         0.5054      0.0903      0.3199    0.6646
          160          7         1         0.4422      0.0986      0.2480    0.6204
posttran=1
           10         16         1         0.9286      0.0688      0.5908    0.9896
           40         43         6         0.7391      0.0900      0.5140    0.8716
           70         45         9         0.6002      0.0841      0.4172    0.7423
          100         40         9         0.4814      0.0762      0.3271    0.6198
          130         38         1         0.4687      0.0752      0.3174    0.6063
          160         36         1         0.4561      0.0742      0.3076    0.5928

Note:  survivor function is calculated over full data and evaluated at
       indicated times; it is not calculated from aggregates shown at left.
```

We specified at(10 40 to 170) when that is not strictly correct; at(10 40 to 160) would make sense and so would at(10 40 to 180), but sts list is not picky.

❑ Technical note

When used with at(), sts list is designed to give you only a snapshot of the full Kaplan–Meier curve. That is, the Beg. Total information is that for the last observed failure time (before the failures occur).

When the at() option is used, the Beg. Total column in the output does not contain the number at risk at the time indicated in the Time column. It shows the number at risk at the time just before the previous failure.

❑

Similar output for the Nelson–Aalen estimated cumulative hazard can be produced by specifying the cumhaz option:

```
. sts list, cumhaz at(10 40 to 170) by(posttran)
                    Beg.                 Nelson-Aalen   Std.
         Time      Total      Fail       Cum. Haz.      Error       [95% Conf. Int.]

posttran=0
           10         74        12         0.1349      0.0391      0.0764    0.2382
           40         31        11         0.3824      0.0871      0.2448    0.5976
           70         17         2         0.4813      0.1124      0.3044    0.7608
          100         11         1         0.5646      0.1400      0.3473    0.9178
          130         10         1         0.6646      0.1720      0.4002    1.1037
          160          7         1         0.7896      0.2126      0.4658    1.3385
posttran=1
           10         16         1         0.0714      0.0714      0.0101    0.5071
           40         43         6         0.2929      0.1176      0.1334    0.6433
           70         45         9         0.4981      0.1360      0.2916    0.8507
          100         40         9         0.7155      0.1542      0.4691    1.0915
          130         38         1         0.7418      0.1564      0.4908    1.1214
          160         36         1         0.7689      0.1587      0.5130    1.1523

Note:  Nelson-Aalen function is calculated over full data and evaluated at
       indicated times; it is not calculated from aggregates shown at left.
```

Here is the result of the survivor functions with the `compare` option:

```
. sts list, at(10 40 to 170) by(posttran) compare
              Survivor Function
posttran              0          1

time      10      0.8724     0.9286
          40      0.6781     0.7391
          70      0.6126     0.6002
         100      0.5616     0.4814
         130      0.5054     0.4687
         160      0.4422     0.4561
```

And here is the result of the cumulative hazard functions with the `compare` option:

```
. sts list, cumhaz at(10 40 to 170) by(posttran) compare
              Nelson-Aalen Cum. Haz.
posttran              0          1

time      10      0.1349     0.0714
          40      0.3824     0.2929
          70      0.4813     0.4981
         100      0.5646     0.7155
         130      0.6646     0.7418
         160      0.7896     0.7689
```

Methods and formulas

See [ST] **sts**.

References

See [ST] **sts** for references.

Also see

[ST] **sts** — Generate, graph, list, and test the survivor and cumulative hazard functions

[ST] **sts generate** — Create variables containing survivor and related functions

[ST] **sts graph** — Graph the survivor and cumulative hazard functions

[ST] **sts test** — Test equality of survivor functions

[ST] **stset** — Declare data to be survival-time data

Title

sts test — Test equality of survivor functions

Syntax

> sts test *varlist* [*if*] [*in*] [, *options*]

options	description
Main	
logrank	perform log-rank test of equality; the default
cox	perform Cox test of equality
wilcoxon	perform Wilcoxon–Breslow–Gehan test of equality
tware	perform Tarone–Ware test of equality
peto	perform Peto–Peto–Prentice test of equality
fh(*p q*)	perform generalized Fleming–Harrington test of equality
trend	test trend of the survivor function across three or more ordered groups
strata(*varlist*)	perform stratified test on *varlist*, displaying overall test results
detail	display individual test results; modifies strata()
Options	
mat(*mname$_1$ mname$_2$*)	store vector **u** in *mname$_1$* and matrix **V** in *mname$_2$*
noshow	do not show st setting information
notitle	suppress title

You must stset your data before using sts test; see [ST] **stset**.

Note that fweights, iweights, and pweights may be specified using stset; see [ST] **stset**.

Menu

Statistics > Survival analysis > Summary statistics, tests, and tables > Test equality of survivor functions

Description

sts test tests the equality of survivor functions across two or more groups. The log-rank, Cox, Wilcoxon–Breslow–Gehan, Tarone–Ware, Peto–Peto–Prentice, and Fleming–Harrington tests are provided, in both unstratified and stratified forms.

sts test also provides a test for trend.

See [ST] **sts** for an introduction to this command.

sts test can be used with single- or multiple-record or single- or multiple-failure st data.

Options

<u>Main</u>

logrank, cox, wilcoxon, tware, peto, and fh(*p q*) specify the test of equality desired. logrank is the default, unless the data are pweighted, in which case cox is the default and is the only possibility.

wilcoxon specifies the Wilcoxon–Breslow–Gehan test; tware, the Tarone–Ware test; peto, the Peto–Peto–Prentice test; and fh(), the generalized Fleming–Harrington test. The Fleming–Harrington test requires two arguments, *p* and *q*. When $p = 0$ and $q = 0$, the Fleming–Harrington test reduces to the log-rank test; when $p = 1$ and $q = 0$, the test reduces to the Mann–Whitney–Wilcoxon test.

trend specifies that a test for trend of the survivor function across three or more ordered groups be performed.

strata(*varlist*) requests that a stratified test be performed.

detail modifies strata(); it requests that, in addition to the overall stratified test, the tests for the individual strata be reported. detail is not allowed with cox.

<u>Options</u>

mat(*mname₁ mname₂*) requests that the vector **u** be stored in *mname₁* and that matrix **V** be stored in *mname₂*. The other tests are rank tests of the form $\mathbf{u}'\mathbf{V}^{-1}\mathbf{u}$. This option may not be used with cox.

noshow prevents sts test from showing the key st variables. This option is seldom used because most people type stset, show or stset, noshow to set whether they want to see these variables mentioned at the top of the output of every st command; see [ST] **stset**.

notitle requests that the title printed above the test be suppressed.

Remarks

Remarks are presented under the following headings:

> The log-rank test
> The Wilcoxon (Breslow–Gehan) test
> The Tarone–Ware test
> The Peto–Peto–Prentice test
> The generalized Fleming–Harrington tests
> The "Cox" test
> The trend test

sts test tests the equality of the survivor function across groups. With the exception of the Cox test, these tests are members of a family of statistical tests that are extensions to censored data of traditional nonparametric rank tests for comparing two or more distributions. A technical description of these tests can be found in the *Methods and formulas* section of this entry. Simply, at each distinct failure time in the data, the contribution to the test statistic is obtained as a weighted standardized sum of the difference between the observed and expected number of deaths in each of the *k* groups. The expected number of deaths is obtained under the null hypothesis of no differences between the survival experience of the *k* groups.

The weights or weight function used determines the test statistic. For example, when the weight is 1 at all failure times, the log-rank test is computed, and when the weight is the number of subjects at risk of failure at each distinct failure time, the Wilcoxon–Breslow–Gehan test is computed.

The following table summarizes the weights used for each statistical test.

Test	Weight at each distinct failure time (t_i)
Log-rank	1
Wilcoxon–Breslow–Gehan	n_i
Tarone–Ware	$\sqrt{n_i}$
Peto–Peto–Prentice	$\widetilde{S}(t_i)$
Fleming–Harrington	$\widehat{S}(t_{i-1})^p \{1 - \widehat{S}(t_{i-1})\}^q$

where $\widehat{S}(t_i)$ is the estimated Kaplan–Meier survivor-function value for the combined sample at failure time t_i, $\widetilde{S}(t_i)$ is a modified estimate of the overall survivor function described in *Methods and formulas*, and n_i is the number of subjects in the risk pool at failure time t_i.

These tests are appropriate for testing the equality of survivor functions across two or more groups. Up to 800 groups are allowed.

The "Cox" test is related to the log-rank test but is performed as a likelihood-ratio test (or, alternatively, as a Wald test) on the results from a Cox proportional hazards regression. The log-rank test should be preferable to what we have labeled the Cox test, but with pweighted data the log-rank test is not appropriate. Whether you perform the log-rank or Cox test makes little substantive difference with most datasets.

sts test, trend can be used to test against the alternative hypothesis that the failure rate increases or decreases as the level of the k groups increases or decreases. This test is appropriate only when there is a natural ordering of the comparison groups, for example, when each group represents an increasing or decreasing level of a therapeutic agent.

trend is not valid when cox is specified.

The log-rank test

sts test, by default, performs the log-rank test, which is, to be clear, the exponential scores test (Savage 1956; Mantel and Haenszel 1959; Mantel 1963; Mantel 1963). This test is most appropriate when the hazard functions are thought to be proportional across the groups if they are not equal.

This test statistic is constructed by giving equal weights to the contribution of each failure time to the overall test statistic.

In *Testing equality of survivor functions* in [ST] **sts**, we demonstrated the use of this command with the heart transplant data, a multiple-record, single-failure st dataset.

(*Continued on next page*)

```
. use http://www.stata-press.com/data/r11/stan3
(Heart transplant data)

. sts test posttran

         failure _d:  died
   analysis time _t:  t1
                 id:  id
```

Log-rank test for equality of survivor functions

posttran	Events observed	Events expected
0	30	31.20
1	45	43.80
Total	75	75.00

```
              chi2(1) =     0.13
              Pr>chi2 =   0.7225
```

We cannot reject the hypothesis that the survivor functions are the same.

sts test, logrank can also perform the stratified log-rank test. Say that it is suggested that calendar year of acceptance also affects survival and that there are three important periods: 1967–1969, 1970–1972, and 1973–1974. Therefore, a stratified test should be performed:

```
. stset, noshow
. generate group = 1 if year <= 69
(117 missing values generated)
. replace group=2 if year>=70 & year<=72
(78 real changes made)
. replace group=3 if year>=73
(39 real changes made)
. sts test posttran, strata(group)
```

Stratified log-rank test for equality of survivor functions

posttran	Events observed	Events expected(*)
0	30	31.51
1	45	43.49
Total	75	75.00

(*) sum over calculations within group

```
              chi2(1) =     0.20
              Pr>chi2 =   0.6547
```

Still finding nothing, we ask Stata to show the within-stratum tests:

```
. sts test posttran, strata(group) detail
```

Stratified log-rank test for equality of survivor functions

-> group = 1

posttran	Events observed	Events expected
0	14	13.59
1	17	17.41
Total	31	31.00

         chi2(1) =   0.03
         Pr>chi2 =   0.8558

-> group = 2

posttran	Events observed	Events expected
0	13	13.63
1	20	19.37
Total	33	33.00

         chi2(1) =   0.09
         Pr>chi2 =   0.7663

-> group = 3

posttran	Events observed	Events expected
0	3	4.29
1	8	6.71
Total	11	11.00

         chi2(1) =   0.91
         Pr>chi2 =   0.3410

-> Total

posttran	Events observed	Events expected(*)
0	30	31.51
1	45	43.49
Total	75	75.00

(*) sum over calculations within group

         chi2(1) =   0.20
         Pr>chi2 =   0.6547

The Wilcoxon (Breslow–Gehan) test

sts test, wilcoxon performs the generalized Wilcoxon test of Breslow (1970) and Gehan (1965). This test is appropriate when hazard functions are thought to vary in ways other than proportionally and when censoring patterns are similar across groups.

The Wilcoxon test statistic is constructed by weighting the contribution of each failure time to the overall test statistic by the number of subjects at risk. Thus it gives heavier weights to earlier failure times when the number at risk is higher. As a result, this test is susceptible to differences in the censoring pattern of the groups.

`sts test, wilcoxon` works the same way as `sts test, logrank`:

```
. sts test posttran, wilcoxon
```

Wilcoxon (Breslow) test for equality of survivor functions

posttran	Events observed	Events expected	Sum of ranks
0	30	31.20	-85
1	45	43.80	85
Total	75	75.00	0

```
              chi2(1) =     0.14
              Pr>chi2 =     0.7083
```

With the `strata()` option, `sts test, wilcoxon` performs the stratified test:

```
. sts test posttran, wilcoxon strata(group)
```

Stratified Wilcoxon (Breslow) test for equality of survivor functions

posttran	Events observed	Events expected(*)	Sum of ranks(*)
0	30	31.51	-40
1	45	43.49	40
Total	75	75.00	0

(*) sum over calculations within group

```
              chi2(1) =     0.22
              Pr>chi2 =     0.6385
```

As with `sts test, logrank`, you can also specify the `detail` option to see the within-stratum tests.

The Tarone–Ware test

`sts test, tware` performs a test suggested by Tarone and Ware (1977), with weights equal to the square root of the number of subjects in the risk pool at time t_i.

Like Wilcoxon's test, this test is appropriate when hazard functions are thought to vary in ways other than proportionally and when censoring patterns are similar across groups. The test statistic is constructed by weighting the contribution of each failure time to the overall test statistic by the square root of the number of subjects at risk. Thus, like the Wilcoxon test, it gives heavier weights, although not as large, to earlier failure times. Although less susceptible to the failure and censoring pattern in the data than Wilcoxon's test, this could remain a problem if large differences in these patterns exist between groups.

`sts test, tware` works the same way as `sts test, logrank`:

```
. sts test posttran, tware
```

Tarone-Ware test for equality of survivor functions

posttran	Events observed	Events expected	Sum of ranks
0	30	31.20	-9.3375685
1	45	43.80	9.3375685
Total	75	75.00	0

```
              chi2(1) =     0.12
              Pr>chi2 =     0.7293
```

With the `strata()` option, `sts test, tware` performs the stratified test:

```
. sts test posttran, tware strata(group)
```

Stratified Tarone-Ware test for equality of survivor functions

posttran	Events observed	Events expected(*)	Sum of ranks(*)
0	30	31.51	-7.4679345
1	45	43.49	7.4679345
Total	75	75.00	0

(*) sum over calculations within group

```
             chi2(1) =     0.21
             Pr>chi2 =   0.6464
```

As with `sts test, logrank`, you can also specify the `detail` option to see the within-stratum tests.

The Peto–Peto–Prentice test

`sts test, peto` performs an alternative to the Wilcoxon test proposed by Peto and Peto (1972) and Prentice (1978). The test uses as the weight function an estimate of the overall survivor function, which is similar to that obtained using the Kaplan–Meier estimator. See *Methods and formulas* for details.

This test is appropriate when hazard functions are thought to vary in ways other than proportionally, but unlike the Wilcoxon–Breslow–Gehan test, it is not affected by differences in censoring patterns across groups.

`sts test, peto` works the same way as `sts test, logrank`:

```
. sts test posttran, peto
```

Peto-Peto test for equality of survivor functions

posttran	Events observed	Events expected	Sum of ranks
0	30	31.20	-.86708453
1	45	43.80	.86708453
Total	75	75.00	0

```
             chi2(1) =     0.15
             Pr>chi2 =   0.6979
```

With the `strata()` option, `sts test, peto` performs the stratified test:

```
. sts test posttran, peto strata(group)
```

Stratified Peto-Peto test for equality of survivor functions

posttran	Events observed	Events expected(*)	Sum of ranks(*)
0	30	31.51	-.96898006
1	45	43.49	.96898006
Total	75	75.00	0

(*) sum over calculations within group

```
             chi2(1) =     0.19
             Pr>chi2 =   0.6618
```

As with the previous tests, you can also specify the `detail` option to see the within-stratum tests.

The generalized Fleming–Harrington tests

sts test, fh(p q) performs the Harrington and Fleming (1982) class of test statistics. The weight function at each distinct failure time, t, is the product of the Kaplan–Meier survivor estimate at time $t-1$ raised to the p power and $1-$ the Kaplan–Meier survivor estimate at time $t-1$ raised to the q power. Thus, when specifying the Fleming and Harrington option, you must specify two nonnegative arguments, p and q.

When $p > q$, the test gives more weights to earlier failures than to later ones. When $p < q$, the opposite is true, and more weight is given to later failures than to earlier ones. When p and q are both zero, the weight is 1 at all failure times and the test reduces to the log-rank test.

sts test, fh(p q) works the same way as sts test, logrank. If we specify $p=0$ and $q=0$ we will get the same results as the log-rank test:

```
. sts test posttran, fh(0 0)
```

Fleming-Harrington test for equality of survivor functions

posttran	Events observed	Events expected	Sum of ranks
0	30	31.20	-1.1995511
1	45	43.80	1.1995511
Total	75	75.00	0

```
        chi2(1) =    0.13
        Pr>chi2 =    0.7225
```

We could, for example, give more weight to later failures than to earlier ones.

```
. sts test posttran, fh(0 3)
```

Fleming-Harrington test for equality of survivor functions

posttran	Events observed	Events expected	Sum of ranks
0	30	31.20	-.09364975
1	45	43.80	.09364975
Total	75	75.00	0

```
        chi2(1) =    0.01
        Pr>chi2 =    0.9139
```

Similarly to the previous tests, with the strata() option, sts test, fh() performs the stratified test:

```
. sts test posttran, fh(0 3) strata(group)
```

Stratified Fleming-Harrington test for equality of survivor functions

posttran	Events observed	Events expected(*)	Sum of ranks(*)
0	30	31.51	-.1623318
1	45	43.49	.1623318
Total	75	75.00	0

(*) sum over calculations within group

```
        chi2(1) =    0.05
        Pr>chi2 =    0.8266
```

As with the other tests, you can also specify the detail option to see the within-stratum tests.

The "Cox" test

The term *Cox test* is our own, and this test is a variation on the log-rank test using Cox regression.

One way of thinking about the log-rank test is as a Cox proportional hazards model on indicator variables for each of the groups. The log-rank test is a test that the coefficients are zero or, if you prefer, that the hazard ratios are one. The log-rank test is, in fact, a score test of that hypothesis performed on a slightly different (partial) likelihood function that handles ties more accurately.

Many researchers think that a (less precise) score test on the precise likelihood function is preferable to a (more precise) likelihood-ratio test on the approximate likelihood function used in Cox regression estimation. In our experience, it makes little difference:

```
. sts test posttran, cox

Cox regression-based test for equality of survival curves

              Events      Events     Relative
posttran    observed    expected       hazard
─────────────────────────────────────────────
0                 30       31.20       0.9401
1                 45       43.80       1.0450
─────────────────────────────────────────────
Total             75       75.00       1.0000
           LR chi2(1) =     0.13
            Pr>chi2  =   0.7222
```

By comparison, `sts test, logrank` also reported $\chi^2 = 0.13$, although the significance level was 0.7225, meaning that the χ^2 values differed in the fourth digit. As mentioned by Kalbfleisch and Prentice (2002, 20), a primary advantage of the log-rank test is the ease with which it can be explained to nonstatisticians, because the test statistic is the difference between the observed and expected number of failures within groups.

Our purpose in offering `sts test, cox` is not to promote its use instead of the log-rank test but to provide a test for researchers with sample-weighted data.

If you have sample weights (if you specified `pweights` when you `stset` the data), you cannot run the log-rank or Wilcoxon tests. The Cox regression model, however, has been generalized to sample-weighted data, and Stata's `stcox` can fit models with such data. In sample-weighted data, the likelihood-ratio statistic is no longer appropriate, but the Wald test based on the robust estimator of variance is.

Thus if we treated these data as sample-weighted data, we would obtain

```
. generate one = 1
. stset t1 [pw=one], id(id) time0(_t0) failure(died) noshow
              id:  id
   failure event:  died != 0 & died < .
obs. time interval: (_t0, t1]
 exit on or before: failure
           weight:  [pweight=one]
─────────────────────────────────────────────
     172  total obs.
       0  exclusions
─────────────────────────────────────────────
     172  obs. remaining, representing
     103  subjects
      75  failures in single failure-per-subject data
 31938.1  total analysis time at risk, at risk from t =         0
                             earliest observed entry t =         0
                              last observed exit t =         1799
```

```
. sts test posttran, cox
```

Cox regression-based test for equality of survival curves

posttran	Events observed	Events expected	Relative hazard
0	30.00	31.20	0.9401
1	45.00	43.80	1.0450
Total	75.00	75.00	1.0000

```
              Wald chi2(1) =      0.13
                  Pr>chi2  =    0.7181
```

sts test, cox now reports the Wald statistic, which is, to two digits, 0.13, just like all the others.

The trend test

When the groups to be compared have a natural order, such as increasing or decreasing age groups or drug dosage, you may want to test the null hypothesis that there is no difference in failure rate among the groups versus the alternative hypothesis that the failure rate increases or decreases as you move from one group to the next.

We illustrate this test with a dataset from a carcinogenesis experiment reprinted in Marubini and Valsecchi (1995, 126). Twenty-nine experimental animals were exposed to three levels (0, 1.5, 2.0) of a carcinogenic agent. The time in days to tumor formation was recorded. Here are a few of the observations:

```
. use http://www.stata-press.com/data/r11/marubini, clear
. list time event group dose in 1/9
```

	time	event	group	dose
1.	67	1	2	1.5
2.	150	1	2	1.5
3.	47	1	3	2
4.	75	0	1	0
5.	58	1	3	2
6.	136	1	2	1.5
7.	58	1	3	2
8.	150	1	2	1.5
9.	43	0	2	1.5

In these data, there are two variables that indicate exposure level. The group variable is coded 1, 2, and 3, indicating a one-unit separation between exposures. The dose variable records the actual exposure dosage. To test the null hypothesis of no difference among the survival experience of the three groups versus the alternative hypothesis that the survival experience of at least one of the groups is different, it does not matter if we use group or dose.

```
. stset time, fail(event) noshow
     failure event:  event != 0 & event < .
obs. time interval:  (0, time]
 exit on or before:  failure
```

```
    29  total obs.
     0  exclusions

    29  obs. remaining, representing
    15  failures in single record/single failure data
  2564  total analysis time at risk, at risk from t =         0
                             earliest observed entry t =         0
                              last observed exit t =          246
. sts test group
```

Log-rank test for equality of survivor functions

group	Events observed	Events expected
1	4	6.41
2	6	6.80
3	5	1.79
Total	15	15.00

```
              chi2(2) =    8.05
              Pr>chi2 =  0.0179
. sts test dose
```

Log-rank test for equality of survivor functions

dose	Events observed	Events expected
0	4	6.41
1.5	6	6.80
2	5	1.79
Total	15	15.00

```
              chi2(2) =    8.05
              Pr>chi2 =  0.0179
```

For the trend test, however, the distance between the values is important, so using group or dose will produce different results.

(Continued on next page)

```
. sts test group, trend
```

Log-rank test for equality of survivor functions

group	Events observed	Events expected
1	4	6.41
2	6	6.80
3	5	1.79
Total	15	15.00

```
                chi2(2) =       8.05
                Pr>chi2 =     0.0179

Test for trend of survivor functions
                chi2(1) =       5.87
                Pr>chi2 =     0.0154

. sts test dose, trend
```

Log-rank test for equality of survivor functions

dose	Events observed	Events expected
0	4	6.41
1.5	6	6.80
2	5	1.79
Total	15	15.00

```
                chi2(2) =       8.05
                Pr>chi2 =     0.0179

Test for trend of survivor functions
                chi2(1) =       3.66
                Pr>chi2 =     0.0557
```

Although the above trend test was constructed using the log-rank test, any of the previously mentioned weight functions can be used. For example, a trend test on the data can be performed using the same weights as the Peto–Peto–Prentice test by specifying the peto option.

```
. sts test dose, trend peto
```

Peto-Peto test for equality of survivor functions

dose	Events observed	Events expected	Sum of ranks
0	4	6.41	-1.2792221
1.5	6	6.80	-1.3150418
2	5	1.79	2.5942639
Total	15	15.00	0

```
                chi2(2) =       8.39
                Pr>chi2 =     0.0150

Test for trend of survivor functions
                chi2(1) =       2.85
                Pr>chi2 =     0.0914
```

Saved results

sts test saves the following in r():

Scalars
r(df)   degrees of freedom   r(chi2) χ^2
r(df_tr) degrees of freedom, trend test r(chi2_tr) χ^2, trend test

Methods and formulas

sts test is implemented as an ado-file.

Let $t_1 < t_2 < \cdots < t_k$ denote the ordered failure times; let d_j be the number of failures at t_j and n_j be the population at risk just before t_j; and let d_{ij} and n_{ij} denote the same things for group i, $i = 1, \ldots, r$.

We are interested in testing the null hypothesis

$$H_0: \lambda_1(t) = \lambda_2(t) = \cdots = \lambda_r(t)$$

where $\lambda(t)$ is the hazard function at time t, against the alternative hypothesis that at least one of the $\lambda_i(t)$ is different for some t_j.

As described in Klein and Moeschberger (2003, 205–216), Kalbfleisch and Prentice (2002, 20–22), and Collett (2003, 48–49), if the null hypothesis is true, the expected number of failures in group i at time t_j is $e_{ij} = n_{ij} d_j / n_j$, and the test statistic

$$\mathbf{u}' = \sum_{j=1}^{k} W(t_j)(d_{1j} - e_{1j}, \ldots, d_{rj} - e_{rj})$$

is formed. $W(t_j)$ is a positive weight function defined as zero when n_{ij} is zero. The various test statistics are obtained by selecting different weight functions, $W(t_j)$. See the table in the *Remarks* section of this entry for a list of these weight functions. For the Peto–Peto–Prentice test,

$$W(t_j) = \widetilde{S}(t_j) = \prod_{\ell : t_\ell \leq t_j} \left(1 - \frac{d_\ell}{n_\ell + 1}\right)$$

The variance matrix $\mathbf{V}$ for $\mathbf{u}$ has elements

$$V_{il} = \sum_{j=1}^{k} \frac{W(t_j)^2 n_{ij} d_j (n_j - d_j)}{n_j (n_j - 1)} \left(\delta_{il} - \frac{n_{ij}}{n_j}\right)$$

where $\delta_{il} = 1$ if $i = l$ and $\delta_{il} = 0$, otherwise.

For the unstratified test, statistic $\mathbf{u}'\mathbf{V}^{-1}\mathbf{u}$ is distributed as χ^2 with $r - 1$ degrees of freedom.

For the stratified test, let $\mathbf{u}_s$ and $\mathbf{V}_s$ be the results of performing the above calculation separately within stratum, and define $\mathbf{u} = \sum_s \mathbf{u}_s$ and $\mathbf{V} = \sum_s \mathbf{V}_s$. The χ^2 test is given by $\mathbf{u}'\mathbf{V}^{-1}\mathbf{u}$ redefined in this way.

The "Cox" test is performed by fitting a (possibly stratified) Cox regression using stcox on $r-1$ indicator variables, one for each group with one of the indicators omitted. The χ^2 test reported is then the likelihood-ratio test (no pweights) or the Wald test (based on the robust estimate of variance); see [ST] **stcox**.

The reported relative hazards are the exponentiated coefficients from the Cox regression renormalized, and the renormalization plays no role in calculating the test statistic. The renormalization is chosen so that the expected-number-of-failures-within-group weighted average of the regression coefficients is 0 (meaning that the hazard is 1). Let b_i, $i = 1, \ldots, r-1$, be the estimated coefficients, and define $b_r = 0$. The constant K is then calculated with

$$K = \sum_{i=1}^{r} e_i b_i / d$$

where $e_i = \sum_j e_{ij}$ is the expected number of failures for group i, d is the total number of failures across all groups, and r is the number of groups. The reported relative hazards are $\exp(b_i - K)$.

The trend test assumes that there is natural ordering of the r groups, $r > 2$. Here we are interested in testing the null hypothesis

$$H_0: \lambda_1(t) = \lambda_2(t) = \cdots = \lambda_r(t)$$

against the alternative hypothesis

$$H_a: \lambda_1(t) \leq \lambda_2(t) \leq \cdots \leq \lambda_r(t)$$

The test uses **u** as previously defined with any of the available weight functions. The test statistic is given by

$$\frac{\left(\sum_{i=1}^{r} a_i u_i\right)^2}{\mathbf{a'Va}}$$

where $a_1 \leq a_2 \leq \cdots \leq a_r$ are scores defining the relationship of interest. A score is assigned to each comparison group, equal to the value of the grouping variable for that group. **a** is the vector of these scores.

References

Breslow, N. E. 1970. A generalized Kruskal–Wallis test for comparing K samples subject to unequal patterns of censorship. *Biometrika* 57: 579–594.

Collett, D. 2003. *Modelling Survival Data in Medical Research*. 2nd ed. London: Chapman & Hall/CRC.

Gehan, E. A. 1965. A generalized Wilcoxon test for comparing arbitrarily singly-censored samples. *Biometrika* 52: 203–223.

Harrington, D. P., and T. R. Fleming. 1982. A class of rank test procedures for censored survival data. *Biometrika* 69: 553–566.

Kalbfleisch, J. D., and R. L. Prentice. 2002. *The Statistical Analysis of Failure Time Data*. 2nd ed. New York: Wiley.

Klein, J. P., and M. L. Moeschberger. 2003. *Survival Analysis: Techniques for Censored and Truncated Data*. 2nd ed. New York: Springer.

Mantel, N. 1963. Chi-square tests with one degree of freedom; extensions of the Mantel–Haenszel procedure. *Journal of the American Statistical Association* 58: 690–700.

——. 1966. Evaluation of survival data and two new rank-order statistics arising in its consideration. *Cancer Chemotherapy Reports* 50: 163–170.

Mantel, N., and W. Haenszel. 1959. Statistical aspects of the analysis of data from retrospective studies of disease. *Journal of the National Cancer Institute* 22: 719–748. Reprinted in *Evolution of Epidemiologic Ideas*, ed. S. Greenland, pp. 112–141. Newton Lower Falls, MA: Epidemiology Resources.

Marubini, E., and M. G. Valsecchi. 1995. *Analysing Survival Data from Clinical Trials and Observational Studies*. New York: Wiley.

Peto, R., and J. Peto. 1972. Asymptotically efficient rank invariant test procedures (with discussion). *Journal of the Royal Statistical Society, Series A* 135: 185–207.

Prentice, R. L. 1978. Linear rank tests with right censored data. *Biometrika* 65: 167–179.

Savage, I. R. 1956. Contributions to the theory of rank-order statistics—the two-sample case. *Annals of Mathematical Statistics* 27: 590–615.

Tarone, R. E., and J. H. Ware. 1977. On distribution-free tests for equality of survival distributions. *Biometrika* 64: 156–160.

White, I. R., S. Walker, and A. Babiker. 2002. strbee: Randomization-based efficacy estimator. *Stata Journal* 2: 140–150.

Wilcoxon, F. 1945. Individual comparisons by ranking methods. *Biometrics* 1: 80–83.

Also see

[ST] **stcox** — Cox proportional hazards model

[ST] **sts** — Generate, graph, list, and test the survivor and cumulative hazard functions

[ST] **sts generate** — Create variables containing survivor and related functions

[ST] **sts graph** — Graph the survivor and cumulative hazard functions

[ST] **sts list** — List the survivor or cumulative hazard function

[ST] **stset** — Declare data to be survival-time data

Title

stset — Declare data to be survival-time data

Syntax

Single-record-per-subject survival data

    stset *timevar* [*if*] [*weight*] [, *single_options*]

    streset [*if*] [*weight*] [, *single_options*]

    st [, no<u>c</u>md <u>not</u>able]

    stset, clear

Multiple-record-per-subject survival data

    stset *timevar* [*if*] [*weight*], id(*idvar*) <u>f</u>ailure(*failvar*[==*numlist*])
        [*multiple_options*]

    streset [*if*] [*weight*] [, *multiple_options*]

    streset, { <u>p</u>ast | <u>f</u>uture | <u>p</u>ast <u>f</u>uture }

    st [, no<u>c</u>md <u>not</u>able]

    stset, clear

single_options	description
Main	
<u>f</u>ailure(*failvar*[==*numlist*])	failure event
<u>nos</u>how	prevent other st commands from showing st setting information
Options	
<u>o</u>rigin(<u>t</u>ime *exp*)	define when a subject becomes at risk
<u>sc</u>ale(*#*)	rescale time value
<u>ent</u>er(<u>t</u>ime *exp*)	specify when subject first enters study
<u>ex</u>it(<u>t</u>ime *exp*)	specify when subject exits study
Advanced	
if(*exp*)	select records for which *exp* is true; recommended rather than if *exp*
time0(*varname*)	mechanical aspect of interpretation about records in dataset; seldom used

multiple_options	description
Main	
* id(*idvar*)	multiple-record ID variable
* <u>f</u>ailure(*failvar*[==*numlist*])	failure event
<u>nosh</u>ow	prevent other st commands from showing st setting information
Options	
<u>o</u>rigin([*varname*==*numlist*] <u>t</u>ime *exp* | min)	define when a subject becomes at risk
<u>scale</u>(#)	rescale time value
<u>enter</u>([*varname*==*numlist*] <u>t</u>ime *exp*)	specify when subject first enters study
<u>ex</u>it(failure | [*varname*==*numlist*] <u>t</u>ime *exp*)	specify when subject exits study
Advanced	
if(*exp*)	select records for which *exp* is true; recommended rather than if *exp*
ever(*exp*)	select objects for which *exp* is ever true
never(*exp*)	select objects for which *exp* is never true
after(*exp*)	select records within subject on or after the first time *exp* is true
<u>before</u>(*exp*)	select records within subject before the first time *exp* is true
time0(*varname*)	mechanical aspect of interpretation about records in dataset; seldom used

* id() and failure() are required with stset multiple-record-per-subject survival data.

fweights, iweights, and pweights are allowed; see [U] **11.1.6 weight**.

Examples

. stset ftime	(time measured from 0, all failed)
. stset ftime, failure(died)	(time measured from 0, censoring)
. stset ftime, failure(died) id(id)	(time measured from 0, censoring & ID)
. stset ftime, failure(died==2,3)	(time measured from 0, failure codes)
. stset ftime, failure(died) origin(time dob)	(time measured from dob, censoring)

You cannot harm your data by using stset, so feel free to experiment.

Menu

Statistics > Survival analysis > Setup and utilities > Declare data to be survival-time data

Description

st refers to survival-time data, which are fully described below.

stset declares the data in memory to be st data, informing Stata of key variables and their roles in a survival-time data analysis. When you stset your data, stset runs various data consistency checks to ensure that what you have declared makes sense. If the data are weighted, you specify the weights when you stset the data, not when you issue the individual st commands.

streset changes how the st dataset is declared. In multiple-record data, streset can also temporarily set the sample to include records from before the time at risk (called the past) and records after failure (called the future). Then typing streset without arguments resets the sample back to the analysis sample.

st displays how the dataset is currently declared.

Whenever you type stset or streset, Stata runs or reruns data consistency checks to ensure that what you are now declaring (or declared in the past) makes sense. Thus if you have made any changes to your data or simply wish to verify how things are, you can type streset with no options.

stset, clear is for use by programmers. It causes Stata to forget the st markers, making the data no longer st data to Stata. The data remain unchanged. It is not necessary to stset, clear before doing another stset.

Options for use with stset and streset

Main

id(*idvar*) specifies the subject-ID variable; observations with equal, nonmissing values of *idvar* are assumed to be the same subject. *idvar* may be string or numeric. Observations for which *idvar* is missing (. or "") are ignored.

When id() is not specified, each observation is assumed to represent a different subject and thus constitutes a one-record-per-subject survival dataset.

When you specify id(), the data are said to be multiple-record data, even if it turns out that there is only one record per subject. Perhaps they would better be called potentially multiple-record data.

If you specify id(), stset requires that you specify failure().

Specifying id() never hurts; we recommend it because a few st commands, such as stsplit, require an ID variable to have been specified when the dataset was stset.

failure(*failvar*[==*numlist*]) specifies the failure event.

If failure() is not specified, all records are assumed to end in failure. This is allowed with single-record data only.

If failure(*failvar*) is specified, *failvar* is interpreted as an indicator variable; 0 and missing mean censored, and all other values are interpreted as representing failure.

If failure(*failvar*==*numlist*) is specified, records with *failvar* taking on any of the values in *numlist* are assumed to end in failure, and all other records are assumed to be censored.

noshow prevents other st commands from showing the key st variables at the top of their output.

Options

origin([*varname*==*numlist*] time *exp* | min) and scale(*#*) define analysis time; i.e., origin() defines when a subject becomes at risk. Subjects become at risk when time = origin(). All analyses are performed in terms of time since becoming at risk, called analysis time.

Let us use the terms *time* for how time is recorded in the data and t for analysis time. Analysis time t is defined

$$t = \frac{time - \text{origin}()}{\text{scale}()}$$

t is time from origin in units of scale.

By default, origin(time 0) and scale(1) are assumed, meaning that $t = $ *time*. Then you must ensure that *time* in your data is measured as time since becoming at risk. Subjects are exposed at $t = $ *time* $= 0$ and later fail. Observations with $t = $ *time* < 0 are ignored because information before becoming at risk is irrelevant.

origin() determines when the clock starts ticking. scale() plays no substantive role, but it can be handy for making t units more readable (such as converting days to years).

origin(time *exp*) sets the origin to *exp*. For instance, if *time* were recorded as dates, such as 05jun1998, in your data and variable expdate recorded the date when subjects were exposed, you could specify origin(time expdate). If instead all subjects were exposed on 12nov1997, you could specify origin(time mdy(11,12,1997)).

origin(time *exp*) may be used with single- or multiple-record data.

origin(*varname*==*numlist*) is for use with multiple-record data; it specifies the origin indirectly. If *time* were recorded as dates in your data, variable obsdate recorded the (ending) date associated with each record, and subjects became at risk upon, say, having a certain operation—and that operation were indicated by code==217—then you could specify origin(code==217). origin(code==217) would mean, for each subject, that the origin time is the earliest time at which code==217 is observed. Records before that would be ignored (because $t < 0$). Subjects who never had code==217 would be ignored entirely.

origin(*varname*==*numlist* time *exp*) sets the origin to the later of the two times determined by *varname*==*numlist* and *exp*.

origin(min) sets origin to the earliest time observed, minus 1. This is an odd thing to do and is described in example 10.

origin() is an important concept; see *Key concepts*, *Two concepts of time*, and *The substantive meaning of analysis time*.

scale() makes results more readable. If you have *time* recorded in days (such as Stata dates, which are really days since 01jan1960), specifying scale(365.25) will cause results to be reported in years.

enter([*varname*==*numlist*] time *exp*) specifies when a subject first comes under observation, meaning that any failures, were they to occur, would be recorded in the data.

Do not confuse enter() and origin(). origin() specifies when a subject first becomes at risk. In many datasets, becoming at risk and coming under observation are coincident. Then it is sufficient to specify origin().

enter(time *exp*), enter(*varname*==*numlist*), and enter(*varname*==*numlist* time *exp*) follow the same syntax as origin(). In multiple-record data, both *varname*==*numlist* and time *exp* are interpreted as the earliest time implied, and if both are specified, the later of the two times is used.

exit(failure | [*varname*==*numlist*] time *exp*) specifies the latest time under which the subject is both under observation and at risk. The emphasis is on latest; obviously, subjects also exit the risk pool when their data run out.

exit(failure) is the default. When the first failure event occurs, the subject is removed from the analysis risk pool, even if the subject has subsequent records in the data and even if some of those subsequent records document other failure events. Specify exit(time .) if you wish to keep all records for a subject after failure. You want to do this if you have multiple-failure data.

exit(*varname*==*numlist*), exit(time *exp*), and exit(*varname*==*numlist* time *exp*) follow the same syntax as origin() and enter(). In multiple-record data, both *varname*==*numlist* and time *exp* are interpreted as the earliest time implied. exit differs from origin() and enter() in that if both are specified, the earlier of the two times is used.

⌐ Advanced ⌐

if(*exp*), ever(*exp*), never(*exp*), after(*exp*), and before(*exp*) select relevant records.

> if(*exp*) selects records for which *exp* is true. We strongly recommend specifying this if() option rather than if *exp* following stset or streset. They differ in that if *exp* removes the data from consideration before calculating beginning and ending times and other quantities. The if() option, on the other hand, sets the restriction after all derived variables are calculated. See *if() versus if exp*.
>
> if() may be specified with single- or multiple-record data. The remaining selection options are for use with multiple-record data only.
>
> ever(*exp*) selects only subjects for which *exp* is ever true.
>
> never(*exp*) selects only subjects for which *exp* is never true.
>
> after(*exp*) selects records within subject on or after the first time *exp* is true.
>
> before(*exp*) selects records within subject before the first time *exp* is true.

time0(*varname*) is seldom specified because most datasets do not contain this information. time0() should be used exclusively with multiple-record data, and even then you should consider whether origin() or enter() would be more appropriate.

time0() specifies a mechanical aspect of interpretation about the records in the dataset, namely, the beginning of the period spanned by each record. See *Intermediate exit and reentry times (gaps)*.

Options unique to streset

past expands the stset sample to include the entire recorded past of the relevant subjects, meaning that it includes observations before becoming at risk and those excluded because of after(), etc.

future expands the stset sample to include the records on the relevant subjects after the last record that previously was included, if any, which typically means to include all observations after failure or censoring.

past future expands the stset sample to include all records on the relevant subjects.

Typing streset without arguments resets the sample to the analysis sample. See *Past and future records* for more information.

Options for st

nocmd suppresses displaying the last stset command.

notable suppresses displaying the table summarizing what has been stset.

Remarks

Remarks are presented under the following headings:

> What are survival-time data?
> Key concepts
> Survival-time datasets
> Using stset
> Two concepts of time
> The substantive meaning of analysis time
> Setting the failure event
> Setting multiple failures
> First entry times
> Final exit times
> Intermediate exit and reentry times (gaps)
> if( ) versus if exp
> Past and future records
> Using streset
> Performance and multiple-record-per-subject datasets
> Sequencing of events within t
> Weights
> Data warnings and errors flagged by stset
> Using survival-time data in Stata

What are survival-time data?

Survival-time data—what we call st data—document spans of time ending in an event. For instance,

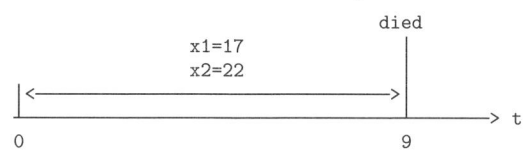

```
                       died
         x1=17          |
         x2=22          |
    |<—————————————————>|
                                 > t
    0                   9
```

which indicates x1 = 17 and x2 = 22 over the time span 0 to 9, and died = 1. More formally, it means x1 = 17 and x2 = 22 for $0 < t \leq 9$, which we often write as $(0, 9]$. However you wish to say it, this information might be recorded by the observation

```
        id          end       x1      x2     died
       101            9       17      22        1
```

and we call this single-record survival data.

The data can be more complicated. For instance, we might have

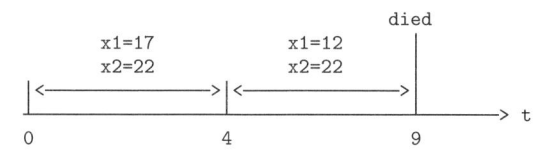

```
                                 died
        x1=17        x1=12        |
        x2=22        x2=22        |
    |<————————>|<————————>|
                                         > t
    0          4          9
```

meaning

> x1 = 17 and x2 = 22 during $(0, 4]$
> x1 = 12 and x2 = 22 during $(4, 9]$, and then died = 1.

and this would be recorded by the data

```
        id     begin      end      x1      x2     died
       101         0        4      17      22        0
       101         4        9      12      22        1
```

We call this multiple-record survival data.

These two formats allow you to record many different possibilities. The last observation on a person need not be failure,

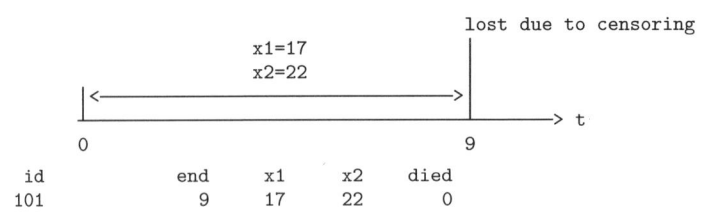

```
                              lost due to censoring
             x1=17            |
             x2=22            |
   |<----------------------->|
   |---------------------------------> t
   0                          9

   id        end      x1      x2     died
   101         9      17      22       0
```

or

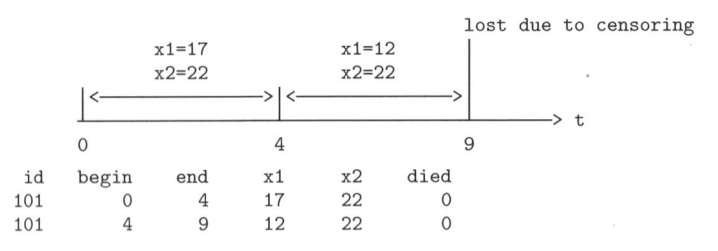

```
                                        lost due to censoring
       x1=17              x1=12         |
       x2=22              x2=22         |
   |<----------->|<--------------->|
   |-------------------------------------> t
   0             4                 9

   id    begin    end     x1      x2     died
   101       0      4     17      22       0
   101       4      9     12      22       0
```

Multiple-record data might have gaps,

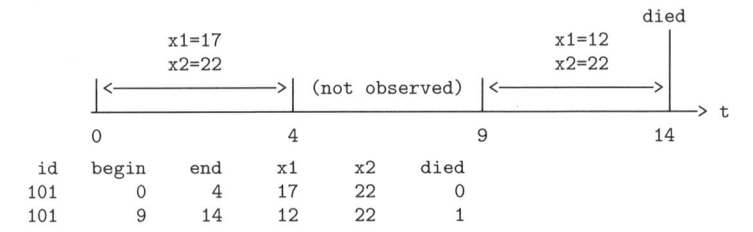

```
                                                     died
       x1=17                            x1=12        |
       x2=22                            x2=22        |
   |<----------->| (not observed) |<----------->|
   |---------------------------------------------------> t
   0             4                9             14

   id    begin    end     x1      x2     died
   101       0      4     17      22       0
   101       9     14     12      22       1
```

or subjects might not be observed from the onset of risk,

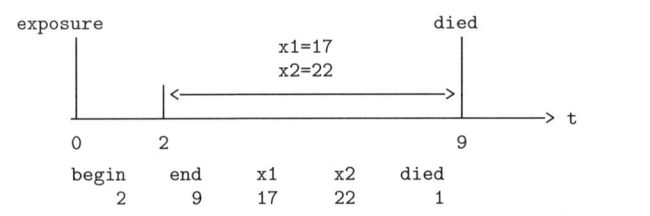

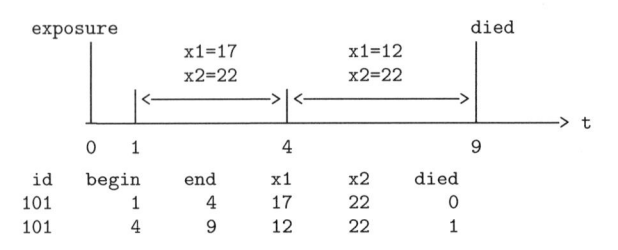

```
   exposure                       died
       |             x1=17        |
       |             x2=22        |
       |      |<--------------->|
       |------------------------------> t
       0      2                 9

          begin    end     x1      x2     died
              2      9     17      22       1
```

and

```
   exposure                              died
       |       x1=17          x1=12      |
       |       x2=22          x2=22      |
       |   |<---------->|<----------->|
       |----------------------------------> t
       0   1            4             9

   id    begin    end     x1      x2     died
   101       1      4     17      22       0
   101       4      9     12      22       1
```

The failure event might not be death but instead something that can be repeated:

```
                              1st                            2nd
                          infarction                     infarction
               x1=17                   x1=12              x1=10
               x2=22                   x2=22              x2=22
       |<----------->|<----------->|<----------->|
                                                                    ----> t
       0             4             9             13

       id    begin    end    x1    x2    infarc
       101      0      4     17    22      1
       101      4      9     12    22      0
       101      9     13     10    22      1
```

Our data may be in different time units; rather than t where $t = 0$ corresponds to the onset of risk, we might have time recorded as age,

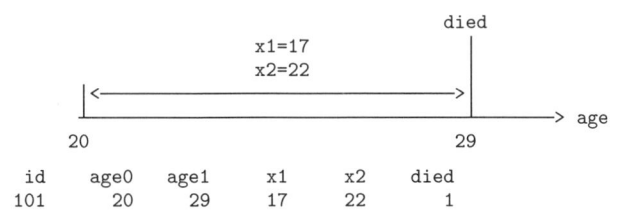

```
                                              died
                         x1=17                 |
                         x2=22
              |<--------------------------->|
                                                          ----> age
              20                            29

              id    age0   age1    x1    x2    died
              101    20     29     17    22     1
```

or time recorded as calendar dates:

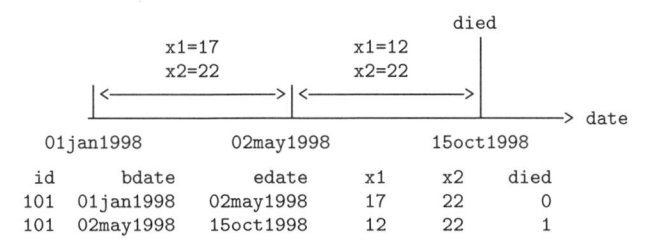

```
                                              died
                 x1=17                  x1=12  |
                 x2=22                  x2=22
              |<--------------->|<------------>|
                                                          ----> date
           01jan1998        02may1998      15oct1998

           id      bdate        edate       x1    x2    died
           101   01jan1998    02may1998     17    22     0
           101   02may1998    15oct1998     12    22     1
```

Finally, we can mix these diagrams however we wish, so we might have time recorded according to the calendar, unobserved periods after the onset of risk, subsequent gaps, and multiple failure events.

The st commands analyze data like these, and the first step is to tell st about your data by using stset. You do not change your data to fit some predefined mold; you describe your data with stset, and the rest of the st commands just do the right thing.

Before we discuss using stset, let's describe one more style of recording time-to-event data because it is common and is inappropriate for use with st. It is inappropriate, but it is easy to convert to the survival-time form. It is called snapshot data, which are data for which you do not know spans of time but you have information recorded at various points in time:

```
           x1=17              x1=12
           x2=22              x2=22            died
           |   (unobserved)   |  (unobserved)   |
           |                  |                 |
                                                        ----> t
           0                  4                 9

                  id     t     x1    x2    died
                  101    0     17    22     0
                  101    4     12    22     0
                  101    9      .     .     1
```

In this snapshot dataset all we know are the values of x1 and x2 at $t = 0$ and $t = 4$, and we know that the subject died at $t = 9$. Snapshot data can be converted to survival-time data if we are willing to assume that x1 and x2 remained constant between times:

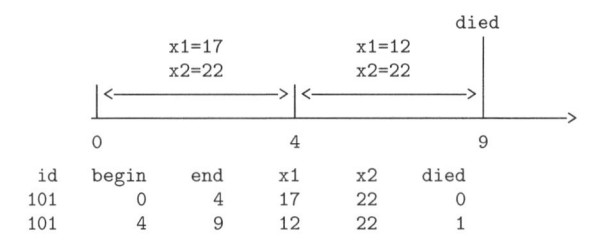

```
                                              died
               x1=17              x1=12        |
               x2=22              x2=22        |
        |<------------->|<------------->|      |
        |                                      |---------->
        0               4                      9

        id    begin    end    x1    x2    died
        101     0       4     17    22     0
        101     4       9     12    22     1
```

The `snapspan` command makes this conversion. If you have snapshot data, first see [ST] **snapspan** to convert it to survival-time data and then use `stset` to tell st about the converted data; see example 10 first.

Key concepts

time, or, better, *time units*, is how time is recorded in your data. It might be numbers (such as 0, 1, 2, ..., with time = 0 corresponding to some exposure event), a subject's age, or calendar time.

events are things that happen at an instant in time, such as being exposed to an environmental hazard, being diagnosed as myopic, becoming employed, being promoted, becoming unemployed, having a heart attack, and dying.

failure event is the event indicating failure as it is defined for analysis. This can be a single or compound event. The failure event might be when variable `dead` is 1, or it might be when variable `diag` is any of 115, 121, or 133.

at risk means the subject is at risk of the failure event occurring. For instance, if the failure event is becoming unemployed, a person must be employed. The subject is not at risk before being employed. Once employed, the subject becomes at risk; once again, the subject is no longer at risk once the failure event occurs. If subjects become at risk upon the occurrence of some event, it is called the exposure event. Gaining employment is the exposure event in our example.

origin is the time when the subject became at risk. If time is recorded as numbers such as 0, 1, 2, ..., with time = 0 corresponding to the exposure event, then origin = 0. Alternatively, origin might be the age of the subject when diagnosed or the date when the subject was exposed. Regardless, origin is expressed in time units.

scale is just a fixed number, typically 1, used in mapping time to analysis time t.

t, or *analysis time*, is (time − origin)/scale, meaning the time since onset of being at risk measured in scale units.

$t = 0$ corresponds to the onset of risk, and scale just provides a way to make the units of t more readable. You might have time recorded in days from 01jan1960 and want t recorded in years, in which case scale would be 365.25.

Time is how time is recorded in your data, and t is how time is reported in the analysis.

under observation means that, should the failure event occur, it would be observed and recorded in the data. Sometimes subjects are under observation only after they are at risk. This would be the case, for instance, if subjects enrolled in a study after being diagnosed with cancer and if, to enroll in the study, subjects were required to be diagnosed with cancer.

Being under observation does not mean that the subject is necessarily at risk. A subject may come under observation before being at risk, and in fact, a subject under observation may never become at risk.

entry time and *exit time* mark when a subject is first and last under observation. The emphasis here is on the words *first* and *last*; entry time and exit time do not record observational gaps; there is only one entry time and one exit time per subject.

Entry time and exit time might be expressed as times (recorded in time units), or they might correspond to the occurrence of some event (such as enrolling in the study).

Often the entry time corresponds to $t = 0$; i.e., because $t = (\text{time} - \text{origin})/\text{scale}$, time = origin, meaning that time equals when the subject became at risk.

Often the exit time corresponds to when the failure event occurs or, failing that, the end of data for the subject.

delayed entry means that entry time corresponds to $t > 0$; the subject became at risk before coming under observation.

ID refers to a subject identification variable; equal values of ID indicate that the records are on the same subject. An ID variable is required for multiple-record data and is optional, but recommended, with single-record data.

time0 refers to the beginning time (recorded in time units) of a record. Some datasets have this variable, but most do not. If the dataset does not contain the beginning time for each record, subsequent records are assumed to begin where previous records ended. A time0 variable may be created for these datasets by using the snapspan command; see [ST] **snapspan**. Do not confuse time0—a mechanical aspect of datasets—with entry time—a substantive aspect of analysis.

gaps refer to gaps in observation between entry time and exit time. During a gap, a subject is not under observation. Gaps can arise only if the data contain a time0 variable, because otherwise subsequent records beginning when previous records end would preclude there being gaps in the data. Gaps are distinct from delayed entry.

past history is a term we use to mean information recorded in the data before the subject was both at risk and under observation. Complex datasets can contain such observations. Say that the dataset contains histories on subjects from birth to death. You might tell st that a subject becomes at risk once diagnosed with a particular kind of cancer. The past history on the subject would then refer to records before the subject was diagnosed.

The word *history* is often dropped, and the term simply becomes *past*. For instance, we might want to know whether the subject smoked in the past.

future history is a term meaning information recorded in the data after the subject is no longer at risk. Perhaps the failure event is not so serious as to preclude the possibility of data after failure.

The word *history* is often dropped, and the term simply becomes *future*. Perhaps the failure event is cardiac infarction, and you want to know whether the subject died soon in the future so that you can exclude the subject.

Survival-time datasets

In survival-time datasets, observations (records) document a span of time. The span might be explicitly indicated, such as

```
     begin    end    x1    x2
         3      9    17    22         <- spans (3,9]
```

or it might be implied that the record begins at 0,

```
                end   x1   x2
                  9   17   22           <- spans (0,9]
```

or it might be implied because there are multiple records per subject:

```
        id     end   x1   x2
         1       4   17   22            <- spans (0,4]
         1       9   12   22            <- spans (4,9]
```

Records may have an event indicator:

```
     begin     end   x1   x2   died
         3       9   17   22      1     <- spans (3,9], died at t=9

                end   x1   x2   died
                  9   17   22      1    <- spans (0,9], died at t=9

        id     end   x1   x2   died
         1       4   17   22      0     <- spans (0,4],
         1       9   12   22      1     <- spans (4,9], died at t=9
```

The first two examples are called single-record survival-time data because there is one record per subject.

The final example is called multiple-record survival-time data. There are two records for the subject with id = 1.

Either way, survival-time data document time spans. Characteristics are assumed to remain constant over the span, and the event is assumed to occur at the end of the span.

Using stset

Once you have stset your data, you can use the other st commands.

If you save your data after stsetting, you will not have to re-stset in the future; Stata will remember.

stset declares your data to be survival-time data. It does not change the data, although it does add a few variables to your dataset.

This means that you can re-stset your data as often as you wish. In fact, the streset command encourages this. Using complicated datasets often requires typing long stset commands, such as

    . stset date, fail(event==27,28) origin(event==15) enter(event==22)

Later, you might want to try fail(event==27). You could retype the stset command, making the substitution, or you could type

    . streset, fail(event==27)

streset takes what you type, merges it with what you have previously declared with stset, and performs the combined stset command.

▷ Example 1: Single-record data

Generators are run until they fail. Here is some of our dataset:

```
. use http://www.stata-press.com/data/r11/kva
(Generator experiment)
. list in 1/3
```

	failtime	load	bearings
1.	100	15	0
2.	140	15	1
3.	97	20	0

The `stset` command for this dataset is

```
. stset failtime
         failure event:  (assumed to fail at time=failtime)
 obs. time interval:  (0, failtime]
 exit on or before:  failure

       12  total obs.
        0  exclusions

       12  obs. remaining, representing
       12  failures in single record/single failure data
      896  total analysis time at risk, at risk from t =         0
                              earliest observed entry t =         0
                                   last observed exit t =       140
```

When you type `stset` *timevar*, *timevar* is assumed to be the time of failure. More generally, you will learn, *timevar* is the time of failure or censoring. Here *timevar* is `failtime`.

◁

▷ Example 2: Single-record data with censoring

Generators are run until they fail, but during the experiment, the room flooded, so some generators were run only until the flood. Here are some of our data:

```
. use http://www.stata-press.com/data/r11/kva2
(Generator experiment)
. list in 1/4
```

	failtime	load	bearings	failed
1.	100	15	0	1
2.	140	15	1	0
3.	97	20	0	1
4.	122	20	1	1

Here the second generator did not fail at time 140; the experiment was merely discontinued then. The `stset` command for this dataset is

```
. stset failtime, failure(failed)
     failure event:  failed != 0 & failed < .
obs. time interval:  (0, failtime]
 exit on or before:  failure
```

```
        12  total obs.
         0  exclusions

        12  obs. remaining, representing
        11  failures in single record/single failure data
       896  total analysis time at risk, at risk from t =          0
                                   earliest observed entry t =     0
                                    last observed exit t =       140
```

When you type stset *timevar*, failure(*failvar*), *timevar* is interpreted as the time of failure or censoring, which is determined by the value of *failvar*. *failvar* = 0 and *failvar* = . (missing) indicate censorings, and all other values indicate failure.

◁

▷ Example 3: Multiple-record data

Assume that we are analyzing survival time of patients with a particular kind of cancer. In this dataset, the characteristics of patients vary over time, perhaps because new readings were taken or because the drug therapy was changed. Some of the data are

```
. list, separator(0)
```

	patid	t	died	x1	x2
1.	90	100	0	1	0
2.	90	150	1	0	0
3.	91	50	1	1	1
4.	92	100	0	0	0
5.	92	150	0	0	1
6.	92	190	0	0	0
7.	93	100	0	0	0

(*output omitted*)

There are two records for patient 90, and died is 0 in the first record but 1 in the second. The interpretation of these two records is

Interval (0,100]: $x1 = 1$ and $x2 = 0$
Interval (100,150]: $x1 = 0$ and $x2 = 0$
at t = 150:   the patient died

Similarly, here is how you interpret the other records:

Patient 91:
Interval (0, 50]: $x1 = 1$ and $x2 = 1$
at t = 50:   the patient died

Patient 92:
Interval (0,100]: $x1 = 0$ and $x2 = 0$
Interval (100,150]: $x1 = 0$ and $x2 = 1$
Interval (150,190]: $x1 = 0$ and $x2 = 0$
at t = 190:   the patient was lost because of censoring

Look again at patient 92's data:

```
    patid        t       died       x1        x2
       92      100          0        0         0
       92      150          0        0         1
       92      190          0        0         0
```

died = 0 for the first event. Mechanically, this removes the subject from the data at t = 100—the patient is treated as censored. The next record, however, adds the patient back into the data (at t = 100) with new characteristics.

The stset command for this dataset is

```
. stset t, id(patid) failure(died)

                id:  patid
     failure event:  died != 0 & died < .
obs. time interval:  (t[_n-1], t]
 exit on or before:  failure

─────────────────────────────────────────────────────────────────────
       126  total obs.
         0  exclusions
─────────────────────────────────────────────────────────────────────
       126  obs. remaining, representing
        40  subjects
        26  failures in single failure-per-subject data
      2989  total analysis time at risk, at risk from t =         0
                                  earliest observed entry t =     0
                                       last observed exit t =   139
```

When you have multiple-record data, you specify stset's id(*idvar*) option. When you type stset *timevar*, id(*idvar*) failure(*failvar*), *timevar* denotes the end of the period (just as it does in single-record data). The first record within *idvar* is assumed to begin at time 0, and later records are assumed to begin where the previous record left off. *failvar* should contain 0 on all but, possibly, the last record within *idvar*, unless your data contain multiple failures (in which case you must specify the exit() option; see *Setting multiple failures* below).

◁

▷ Example 4: Multiple-record data with multiple events

We have the following data on hospital patients admitted to a particular ward:

```
    patid     day     sex      x1      x2     code
      101       5       1      10      10      177
      101      13       1      20       8      286
      101      21       1      16      11      208
      101      24       1      11      17      401
      102       8       0      20      19      204
      102      18       0      19       1      401
      103     etc.
```

Variable code records various actions; code 401 indicates being discharged alive, and 402 indicates death. We stset this dataset by typing

```
. stset day, id(patid) fail(code==402)
              id:  patid
   failure event:  code == 402
obs. time interval: (day[_n-1], day]
exit on or before:  failure

   243  total obs.
     0  exclusions

   243  obs. remaining, representing
    40  subjects
    15  failures in single failure-per-subject data
  1486  total analysis time at risk, at risk from t =         0
                            earliest observed entry t =       0
                             last observed exit t =          62
```

When you specify failure(*eventvar* == #), the failure event is as specified. You may include a list of numbers following the equal signs. If failure were codes 402 and 403, you could specify failure(code == 402 403). If failure were codes 402, 403, 404, 405, 406, 407, and 409, you could specify failure(code == 402/407 409).

◁

▷ Example 5: Multiple-record data recording time rather than t

More reasonably, the hospital data in the above example would not contain days since admission but would contain admission and current dates. In the dataset below, adday contains the day of admission, and curdate contains the ending date of the record, both recorded as number of days since the ward opened:

patid	adday	curday	sex	x1	x2	code
101	287	292	1	10	10	177
101	.	300	1	20	8	286
101	.	308	1	16	11	208
101	.	311	1	11	17	401
102	289	297	0	20	19	204
102	.	307	0	19	1	401
103	etc.					

This is the same dataset as shown in example 4. Previously, the first record on patient 101 was recorded 5 days after admission. In this dataset, $292 - 287 = 5$. We would stset this dataset by typing

```
. stset curday, id(patid) fail(code==402) origin(time adday)
              id:  patid
   failure event:  code == 402
obs. time interval: (curday[_n-1], curday]
exit on or before:  failure
   t for analysis:  (time-origin)
          origin:  time adday

   243  total obs.
     0  exclusions

   243  obs. remaining, representing
    40  subjects
    15  failures in single failure-per-subject data
  1486  total analysis time at risk, at risk from t =         0
                            earliest observed entry t =       0
                             last observed exit t =          62
```

origin() sets when a subject becomes at risk. It does this by defining analysis time.

When you specify stset *timevar*, ... origin(time *originvar*), analysis time is defined as $t = (timevar - originvar)/\text{scale}()$. In analysis-time units, subjects become at risk at $t = 0$. See *Two concepts of time* and *The substantive meaning of analysis time* below.

◁

▷ Example 6: Multiple-record data with time recorded as a date

Even more reasonably, dates would not be recorded as integers 428, 433, and 453, meaning the number of days since the ward opened. The dates would be recorded as dates:

```
patid    addate    curdate    sex    x1    x2    code
 101   18aug1998  23aug1998    1     10    10    177
 101       .      31aug1998    1     20     8    286
 101       .      08sep1998    1     16    11    208
 101       .      11sep1998    1     11    17    401
 102   20aug1998  28aug1998    0     20    19    204
 102       .      07sep1998    0     19     1    401
 103     etc.
```

That, in fact, changes nothing. We still type what we previously typed:

. stset curdate, id(patid) fail(code==402) origin(time addate)

Stata dates are, in fact, integers—they are the number of days since 01jan1960—and it is merely Stata's %td display format that makes them display as dates.

◁

▷ Example 7: Multiple-record data with extraneous information

Perhaps we wish to study the outcome after a certain operation, said operation being indicated by code 286. Subjects become at risk when the operation is performed. Here we do not type

. stset curdate, id(patid) fail(code==402) origin(time addate)

We instead type

. stset curdate, id(patid) fail(code==402) origin(code==286)

The result of typing this would be to set analysis time t to

$$t = \text{curdate} - (\text{the value of curdate when code==286})$$

Let's work through this for the first patient:

```
patid    addate    curdate    sex    x1    x2    code
 101   18aug1998  23aug1998    1     10    10    177
 101       .      31aug1998    1     20     8    286
 101       .      08sep1998    1     16    11    208
 101       .      11sep1998    1     11    17    401
```

The event 286 occurred on 31aug1998, and thus the values of t for the four records are

$$t_1 = \text{curdate}_1 - 31\text{aug}1998 = 23\text{aug}1998 - 31\text{aug}1998 = -8$$
$$t_2 = \text{curdate}_2 - 31\text{aug}1998 = 31\text{aug}1998 - 31\text{aug}1998 = 0$$
$$t_3 = \text{curdate}_3 - 31\text{aug}1998 = 08\text{sep}1998 - 31\text{aug}1998 = 8$$
$$t_4 = \text{curdate}_4 - 31\text{aug}1998 = 11\text{sep}1998 - 31\text{aug}1998 = 11$$

Information prior to $t = 0$ is not relevant because the subject is not yet at risk. Thus the relevant data on this subject are

t in $(0, 8]$ sex = 1, x1 = 16, x2 = 11
t in $(8, 11]$ sex = 1, x1 = 11, x2 = 17, and the subject is censored (code $\neq$ 402)

That is precisely the logic that `stset` went through. For your information, `stset` quietly creates the variables

_st	1 if the record is to be used, 0 if ignored
_t0	analysis time when record begins
_t	analysis time when record ends
_d	1 if failure, 0 if censored

You can examine these variables after issuing the `stset` command:

. list _st _t0 _t _d

	_st	_t0	_t	_d
1.	0	.	.	.
2.	0	.	.	.
203.	1	0	8	0
204.	1	8	11	0

Results are just as we anticipated. Do not let the observation numbers bother you; `stset` sorts the data in a way it finds convenient. Feel free to re-sort the data; if any of the st commands need the data in a different order, they will sort it themselves.

There are two ways of specifying `origin()`:

origin(time *timevar*)   or   origin(time *exp*)
origin(*eventvar* == *numlist*)

In the first syntax—which is denoted by typing the word `time`—you directly specify when a subject becomes at risk. In the second syntax—which is denoted by typing a variable name and equal signs—you specify the same thing indirectly. The subject becomes at risk when the specified event occurs (which may be never).

Information prior to `origin()` is ignored. That information composes what we call the past history.

◁

▷ Example 8: Multiple-record data with delayed entry

In another analysis, we want to use the above data to analyze all patients, not just those undergoing a particular operation. In this analysis, subjects become at risk when they enter the ward. For this analysis, however, we need information from a particular test, and that information is available only if the test is administered to the patient. Even if the test is administered, some amount of time passes before that. Assume that when the test is administered, `code==152` is inserted into the patient's hospital record.

To summarize, we want `origin(time addate)`, but patients do not enter our sample until `code==152`. The way to `stset` these data is

. stset curdate, id(patid) fail(code==402) origin(time addate) enter(code==152)

Patient 107 has code 152:

```
   patid    addate    curdate   sex   x1   x2   code
     107 22aug1998  25aug1998     1    9   13    274
     107         .  28aug1998     1   19   19    152
     107         .  30aug1998     1   18   12    239
     107         .  07sep1998     1   12   11    401
```

In analysis time, $t = 0$ corresponds to 22aug1998. The test was not administered, however, until 6 days later. The analysis times for these records are

$$t_1 = curdate_1 - 22\text{aug}1998 = 25\text{aug}1998 - 22\text{aug}1998 = 3$$
$$t_2 = curdate_2 - 22\text{aug}1998 = 28\text{aug}1998 - 22\text{aug}1998 = 6$$
$$t_3 = curdate_3 - 22\text{aug}1998 = 30\text{aug}1998 - 22\text{aug}1998 = 8$$
$$t_4 = curdate_4 - 22\text{aug}1998 = 07\text{sep}1998 - 22\text{aug}1998 = 16$$

and the data we want in our sample are

t in $(6, 8]$ sex $= 1$, x1 $= 18$, x2 $= 12$
t in $(8, 16]$ sex $= 1$, x1 $= 12$, x2 $= 11$, and patient was censored (code $\neq 402$)

The above stset command produced this:

. list _st _t0 _t _d

	_st	_t0	_t	_d
1.	0	.	.	.
2.	0	.	.	.
39.	1	6	8	0
40.	1	8	16	0

◁

▷ Example 9: Multiple-record data with extraneous information and delayed entry

The origin() and enter() options can be combined. For instance, we want to analyze patients receiving a particular operation (time at risk begins upon code == 286, but patients may not enter the sample before a test is administered, denoted by code == 152). We type

. stset curdate, id(patid) fail(code==402) origin(code==286) enter(code==152)

If we typed the above commands, it would not matter whether the test was performed before or after the operation.

A patient who had the test and then the operation would enter at analysis time $t = 0$.

A patient who had the operation and then the test would enter at analysis time $t > 0$, the analysis time being the time the test was performed.

If we wanted to require that the operation be performed after the test, we could type

. stset curdate, id(patid) fail(code==402) origin(code==286) after(code==152)

Admittedly, this can be confusing. The way to proceed is to find a complicated case in your data and then list _st _t0 _t _d for that case after you stset the data.

◁

▷ Example 10: Real data

All of our hospital ward examples are artificial in one sense: it is unlikely the data would have come to us in survival-time form:

```
patid    addate   curdate  sex  x1  x2  code
  101 18aug1998 23aug1998    1  10  10   177
  101         . 31aug1998    1  20   8   286
  101         . 08sep1998    1  16  11   208
  101         . 11sep1998    1  11  17   401
  102 20aug1998 28aug1998    0  20  19   204
  102         . 07sep1998    0  19   1   401
  103  etc.
```

Rather, we would have received a snapshot dataset:

```
patid      date  sex  x1  x2  code
  101 18aug1998    1  10  10    22
  101 23aug1998    .  20   8   177
  101 31aug1998    .  16  11   286
  101 08sep1998    .  11  17   208
  101 11sep1998    .   .   .   401
  102 20aug1998    0  20  19    22
  102 28aug1998    .  19   1   204
  102 07sep1998    .   .   .   401
  103  etc.
```

In a snapshot dataset, we have a time (here a date) and values of the variables as of that instant.

This dataset can be converted to the appropriate form by typing

```
. snapspan patid date code
```

The result would be as follows:

```
patid      date  sex  x1  x2  code
  101 18aug1998    .   .   .    22
  101 23aug1998    1  10  10   177
  101 31aug1998    .  20   8   286
  101 08sep1998    .  16  11   208
  101 11sep1998    .  11  17   401
  102 20aug1998    .   .   .    22
  102 28aug1998    0  20  19   204
  102 07sep1998    .  19   1   401
```

This is virtually the same dataset with which we have been working, but it differs in two ways:

1. The variable sex is not filled in for all the observations because it was not filled in on the original form. The hospital wrote down the sex on admission and then never bothered to document it again.

2. We have no admission date (addate) variable. Instead, we have an extra first record for each patient with code = 22 (22 is the code the hospital uses for admissions).

The first problem is easily fixed, and the second, it turns out, is not a problem because we can vary what we type when we stset the data.

First, let's fix the problem with variable sex. There are two ways to proceed. One would be simply to fill in the variable ourselves:

```
. by patid (date), sort: replace sex = sex[_n-1] if sex>=.
```

We could also perform a phony `stset` that is good enough to set all the data and then use `stfill` to fill in the variable for us. Let's begin with the phony `stset`:

```
. stset date, id(patid) origin(min) fail(code==-1)
                 id:  patid
      failure event:  code == -1
 obs. time interval:  (date[_n-1], date]
  exit on or before:  failure
     t for analysis:  (time-origin)
             origin:  min
 ───────────────────────────────────────────────────────────────
      283  total obs.
        0  exclusions
 ───────────────────────────────────────────────────────────────
      283  obs. remaining, representing
       40  subjects
        0  failures in single failure-per-subject data
     2224  total analysis time at risk, at risk from t =         0
                                    earliest observed entry t =         0
                                     last observed exit t =        89
```

Typing `stset date, id(patid) origin(min) fail(code == -1)` does not produce anything we would want to use for analysis. This is a trick to get the dataset temporarily `stset` so that we can use some st data-management commands on it.

The first part of the trick is to specify `origin(min)`. This defines the analysis time as $t = 0$, corresponding to the minimum observed value of the time variable minus 1. The time variable is date here. Why the minimum minus 1? Because st ignores observations for which analysis time $t < 0$. `origin(min)` provides a phony definition of t that ensures $t > 0$ for all observations.

The second part of the trick is to specify `fail(code == -1)`, and you might have to vary what you type. We just wanted to choose an event that we know never happens, thus ensuring that no observations are ignored after failure.

Now that we have the dataset `stset`, we can use the other st commands. Do not use the st analysis commands unless you want ridiculous results, but one of the st data-management commands is just what we want:

```
. stfill sex, forward
         failure _d:  code == -1
    analysis time _t:  (date-origin)
             origin:  min
                 id:  patid
replace missing values with previously observed values:
        sex:    203 real changes made
```

Problem one solved.

The second problem concerns the lack of an admission-date variable. That is not really a problem because we have a new first observation with code = 22 recording the date of admission, so every place we previously coded `origin(time addate)`, we substitute `origin(code == 22)`.

Problem two solved.

We also solved the big problem—converting a snapshot dataset into a survival-time dataset; see [ST] **snapspan**.

◁

Two concepts of time

The st system has two concepts of time. The first is *time* in italics, which corresponds to how time is recorded in your data. The second is analysis time, which we write as t. Substantively, analysis time is time at risk. stset defines analysis time in terms of *time* via

$$t = \frac{time - \texttt{origin()}}{\texttt{scale()}}$$

t and *time* can be the same thing, and by default they are because, by default, origin() is 0 and scale() is 1.

All the st analysis commands work with analysis time, t.

By default, if you do not specify the origin() and scale() options, your time variables are expected to be the analysis-time variables. This means that *time* = 0 corresponds to when subjects became at risk, and that means, among other things, that observations for which *time* < 0 are ignored because survival analysis concerns persons who are at risk, and no one is at risk before $t = 0$.

origin determines when the clock starts ticking. If you do not specify origin(), origin(time 0) is assumed, meaning that $t = $ *time* and that persons are at risk from $t = $ *time* $ = 0$.

time and t will often differ. *time* might be calendar time and t the length of time since some event, such as being born or being exposed to some risk factor. origin() sets when $t = 0$. scale() merely sets a constant that makes t more readable.

The syntax for the origin() option makes it look more complicated than it really is

origin([*varname* == *numlist*] time *exp* | min)

This says that there are four different ways to specify origin():

origin(time *exp*)
origin(*varname* == *numlist*)
origin(*varname* == *numlist* time *exp*)
origin(min)

The first syntax can be used with single- or multiple-record data. It states that the origin is given by *exp*, which can be a constant for all observations, a variable (and hence may vary subject by subject), or even an expression composed of variables and constants. Perhaps the origin is a fixed date or a date recorded in the data when the subject was exposed or when the subject turned 18.

The second and third syntaxes are for use with multiple-record data. The second states that the origin corresponds to the (earliest) time when the designated event occurred. Perhaps the origin is when an operation was performed. The third syntax calculates the origin both ways and then selects the later one.

The fourth syntax does something odd; it sets the origin to the minimum time observed minus 1. This is not useful for analysis but is sometimes useful for playing data-management tricks; see example 10 above.

Let's start with the first syntax. Say that you had the data

```
      faildate      x1      x2
      28dec1997     12      22
      12nov1997     15      22
      03feb1998     55      22
```

and that all the observations came at risk on the same date, 01nov1997. You could type

. stset faildate, origin(time mdy(11,1,1997))

Remember that stset adds _t0 and _t to your dataset and that they contain the time span for each record, documented in analysis-time units. After typing stset, you can list the results:

. list faildate x1 x2 _t0 _t

	faildate	x1	x2	_t0	_t
1.	28dec1997	12	22	0	57
2.	12nov1997	15	22	0	11
3.	03feb1998	55	22	0	94

Record 1 reflects the period $(0, 57]$ in analysis-time units, which are days here. stset calculated the 57 from 28dec1997 − 01nov1997 = 13,876 − 13,819 = 57. (Dates such as 28dec1997 are really just integers containing the number of days from 01jan1960, and Stata's %td display format makes them display nicely. 28dec1997 is really the number 13,876.)

As another example, we might have data recording exposure and failure dates:

expdate	faildate	x1	x2
07may1998	22jun1998	12	22
02feb1998	11may1998	11	17

The way to stset this dataset is

. stset faildate, origin(time expdate)

and the result, in analysis units, is

. list expdate faildate x1 x2 _t0 _t

	expdate	faildate	x1	x2	_t0	_t
1.	07may1998	22jun1998	12	22	0	46
2.	02feb1998	11may1998	11	17	0	98

There is nothing magical about dates. Our original data could just as well have been

expdate	faildate	x1	x2
32	78	12	22
12	110	11	17

and the result would still be the same because 78 − 32 = 46 and 110 − 12 = 98.

Specifying an expression can sometimes be useful. Suppose that your dataset has the variable date recording the date of event and variable age recording the subject's age as of date. You want to make $t = 0$ correspond to when the subject turned 18. You could type origin(time date-int((age-18)*365.25)).

origin(*varname* == *numlist*) is for use with multiple-record data. It states when each subject became at risk indirectly; the subject became at risk at the earliest time that *varname* takes on any of the enumerated values. Say that you had

```
patid      date       x1    x2    event
101    12nov1997      15    22     127
101    28dec1997      12    22     155
101    03feb1998      55    22     133
101    05mar1998      14    22     127
101    09apr1998      12    22     133
101    03jun1998      13    22     101
102    22nov1997       .     .       .
```

and assume event = 155 represents the onset of exposure. You might stset this dataset by typing

```
. stset date, id(patid) origin(event==155) ...
```

If you did that, the information for patient 101 before 28dec1997 would be ignored in subsequent analysis. The prior information would not be removed from the dataset; it would just be ignored. Probably something similar would happen for patient 102, or if patient 102 has no record with event = 155, all the records on the patient would be ignored.

For analysis time, $t = 0$ would correspond to when event 155 occurred. Here are the results in analysis-time units:

```
patid      date       x1    x2    event    _t0      _t
101    12nov1997      15    22     127      .        .
101    28dec1997      12    22     155      .        .
101    03feb1998      55    22     133      0       37
101    05mar1998      14    22     127     37       67
101    09apr1998      12    22     133     67      102
101    03jun1998      13    22     101    102      157
102    22nov1997       .     .       .      .        .
```

Patient 101's second record is excluded from the analysis. That is not a mistake. Records document durations, date reflects the end of the period, and events occur at the end of periods. Thus event 155 occurred at the instant date = 28dec1997, and the relevant first record for the patient is (28dec1997, 03feb1998] in time units, which is (0, 37] in t units.

The substantive meaning of analysis time

In specifying origin(), you must ask yourself whether two subjects with identical characteristics face the same risk of failure. The answer is that they face the same risk when they have the same value of $t = (time - \text{origin}())/\text{scale}()$ or, equivalently, when the same amount of time has elapsed from origin().

Say that we have the following data on smokers who have died:

```
    ddate      x1    x2    reason
11mar1984      23    11      2
15may1994      21     9      1
22nov1993      22    13      2
   etc.
```

We wish to analyze death due to reason==2. However, typing

```
. stset ddate, fail(reason==2)
```

would probably not be adequate. We would be saying that smokers were at risk of death from 01jan1960. Would it matter? It would if we planned on doing anything parametric because parametric hazard functions, except for the exponential, are functions of analysis time, and the location of 0 makes a difference.

Even if we were thinking of performing nonparametric analysis, there would probably be difficulties. We would be asserting that two "identical" persons (in terms of x1 and x2) face the same risk on the same calendar date. Does the risk of death due to smoking really change as the calendar changes?

It would be more reasonable to assume that the risk changes with how long a subject has been smoking and that our data would probably include that date. We would type

        . stset ddate, fail(reason==2) origin(time smdate)

if smdate were the name of the date-started-smoking variable. We would now be saying that the risk is equal when the number of days smoked is the same. We might prefer to see t in years,

        . stset ddate, fail(reason==2) origin(time smdate) scale(365.25)

but that would make no substantive difference.

Consider single-record data on firms that went bankrupt:

```
        incorp    bankrupt    x1    x2   btype
     22jan1983   11mar1984    23    11     2
     17may1992   15may1994    21     9     1
     03nov1991   22nov1993    22    13     2
     etc.
```

Say that we wish to examine the risk of a particular kind of bankruptcy, btype == 2, among firms that become bankrupt. Typing

        . stset bankrupt, fail(btype==2)

would be more reasonable than it was in the smoking example. It would not be reasonable if we were thinking of performing any sort of parametric analysis, of course, because then location of $t = 0$ would matter, but it might be reasonable for semiparametric analysis. We would be asserting that two "identical" firms (with respect to the characteristics we model) have the same risk of bankruptcy when the calendar dates are the same. We would be asserting that the overall state of the economy matters.

It might be reasonable to instead measure time from the date of incorporation:

        . stset bankrupt, fail(btype==2) origin(time incorp)

Understand that the choice of origin() is a substantive decision.

Setting the failure event

You set the failure event by using the failure() option.

In single-record data, if failure() is not specified, every record is assumed to end in a failure. For instance, with

```
          failtime    load    bearings
     1.        100      15           0
     2.        140      15           1
     etc.
```

you would type stset failtime, and the first observation would be assumed to fail at time = 100; the second, at time = 140; and so on.

failure(*varname*) specifies that a failure occurs whenever *varname* is not zero and is not missing. For instance, with

	failtime	load	bearings	burnout
1.	100	15	0	1
2.	140	15	1	0
3.	97	20	0	1
4.	122	20	1	0
5.	84	25	0	1
6.	100	25	1	1
etc.				

you might type stset failtime, failure(burnout). Observations 1, 3, 5, and 6 would be assumed to fail at times 100, 97, 84, and 100, respectively; observations 2 and 4 would be assumed to be censored at times 140 and 122.

Similarly, if the data were

	failtime	load	bearings	burnout
1.	100	15	0	1
2.	140	15	1	0
3.	97	20	0	2
4.	122	20	1	.
5.	84	25	0	2
6.	100	25	1	3
etc.				

the result would be the same. Nonzero, nonmissing values of the failure variable are assumed to represent failures. (Perhaps burnout contains a code on how the burnout occurred.)

failure(*varname* == *numlist*) specifies that a failure occurs whenever *varname* takes on any of the values of *numlist*. In the above example, specifying

```
. stset failtime, failure(burnout==1 2)
```

would treat observation 6 as censored.

```
. stset failtime, failure(burnout==1 2 .)
```

would also treat observation 4 as a failure.

```
. stset failtime, failure(burnout==1/3 6 .)
```

would treat burnout==1, burnout==2, burnout==3, burnout==6, and burnout==. as representing failures and all other values as representing censorings. (Perhaps we want to examine "failure due to meltdown", and these are the codes that represent the various kinds of meltdown.)

failure() is treated the same way in both single- and multiple-record data. Consider

	patno	t	x1	x2	died
1.	1	4	23	11	1
2.	2	5	21	9	0
3.	2	8	22	13	1
4.	3	7	20	5	0
5.	3	9	22	5	0
6.	3	11	21	5	0
7.	4	...			

Typing

```
. stset t, id(patno) failure(died)
```

would treat

patno==1	as dying	at t==4
patno==2	as dying	at t==8
patno==3	as being censored	at t==11

Intervening records on the same subject are marked as "censored". Technically, they are not really censored if you think about it carefully; they are simply marked as not failing. Look at the data for subject 3:

```
         patno          t           x1          x2        died
             3          9           22           5           0
             3         11           21           5           0
```

The subject is not censored at t = 9 because there are more data on the subject; it is merely the case that the subject did not die at that time. At t = 9, x1 changed from 22 to 21. The subject is really censored at t = 11 because the subject did not die and there are no more records on the subject.

Typing `stset t, id(patno) failure(died)` would mark the same persons as dying and the same persons as censored, as in the previous case. If `died` contained not 0 and 1, but 0 and nonzero, nonmissing codes for the reason for death would be

```
         patno          t           x1          x2        died
    1.       1          4           23          11         103
    2.       2          5           21           9           0
    3.       2          8           22          13         207
    4.       3          7           20           5           0
    5.       3          9           22           5           0
    6.       3         11           21           5           0
    7.       4        ...
```

Typing

    . stset t, id(patno) failure(died)

or

    . stset t, id(patno) failure(died==103 207)

would yield the same results; subjects 1 and 2 would be treated as dying and subject 3 as censored.

Typing

    . stset t, id(patno) failure(died==207)

would treat subject 2 as dying and subjects 1 and 3 as censored. Thus when you specify the values for the code, the code variable need not ever contain 0. In

```
         patno          t           x1          x2        died
    1.       1          4           23          11         103
    2.       2          5           21           9          13
    3.       2          8           22          13         207
    4.       3          7           20           5          11
    5.       3          9           22           5          12
    6.       3         11           21           5          12
    7.       4        ...
```

typing

    . stset t, id(patno) failure(died==207)

treats patient 2 as dying and 1 and 3 as censored. Typing

    . stset t, id(patno) failure(died==103 207)

treats patients 1 and 2 as dying and 3 as censored.

Setting multiple failures

In multiple-record data, records after the first failure event are ignored unless you specify the exit() option. Consider the following data:

	patno	t	x1	x2	code
1.	1	4	21	7	14
2.	1	5	21	7	11
3.	1	7	20	7	17
4.	1	8	22	7	22
5.	1	9	22	7	22
6.	1	11	21	7	29
7.	2	...			

Perhaps code 22 represents the event of interest—say, the event "visited the doctor". Were you to type stset t, id(patno) failure(code == 22), the result would be as if the data contained

	patno	t	x1	x2	code
1.	1	4	21	7	14
2.	1	5	21	7	11
3.	1	7	20	7	17
4.	1	8	22	7	22

Records after the first occurrence of the failure event are ignored. If you do not want this, you must specify the exit() option. Probably you would want to specify exit(time .), here meaning that subjects are not to exit the risk group until their data run out. Alternatively, perhaps code 142 means "entered the nursing home" and, once that event happens, you no longer want them in the risk group. Then you would code exit(code == 142); see *Final exit times* below.

First entry times

Do not confuse enter() with origin(). origin() specifies when a subject first becomes at risk. enter() specifies when a subject first comes under observation. In most datasets, becoming at risk and coming under observation are coincident. Then it is sufficient to specify origin() alone, although you could specify both options.

Some persons enter the data after they have been at risk of failure. Say that we are studying deaths due to exposure to substance X and we know the date at which a person was first exposed to the substance. We are willing to assume that persons are at risk from the date of exposure forward. A person arrives at our door who was exposed 15 years ago. Can we add this person to our data? The statistical issue is labeled *left-truncation*, and the problem is that had the person died before arriving at our door, we would never have known about her. We can add her to our data, but we must be careful to treat her subsequent survival time as conditional on having already survived 15 years.

Say that we are examining visits to the widget repair facility, "failure" being defined as a visit (so failures can be repeated). The risk begins once a person buys a widget. We have a woman who bought a widget 3 years ago, and she has no records on when she has visited the facility in the last 3 years. Can we add her to our data? Yes, as long as we are careful to treat her subsequent behavior as already being 3 years after she first became at risk.

The jargon for this is "under observation". All this means is that any failures would be observed. Before being under observation, failure would not be observed.

If enter() is not specified, we assume that subjects are under observation at the time they enter the risk group as specified by origin(), 0 if origin() is not specified, or possibly time0(). To be precise, subject i is assumed to first enter the analysis risk pool at

$$time_i = \max\left(\text{earliest time0() for } i, \text{enter()}, \text{origin()}\right)$$

Say that we have multiple-record data recording "came at risk" (mycode == 1), "enrolled in our study" (mycode == 2), and "failed due to risk" (mycode == 3). We stset this dataset by typing

        . stset time, id(id) origin(mycode==1) enter(mycode==2) failure(mycode==3)

The above stset correctly handles the came at risk/came under observation problem regardless of the order of events 1, 2, and 3. For instance, if the subject comes under observation before he or she becomes at risk, the subject will be treated as entering the analysis risk pool at the time he or she came at risk.

Say that we have the same data in single-record format: variable riskdate documents becoming at risk and variable enr_date the date of enrollment in our study. We would stset this dataset by typing

        . stset time, origin(time riskdate) enter(time enr_date) failure(mycode==3)

For a final example, let's return to the multiple-record way of recording our data and say that we started enrolling people in our study on 12jan1998 but that, up until 16feb1998, we do not trust that our records are complete (we had start-up problems). We would stset that dataset by typing

        . stset time, origin(mycode==1) enter(mycode==2 time mdy(2,16,1998))
        > fail(mycode==3)

enter(*varname*==*numlist* time *exp*) is interpreted as

$$\max(\text{time of earliest event in } numlist, exp)$$

Thus persons having mycode == 2 occurring before 16feb1998 are assumed to be under observation from 16feb1998, and those having mycode == 2 thereafter are assumed to be under observation from the time of mycode == 2.

Final exit times

exit() specifies the latest time under which the subject is both under observation and at risk of the failure event. The emphasis is on latest; obviously subjects also exit the data when their data run out.

When you type

        . stset ..., ... failure(outcome==1/3 5) ...

the result is as if you had typed

        . stset ..., ... failure(outcome==1/3 5) exit(failure) ...

which, in turn, is the same as

        . stset ..., ... failure(outcome==1/3 5) exit(outcome==1/3 5) ...

When are people to be removed from the analysis risk pool? When their data end, of course, and when the event 1, 2, 3, or 5 first occurs. How are they to be removed? According to their status at that time. If the event is 1, 2, 3, or 5 at that instant, they exit as a failure. If the event is something else, they exit as censored.

Perhaps events 1, 2, 3, and 5 represent death due to heart disease, and that is what we are studying. Say that outcome == 99 represents death for some other reason. Obviously, once the person dies, she is no longer at risk of dying from heart disease, so we would want to specify

        . stset ..., ... failure(outcome==1/3 5) exit(outcome==1/3 5 99) ...

When we explicitly specify `exit()`, we must list all the reasons for which a person is to be removed other than simply running out of data. Here it would have been a mistake to specify just `exit(99)` because that would have left persons in the analysis risk pool who died for reasons 1, 2, 3, and 5. We would have treated those people as if they were still at risk of dying.

In fact, it probably would not have mattered had we specified `exit(99)` because, once a person is dead, he or she is unlikely to have any subsequent records anyway. By that logic, we did not even have to specify `exit(99)` because death is death and there should be no records following it.

For other kinds of events, however, `exit()` becomes important. Let's assume that the failure event is to be diagnosed with heart disease. A person may surely have records following diagnosis, but even so,

```
. stset ..., ... failure(outcome==22) ...
```

would be adequate because, by not specifying `exit()`, we are accepting the default that `exit()` is equivalent to `failure()`. Once outcome 22 occurs, subsequent records on the subject will be ignored—they constitute the future history of the subject.

Say, however, that we wish to treat as censored persons diagnosed with kidney disease. We would type

```
. stset ..., ... failure(outcome==22) exit(outcome==22 29) ...
```

assuming that outcome = 29 is "diagnosed with kidney disease". It is now of great importance that we specified `exit(outcome==22 29)` and not just `exit(outcome==29)` because, had we omitted code 22, persons would have remained in the analysis risk pool even after the failure event, i.e., being diagnosed with heart disease.

If, in addition, our data were untrustworthy after 22nov1998 (perhaps not all the data have been entered yet), we would type

```
. stset ..., ... failure(outcome==22) exit(outcome==22 29 time
> mdy(11,22,1998)) ...
```

If we type `exit(varname==numlist time exp)`, the exit time is taken to be

$$\min(\text{time of earliest event in } \textit{numlist}, \textit{exp})$$

For some analyses, repeated failures are possible. If you have repeated failure data, you specify the `exit()` option and include whatever reasons, if any, that would cause the person to be removed. If there are no such reasons and you wish to retain all observations for the person, you type

```
. stset ..., ... exit(time .) ...
```

`exit(time .)` specifies that the maximum time a person can be in the risk pool is infinite; thus subjects will not be removed until their data run out.

Intermediate exit and reentry times (gaps)

Gaps arise when a subject is temporarily not under observation. The statistical importance of gaps is that, if failure is death and if the person died during such a gap, he would not have been around to be found again. The solution to this is to remove the person from the risk pool during the observational gap.

To determine that you have gaps, your data must provide starting and ending times for each record. Most datasets provide only ending times, making it impossible to know that you have gaps.

You use time0() to specify the beginning times of records. time0() specifies a mechanical aspect of interpreting the records in the dataset, namely, the beginning of the period spanned by each record. Do not confuse time0() with origin(), which specifies when a subject became at risk, or with enter(), which specifies when a subject first comes under observation.

time0() merely identifies the beginning of the time span covered by each record. Say that we had two records on a subject, the first covering the span (40,49] and the second, (49,57]:

```
         |<— record 1 —>|<- record 2 ->|
                                        -> time
         40             49             57
```

A time0() variable would contain

```
                40 in record 1
                49 in record 2
```

and not, for instance, 40 and 40. A time0() variable varies record by record for a subject.

Most datasets merely provide an end-of-record time value, *timevar*, which you specify by typing stset *timevar*, ... When you have multiple records per subject and you do not specify a time0() variable, stset assumes that each record begins where the previous one left off.

if( ) versus if exp

Both the if *exp* and if(*exp*) options select records for which *exp* is true. We strongly recommend specifying the if() option in preference to if *exp*. They differ in that if *exp* removes data from consideration before calculating beginning and ending times, and other quantities as well. The if() option, on the other hand, sets the restriction after all derived variables are calculated. To appreciate this difference, consider the following multiple-record data:

```
    patno       t        x1       x2      code
       3        7        20        5        14
       3        9        22        5        23
       3       11        21        5        29
```

Consider the difference in results between typing

```
. stset t if x1!=22, failure(code==14)
```

and

```
. stset t,   if(x1!=22) failure(code==14)
```

The first would remove record 2 from consideration at the outset. In constructing beginning and ending times, stset and streset would see

```
    patno       t        x1       x2      code
       3        7        20        5        14
       3       11        21        5        29
```

and would construct the result:

```
                   x1=20              x1=21
                   x2= 5              x2= 5
         |<—————————————>|<——————————————>|
         0               7                11
```

In the second case, the result would be

```
                   x1=20                         x1=21
```

```
              x2= 5                        x2= 5
        |<————————————>|  (gap)  |<————>|
        0              7         9      11
```

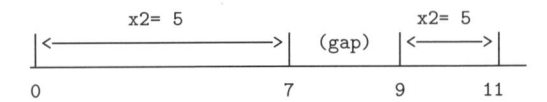

The latter result is correct and the former incorrect because x1 = 21 is not true in the interval $(7, 9)$.

The only reason to specify if *exp* is to ignore errors in the data—observations that would confuse stset and streset—without actually dropping the offending observations from the dataset.

You specify the if() option to ignore information in the data that are not themselves errors. Specifying if() yields the same result as specifying if *exp* on the subsequent st commands after the dataset has been stset.

Past and future records

Consider the hospital ward data that we have seen before:

```
patid    addate    curdate   sex   x1   x2   code
  101  18aug1998  23aug1998    1   10   10    177
  101      .     31aug1998    1   20    8    286
  101      .     08sep1998    1   16   11    208
  101      .     11sep1998    1   11   17    401
  etc.
```

Say that you stset this dataset such that you selected the middle two records. Perhaps you typed

    . stset curdate, id(patid) origin(time addate) enter(code==286) failure(code==208)

The first record for the subject, because it was not selected, is called a *past history record*. Any earlier records that were not selected would also be called past history records.

The last record for the subject, because it was not selected, is called a *future history record*. Any later records that were not selected would also be called future history records.

If you typed

    . streset, past

the first three records for this subject would be selected.

Typing

    . streset, future

would select the last three records for this subject.

If you typed

    . streset, past future

all four records for this subject would be selected.

If you then typed

    . streset

the original two records would be selected, and things would be back just as they were before.

After typing streset, past; streset, future; or streset, past future, you would not want to use any analysis commands. streset did some strange things, especially with the analysis-time variable, to include the extra records. It would be the wrong sample, anyway.

You might, however, want to use certain data-management commands on the data, especially those for creating new variables.

Typically, streset, past is of greater interest than the other commands. Past records—records prior to being at risk or excluded for other reasons—are not supposed to play a role in survival analysis. stset makes sure they do not. But it is sometimes reasonable to ask questions about them such as, was the subject ever on the drug cisplatin? Has the subject ever been married? Did the subject ever have a heart attack?

To answer questions like that, you sometimes want to dig into the past. Typing streset, past makes that easy, and once the past is set, the data can be used with stgen and st_is 2. You might well type the following:

```
. stset curdate, id(patid) origin(addate) enter(code==286) failure(code==208)
. streset, past
. stgen attack = ever(code==177)
. streset
. stcox attack ...
```

Do not be concerned about doing something inappropriate while having the past or future set; st will not let you:

```
. stset curdate, id(patid) origin(time addate) enter(code==286) failure(code==208)
  (output omitted)
. streset, past
  (output omitted)
. stcox x1
you last "streset, past"
you must type "streset" to restore the analysis sample
r(119);
```

Using streset

streset is a useful tool for gently modifying what you have previously stset. Rather than typing the whole stset command, you can type streset followed just by what has changed.

For instance, you might type

```
. stset curdate, id(patid) origin(time addate) enter(code==286) failure(code==208)
```

and then later want to restrict the analysis to subjects who ever have x1>20. You could retype the whole stset command and add ever(x1>20), but it would be easier to type

```
. streset, ever(x1>20)
```

If later you decide you want to remove the restriction, type

```
. streset, ever(.)
```

That is the general rule for resetting options to the default: type '.' as the option's argument.

Be careful using streset because you can make subtle mistakes. In another analysis with another dataset, consider the following:

```
. stset date, fail(code==2) origin(code==1107)
. ...
```

```
. streset date, fail(code==9) origin(code==1422) after(code==1423)
. ...
. streset, fail(code==2) origin(code==1107)
```

If, in the last step, you are trying to get back to the results of the first stset, you will fail. The last streset is equivalent to

```
. stset date, fail(code==9) origin(code==1107) after(code==1423)
```

streset() remembers the previously specified options and uses them if you do not override them. Both stset and streset display the current command line. Make sure that you verify that the command is what you intended.

Performance and multiple-record-per-subject datasets

stset and streset do not drop data; they simply mark data to be excluded from consideration. Some survival-time datasets can be large, although the relevant subsamples are small. In such cases, you can reduce memory requirements and speed execution by dropping the irrelevant observations.

stset and streset mark the relevant observations by creating a variable named _st (it is always named this) containing 1 and 0; _st = 1 marks the relevant observations and _st = 0 marks the irrelevant ones. If you type

```
. drop if _st==0
```

or equivalently

```
. keep if _st==1
```

or equivalently

```
. keep if _st
```

you will drop the irrelevant observations. All st commands produce the same results whether you do this or not. Be careful, however, if you are planning future stsets or stresets. Observations that are irrelevant right now might be relevant later.

One solution to this conundrum is to keep only those observations that are relevant after setting the entire history:

```
. stset date, fail(code==9) origin(code==1422) after(code==1423)
. streset, past future
. keep if _st
. streset
```

As a final note, you may drop the irrelevant observations as marked by _st = 0, but do not drop the _st variable itself. The other st commands expect to find variable _st.

Sequencing of events within t

Consider the following bit of data:

etime	failtime	fail
0	5	1
0	5	0
5	7	1

Note all the different events happening at time 5: the first observation fails, the second is censored, and the third enters.

What does it mean for something to happen at time 5? In particular, is it at least potentially possible for the second observation to have failed at time 5; i.e., was it in the risk group when the first observation failed? How about the third observation? Was it in the risk group, and could it have potentially failed at time 5?

Stata sequences events within a time as follows:

    first,   at time t the failures occur
    then,   at time $t + 0$ the censorings are removed from the risk group
    finally, at time $t + 0 + 0$ the new entries are added to the risk group

Thus, to answer the questions:

Could the second observation have potentially failed at time 5? Yes.

Could the third observation have potentially failed at time 5? No, because it was not yet in the risk group.

By this logic, the following makes no sense:

    etime   failtime   fail
      5   5   1

This would mark a subject as failing before being at risk. It would make no difference if `fail` were 0—the subject would then be marked as being censored too soon. Either way, `stset` would flag this as an error. If you had a subject who entered and immediately exited, you would code this as

    etime   failtime   fail
    4.99   5   1

Weights

`stset` allows you to specify `fweights`, `pweights`, and `iweights`.

`fweights` are Stata's frequency or replication weights. Consider the data

    failtime   load   bearings   count
      100   15   0   3
      140   15   1   2
       97   20   0   1

and the `stset` command

```
. stset failtime [fw=count]

     failure event:  (assumed to fail at time=failtime)
obs. time interval:  (0, failtime]
 exit on or before:  failure
            weight:  [fweight=count]

        3  total obs.
        0  exclusions

        3  physical obs. remaining, equal to
        6  weighted obs., representing
        6  failures in single record/single failure data
      677  total analysis time at risk, at risk from t =         0
                                earliest observed entry t =         0
                                     last observed exit t =       140
```

This combination is equivalent to the expanded data

failtime	load	bearings
100	15	0
100	15	0
100	15	0
140	15	1
140	15	1
97	20	0

and the command

. stset failtime

pweights are Stata's sampling weights—the inverse of the probability that the subject was chosen from the population. pweights are typically integers, but they do not have to be. For instance, you might have

time0	time	died	sex	reps
0	300	1	0	1.50
0	250	0	1	4.50
30	147	1	0	2.25

Here reps is how many patients each observation represents in the underlying population—perhaps when multiplied by 10. The stset command for these data is

. stset time [pw=reps], origin(time time0) failure(died)

For variance calculations, the scale of the pweights does not matter. reps in the 3 observations shown could just as well be 3, 9, and 4.5. Nevertheless, the scale of the pweights is used when you ask for counts. For instance, stsum would report the person-time at risk as

$$(300 - 0)\,1.5 + (250 - 0)\,4.5 + (147 - 30)\,2.25 = 1{,}838.25$$

for the 3 observations shown. stsum would count that $1.5 + 2.25 = 3.75$ persons died, and so the incidence rate for these 3 observations would be $3.75/1{,}838.25 = 0.0020$. The incidence rate is thus unaffected by the scale of the weights. Similarly, the coefficients and confidence intervals reported by, for instance, streg, dist(exponential) would be unaffected. The 95% confidence interval for the incidence rate would be $[0.0003, 0.0132]$, regardless of the scale of the weights.

If these 3 observations were examined unweighted, the incidence rate would be 0.0030 and the 95% confidence interval would be $[0.0007, 0.0120]$.

Finally, stset allows you to set iweights, which are Stata's "importance" weights, but we recommend that you do not. iweights are provided for those who wish to create special effects by manipulating standard formulas. The st commands treat iweights just as they would fweights, although they do not require that the weights be integers, and push their way through conventional variance calculations. Thus results—counts, rates, and variances—depend on the scale of these weights.

Data warnings and errors flagged by stset

When you stset your data, stset runs various checks to verify that what you are setting makes sense. stset refuses to set the data only if, in multiple-record, weighted data, weights are not constant within ID. Otherwise, stset merely warns you about any inconsistencies that it identifies.

Although stset will set the data, it will mark out records that it cannot understand; for instance,

```
. stset curdate, origin(time addate) failure(code==402) id(patid)
              id:  patid
   failure event:  code == 402
obs. time interval: (curdate[_n-1], curdate]
exit on or before: failure
   t for analysis: (time-origin)
           origin: time addate
```

243	total obs.	
1	event time missing (curdate>=.)	PROBABLE ERROR
4	multiple records at same instant (curdate[_n-1]==curdate)	PROBABLE ERROR
238	obs. remaining, representing	
40	subjects	
15	failures in single failure-per-subject data	
1478	total analysis time at risk, at risk from t =	0
	earliest observed entry t =	0
	last observed exit t =	62

You must ensure that the result, after exclusions, is correct.

The warnings stset might issue include

ignored because patid missing	
event time missing	PROBABLE ERROR
entry time missing	PROBABLE ERROR
entry on or after exit (etime>t)	PROBABLE ERROR
obs. end on or before enter()	
obs. end on or before origin()	
multiple records at same instant (t[_n-1]==t)	PROBABLE ERROR
overlapping records (t[_n-1]>entry time)	PROBABLE ERROR
weights invalid	PROBABLE ERROR

stset sets $\_st = 0$ when observations are excluded for whatever reason. Thus observations with any of the above problems can be found among the $\_st = 0$ observations.

Using survival-time data in Stata

In the examples above, we have shown you how Stata expects survival-time data to be recorded. To summarize:

- Each subject's history is represented by 1 or more observations in the dataset.
- Each observation documents a span of time. The observation must contain when the span ends (exit time) and may optionally contain when the span begins (entry time). If the entry time is not recorded, it is assumed to be 0 or, in multiple-record data, the exit time of the subject's previous observation, if there is one. By *previous*, we mean that the data have already been temporally ordered on exit times within subject. The physical order of the observations in your dataset does not matter.
- Each observation documents an outcome associated with the exit time. Unless otherwise specified with failure(), 0 and missing mean censored, and nonzero means failed.
- Each observation contains other variables (called covariates) that are assumed to be constant over the span of time recorded by the observation.

Data rarely arrive in this neatly organized form. For instance, Kalbfleisch and Prentice (2002, 4–5) present heart transplant survival data from Stanford (Crowley and Hu 1977). These data can be converted into the correct st format in at least two ways. The first method is shown in example 11. A second, shorter, method using the st commands is described as an example in the `stsplit` entry.

▷ Example 11

Here we will describe the process that uses the standard Stata commands.

```
. use http://www.stata-press.com/data/r11/stan2
(Heart transplant data)

. describe

Contains data from http://www.stata-press.com/data/r11/stan2.dta
  obs:           103                          Heart transplant data
 vars:             5                          26 Mar 2009 17:48
 size:         1,442 (99.9% of memory free)
```

variable name	storage type	display format	value label	variable label
id	int	%8.0g		Patient Identifier
died	byte	%8.0g		Survival Status (1=dead)
stime	float	%8.0g		Survival Time (Days)
transplant	byte	%8.0g		Heart Transplant
wait	int	%8.0g		Waiting Time

Sorted by:

The data are from 103 patients selected as transplantation candidates. There is one record on each patient, and the important variables, from an st-command perspective, are

id	the patient's ID number
transplant	whether the patient received a transplant
wait	when (after acceptance) the patient received the transplant
stime	when (after acceptance) the patient died or was censored
died	the patient's status at `stime`

To better understand, let's examine two records from this dataset:

```
. list id transplant wait stime died if id==44 | id==16
```

	id	transp~t	wait	stime	died
33.	44	0	0	40	1
34.	16	1	20	43	1

Patient 44 never did receive a new heart; he or she died 40 days after acceptance while still on the waiting list. Patient 16 did receive a new heart—20 days after acceptance—yet died 43 days after acceptance.

Our goal is to turn this into st data that contain the histories of each of these patients. That is, we want records that appear as

id	t1	died	posttran
16	20	0	0
16	43	1	1
44	40	1	0

or, even more explicitly, as

```
    id        t0        t1     died  posttran
    16         0        20       0       0
    16        20        43       1       1
    44         0        40       1       0
```

The new variable, posttran, would be 0 before transplantation and 1 afterward.

Patient 44 would have one record in this new dataset recording that he or she died at time 40 and that posttran was 0 over the entire interval.

Patient 16, however, would have two records: one documenting the duration $(0, 20]$, during which posttran was 0, and one documenting the duration $(20, 43]$, during which posttran was 1.

Our goal is to take the first dataset and convert it into the second, which we can then stset. We make the transformation by using Stata's other data-management commands. One way we could do this is by typing

```
. expand 2 if transplant
(69 observations created)
. by id, sort: gen byte posttran = (_n==2)
. by id: gen t1 = stime if _n==_N
(69 missing values generated)
. by id: replace t1 = wait if _n==1 & transplant
(69 real changes made)
. by id: replace died=0 if _n==1 & transplant
(45 real changes made)
```

expand 2 if transplant duplicated the observations for patients who had transplant $\neq 0$. Considering our two sample patients, we would now have the following data:

```
    id    transp~t     wait     stime     died
    44         0         0        40        1
    16         1        20        43        1
    16         1        20        43        1
```

We would have 1 observation for patient 44 and 2 identical observations for patient 16.

We then by id, sort: gen posttran = (_n==2), resulting in

```
    id    transp~t     wait     stime     died   posttran
    16         1        20        43        1         0
    16         1        20        43        1         1
    44         0         0        40        1         0
```

This type of trickiness is discussed in [U] **13.7 Explicit subscripting**. Statements such as _n==2 produce values 1 (meaning true) and 0 (meaning false), so new variable posttran will contain 1 or 0 depending on whether _n is or is not 2. _n is the observation counter and, combined with by id:, becomes the observation-within-ID counter. Thus we set posttran to 1 on second records and to 0 on all first records.

Finally, we produce the exit-time variable. Final exit time is just stime, and that is handled by the command by id: gen t1 = stime if _n==_N. _n is the observation-within-ID counter and _N is the total number of observations within id, so we just set the last observation on each patient to stime. Now we have

```
    id    transp~t     wait     stime     died   posttran        t1
    16         1        20        43        1         0          .
    16         1        20        43        1         1         43
    44         0         0        40        1         0         40
```

All that is left to do is to fill in `t1` with the value from `wait` on the interim records, meaning `replace t1=wait if it is an interim record`.

There are many ways we could identify the interim records. In the output above, we did it by

```
. by id: replace t1 = wait if _n==1 & transplant
```

meaning that an interim record is a first record of a person who did receive a transplant. More easily, but with more trickery, we could have just said

```
. replace t1=wait if t1>=.
```

because the only values of `t1` left to be filled in are the missing ones. Another alternative would be

```
. by id: replace t1 = wait if _n==1 & _N==2
```

which would identify the first record of two-record pairs. There are many alternatives, but they would all produce the same thing:

id	transp~t	wait	stime	died	posttran	t1
16	1	20	43	1	0	20
16	1	20	43	1	1	43
44	0	0	40	1	0	40

There is one more thing we must do, which is to reset `died` to contain 0 on the interim records:

```
. by id: replace died=0 if _n==1 & transplant
```

The result is

id	transp~t	wait	stime	died	posttran	t1
16	1	20	43	0	0	20
16	1	20	43	1	1	43
44	0	0	40	1	0	40

We now have the desired result and are ready to `stset` our data:

```
. stset t1, failure(died) id(id)
                id:  id
     failure event:  died != 0 & died < .
obs. time interval:  (t1[_n-1], t1]
 exit on or before:  failure

     172  total obs.
       2  multiple records at same instant           PROBABLE ERROR
          (t1[_n-1]==t1)

     170  obs. remaining, representing
     102  subjects
      74  failures in single failure-per-subject data
   31933  total analysis time at risk, at risk from t =         0
                              earliest observed entry t =         0
                               last observed exit t =         1799
```

Well, something went wrong. Two records were excluded. There are few enough data here that we could just list the dataset and look for the problem, but let's pretend otherwise. We want to find the records that, within patient, are marked as exiting at the same time:

```
. by id: gen problem = t1==t1[_n-1]
. sort id died
```

. list id if problem

	id
61.	38

. list id transplant wait stime died posttran t1 if id==38

	id	transp~t	wait	stime	died	posttran	t1
60.	38	1	5	5	0	0	5
61.	38	1	5	5	1	1	5

There is no typographical error in these data—we checked that variables `transplant`, `wait`, and `stime` contain what the original source published. Those variables indicate that patient 38 waited 5 days for a heart transplant, received one on the fifth day, and then died on the fifth day, too.

That makes perfect sense, but not to Stata, which orders events within t as failures, followed by censorings, followed by entries. Reading `t1`, Stata went for this literal interpretation: patient 38 was censored at time 5 with `posttran` $= 0$; then, at time 5, patient 38 died; and then, at time 5, patient 38 reentered the data, but this time with `posttran` $= 1$. That made no sense to Stata.

Stata's sequencing of events may surprise you, but trust us, there are good reasons for it, and really, the ordering convention does not matter. To fix this problem, we just have to put a little time between the implied entry at time 5 and the subsequent death:

```
. replace t1 = 5.1 in 61
(1 real change made)
. list id transplant wait stime died posttran t1 if id==38
```

	id	transp~t	wait	stime	died	posttran	t1
60.	38	1	5	5	0	0	5
61.	38	1	5	5	1	1	5.1

Now the data make sense both to us and to Stata: until time 5, the patient had `posttran` $= 0$; then, at time 5, the value of `posttran` changed to 1; and then, at time 5.1, the patient died.

```
. stset t1, id(id) failure(died)
                id:  id
     failure event:  died != 0 & died < .
obs. time interval:  (t1[_n-1], t1]
 exit on or before:  failure

        172  total obs.
          0  exclusions

        172  obs. remaining, representing
        103  subjects
         75  failures in single failure-per-subject data
    31938.1  total analysis time at risk, at risk from t =         0
                                 earliest observed entry t =         0
                                      last observed exit t =      1799
```

This dataset is now ready for use with all the other st commands. Here is an illustration:

```
. use http://www.stata-press.com/data/r11/stan3, clear
(Heart transplant data)
. stset, noshow
. stsum, by(posttran)
```

		incidence	no. of	Survival time		
posttran	time at risk	rate	subjects	25%	50%	75%
0	5936	.0050539	103	36	149	340
1	26002.1	.0017306	69	39	96	979
total	31938.1	.0023483	103	36	100	979

```
. stcox age posttran surgery year
Iteration 0:   log likelihood = -298.31514
Iteration 1:   log likelihood =  -289.7344
Iteration 2:   log likelihood = -289.53498
Iteration 3:   log likelihood = -289.53378
Iteration 4:   log likelihood = -289.53378
Refining estimates:
Iteration 0:   log likelihood = -289.53378

Cox regression -- Breslow method for ties

No. of subjects =          103                 Number of obs   =       172
No. of failures =           75
Time at risk    =      31938.1
                                               LR chi2(4)      =     17.56
Log likelihood  =   -289.53378                 Prob > chi2     =    0.0015
```

_t	Haz. Ratio	Std. Err.	z	P>|z|	[95% Conf. Interval]	
age	1.030224	.0143201	2.14	0.032	1.002536	1.058677
posttran	.9787243	.3032597	-0.07	0.945	.5332291	1.796416
surgery	.3738278	.163204	-2.25	0.024	.1588759	.8796
year	.8873107	.059808	-1.77	0.076	.7775022	1.012628

◁

Methods and formulas

stset is implemented as an ado-file.

References

Cleves, M. A. 1999. ssa13: Analysis of multiple failure-time data with Stata. *Stata Technical Bulletin* 49: 30–39. Reprinted in *Stata Technical Bulletin Reprints*, vol. 9, pp. 338–349. College Station, TX: Stata Press.

Cleves, M. A., W. W. Gould, R. G. Gutierrez, and Y. Marchenko. 2008. *An Introduction to Survival Analysis Using Stata*. 2nd ed. College Station, TX: Stata Press.

Crowley, J., and M. Hu. 1977. Covariance analysis of heart transplant survival data. *Journal of the American Statistical Association* 72: 27–36.

Hills, M., and B. L. De Stavola. 2006. *A Short Introduction to Stata for Biostatistics*. London: Timberlake.

Kalbfleisch, J. D., and R. L. Prentice. 2002. *The Statistical Analysis of Failure Time Data*. 2nd ed. New York: Wiley.

Also see

[ST] **snapspan** — Convert snapshot data to time-span data

[ST] **stdescribe** — Describe survival-time data

Title

stsplit — Split and join time-span records

Syntax

Split at designated times

    stsplit *newvar* [*if*], {at(*numlist*) | every(*#*)} [*stsplitDT_options*]

Split at failure times

    stsplit [*if*], at(failures) [*stsplitFT_options*]

Join episodes

    stjoin [, censored(*numlist*)]

stsplitDT_options	description
Main	
* at(*numlist*)	split records at specified analysis times
* every(*#*)	split records when analysis time is a multiple of #
after(*spec*)	use time since *spec* for at() or every() rather than time since onset of risk; see *Options*
trim	exclude observations outside of range
† nopreserve	do not save original data; programmer's option

*Either at(*numlist*) or every(*#*) is required with stsplit at designated times.

stsplitFT_options	description
Main	
* at(failures)	split at observed failure times
strata(*varlist*)	restrict splitting to failures within stratum defined by *varlist*
riskset(*newvar*)	create a risk-set ID variable named *newvar*
† nopreserve	do not save original data; programmer's option

*at(failures) is required with stsplit at failure times.

† nopreserve is not shown in the dialog box.
You must stset your dataset using the id() option before using stsplit and stjoin; see [ST] **stset**.

Menu

stsplit

Statistics > Survival analysis > Setup and utilities > Split time-span records

stjoin

Statistics > Survival analysis > Setup and utilities > Join time-span records

Description

stsplit with the at(*numlist*) or every(*#*) option splits episodes into two or more episodes at the implied time points since being at risk or after a time point specified by after(). Each resulting record contains the follow-up on one subject through one time band. Expansion on multiple time scales may be obtained by repeatedly using stsplit. *newvar* specifies the name of the variable to be created containing the observation's category. The new variable records the interval to which each new observation belongs and is bottom coded.

stsplit, at(failures) performs episode splitting at the failure times (per stratum).

stjoin performs the reverse operation, namely, joining episodes back together when such can be done without losing information.

Options for stsplit

Main

at(*numlist*) or every(*#*) is required in syntax one; at(failures) is required for syntax two. These options specify the analysis times at which the records are to be split.

at(5(5)20) splits records at $t = 5$, $t = 10$, $t = 15$, and $t = 20$.

If at([...] max) is specified, max is replaced by a suitably large value. For instance, to split records every five analysis-time units from time zero to the largest follow-up time in our data, we could find out what the largest time value is by typing summarize _t and then explicitly typing it into the at() option, or we could just specify at(0(5)max).

every(*#*) is a shorthand for at(*#*(*#*)max); i.e., episodes are split at each positive multiple of *#*.

after(*spec*) specifies the reference time for at() or every(). Syntax one can be thought of as corresponding to after(*time of onset of risk*), although you cannot really type this. You could type, however, after(time=birthdate) or after(time=marrydate).

spec has syntax

$$\{\texttt{time} \,|\, \texttt{t} \,|\, \texttt{\_t}\} = \{exp \,|\, \min(exp) \,|\, \texttt{asis}(exp)\}$$

where

time specifies that the expression be evaluated in the same time units as *timevar* in stset *timevar*, ... This is the default.

t and _t specify that the expression be evaluated in units of analysis time. t and _t are synonyms; it makes no difference whether you specify one or the other.

exp specifies the reference time. For multiepisode data, *exp* should be constant within subject ID.

min(*exp*) specifies that for multiepisode data, the minimum of *exp* be taken within ID.

asis(*exp*) specifies that for multiepisode data, *exp* be allowed to vary within ID.

trim specifies that observations with values less than the minimum or greater than the maximum value listed in at() be excluded from subsequent analysis. Such observations are not dropped from the data; trim merely sets their value of variable _st to 0 so that they will not be used, yet they can still be retrieved the next time the dataset is stset.

strata(*varlist*) specifies up to five strata variables. Observations with equal values of the variables are assumed to be in the same stratum. strata() restricts episode splitting to failures that occur within the stratum, and memory requirements are reduced when strata are specified.

riskset(*newvar*) specifies the name for a new variable recording the unique risk set in which an episode occurs, and missing otherwise.

The following option is available with stsplit but is not shown in the dialog box:

nopreserve is intended for use by programmers. It speeds the transformation by not saving the original data, which can be restored should things go wrong or if you press *Break*. Programmers often specify this option when they have already preserved the original data. nopreserve does not affect the transformation.

Option for stjoin

censored(*numlist*) specifies values of the failure variable, *failvar*, from stset ..., failure(*failvar*=...) that indicate "no event" (censoring).

If you are using stjoin to rejoin records after stsplit, you do not need to specify censored(). Just do not forget to drop the variable created by stsplit before typing stjoin. See example 4 below.

Neither do you need to specify censored() if, when you stset your dataset, you specified failure(*failvar*) and not failure(*failvar*=...). Then stjoin knows that *failvar* = 0 and *failvar* = . (missing) correspond to no event. Two records can be joined if they are contiguous and record the same data and the first record has *failvar* = 0 or *failvar* = ., meaning no event at that time.

You may need to specify censored(), and you probably do if, when you stset the dataset, you specified failure(*failvar*=...). If stjoin is to join records, it needs to know what events do not count and can be discarded. If the only such event is *failvar* = ., you do not need to specify censored().

Remarks

Remarks are presented under the following headings:

> What stsplit does and why
> Using stsplit to split at designated times
> Time versus analysis time
> Splitting data on recorded ages
> Using stsplit to split at failure times

What stsplit does and why

stsplit splits records into two or more records on the basis of analysis time or on a variable that depends on analysis time, such as age. stsplit takes data like

```
       id     _t0     _t     x1    x2    _d
        1       0     18     12    11     1
```

and produces

id	_t0	_t	x1	x2	_d	tcat
1	0	5	12	11	0	0
1	5	10	12	11	0	5
1	10	18	12	11	1	10

or

id	_t0	_t	x1	x2	_d	agecat
1	0	7	12	11	0	30
1	7	17	12	11	0	40
1	17	18	12	11	1	50

The above alternatives record the same underlying data: subject 1 had x1 = 12 and x2 = 11 during $0 < t \leq 18$, and at $t = 18$, the subject failed.

The difference between the two alternatives is that the first breaks out the analysis times 0–5, 5–10, and 10–20 (although subject 1 failed before $t = 20$). The second breaks out age 30–40, 40–50, and 50–60. You cannot tell from what is presented above, but at $t = 0$, subject 1 was 33 years old.

In our example, that the subject started with one record is not important. The original data on the subject might have been

id	_t0	_t	x1	x2	_d
1	0	14	12	11	0
1	14	18	12	9	1

and then we would have obtained

id	_t0	_t	x1	x2	_d	tcat
1	0	5	12	11	0	0
1	5	10	12	11	0	5
1	10	14	12	11	0	10
1	14	18	12	9	1	10

or

id	_t0	_t	x1	x2	_d	agecat
1	0	7	12	11	0	30
1	7	14	12	11	0	40
1	14	17	12	9	0	40
1	17	18	12	9	1	50

Also we could just as easily have produced records with analysis time or age recorded in single-year categories. That is, we could start with

id	_t0	_t	x1	x2	_d
1	0	14	12	11	0
1	14	18	12	9	1

and produce

id	_t0	_t	x1	x2	_d	tcat
1	0	1	12	11	0	0
1	1	2	12	11	0	1
1	2	3	12	11	0	2
...						

or

id	_t0	_t	x1	x2	_d	agecat
1	0	1	12	11	0	30
1	1	2	12	11	0	31
1	2	3	12	11	0	32
...						

Moreover, we can even do this splitting on more than one variable. Let's go back and start with

```
    id     _t0     _t     x1     x2     _d
     1       0     18     12     11      1
```

Let's split it into the analysis-time intervals 0–5, 5–10, and 10–20, *and* let's split it into 10-year age intervals 30–40, 40–50, and 50–60. The result would be

```
    id     _t0     _t     x1     x2     _d    tcat    agecat
     1       0      5     12     11      0       0        30
     1       5      7     12     11      0       5        30
     1       7     10     12     11      0       5        40
     1      10     17     12     11      0      10        40
     1      17     18     12     11      1      10        50
```

Why would we want to do any of this?

We might want to split on a time-dependent variable, such as age, if we want to estimate a Cox proportional hazards model and include current age among the regressors (although we could instead use `stcox`'s `tvc()` option) or if we want to make tables by age groups (see [ST] **strate**).

Using stsplit to split at designated times

`stsplit`'s syntax to split at designated times is, ignoring other options,

> `stsplit` *newvar* [*if*] , `at`(*numlist*)
>
> `stsplit` *newvar* [*if*] , `at`(*numlist*) `after`(*spec*)

`at()` specifies the analysis times at which records are to be split. Typing `at(5 10 15)` splits records at the indicated analysis times and separates records into the four intervals 0–5, 5–10, 10–15, and 15+.

In the first syntax, the splitting is done on analysis time, t. In the second syntax, the splitting is done on 5, 10, and 15 analysis-time units after the time given by `after(`*spec*`)`.

In either case, `stsplit` also creates *newvar* containing the interval to which each observation belongs. Here *newvar* would contain 0, 5, 10, and 15; it would contain 0 if the observation occurred in the interval 0–5, 5 if the observation occurred in the interval 5–10, and so on. To be precise,

Category	Precise meaning	*newvar* value
0–5	$(-\infty, 5]$	0
5–10	$(5, 10]$	5
10–15	$(10, 15]$	10
15+	$(15, \infty)$	15

If any of the `at()` numbers are negative (which would be allowed only by specifying the `after()` option and would be unusual), the first category is labeled one less than the minimum value specified by `at()`.

Consider the data

```
    id     yr0     yr1    yrborn     x1    event
     1    1990    1995      1960      5       52
     2    1993    1997      1964      3       47
```

In these data, subjects became at risk in yr0. The failure event of interest is event = 47, so we stset our dataset by typing

```
. stset yr1, id(id) origin(time yr0) failure(event==47)
(output omitted)
```

and that results in

id	_t0	_t	yr0	yr1	yrborn	x1	event	_d
1	0	5	1990	1995	1960	5	52	0
2	0	4	1993	1997	1964	3	47	1

In the jargon of st, variables _t0 and _t record the span of each record in analysis-time (t) units. Variables yr0 and yr1 also record the time span, but in time units. Variable _d records 1 for failure and 0 otherwise.

Typing stsplit cat, at(2 4 6 8) would split the records on the basis of analysis time:

```
. stsplit cat, at(2 4 6 8)
(3 observations (episodes) created)
. order id _t0 _t yr0 yr1 yrborn x1 event _d cat
. list id-cat
```

	id	_t0	_t	yr0	yr1	yrborn	x1	event	_d	cat
1.	1	0	2	1990	1992	1960	5	.	0	0
2.	1	2	4	1990	1994	1960	5	.	0	2
3.	1	4	5	1990	1995	1960	5	52	0	4
4.	2	0	2	1993	1995	1964	3	.	0	0
5.	2	2	4	1993	1997	1964	3	47	1	2

The first record, which represented the analysis-time span $(0, 5]$, was split into three records: $(0, 2]$, $(2, 4]$, and $(4, 5]$. The yrborn and x1 values from the single record were duplicated in $(0, 2]$, $(2, 4]$, and $(4, 5]$. The original event variable was changed to missing at $t = 2$ and $t = 4$ because we do not know the value of event; all we know is that event is 52 at $t = 5$. The _d variable was correspondingly set to 0 for $t = 2$ and $t = 4$ because we do know, at least, that the subject did not fail.

stsplit also keeps your original time variables up to date in case you want to streset or re-stset your dataset. yr1 was updated, too.

Now let's go back to our original dataset after we stset it but before we split it,

id	_t0	_t	yr0	yr1	yrborn	x1	event	_d
1	0	5	1990	1995	1960	5	52	0
2	0	4	1993	1997	1964	3	47	1

and consider splitting on age. In 1990, subject 1 is age $1990 - \text{yrborn} = 1990 - 1960 = 30$, and subject 2 is 29. If we type

```
. stsplit acat, at(30 32 34) after(time=yrborn)
```

we will split the data according to

```
         age <= 30      (called acat=0)
    30 < age <= 32      (called acat=30)
    32 < age <= 34      (called acat=32)
    34 < age            (called acat=34)
```

The result will be

```
    id   _t0   _t   yr0   yr1   yrborn   x1   event   _d   acat
    1     0    2    1990  1992  1960     5      .     0    30
    1     2    4    1990  1994  1960     5      .     0    32
    1     4    5    1990  1995  1960     5     52     0    34
    2     0    1    1993  1994  1964     3      .     0     0
    2     1    3    1993  1996  1964     3      .     0    30
    2     3    4    1993  1997  1964     3     47     1    32
```

The original record on subject 1 corresponding to $(0,5]$ was split into $(0,2]$, $(2,4]$, and $(4,5]$ because those are the t values at which age becomes 32 and 34.

You can stsplit the data more than once. Now having these data, if we typed

. stsplit cat, at(2 4 6 8)

the result would be

```
    id   _t0   _t   yr0   yr1   yrborn   x1   event   _d   acat   cat
    1     0    2    1990  1992  1960     5      .     0    30     0
    1     2    4    1990  1994  1960     5      .     0    32     2
    1     4    5    1990  1995  1960     5     52     0    34     4
    2     0    1    1993  1994  1964     3      .     0     0     0
    2     1    2    1993  1995  1964     3      .     0    30     0
    2     2    3    1993  1996  1964     3      .     0    30     2
    2     3    4    1993  1997  1964     3     47     1    32     2
```

Whether we typed

. stsplit acat, at(30 32 34) after(time=yrborn)
. stsplit cat, at(2 4 6 8)

or

. stsplit cat, at(2 4 6 8)
. stsplit acat, at(30 32 34) after(time=yrborn)

would make no difference.

Time versus analysis time

Be careful using the after() option if, when you stset your dataset, you specified stset's scale() option. We say be careful, but actually we mean be appreciative, because stsplit will do just what you would expect if you did not think too hard.

When you split a record on a time-dependent variable, at() is still specified in analysis-time units, meaning units of time/scale().

For instance, if your original data recorded time as Stata dates, i.e., number of days since 1960,

```
    id    date0      date1     birthdate   x1   event
    1    14apr1993  27mar1995  12jul1959    5    52
    ...
```

and you previously stset your dataset by typing

. stset date1, id(id) origin(time date0) scale(365.25) ...

and you now wanted to split on the age implied by birthdate, you would specify the split points in *years* since birth:

. stsplit agecat, at(20(5)60) after(time=birth)

at(20(5)60) means to split the records at the ages, measured in years, of 20, 25, ..., 60.

When you `stset` your dataset, you basically told st how you recorded times (you recorded them as dates) and how to map such times (dates) into analysis time. That was implied by what you typed, and all of st remembers that. Typing

```
. stsplit agecat, at(20(5)60) after(time=birth)
```

tells `stsplit` to split the data on 20, 25, ..., 60 analysis-time units after `birthdate` for each subject.

Splitting data on recorded ages

Recorded ages can sometimes be tricky. Consider the data

id	yr0	yr1	age	x1	event
1	1980	1996	30	5	52
...					

When was age = 30 recorded—1980 or 1996? Put aside that question because things are about to get worse. Say that you `stset` this dataset so that yr0 is the `origin()`,

id	_t0	_t	yr0	yr1	age	x1	event
1	0	16	1980	1996	30	5	52
...							

and then split on analysis time by typing `stsplit cat, at(5(5)20)`. The result would be

id	_t0	_t	yr0	yr1	age	x1	event
1	0	5	1980	1985	30	5	.
1	5	10	1980	1990	30	5	.
1	10	15	1980	1995	30	5	.
1	15	16	1980	1996	30	5	52

Regardless of the answer to the question on when age was measured, age is most certainly not 30 in the newly created records, although you might argue that age at baseline was 30 and that is what you wanted, anyway.

The only truly safe way to deal with ages is to convert them back to birthdates at the outset. Here we would, early on, type

```
. gen bdate = yr1 - age
```
(if age was measured at yr1)

or

```
. gen bdate = yr0 - age
```
(if age was measured at yr0)

In fact, `stsplit` tries to protect you from making age errors. Suppose that you did not do as we just recommended. Say that age was measured at yr1, and you typed, knowing that `stsplit` wants a date,

```
. stsplit acat, at(20(5)50) after(time= yr1-age)
```

on these already `stsplit` data. `stsplit` will issue the error message "after() should be constant within id". To use the earliest date, you need to type

```
. stsplit acat, at(20(5)50) after(time= min(yr1-age))
```

Nevertheless, be aware that when you `stsplit` data, if you have recorded ages in your data, and if the records were not already split to control for the range of those ages, then age values, just like all the other variables, are carried forward and no longer reflect the age of the newly created record.

▷ Example 1: Splitting on age

Consider the data from a heart disease and diet survey. The data arose from a study described more fully in Morris, Marr, and Clayton (1977) and analyzed in Clayton and Hills (1993). (Their results differ slightly from ours because the dataset has been updated.)

```
. use http://www.stata-press.com/data/r11/diet
(Diet data with dates)
. describe
Contains data from http://www.stata-press.com/data/r11/diet.dta
  obs:           337                          Diet data with dates
 vars:            11                          1 May 2009 19:01
 size:        17,861 (99.8% of memory free)
```

variable name	storage type	display format	value label	variable label
id	float	%9.0g		Subject identity number
fail	int	%8.0g		Outcome (CHD = 1 3 13)
job	int	%8.0g		Occupation
month	byte	%8.0g		month of survey
energy	float	%9.0g		Total energy (1000kcals/day)
height	float	%9.0g		Height (cm)
weight	float	%9.0g		Weight (kg)
hienergy	float	%9.0g		Indicator for high energy
doe	double	%td		Date of entry
dox	double	%td		Date of exit
dob	double	%td		Date of birth

```
Sorted by: id
```

In this dataset, the outcome variable, fail, has been coded as 0, 1, 3, 5, 12, 13, 14, and 15. Codes 1, 3, and 13 indicated coronary heart disease (CHD), other nonzero values code other events such as cancer, and 0 is used to mean "no event" at the end of the study.

The variable hienergy is coded 1 if the total energy consumption is more than 2.75 Mcal and 0 otherwise.

We would like to expand the data, using age as the time scale with 10-year age bands. We do this by first stsetting the dataset, specifying the date of birth as the origin.

```
. stset dox, failure(fail) origin(time dob) enter(time doe) scale(365.25) id(id)
                id:  id
     failure event:  fail != 0 & fail < .
obs. time interval:  (dox[_n-1], dox]
 enter on or after:  time doe
 exit on or before:  failure
    t for analysis:  (time-origin)/365.25
            origin:  time dob

      337  total obs.
        0  exclusions

      337  obs. remaining, representing
      337  subjects
       80  failures in single failure-per-subject data
 4603.669  total analysis time at risk, at risk from t =         0
                              earliest observed entry t =  30.07529
                                   last observed exit t =  69.99863
```

The origin is set to date of birth, making time-since-birth analysis time, and the scale is set to 365.25, so that time since birth is measured in years.

Let's list a few records and verify that the analysis-time variables _t0 and _t are indeed recorded as we expect:

```
. list id dob doe dox fail _t0 _t if id==1 | id==34
```

	id	dob	doe	dox	fail	_t0	_t
1.	1	04jan1915	16aug1964	01dec1976	0	49.615332	61.908282
34.	34	12jun1899	16apr1959	31dec1966	3	59.841205	67.550992

We see that patient 1 was 49.6 years old at time of entry into our study and left at age 61.9. Patient 34 entered the study at age 59.8 and exited the study with CHD at age 67.6.

Now we can split the data by age:

```
. stsplit ageband, at(40(10)70)
(418 observations (episodes) created)
```

stsplit added 418 observations to the dataset in memory and generated a new variable, ageband, which identifies each observation's age group.

```
. list id _t0 _t ageband fail height if id==1 | id==34
```

	id	_t0	_t	ageband	fail	height
1.	1	49.615332	50	40	.	175.387
2.	1	50	60	50	.	175.387
3.	1	60	61.908282	60	0	175.387
61.	34	59.841205	60	50	.	177.8
62.	34	60	67.550992	60	3	177.8

The single record for the subject with id = 1 has expanded to three records. The first refers to the age band 40–49, coded 40, and the subject spends _t − _t0 = 0.384668 years in this band. The second refers to the age band 50–59, coded 50, and the subject spends 10 years in this band, and so on. The follow-up in each of the three bands is censored (fail = .). The single record for the subject with id = 34 is expanded to two age bands; the follow-up for the first band was censored (fail = .), and the follow-up for the second band ended in CHD (fail = 3).

The values for variables that do not change with time, such as height, are simply repeated in the new records. This can lead to much larger datasets after expansion. Dropping unneeded variables before using split may be necessary.

◁

▷ Example 2: Splitting on age and time in study

To use stsplit to expand the records on two time scales simultaneously, such as age and time in study, we can first expand on the age scale as described in example 1, and then on the time-in-study scale with the command

(Continued on next page)

```
. stsplit timeband, at(0(5)25) after(time=doe)
(767 observations (episodes) created)

. list id _t0 _t ageband fail if id==1 | id==34
```

	id	_t0	_t	ageband	fail
1.	1	49.615332	50	40	.
2.	1	50	54.615332	50	.
3.	1	54.615332	59.615332	50	.
4.	1	59.615332	60	50	.
5.	1	60	61.908282	60	0
111.	34	59.841205	60	50	.
112.	34	60	64.841205	60	.
113.	34	64.841205	67.550992	60	3

By splitting the data by using two time scales, we partitioned the data into time cells corresponding to a *Lexis diagram* as described, for example, in Clayton and Hills (1993). Also see Keiding (1998) for an overview of Lexis diagrams. Each new observation created by splitting the data records the time that the individual spent in a Lexis cell. We can obtain the time spent in the cell by calculating the difference _t − _t0. For example, the subject with id = 1 spent 0.384668 years (50 − 49.615332) in the cell corresponding to age 40–49 and study time 0–5, and 4.615332 years (54.615332 − 50) in the cell for age 50–59 and study time 0–5.

We can also do these expansions in reverse order, that is, split first on study time and then on age.

◁

▷ Example 3: Explanatory variables that change with time

In the previous examples, time, in the form of age or time in study, is the explanatory variable to be studied or controlled for, but in some studies other explanatory variables also vary with time. The stsplit command can sometimes be used to expand the records so that in each new record such an explanatory variable is constant over time. For example, in the Stanford heart data (see [ST] stset), we would like to split the data and generate the explanatory variable posttran, which takes the value 0 before transplantation and 1 thereafter. The follow-up must therefore be divided into time before transplantation and time after.

We first generate for each observation an entry time and an exit time that preserve the correct follow-up time in such a way that the time of transplants is the same for all individuals. By summarizing wait, the time to transplant, we obtain its maximum value of 310. By selecting a value greater than this maximum, say, 320, we now generate two new variables:

```
. use http://www.stata-press.com/data/r11/stanford, clear
(Heart transplant data)

. generate enter = 320 - wait

. generate exit = 320 + stime
```

We have created a new artificial time scale where all transplants are coded as being performed at time 320. By defining enter and exit in this manner, we maintain the correct total follow-up time for each patient. We now stset and stsplit the data:

```
. stset exit, enter(time enter) failure(died) id(id)

                 id:  id
      failure event:  died != 0 & died < .
obs. time interval:  (exit[_n-1], exit]
 enter on or after:  time enter
 exit on or before:  failure
```

```
      103  total obs.
        0  exclusions

      103  obs. remaining, representing
      103  subjects
       75  failures in single failure-per-subject data
  34589.1  total analysis time at risk, at risk from t =         0
                                  earliest observed entry t =   10
                                      last observed exit t = 2119
```

```
. stsplit posttran, at(0,320)
(69 observations (episodes) created)
. replace posttran=0 if transplant==0
(34 real changes made)
. replace posttran=1 if posttran==320
(69 real changes made)
```

We replaced posttran in the last command so that it is now a 0/1 indicator variable. We can now generate our follow-up time, t1, as the difference between our analysis-time variables, list the data, and stset the dataset.

```
. generate  t1 =_t - _t0
. list id enter exit _t0 _t posttran if id==16 | id==44
```

	id	enter	exit	_t0	_t	posttran
41.	44	320	360	320	360	0
42.	16	300	320	300	320	0
43.	16	300	363	320	363	1

```
. stset t1, failure(died) id(id)

                 id:  id
      failure event:  died != 0 & died < .
obs. time interval:  (t1[_n-1], t1]
 exit on or before:  failure
```

```
      172  total obs.
        0  exclusions

      172  obs. remaining, representing
      103  subjects
       75  failures in single failure-per-subject data
  31938.1  total analysis time at risk, at risk from t =         0
                                  earliest observed entry t =    0
                                      last observed exit t = 1799
```

Using stsplit to split at failure times

To split data at failure times, you would use stsplit with the following syntax, ignoring other options:

$$\text{stsplit } [\textit{if}] , \text{ at(failures)}$$

This form of episode splitting is useful for Cox regression with time-varying covariates. Splitting at the failure times is useful because of a property of the maximum partial-likelihood estimator for a Cox regression model: the likelihood is evaluated only at the times at which failures occur in the data, and the computation depends only on the risk pools at those failure times. Changes in covariates between failure times do not affect estimates for a Cox regression model. Thus, to fit a model with time-varying covariates, all you have to do is define the values of these time-varying covariates at all failure times at which a subject was at risk (Collett 2003, chap. 8). After splitting at failure times, you define time-varying covariates by referring to the system variable _t (analysis time) or the *timevar* variable used to stset the data.

After splitting at failure times, all st commands still work fine and produce the same results as before splitting. To fit parametric models with time-varying covariates, it does not suffice to specify covariates at failure times. Stata can fit "piecewise constant" models by manipulating data using stsplit, {at()|every()}.

◁

▷ Example 4: Splitting on failure times to test the proportional-hazards assumption

Collett (2003, 187–190) presents data on 26 ovarian cancer patients who underwent two different chemotherapy protocols after a surgical intervention. Here are a few of the observations:

```
. use http://www.stata-press.com/data/r11/ocancer, clear
. list patient time cens treat age rdisea in 1/6, separator(0)
```

	patient	time	cens	treat	age	rdisea
1.	1	156	1	1	66	2
2.	2	1040	0	1	38	2
3.	3	59	1	1	72	2
4.	4	421	0	2	53	2
5.	5	329	1	1	43	2
6.	6	769	0	2	59	2

The treat variable indicates the chemotherapy protocol administered, age records the age of the patient at the beginning of the treatment, and rdisea records each patient's residual disease after surgery. After stsetting this dataset, we fit a Cox proportional-hazards regression model on age and treat to ascertain the effect of treatment, controlling for age.

```
. stset time, failure(cens) id(patient)
                id:  patient
     failure event:  cens != 0 & cens < .
obs. time interval:  (time[_n-1], time]
 exit on or before:  failure
```

```
       26  total obs.
        0  exclusions

       26  obs. remaining, representing
       26  subjects
       12  failures in single failure-per-subject data
    15588  total analysis time at risk, at risk from t =         0
                                   earliest observed entry t =   0
                                    last observed exit t =    1227
```

```
. stcox age treat, nolog nohr
         failure _d:  cens
   analysis time _t:  time
                id:   patient

Cox regression -- no ties

No. of subjects  =         26                 Number of obs   =        26
No. of failures  =         12
Time at risk     =      15588
                                              LR chi2(2)      =     15.82
Log likelihood   =  -27.073767                Prob > chi2     =    0.0004
```

_t	Coef.	Std. Err.	z	P>|z|	[95% Conf. Interval]	
age	.1465698	.0458537	3.20	0.001	.0566982	.2364415
treat	-.7959324	.6329411	-1.26	0.209	-2.036474	.4446094

One way to test the proportional-hazards assumption is to include in the model a term for the interaction between age and time at risk, which is a continuously varying covariate. This can be easily done by first splitting the data at the failure times and then generating the interaction term.

```
. stsplit, at(failures)
(12 failure times)
(218 observations (episodes) created)

. generate tage = age * _t

. stcox age treat tage, nolog nohr
         failure _d:  cens
   analysis time _t:  time
                id:   patient

Cox regression -- no ties

No. of subjects  =         26                 Number of obs   =       244
No. of failures  =         12
Time at risk     =      15588
                                              LR chi2(3)      =     16.36
Log likelihood   =  -26.806607                Prob > chi2     =    0.0010
```

_t	Coef.	Std. Err.	z	P>|z|	[95% Conf. Interval]	
age	.2156499	.1126093	1.92	0.055	-.0050602	.43636
treat	-.6635945	.6695492	-0.99	0.322	-1.975887	.6486978
tage	-.0002031	.0002832	-0.72	0.473	-.0007582	.000352

Other time-varying interactions of age and time at risk could be generated. For instance,

        . generate lntage = age * ln(_t)
        . generate dage = age * (_t >= 500)

Although in most analyses in which we include interactions we also include main effects, if we include in a Cox regression a multiplicative interaction between analysis time (or any transformation) and some covariate, we should not include the analysis time as a covariate in stcox. The analysis time is constant within each risk set, and hence, its effect is not identified.

❑ Technical note

If our interest really were just in performing this test of the proportional-hazards assumption, we would not have had to use stsplit at all. We could have just typed

        . stcox age treat, tvc(age)

to have fit a model including $t*$age, and if we wanted instead to include $\ln(t)*$age or age$*t \geq 500$, we could have typed

        . stcox age treat, tvc(age) texp(ln(_t))
        . cstoc age treat, tvc(age) texp(_t >= 500)

Still, it is worth understanding how stsplit could be used to obtain the same results for instances when stcox's tvc() and texp() options are not rich enough to handle the desired specification.

❑

Assume that we want to control for rdisea as a stratification variable. If the data are already split at all failure times, we can proceed with

        . stcox age treat tage, strata(rdisea)

If the data are not yet split, and memory is scarce, then we could just split the data at the failure times within the respective stratum. That is, with the original data in memory, we could type

        . stset time, failure(cens) id(patient)
        . stsplit, at(failures) strata(rdisea)
        . generate tage = age * _t
        . stcox treat age tage, strata(rdisea)

This would save memory by reducing the size of the split dataset.

◁

❑ Technical note

Of course, the above model could also be obtained by typing

        . stcox age treat, tvc(age) strata(rdisea)

without splitting the data.

❑

▷ Example 5: Cox regression versus conditional logistic regression

Cox regression with the "exact partial" method of handling ties is tightly related to conditional logistic regression. In fact, we can perform Cox regression via `clogit`, as illustrated in the following example using Stata's cancer data. First, let's fit the Cox model.

```
. use http://www.stata-press.com/data/r11/cancer, clear
(Patient Survival in Drug Trial)
. generate id =_n
. stset studytime, failure(died) id(id)
                id:  id
     failure event:  died != 0 & died < .
obs. time interval:  (studytime[_n-1], studytime]
 exit on or before:  failure

       48  total obs.
        0  exclusions

       48  obs. remaining, representing
       48  subjects
       31  failures in single failure-per-subject data
      744  total analysis time at risk, at risk from t =         0
                              earliest observed entry t =         0
                                   last observed exit t =        39
. stcox age drug, nolog nohr exactp
         failure _d:  died
   analysis time _t:  studytime
                 id:  id
Cox regression -- exact partial likelihood
No. of subjects =           48                Number of obs   =        48
No. of failures =           31
Time at risk    =          744
                                              LR chi2(2)      =     38.13
Log likelihood  =    -73.10556                Prob > chi2     =    0.0000
```

_t _d	Coef.	Std. Err.	z	P>|z|	[95% Conf. Interval]
age	.1169906	.0374955	3.12	0.002	.0435008 .1904805
drug	-1.664873	.3437487	-4.84	0.000	-2.338608 -.9911376

We will now perform the same analysis by using `clogit`. To do this, we first split the data at failure times, specifying the `riskset()` option so that a risk set identifier is added to each observation. We then fit the conditional logistic regression, using _d as the outcome variable and the risk set identifier as the grouping variable.

(Continued on next page)

```
. stsplit, at(failures) riskset(RS)
(21 failure times)
(534 observations (episodes) created)
. clogit _d age drug, group(RS) nolog
note: multiple positive outcomes within groups encountered.
```

Conditional (fixed-effects) logistic regression   Number of obs   =   573
                                                    LR chi2(2)   =   38.13
                                                    Prob > chi2   =   0.0000
Log likelihood = -73.10556   Pseudo R2   =   0.2069

_d	Coef.	Std. Err.	z	P>|z|	[95% Conf. Interval]	
age	.1169906	.0374955	3.12	0.002	.0435008	.1904805
drug	-1.664873	.3437487	-4.84	0.000	-2.338608	-.9911376

◁

> Example 6: Joining data that have been split with stsplit

Let's return to the first example. We split the diet data into age bands, using the following commands:

```
. use http://www.stata-press.com/data/r11/diet, clear
(Diet data with dates)
. stset dox, failure(fail) origin(time dob) enter(time doe) scale(365.25) id(id)
 (output omitted)
. stsplit ageband, at(40(10)70)
(418 observations (episodes) created)
```

We can rejoin the data by typing stjoin:

```
. stjoin
(option censored(0) assumed)
(0 obs. eliminated)
```

Nothing happened! stjoin will combine records that are contiguous and record the same data. Here, when we split the data, stsplit created the new variable ageband, and that variable takes on different values across the split observations. Remember to drop the variable that stsplit creates:

```
. drop ageband
. stjoin
(option censored(0) assumed)
(418 obs. eliminated)
```

◁

Methods and formulas

stsplit and stjoin are implemented as ado-files.

> Wilhelm Lexis (1837–1914) was born near Aachen in Germany. He studied law, mathematics, and science at the University of Bonn and developed interests in the social sciences during a period in Paris. Lexis held posts at universities in Strassburg (now Strasbourg, in France), Dorpat (now Tartu, in Estonia), Freiburg, Breslau (now Wroclaw, in Poland), and Göttingen. During this peripatetic career, he carried out original work in statistics on the analysis of dispersion, foreshadowing the later development of chi-squared and analysis of variance.

Acknowledgments

stsplit and stjoin are extensions of lexis by David Clayton, Cambridge Institute for Medical Research, and Michael Hills, London School of Hygiene and Tropical Medicine (retired) (Clayton and Hills 1995). The original stsplit and stjoin commands were written by Jeroen Weesie, Utrecht University, The Netherlands (Weesie 1998a, 1998b), as was the revised stsplit command in this release.

References

Clayton, D. G., and M. Hills. 1993. *Statistical Models in Epidemiology*. Oxford: Oxford University Press.

———. 1995. ssa7: Analysis of follow-up studies. *Stata Technical Bulletin* 27: 19–26. Reprinted in *Stata Technical Bulletin Reprints*, vol. 5, pp. 219–227. College Station, TX: Stata Press.

Cleves, M. A., W. W. Gould, R. G. Gutierrez, and Y. Marchenko. 2008. *An Introduction to Survival Analysis Using Stata*. 2nd ed. College Station, TX: Stata Press.

Collett, D. 2003. *Modelling Survival Data in Medical Research*. 2nd ed. London: Chapman & Hall/CRC.

Hertz, S. 2001. Wilhelm Lexis. In *Statisticians of the Centuries*, ed. C. C. Heyde and E. Seneta, 204–207. New York: Springer.

Keiding, N. 1998. Lexis diagrams. In *Encyclopedia of Biostatistics*, ed. P. Armitage and T. Colton, 2844–2850. New York: Wiley.

Lexis, W. H. 1875. *Einleitung in die Theorie der Bevölkerungsstatistik*. Strassburg: Trübner.

Mander, A. 1998. gr31: Graphical representation of follow-up by time bands. *Stata Technical Bulletin* 45: 14–17. Reprinted in *Stata Technical Bulletin Reprints*, vol. 8, pp. 50–53. College Station, TX: Stata Press.

Morris, J. N., J. W. Marr, and D. G. Clayton. 1977. Diet and heart: A postscript. *British Medical Journal* 19: 1307–1314.

Weesie, J. 1998a. ssa11: Survival analysis with time-varying covariates. *Stata Technical Bulletin* 41: 25–43. Reprinted in *Stata Technical Bulletin Reprints*, vol. 7, pp. 268–292. College Station, TX: Stata Press.

———. 1998b. dm62: Joining episodes in multi-record survival time data. *Stata Technical Bulletin* 45: 5–6. Reprinted in *Stata Technical Bulletin Reprints*, vol. 8, pp. 27–28. College Station, TX: Stata Press.

Also see

[ST] **stset** — Declare data to be survival-time data

Title

stsum — Summarize survival-time data

Syntax

stsum [*if*] [*in*] [, by(*varlist*) <u>noshow</u>]

You must stset your data before using stsum; see [ST] **stset**.

by is allowed; see [D] **by**.

fweights, iweights, and pweights may be specified using stset; see [ST] **stset**.

Menu

Statistics > Survival analysis > Summary statistics, tests, and tables > Summarize survival-time data

Description

stsum presents summary statistics: time at risk; incidence rate; number of subjects; and the 25th, 50th, and 75th percentiles of survival time.

stsum can be used with single- or multiple-record or single- or multiple-failure st data.

Options

\[Main \]

by(*varlist*) requests separate summaries for each group along with an overall total. Observations are in the same group if they have equal values of the variables in *varlist*. *varlist* may contain any number of string or numeric variables.

noshow prevents stsum from showing the key st variables. This option is seldom used because most people type stset, show or stset, noshow to set whether they want to see these variables mentioned at the top of the output of every st command; see [ST] **stset**.

Remarks

Remarks are presented under the following headings:

> *Single-failure data*
> *Multiple-failure data*

Single-failure data

Here is an example of stsum with single-record survival data:

```
. use http://www.stata-press.com/data/r11/page2
. stset, noshow
. stsum
```

	time at risk	incidence rate	no. of subjects	Survival time 25%	50%	75%
total	9118	.0039482	40	198	232	261

```
. stsum, by(group)
```

group	time at risk	incidence rate	no. of subjects	Survival time 25%	50%	75%
1	4095	.0041514	19	190	216	234
2	5023	.0037826	21	232	233	280
total	9118	.0039482	40	198	232	261

stsum works equally well with multiple-record survival data. Here is a summary of the multiple-record Stanford heart transplant data introduced in [ST] **stset**:

```
. use http://www.stata-press.com/data/r11/stan3
(Heart transplant data)
. stsum
       failure _d:  died
   analysis time _t:  t1
              id:  id
```

	time at risk	incidence rate	no. of subjects	Survival time 25%	50%	75%
total	31938.1	.0023483	103	36	100	979

stsum with the **by()** option may produce results with multiple-record data that, at first, you may think are in error.

```
. stsum, by(posttran) noshow
```

posttran	time at risk	incidence rate	no. of subjects	Survival time 25%	50%	75%
0	5936	.0050539	103	36	149	340
1	26002.1	.0017306	69	39	96	979
total	31938.1	.0023483	103	36	100	979

For the time at risk, $5,936 + 26,002.1 = 31,938.1$, but, for the number of subjects, $103 + 69 \neq 103$. The **posttran** variable is not constant for the subjects in this dataset:

```
. stset, noshow
. stvary posttran
                  subjects for whom the variable is
```

variable	constant	varying	never missing	always missing	sometimes missing
posttran	34	69	103	0	0

In this dataset, subjects have one or two records. All subjects were eligible for heart transplantation. They have one record if they die or are lost because of censoring before transplantation, and they have two records if the operation was performed. Then the first record records their survival up to transplantation and the second records their subsequent survival. **posttran** is 0 in the first record and 1 in the second.

Thus all 103 subjects have records with **posttran** = 0, and when **stsum** reported results for this group, it summarized the pretransplantation survival. The incidence of death was 0.005, and median survival time was 149 days.

The `posttran = 1` line of `stsum`'s output summarizes the posttransplantation survival: 69 patients underwent transplantation, incidence of death was 0.002, and median survival time was 96 days. For these data, this is not 96 more days, but 96 days in total. That is, the clock was not reset at transplantation. Thus, without attributing cause, we can describe the differences between the groups as an increased hazard of death at early times followed by a decreased hazard later.

Multiple-failure data

If you simply type `stsum` with multiple-failure data, the reported survival time is the survival time to the first failure, assuming that the hazard function is not indexed by number of failures.

Here we have some multiple-failure data:

```
. use http://www.stata-press.com/data/r11/mfail2
. st
-> stset t, id(id) failure(d) time0(t0) exit(time .) noshow
             id:  id
  failure event:  d != 0 & d < .
obs. time interval:  (t0, t]
  exit on or before:  time .
. stsum
```

	time at risk	incidence rate	no. of subjects	Survival time 25%	50%	75%
total	435444	.0018556	926	201	420	703

To understand this output, let's also obtain output for each failure separately:

```
. stgen nf = nfailures()
. stsum, by(nf)
```

nf	time at risk	incidence rate	no. of subjects	Survival time 25%	50%	75%
0	263746	.0020057	926	196	399	604
1	121890	.0018131	529	252	503	816
2	38807	.0014946	221	415	687	.
3	11001	0	58	.	.	.
total	435444	.0018556	926	201	420	703

The `stgen` command added, for each subject, a variable containing the number of previous failures. For a subject, up to and including the first failure, `nf` is 0. Then `nf` is 1 up to and including the second failure, and then it is 2, and so on; see [ST] **stgen**.

The first line of the output, corresponding to $nf = 0$, states that among those who had experienced no failures yet, the incidence rate for (first) failure is 0.0020. The distribution of the time to the first failure is as shown.

Similarly, the second line, corresponding to $nf = 1$, is for those who have already experienced one failure. The incidence rate for (second) failures is 0.0018, and the distribution of time of (second) failures is as shown.

When we simply typed `stsum`, we obtained the same information shown as the total line of the more detailed output. The total incidence rate is easy to interpret, but what is the "total" survival-time distribution? It is an estimate of the distribution of the time to the first failure assuming that the hazard function $h(t)$ is the same across failures—that the second failure is no different from the first failure. This is an odd definition of "same" because the clock, t, is not reset in $h(t)$. What is the hazard of a failure—any failure—at time t? The answer is $h(t)$.

Another definition of "same" would have it that the hazard of a failure is given by $h(\tau)$, where τ is the time since last failure—that the process repeats. These definitions are different unless $h()$ is a constant function of t (τ).

So let's examine these multiple-failure data under the process-replication idea. The key variables in these st data are id, t0, t, and d:

```
. st
-> stset t, id(id) failure(d) time0(t0) exit(time .) noshow
                id:  id
     failure event:  d != 0 & d < .
obs. time interval:  (t0, t]
 exit on or before:  time .
```

Our goal is, for each subject, to reset t0 and t to 0 after every failure event. We are going to have to trick Stata, or at least trick stset, because it will not let us set data where the same subject has multiple records summarizing the overlapping periods. So, the trick is to create a new id variable that is different for every ID–nf combination (remember, nf is the variable we previously created that records the number of prior failures). Then all the "new" subjects can have their clocks start at time 0:

```
. egen newid = group(id nf)
. sort newid t
. by newid: replace t = t - t0[1]
(808 real changes made)
. by newid: gen newt0 = t0 - t0[1]
. stset t, failure(d) id(newid) time0(newt0)
                id:  newid
     failure event:  d != 0 & d < .
obs. time interval:  (newt0, t]
 exit on or before:  failure

     1734  total obs.
        0  exclusions

     1734  obs. remaining, representing
     1734  subjects
      808  failures in single failure-per-subject data
   435444  total analysis time at risk, at risk from t =         0
                             earliest observed entry t =         0
                                  last observed exit t =       797
```

stset no longer thinks that we have multiple-failure data. Whereas with id, subjects had multiple failures, newid gives a unique identity to each ID–nf combination. Each "new" subject has at most one failure.

```
. stsum, by(nf)
         failure _d:  d
   analysis time _t:  t
                id:  newid
```

nf	time at risk	incidence rate	no. of subjects	Survival time		
				25%	50%	75%
0	263746	.0020057	926	196	399	604
1	121890	.0018131	529	194	384	580
2	38807	.0014946	221	210	444	562
3	11001	0	58	.	.	.
total	435444	.0018556	1734	201	404	602

Compare this table with the one we previously obtained. The incidence rates are the same, but the survival times differ because now we measure the times from one failure to the next. Previously we measured the time from a fixed point. The time between events in these data appears to be independent of event number.

❑ Technical note

The method shown for converting multiple-failure data to replicated-process single-event failure data is completely general. The generic outline of the conversion process is

```
. stgen nf = nfailures()
. egen newid = group(id nf)
. sort newid t
. by newid: replace t = t - t0[1]
. by newid: gen newt0 = t0 - t0[1]
. stset t, failure(d) id(newid) t0(newt0)
```

where *id*, *t*, *t0*, and *d* are the names of your key survival-time variables.

Once you have done this to your data, you need exercise only one caution. If, in fitting models with stcox, streg, etc., you wish to obtain robust estimates of variance, you should include the vce(cluster *id*) option.

When you specify the vce(robust) option, stcox, streg, etc., assume that you mean vce(cluster *stset_id_variable*), which, here, will be vce(cluster newid). The data, however, are really more clustered than that. Two "subjects" with different newid values may, in fact, be the same real subject. vce(cluster *id*) is what is appropriate.

❑

Saved results

stsum saves the following in r():

Scalars
r(p25)	25th percentile	r(risk)	time at risk
r(p50)	50th percentile	r(ir)	incidence rate
r(p75)	75th percentile	r(N_sub)	number of subjects

Methods and formulas

stsum is implemented as an ado-file.

The 25th, 50th, and 75th percentiles of survival times are obtained from $S(t)$, the Kaplan–Meier product-limit estimate of the survivor function. The 25th percentile, for instance, is obtained as the minimum value of t such that $S(t) \leq 0.75$.

Also see

[ST] **stdescribe** — Describe survival-time data

[ST] **stgen** — Generate variables reflecting entire histories

[ST] **stir** — Report incidence-rate comparison

[ST] **sts** — Generate, graph, list, and test the survivor and cumulative hazard functions

[ST] **stset** — Declare data to be survival-time data

[ST] **stvary** — Report whether variables vary over time

[ST] **stci** — Confidence intervals for means and percentiles of survival time

[ST] **stptime** — Calculate person-time, incidence rates, and SMR

Title

> **sttocc** — Convert survival-time data to case–control data

Syntax

> sttocc [*varlist*] [, *options*]

options	description
Main	
match(*matchvarlist*)	match cases and controls on analysis time and specified categorical variables; default is to match on analysis time only
number(*#*)	use *#* controls for each case; default is **number(1)**
nodots	suppress displaying dots during calculation
Options	
generate(*case set time*)	new variable names; default is _case, _set, and _time

You must stset your data before using sttocc; see [ST] **stset**.
fweights, iweights, and pweights may be specified using stset; see [ST] **stset**.

Menu

Statistics > Survival analysis > Setup and utilities > Convert survival-time data to case-control data

Description

sttocc generates a nested case–control study dataset from a cohort-study dataset by sampling controls from the risk sets. For each case, the controls are chosen randomly from those members of the cohort who are at risk at the failure time of the case. That is, the resulting case–control sample is matched with respect to analysis time—the time scale used to compute risk sets. The following variables are added to the dataset:

_case	coded 0 for controls, 1 for cases
_set	case–control ID; matches cases and controls that belong together
_time	analysis time of the case's failure

The names of these three variables can be changed by specifying the **generate()** option. *varlist* defines variables that, in addition to those used in the creation of the case–control study, will be retained in the final dataset. If *varlist* is not specified, all variables are carried over into the resulting dataset.

When the resulting dataset is analyzed as a matched case–control study, odds ratios will estimate corresponding rate-ratio parameters in the proportional hazards model for the cohort study.

Randomness in the matching is obtained using Stata's **runiform()** function. To ensure that the sample truly is random, you should set the random-number seed; see [R] **set seed**.

Options

[Main]

match(*matchvarlist*) specifies more categorical variables for matching controls to cases. When match() is not specified, cases and controls are matched with respect to time only. If match(*matchvarlist*) is specified, the cases will also be matched by *matchvarlist*.

number(#) specifies the number of controls to draw for each case. The default is 1, even though this is not a sensible choice.

nodots requests that dots not be placed on the screen at the beginning of each case–control group selection. By default, dots are displayed to show progress.

[Options]

generate(*case set time*) specifies variable names for the three new variables; the default is _case, _set, and _time.

Remarks

Nested case–control studies are an attractive alternative to full Cox regression analysis, particularly when time-varying explanatory variables are involved. They are also attractive when some explanatory variables involve laborious coding. For example, you can create a file with a subset of variables for all subjects in the cohort, generate a nested case–control study, and go on to code the remaining data only for those subjects selected.

In the same way as with Cox regression, the results of the analysis are critically dependent on the choice of analysis time (time scale). The choice of analysis time may be calendar time—so that controls would be chosen from subjects still being monitored on the date that the case fails—but other time scales, such as age or time in study, may be more appropriate in some studies. Remember that the analysis time set in selecting controls is implicitly included in the model in subsequent analysis.

match() requires that controls also be matched to the case with respect to other categorical variables, such as sex. This produces an analysis closely mirroring stratified Cox regression. If we wanted to match on calendar time and 5-year age bands, we could first type stsplit ageband ... to create the age bands and then specify match(ageband) on the sttocc command. Analyzing the resulting data as a matched case–control study would estimate rate ratios in the underlying cohort that are controlled for calendar time (very finely) and age (less finely). Such analysis could be carried out by Mantel–Haenszel (odds ratio) calculations, for example, using mhodds, or by conditional logistic regression using clogit.

When ties occur between entry times, censoring times, and failure times, the following convention is adopted:

$$\text{Entry time} < \text{Failure time} < \text{Censoring time}$$

Thus censored subjects and subjects entering at the failure time of the case are included in the risk set and are available for selection as controls. Tied failure times are broken at random. See Clayton and Hills (1997) for more information.

▷ Example 1: Creating a nested case–control study

Using the diet data introduced in [ST] **stsplit**, we will illustrate the use of sttocc, letting age be analysis time. Controls are chosen from subjects still being monitored at the age at which the case fails.

```
. use http://www.stata-press.com/data/r11/diet
(Diet data with dates)

. stset dox, failure(fail) enter(time doe) id(id) origin(time dob) scale(365.25)

                id:  id
     failure event:  fail != 0 & fail < .
obs. time interval:  (dox[_n-1], dox]
 enter on or after:  time doe
 exit on or before:  failure
    t for analysis:  (time-origin)/365.25
            origin:  time dob

       337  total obs.
         0  exclusions

       337  obs. remaining, representing
       337  subjects
        80  failures in single failure-per-subject data
  4603.669  total analysis time at risk, at risk from t =         0
                                earliest observed entry t =  30.07529
                                     last observed exit t =  69.99863

. set seed 9123456

. sttocc, match(job) n(5) nodots
         failure _d:  fail
   analysis time _t:  (dox-origin)/365.25
             origin:  time dob
  enter on or after:  time doe
                 id:  id
        matching for: job
There were 2 tied times involving failure(s)
  - failures assumed to precede censorings,
  - tied failure times split at random
There are 80 cases
Sampling 5 controls for each case
```

The above two commands create a new dataset in which there are five controls per case, matched on job, with the age of the subjects when the case failed recorded in the variable _time. The case indicator is given in _case and the matched set number, in _set. Because we did not specify the optional *varlist*, all variables are carried over into the new dataset.

We can verify that the controls were correctly selected:

```
. gen ageentry=(doe-dob)/365.25

. gen ageexit=(dox-dob)/365.25

. sort _set _case id
```

```
. list _set id _case _time ageentry ageexit job, sepby(_set)
```

	_set	id	_case	_time	ageentry	ageexit	job
1.	1	37	0	42.57358	35.2115	52.67351	0
2.	1	57	0	42.57358	40.04107	56.5859	0
3.	1	86	0	42.57358	38.14921	54.10815	0
4.	1	92	0	42.57358	36.67625	52.38877	0
5.	1	100	0	42.57358	36.15332	46.86653	0
6.	1	90	1	42.57358	31.4141	42.57358	0
7.	2	203	0	47.8987	41.26215	61.22108	2
8.	2	213	0	47.8987	47.23614	67.02532	2
9.	2	219	0	47.8987	41.86721	61.74127	2
10.	2	313	0	47.8987	41.50582	57.05133	2
11.	2	316	0	47.8987	40.68994	56.23545	2
12.	2	196	1	47.8987	45.46475	47.8987	2
13.	3	57	0	47.964408	40.04107	56.5859	0
14.	3	71	0	47.964408	47.52909	64.57221	0
				(output omitted)			
479.	80	136	0	68.596851	58.41205	69.99863	1
480.	80	108	1	68.596851	55.72074	68.59686	1

The controls do indeed belong to the appropriate risk set. The controls in each set enter at an age that is less than the age of the case at failure, and they exit at an age that is greater than the age of the case at failure. To estimate the effect of high energy, use clogit, just as you would for any matched case–control study:

```
. clogit _case hienergy, group(_set) or
Iteration 0:   log likelihood = -142.31278
Iteration 1:   log likelihood = -142.31276
Iteration 2:   log likelihood = -142.31276
Conditional (fixed-effects) logistic regression    Number of obs   =      480
                                                   LR chi2(1)      =     2.06
                                                   Prob > chi2     =   0.1516
Log likelihood = -142.31276                        Pseudo R2       =   0.0072
```

_case	Odds Ratio	Std. Err.	z	P>|z|	[95% Conf. Interval]	
hienergy	.7026801	.1734294	-1.43	0.153	.433183	1.13984

◁

Methods and formulas

sttocc is implemented as an ado-file.

Acknowledgments

The original version of sttocc was written by David Clayton, Cambridge Institute for Medical Research, and Michael Hills, London School of Hygiene and Tropical Medicine (retired).

References

Clayton, D. G., and M. Hills. 1993. *Statistical Models in Epidemiology*. Oxford: Oxford University Press.

———. 1995. ssa7: Analysis of follow-up studies. *Stata Technical Bulletin* 27: 19–26. Reprinted in *Stata Technical Bulletin Reprints*, vol. 5, pp. 219–227. College Station, TX: Stata Press.

———. 1997. ssa10: Analysis of follow-up studies with Stata 5.0. *Stata Technical Bulletin* 40: 27–39. Reprinted in *Stata Technical Bulletin Reprints*, vol. 7, pp. 253–268. College Station, TX: Stata Press.

Coviello, V. 2001. sbe41: Ordinary case–cohort design and analysis. *Stata Technical Bulletin* 59: 12–18. Reprinted in *Stata Technical Bulletin Reprints*, vol. 10, pp. 121–129. College Station, TX: Stata Press.

Langholz, B., and D. C. Thomas. 1990. Nested case-control and case-cohort methods of sampling from a cohort: A critical comparison. *American Journal of Epidemiology* 131: 169–176.

Also see

[ST] **stbase** — Form baseline dataset

[ST] **stdescribe** — Describe survival-time data

[ST] **stsplit** — Split and join time-span records

Title

sttoct — Convert survival-time data to count-time data

Syntax

sttoct *newfailvar newcensvar* [*newentvar*] [, *options*]

options	description
by(*varlist*)	reflect counts by group, where groups are defined by observations with equal values of *varlist*
replace	proceed with transformation, even if current data are not saved
<u>nosh</u>ow	do not show st setting information

You must stset your data before using sttoct; see [ST] **stset**.
fweights, iweights, and pweights may be specified using stset; see [ST] **stset**.
There is no dialog-box interface for sttoct.

Description

sttoct converts survival-time (st) data to count-time (ct) data; see [ST] **ct**.

At present, there is absolutely no reason that you would want to do this.

Options

by(*varlist*) specifies that counts reflect counts by group where the groups are defined by observations with equal values of *varlist*.

replace specifies that it is okay to proceed with the transformation, even though the current dataset has not been saved on disk.

noshow prevents sttoct from showing the key st variables. This option is seldom used because most people type stset, show or stset, noshow to set whether they want to see these variables mentioned at the top of every st command; see [ST] **stset**.

Remarks

sttoct is a never-used command that is included only for completeness. The definition of ct data is found in [ST] **ct**. In the current version of Stata, all you can do with ct data is convert it to st data (which thus provides access to Stata's survival analysis capabilities to those with ct data), so there is little point in converting st data to ct data.

The converted dataset will contain

varlist	from by(*varlist*), if specified
t	the exit-time variable previously stset
newfailvar	number of failures at *t*
newcensvar	number of censored at *t* (after failures)
newentvar	if specified, number of entries at *t* (after censorings)

The resulting dataset will be ctset automatically.

There are two forms of the sttoct command:

1. sttoct *failvar censvar*, ...

2. sttoct *failvar censvar entvar*, ...

That is, specifying *entvar* makes a difference.

Case 1: entvar not specified

This is possible only if

  a. the risk is not recurring;

  b. the original st data is single-record data, or if it is multiple-record data, all subjects enter at time 0 and have no gaps thereafter; and

  c. if by(*varlist*) is specified, subjects do not have changing values of the variables in *varlist* over their histories.

If you do not specify *entvar*, sttoct verifies that (a), (b), and (c) are true. If the assumptions are true, sttoct converts your data and counts each subject only once. That is, in multiple-record data, all thrashing (censoring followed by immediate reenter with different covariates) is removed.

Case 2: entvar specified

Any kind of survival-time data can be converted to count-time data with an entry variable. You can convert your data in this way whether assumptions (a), (b), and (c) are true or not.

When you specify a third variable, thrashing is not removed, even if it could be—even if assumptions (a), (b), and (c) are true.

Methods and formulas

sttoct is implemented as an ado-file.

Also see

[ST] **ct** — Count-time data

[ST] **st_is** — Survival analysis subroutines for programmers

[ST] **sttocc** — Convert survival-time data to case–control data

Title

stvary — Report whether variables vary over time

Syntax

stvary [*varlist*] [*if*] [*in*] [, <u>nosh</u>ow]

You must stset your data before using stvary; see [ST] **stset**.

by is allowed; see [D] **by**.

fweights, iweights, and pweights may be specified using stset; see [ST] **stset**.

Menu

Statistics > Survival analysis > Setup and utilities > Report variables that vary over time

Description

stvary is for use with multiple-record datasets, for which id() has been stset. It reports whether values of variables within subject vary over time and reports their pattern of missing values. Although stvary is intended for use with multiple-record st data, it may be used with single-record data as well, but this produces little useful information.

stvary ignores weights, even if you have set them. stvary summarizes the variables in the computer or data sense of the word.

Option

> Main

noshow prevents stvary from showing the key st variables. This option is seldom used because most people type stset, show or stset, noshow to set whether they want to see these variables mentioned at the top of the output of every st command; see [ST] **stset**.

Remarks

Consider a multiple-record dataset. A subject's sex, presumably, does not change, but his or her age might. stvary allows you to verify that values vary in the way that you expect:

```
. use http://www.stata-press.com/data/r11/stan3
(Heart transplant data)

. stvary
        failure _d:  died
   analysis time _t:  t1
              id:  id
```

```
            subjects for whom the variable is
                                    never    always   sometimes
  variable |  constant   varying  missing   missing    missing
-----------+----------------------------------------------------
      year |      103         0       103         0          0
       age |      103         0       103         0          0
     stime |      103         0       103         0          0
   surgery |      103         0       103         0          0
transplant |      103         0       103         0          0
      wait |      103         0       103         0          0
  posttran |       34        69       103         0          0
```

That 103 values for year are "constant" does not mean that year itself is a constant—it means merely that, for each subject, the value of year does not change across the records. Whether the values of year vary across subjects is still an open question.

Now look at the bottom of the table: posttran is constant over time for 34 subjects and varies for the remaining 69.

Below we have another dataset, and we will examine just two of the variables:

```
. use http://www.stata-press.com/data/r11/stvaryex
. stvary sex drug
            subjects for whom the variable is
                                    never    always   sometimes
  variable |  constant   varying  missing   missing    missing
-----------+----------------------------------------------------
       sex |      119         1       119         3          1
      drug |      121         2       123         0          0
```

Clearly, there are errors in the sex variable; for 119 of the subjects, sex does not change over time, but for one, it does. Also we see that we do not know the sex of three of the patients, but for another, we sometimes know it and sometimes do not. The latter must be a simple data-construction error. As for drug, we see that for two of our patients, the drug administered varied over time. Perhaps this is an error, or perhaps those two patients were treated differently from all the rest.

Saved results

stvary saves the following in r():

Scalars
    r(cons)   number of subjects for whom variable is constant when not missing
    r(varies)   number of subjects for whom nonmissing values vary
    r(never)   number of subjects for whom variable is never missing
    r(always)   number of subjects for whom variable is always missing
    r(miss)   number of subjects for whom variable is sometimes missing

Methods and formulas

stvary is implemented as an ado-file.

Reference

Cleves, M. A., W. W. Gould, R. G. Gutierrez, and Y. Marchenko. 2008. *An Introduction to Survival Analysis Using Stata*. 2nd ed. College Station, TX: Stata Press.

Also see

[ST] **stdescribe** — Describe survival-time data

[ST] **stfill** — Fill in by carrying forward values of covariates

[ST] **stset** — Declare data to be survival-time data

Glossary

accelerated failure-time model. A model in which everyone has, in a sense, the same survival function, $S(\tau)$, and an individual's τ_j is a function of his or her characteristics and of time, such as $\tau_j = t * \exp(\beta_0 + \beta_1 x_{1j} + \beta_2 x_{2j})$.

accrual period or **recruitment period** or **accrual**. The accrual period (or recruitment period) is the period during which subjects are being enrolled (recruited) into a study. Also see *follow-up period*.

administrative censoring. Administrative censoring is the right-censoring that occurs when the study observation period ends. All subjects complete the course of the study and are known to have experienced either of two outcomes at the end of the study: survival or failure. This type of censoring should not be confused with *withdrawal* and *loss to follow-up*. Also see *censoring*.

analysis time. Analysis time is like time, except that 0 has a special meaning: $t = 0$ is the time of onset of risk, the time when failure first became possible.

Analysis time is usually not what is recorded in a dataset. A dataset of patients might record calendar time. Calendar time must then be mapped to analysis time.

The letter t is reserved for time in analysis-time units. The term *time* is used for time measured in other units.

The *origin* is the *time* corresponding to $t = 0$, which can vary subject to subject. Thus $t =$ time $-$ origin.

at risk. A subject is at risk from the instant the first failure event becomes possible and usually stays that way until failure, but a subject can have periods of being at risk and not at risk.

attributable fraction. An attributable fraction is the reduction in the risk of a disease or other condition of interest when a particular risk factor is removed.

baseline. In survival analysis, baseline is the state at which the covariates, usually denoted by the row vector **x**, are zero. For example, if the only measured covariate is systolic blood pressure, the baseline survival function would be the survival function for someone with zero systolic blood pressure. This may seem ridiculous, but covariates are usually centered so that the mathematical definition of baseline (covariate is zero) translates into something meaningful (mean systolic blood pressure).

case–control studies. In case–control studies, cases meeting a fixed criterion are matched to noncases ex post to study differences in possible covariates. Relative sample sizes are usually fixed at 1:1 or 1:2 but sometimes vary once the survey is complete. In any case, sample sizes do not reflect the distribution in the underlying population.

cause-specific hazard. In a competing-risks analysis, the cause-specific hazard is the hazard function that generates the events of a given type. For example, if heart attack and stroke are competing events, then the cause-specific hazard for heart attacks describes the biological mechanism behind heart attacks independently of that for strokes. Cause-specific hazards can be modeled using Cox regression, treating the other events as censored.

censored, censoring, left-censoring, and **right-censoring.** An observation is left-censored when the exact time of the onset of risk is not known; it is merely known that the onset of risk occurred before t_l.

An observation is right-censored when the time of failure is not known; it is merely known that the failure occurred after t_r.

In common usage, *censored* without a modifier means right-censoring.

Also see *truncation*.

CIF. See *cumulative incidence function*.

cohort studies. In cohort studies, a group that is well defined is monitored over time to track the transition of noncases to cases. Cohort studies differ from incidence studies in that they can be retrospective as well as prospective.

competing risks. Competing risks models are survival-data models in which the failures are generated by more than one underlying process. For example, death may be caused by either heart attack or stroke. There are various methods for dealing with competing risks. One direct way is to duplicate failures for one competing risk as censored observations for the other risk and stratify on the risk type. Another is to directly model the cumulative incidence of the event of interest in the presence of competing risks. The former method uses `stcox` and the latter, `stcrreg`.

confounding. In the analysis of epidemiological tables, factor or interaction effects are said to be confounded when the effect of one factor is combined with that of another. For example, the effect of alcohol consumption on esophageal cancer may be confounded with the effects of age, smoking, or both. In the presence of confounding, it is often useful to stratify on the confounded factors that are not of primary interest, in the above example, age and smoking.

count-time data. See *ct data*.

covariates. Covariates are the explanatory variables that appear in a model. For instance, if survival time were to be explained by age, sex, and treatment, then those variables would be the covariates. Also see *time-varying covariates*.

cross-sectional or **prevalence studies**. Cross-sectional studies sample distributions of healthy and diseased subjects in the population at one point in time.

crude estimate. A crude estimate is an estimate that is chosen not for accuracy or precision but instead for ease of calculation.

ct data. ct stands for count time. ct data is an aggregate organized like a life table. Each observation records a time, the number known to fail at that time, the number censored, and the number of new entries. See [ST] **ctset**.

cumulative incidence estimator. In a competing-risks analysis, the cumulative incidence estimator estimates the cumulative incidence function (CIF). Assume for now that you have one event of interest (type 1) and one competing event (type 2). The cumulative incidence estimator for type 1 failures is then obtained by

$$\widehat{\text{CIF}}_1(t) = \sum_{j:t_j \leq t} \widehat{h}_1(t_j) \widehat{S}(t_{j-1})$$

with

$$\widehat{S}(t) = \prod_{j:t_j \leq t} \left\{ 1 - \widehat{h}_1(t_j) - \widehat{h}_2(t_j) \right\}$$

The t_j index the times at which events (of any type) occur, and $\widehat{h}_1(t_j)$ and $\widehat{h}_2(t_j)$ are the cause-specific hazard contributions for type 1 and type 2, respectively. $\widehat{S}(t)$ estimates the probability that you are event free at time t.

The above generalizes to multiple competing events in the obvious way.

cumulative incidence function. In a competing-risks analysis, the cumulative incidence function, or CIF, is the probability that will you will observe the event of primary interest before a given time. Formally,
$$\text{CIF}(t) = P(T \leq t \text{ and event type of interest})$$
for time-to-failure, T.

DFBETA. A DFBETA measures the change in the regressor's coefficient because of deletion of that subject. Also see *partial DFBETA*.

effect size. The effect size is the size of the clinically significant difference between the treatments being compared, often expressed as the hazard ratio (or the log of the hazard ratio) in survival analysis.

event. An event is something that happens at an instant in time, such as being exposed to an environmental hazard, being diagnosed as myopic, or becoming employed.

The failure event is of special interest in survival analysis, but there are other equally important events, such as the exposure event, from which analysis time is defined.

In st data, events occur at the end of the recorded time span.

event of interest. In a competing-risks analysis, the event of interest is the event that is the focus of the analysis, that for which the cumulative incidence in the presence of competing risks is estimated.

exponential test. The exponential test is the parametric test comparing the hazard rates, λ_1 and λ_2, (or log hazards) from two independent exponential (constant only) regression models with the null hypothesis H_0: $\lambda_2 - \lambda_1 = 0$ (or H_0: $\ln(\lambda_2) - \ln(\lambda_1) = \ln(\lambda_2/\lambda_1) = 0$).

failure event. Survival analysis is really time-to-failure analysis, and the failure event is the event under analysis. The failure event can be death, heart attack, myopia, or finding employment. Many authors—including Stata—write as if the failure event can occur only once per subject, but when we do, we are being sloppy. Survival analysis encompasses repeated failures, and all of Stata's survival analysis features can be used with repeated-failure data.

follow-up period or **follow-up**. The (minimum) follow-up period is the period after the last subject entered the study until the end of the study. The follow-up defines the phase of a study during which subjects are under observation and no new subjects enter the study. If T is the total duration of a study, and R is the accrual period of the study, then follow-up period f is equal to $T - R$. Also see *accrual period*.

frailty. In survival analysis, it is often assumed that subjects are alike—homogeneous—except for their observed differences. The probability that subject j fails at time t may be a function of j's covariates and random chance. Subjects j and k, if they have equal covariate values, are equally likely to fail.

Frailty relaxes that assumption. The probability that subject j fails at time t becomes a function of j's covariates and j's unobserved frailty value, ν_j. Frailty ν is assumed to be a random variable. Parametric survival models can be fit even in the presence of such heterogeneity.

Shared frailty refers to the case in which groups of subjects share the same frailty value. For instance, subjects 1 and 2 may share frailty value ν because they are genetically related. Both parametric and semiparametric models can be fit under the shared-frailty assumption.

future history. Future history is information recorded after a subject is no longer at risk. The word *history* is often dropped, and the term becomes simply *future*. Perhaps the failure event is cardiac infarction, and you want to know whether the subject died soon in the *future*, in which case you might exclude the subject from analysis.

Also see *past history*.

gaps. Gaps refers to gaps in observation between entry time and exit time; see *under observation*.

hazard, cumulative hazard, and **hazard ratio**. The hazard or hazard rate at time t, $h(t)$, is the instantaneous rate of failure at time t conditional on survival until time t. Hazard rates can exceed 1. Say that the hazard rate were 3. If an individual faced a constant hazard of 3 over a unit interval and if the failure event could be repeated, the individual would be expected to experience three failures during the time span.

The cumulative hazard, $H(t)$, is the integral of the hazard function $h(t)$, from 0 (the onset of risk) to t. It is the total number of failures that would be expected to occur up until time t, if the failure event could be repeated. The relationship between the cumulative hazard function, $H(t)$, and the survivor function, $S(t)$, is

$$S(t) = \exp\{-H(t)\}$$
$$H(t) = -\ln\{S(t)\}$$

The hazard ratio is the ratio of the hazard function evaluated at two different values of the covariates: $h(t\,|\,\mathbf{x})/h(t\,|\,\mathbf{x}_0)$. The hazard ratio is often called the relative hazard, especially when $h(t\,|\,\mathbf{x}_0)$ is the baseline hazard function.

hazard contributions. Hazard contributions are the increments of the estimated cumulative hazard function obtained through either a nonparametric or semiparametric analysis. For these analysis types, the estimated cumulative hazard is a step function that increases every time a failure occurs. The hazard contribution for that time is the magnitude of that increase.

Because the time between failures usually varies from failure to failure, hazard contributions do not directly estimate the hazard. However, one can use the hazard contributions to formulate an estimate of the hazard function based on the method of smoothing.

ID variable. An ID variable identifies groups; equal values of an ID variable indicate that the observations are for the same group. For instance, a stratification ID variable would indicate the strata to which each observation belongs.

When an ID variable is referred to without modification, it means subjects, and usually this occurs in multiple-record st data. In multiple-record data, each physical observation in the dataset represents a time span, and the ID variable ties the separate observations together:

idvar	t0	t
1	0	5
1	5	7

ID variables are usually numbered 1, 2, ..., but that is not required. An ID variable might be numbered 1, 3, 7, 22, ..., or -5, -4, ..., or even 1, 1.1, 1.2, ...

incidence and **incidence rate**. Incidence is the number of new failures (e.g., number of new cases of a disease) that occur during a specified period in a population at risk (e.g., of the disease).

Incidence rate is incidence divided by the sum of the length of time each individual was exposed to the risk.

Do not confuse incidence with prevalence. Prevalence is the fraction of a population that has the disease. Incidence refers to the rate at which people contract a disease, whereas prevalence is the total number actually sick at a given time.

incidence studies, **longitudinal studies**, and **follow-up studies**. Whichever word is used, these studies monitor a population for a time to track the transition of noncases into cases. Incidence studies are prospective. Also see *cohort studies*.

Kaplan–Meier product-limit estimate. This is an estimate of the survivor function, which is the product of conditional survival to each time at which an event occurs. The simple form of the calculation, which requires tallying the number at risk and the number who die and at each time, makes accounting for censoring easy. The resulting estimate is a step function with jumps at the event times.

life table. Also known as a mortality table or actuarial table, a life table is a table that shows for each analysis time the fraction that survive to that time. In mortality tables, analysis time is often age.

likelihood displacement value. A likelihood displacement value is an influence measure of the effect of deleting a subject on the overall coefficient vector. Also see *partial likelihood displacement value*.

LMAX value. An LMAX value is an influence measure of the effect of deleting a subject on the overall coefficient vector and is based on an eigensystem analysis of efficient score residuals. Also see *partial LMAX value*.

loss to follow-up. Subjects are lost to follow-up if they do not complete the course of the study for reasons unrelated to the event of interest. For example, loss to follow-up occurs if subjects move to a different area or decide to no longer participate in a study. Loss to follow-up should not be confused with administrative censoring. If subjects are lost to follow-up, the information about the outcome these subjects would have experienced at the end of the study, had they completed the study, is unavailable. Also see *withdrawal*, *administrative censoring*, and *follow-up*.

matched case–control study. Also known as a retrospective study, a matched case–control study is a study in which persons with positive outcomes are each matched with one or more persons with negative outcomes but with similar characteristics.

multiarm trial. A multiarm trial is a trial comparing survivor functions of more than two groups.

odds and **odds ratio**. The odds in favor of an event are $o = p/(1-p)$, where p is the probability of the event. Thus if $p = 0.2$, the odds are 0.25, and if $p = 0.8$, the odds are 4.

The log of the odds is $\ln(o) = \text{logit}(p) = \ln\{p/(1-p)\}$, and logistic-regression models, for instance, fit $\ln(o)$ as a linear function of the covariates.

The odds ratio is a ratio of two odds: o_1/o_0. The individual odds that appear in the ratio are usually for an experimental group and a control group, or two different demographic groups.

offset variable and **exposure variable**. An offset variable is a variable that is to appear on the right-hand side of a model with coefficient 1:

$$y_j = \text{offset}_j + b_0 + b_1 x_j + \cdots$$

In the above, b_0 and b_1 are to be estimated. The offset is not constant. Offset variables are often included to account for the amount of exposure. Consider a model where the number of events observed over a period is the length of the period multiplied by the number of events expected in a unit of time:

$$n_j = T_j e(X_j)$$

When we take logs, this becomes

$$\log(n_j) = \log(T_j) + \log\{e(X_j)\}$$

$\ln(T_j)$ is an offset variable in this model.

When the log of a variable is an offset variable, the variable is said to be an exposure variable. In the above, T_j is an exposure variable.

partial DFBETA. A partial DFBETA measures the change in the regressor's coefficient because of deletion of that individual record. In single-record data, the partial DFBETA is equal to the DFBETA. Also see *DFBETA*.

partial likelihood displacement value. A partial likelihood displacement value is an influence measure of the effect of deleting an individual record on the coefficient vector. For single-record data, the partial likelihood displacement value is equal to the likelihood displacement value. Also see *likelihood displacement value*.

partial LMAX value. A partial LMAX value is an influence measure of the effect of deleting an individual record on the overall coefficient vector and is based on an eigensystem analysis of efficient score residuals. In single-record data, the partial LMAX value is equal to the LMAX value. Also see *LMAX value*.

past history. Past history is information recorded about a subject before the subject was both *at risk* and *under observation*. Consider a dataset that contains information on subjects from birth to death and an analysis in which subjects became at risk once diagnosed with a particular kind of cancer. The past history on the subject would then refer to records before the subjects were diagnosed.

The word *history* is often dropped, and the term becomes simply *past*. For instance, we might want to know whether a subject smoked in the past.

Also see *future history*.

penalized log-likelihood function. This is a log-likelihood function that contains an added term, usually referred to as a roughness penalty, that reduces its value when the model overfits the data. In Cox models with frailty, such functions are used to prevent the variance of the frailty from growing too large, which would allow the individual frailty values to perfectly fit the data.

power. The power of a test is the probability of correctly rejecting the null hypothesis when it is false. It is often denoted as $1 - \beta$ in statistical literature, where β is the type II error probability. Commonly used values for power are 80% and 90%. Also see *type II error* and *type I error*.

prevented fraction. A prevented fraction is the reduction in the risk of a disease or other condition of interest caused by including a protective risk factor or public-health intervention.

proportional hazards model. This is a model in which, between individuals, the ratio of the instantaneous failure rates (the hazards) is constant over time.

prospective study. Also known as a prospective longitudinal study, a prospective study is a study based on observations over the same subjects for a given period.

risk factor. This is a variable associated with an increased or decreased risk of failure.

risk pool. At a particular point in time, this is the subjects at risk of failure.

risk ratio. In a log-linear model, this is the ratio of probability of survival associated with a one-unit increase in a risk factor relative to that calculated without such an increase, that is, $R(x+1)/R(x)$. Given the exponential form of the model, $R(x+1)/R(x)$ is constant and is given by the exponentiated coefficient.

semiparametric model. This is a model that is not fully parameterized. The Cox proportional hazards model is such a model:

$$h(t) = h_0(t) \exp(\beta_1 x_1 + \cdots + \beta_k x_k)$$

In the Cox model, $h_o(t)$ is left unparameterized and not even estimated. Meanwhile, the relative effects of covariates are parameterized as $\exp(\beta_1 x_1 + \cdots + \beta_k x_k)$.

singleton-group data. A singleton is a frailty group that contains only 1 observation. A dataset containing only singletons is known as singleton-group data.

SMR. SMR stands for standardized mortality (morbidity) ratio and is the observed number of deaths divided by the expected number of deaths. It is calculated using indirect standardization: you take the population of the group of interest—say, by age, sex, and other factors—and calculate the expected number of deaths in each cell (expected being defined as the number of deaths that would have been observed if those in the cell had the same mortality as some other population). You then take the ratio to compare the observed with the expected number of deaths. For instance,

Age	(1) Population of group	(2) Deaths per 100,000 in general pop.	(1)×(2) Expected # of deaths	(4) Observed deaths
25–34	95,965	105.2	100.9	92
34–44	78,280	203.6	159.4	180
44–54	52,393	428.9	224.7	242
55–64	28,914	964.6	278.9	312
Total			763.9	826

$$\text{SMR} = 826/763.9 = 1.08$$

snapshot data. Snapshot data are those in which each record contains the values of a set of variables for a subject at an instant in time. The name arises because each observation is like a snapshot of the subject.

In snapshot datasets, one usually has a group of observations (snapshots) for each subject.

Snapshot data must be converted to st data before they can be analyzed. This requires making assumptions about what happened between the snapshots. See [ST] **snapspan**.

spell data. Spell data are survival data in which each record represents a fixed period, consisting of a begin time, an end time, possibly a censoring/failure indicator, and other measurements (covariates) taken during that specific period.

st data. st stands for survival time. In survival-time data, each observation represents a span of survival, recorded in variables $t0$ and t. For instance, if in an observation $t0$ were 3 and t were 5, the span would be $(t0, t]$, meaning from just after $t0$ up to and including t.

Sometimes variable $t0$ is not recorded; $t0$ is then assumed to be 0. In such a dataset, an observation that had $t = 5$ would record the span $(0, 5]$.

Each observation also includes a variable d, called the failure variable, which contains 0 or nonzero (typically, 1). The failure variable records what happened at the end of the span: 0, the subject was still alive (had not yet failed) or 1, the subject died (failed).

Sometimes variable d is not recorded; d is then assumed to be 1. In such a dataset, all time-span observations would be assumed to end in failure.

Finally, each observation in an st dataset can record the entire history of a subject or each can record a part of the history. In the latter case, groups of observations record the full history. One observation might record the period $(0, 5]$ and the next, $(5, 8]$. In such cases, there is a variable ID that records the subject for which the observation records a time span. Such data are called multiple-record st data. When each observation records the entire history of a subject, the data are called single-record st data. In the single-record case, the ID variable is optional.

See [ST] **stset**.

subhazard, **cumulative subhazard**, and **subhazard ratio**. In a competing risks analysis, the hazard of the subdistribution (or subhazard for short) for the event of interest (type 1) is defined formally as

$$\overline{h}_1(t) = \lim_{\delta \to 0} \left\{ \frac{P(t < T \leq t + \delta \text{ and event type 1 })|\ T > t \text{ or } (T \leq t \text{ and not event type 1})}{\delta} \right\}$$

Less formally, think of this hazard as that which generates failure events of interest while keeping subjects who experience competing events "at risk" so that they can be adequately counted as not having any chance of failing.

The cumulative subhazard $\overline{H}_1(t)$ is the integral of the subhazard function $\overline{h}_1(t)$, from 0 (the onset of risk) to t. The cumulative subhazard plays a very important role in competing-risks analysis. The cumulative incidence function (CIF) is a direct function of the cumulative subhazard:

$$\text{CIF}_1(t) = 1 - \exp\{-\overline{H}_1(t)\}$$

The subhazard ratio is the ratio of the subhazard function evaluated at two different values of the covariates: $\overline{h}_1(t|\mathbf{x})/\overline{h}_1(t|\mathbf{x}_0)$. The subhazard ratio is often called the relative subhazard, especially when $\overline{h}_1(t|\mathbf{x}_0)$ is the baseline subhazard function.

survival-time data. See *st data*.

survivor function. Also known as the survivorship function and the survival function, the survivor function, $S(t)$, is 1) the probability of surviving beyond time t, or equivalently, 2) the probability that there is no failure event prior to t, 3) the proportion of the population surviving to time t, or equivalently, 4) the reverse cumulative distribution function of T, the time to the failure event: $S(t) = \Pr(T > t)$. Also see *hazard*.

time-varying covariates. Time-varying covariates appear in a survival model whose values vary over time. The values of the covariates vary, not the effect. For instance, in a proportional hazards model, the log hazard at time t might be $b \times \text{age}_t + c \times \text{treatment}_t$. Variable age might be time varying, meaning that as the subject ages, the value of age changes, which correspondingly causes the hazard to change. The effect b, however, remains constant.

Time-varying variables are either continuously varying or discretely varying.

In the continuously varying case, the value of the variable x at time t is $x_t = x_o + f(t)$, where $f()$ is some function and often is the identity function, so that $x_t = x_o + t$.

In the discretely varying case, the value of x changes at certain times and often in no particular pattern:

idvar	t0	t	bp
1	0	5	150
1	5	7	130
1	7	9	135

In the above data, the value of bp is 150 over the period $(0, 5]$, then 130 over $(5, 7]$, and 135 over $(7, 9]$.

truncation. In statistics, truncation refers to cases where subjects are unobserved (do not appear in the data). Censoring, on the other hand, refers to cases when subjects are known to fail within a certain time span, except that the exact failure time is unknown.

In survival analysis, truncation—sometimes called left-truncation—occurs when subjects are not observed before time t_c, only to appear in the study at time t_c. It is only because they didn't fail before t_c that we even knew about their existence. Left-truncation differs from left-censoring in that, in the censored case, we know that the subject failed before time t_c, but we just do not know exactly when.

Given these conventions, right-truncation is equivalent to right-censoring, because failure (at some point before time infinity) is inevitable.

type I error or **false-positive result**. The type I error of a test is the error of rejecting the null hypothesis when it is true. The probability of committing a type I error, significance level of a test, is often denoted as α in statistical literature. One traditionally used value for α is 5%. Also see *type II error* and *power*.

type I study. A type I study is a study in which all subjects fail (or experience an event) by the end of the study; that is, no censoring of subjects occurs.

type II error or **false-negative result**. The type II error of a test is the error of not rejecting the null hypothesis when it is false. The probability of committing a type II error is often denoted as β in statistical literature. Commonly used values for β are 20% or 10%. Also see *type I error* and *power*.

type II study. A type II study is a study in which there are subjects who do not fail (or do not experience an event) by the end of the study. These subjects are known to be censored.

under observation. A subject is under observation when failure events, should they occur, would be observed (and so recorded in the dataset). Being under observation does not mean that a subject is necessarily at risk. Subjects usually come under observation before they are at risk. The statistical concern is with periods when subjects are at risk but not under observation, even when the subject is (later) known not to have failed during the hiatus.

In such cases, since failure events would not have been observed, the subject necessarily had to survive the observational hiatus, and that leads to bias in statistical results unless the hiatus is accounted for properly.

Entry time and exit time record when a subject first and last comes under observation, between which there may be observational gaps, but usually there are not. There is only one entry time and one exit time for each subject. Often, entry time corresponds to analysis time $t = 0$, or before, and exit time corresponds to the time of failure.

Delayed entry means that the entry time occurred after $t = 0$.

withdrawal. Withdrawal is the process under which subjects withdraw from a study for reasons unrelated to the event of interest. For example, withdrawal occurs if subjects move to a different area or decide to no longer participate in a study. Withdrawal should not be confused with administrative censoring. If subjects withdraw from the study, the information about the outcome those subjects would have experienced at the end of the study, had they completed the study, is unavailable. Also see *loss to follow-up* and *administrative censoring*.

Subject and author index

This is the subject and author index for the *Survival Analysis and Epidemiological Tables Reference Manual*. Readers interested in topics other than survival analysis should see the combined subject index (and the combined author index) in the *Quick Reference and Index*. The combined index indéxes the *Getting Started* manuals, the *User's Guide*, and all the reference manuals except the *Mata Reference Manual*.

Semicolons set off the most important entries from the rest. Sometimes no entry will be set off with semicolons, meaning that all entries are equally important.

A

Aalen, O. O., [ST] **stcrreg postestimation**, [ST] **sts**
Aalen–Nelson cumulative hazard, *see* Nelson–Aalen cumulative hazard
Abraira-García, L., [ST] **epitab**
Abramson, J. H., [ST] **epitab**
Abramson, Z. H., [ST] **epitab**
accelerated failure-time model, [ST] **Glossary**, [ST] **streg**
accrual period, [ST] **Glossary**, [ST] **stpower exponential**, [ST] **stpower logrank**
actuarial tables, *see* life tables
adjusted Kaplan–Meier survivor function, [ST] **sts**
administrative censoring, [ST] **Glossary**
Agresti, A., [ST] **epitab**
AIC (Akaike information criterion), [ST] **streg**
Aisbett, C. W., [ST] **stcox**, [ST] **streg**
Akaike, H., [ST] **streg**
Allison, P. D., [ST] **discrete**
analysis time, [ST] **Glossary**
Andersen, P. K., [ST] **stcrreg**
Anderson, M. L., [ST] **stcrreg**
association test, [ST] **epitab**
at risk, [ST] **Glossary**
at-risk table, [ST] **sts graph**
attributable fraction, [ST] **epitab**, [ST] **Glossary**
attributable proportion, [ST] **epitab**

B

Babiker, A., [ST] **epitab**, [ST] **stpower cox**, [ST] **stpower**, [ST] **sts test**
Baldus, W. P., [ST] **stcrreg**
Barthel, F. M.-S., [ST] **stcox PH-assumption tests**, [ST] **stpower**, [ST] **stpower cox**
baseline, [ST] **Glossary**
baseline dataset, [ST] **stbase**
baseline hazard and survivor functions, [ST] **stcox**, [ST] **stcox PH-assumption tests**, [ST] **stcrreg**
Bellocco, R., [ST] **epitab**
Beyersman, J., [ST] **stcrreg**
Boggess, M., [ST] **stcrreg**, [ST] **stcrreg postestimation**
Boice Jr., J. D., [ST] **epitab**
Borgan, Ø., [ST] **stcrreg**
Bottai, M., [ST] **epitab**
Box-Steffensmeier, J. M., [ST] **stcox**, [ST] **streg**
Breslow, N. E., [ST] **epitab**, [ST] **stcox**, [ST] **stcox PH-assumption tests**, [ST] **sts test**, [ST] **sts**
Brown, B. W., [ST] **sts graph**
Brown, C. C., [ST] **epitab**
Buchholz, A., [ST] **stcrreg**

C

Califf, R. M., [ST] **stcox postestimation**
Campbell, M. J., [ST] **stpower**, [ST] **stpower cox**, [ST] **stpower logrank**
Carlin, J., [ST] **epitab**
Carnes, B. A., [ST] **streg**
Carter, S., [ST] **stcox**, [ST] **streg**
case–cohort data, [ST] **sttocc**
case–control data, [ST] **epitab**, [ST] **sttocc**
case–control studies, [ST] **Glossary**
categorical data, [ST] **epitab**
cause-specific hazard, [ST] **Glossary**
cc command, [ST] **epitab**
cci command, [ST] **epitab**
Chiang, C. L., [ST] **ltable**
chi-squared, test of independence, [ST] **epitab**
Chow, S.-C., [ST] **stpower**, [ST] **stpower exponential**
CIF, [ST] **Glossary**, *also see* cumulative incidence function
Clark, V. A., [ST] **ltable**
Clayton, D. G., [ST] **epitab**, [ST] **stptime**, [ST] **strate**, [ST] **stsplit**, [ST] **sttocc**
Cleves, M. A., [ST] **st**, [ST] **stcox**, [ST] **stcrreg**, [ST] **stcurve**, [ST] **stdescribe**, [ST] **streg**, [ST] **sts**, [ST] **stset**, [ST] **stsplit**, [ST] **stvary**, [ST] **survival analysis**
clinical trial, [ST] **stpower**
cluster estimator of variance,
    Cox proportional hazards model, [ST] **stcox**, [ST] **stcrreg**
    parametric survival models, [ST] **streg**
cohort studies, [ST] **Glossary**
Collett, D., [ST] **stci**, [ST] **stcox postestimation**, [ST] **stcrreg postestimation**, [ST] **stpower**, [ST] **stpower logrank**, [ST] **streg postestimation**, [ST] **sts test**, [ST] **stsplit**
competing risks, [ST] **Glossary**, [ST] **stcrreg**
concordance, estat subcommand, [ST] **stcox postestimation**
confidence intervals for odds and risk ratios, [ST] **epitab**, [ST] **stci**
confounding, [ST] **Glossary**
constrained estimation
    competing risks, [ST] **stcrreg**
    parametric survival models, [ST] **streg**
contingency tables, [ST] **epitab**
Cornfield, J., [ST] **epitab**
Cornfield confidence intervals, [ST] **epitab**

count-time data, [ST] **ct**, [ST] **Glossary**; [ST] **ctset**, [ST] **cttost**, [ST] **sttoct**
covariates, [ST] **Glossary**
Coviello, V., [ST] **stcrreg**, [ST] **stcrreg postestimation**, [ST] **sttocc**
Cox, D. R., [ST] **ltable**, [ST] **stcox**, [ST] **stcox PH-assumption tests**, [ST] **stcrreg**, [ST] **stpower**, [ST] **stpower cox**, [ST] **streg postestimation**, [ST] **streg**, [ST] **sts**
cox, stpower subcommand, [ST] **stpower cox**
Cox proportional hazards model, [ST] **stcox**
    power, [ST] **stpower cox**
    sample size, [ST] **stpower cox**
    test of assumption, [ST] **stcox**, [ST] **stcox PH-assumption tests**, [ST] **stcox postestimation**, [ST] **stsplit**
    Wald test, power, [ST] **stpower cox**
Cox–Snell residual, [ST] **stcox postestimation**, [ST] **streg postestimation**
cross-sectional studies, [ST] **Glossary**
Crowder, M. J., [ST] **stcrreg**, [ST] **streg**
Crowley, J., [ST] **stcox**, [ST] **stcrreg**, [ST] **stset**
crude estimate, [ST] **epitab**, [ST] **Glossary**
cs command, [ST] **epitab**
csi command, [ST] **epitab**
ct data, [ST] **Glossary**, *also see* count-time data
ctset command, [ST] **ctset**
cttost command, [ST] **cttost**
Cui, J., [ST] **stcox**, [ST] **streg**
cumulative hazard function, [ST] **stcurve**, [ST] **sts generate**, [ST] **sts graph**, [ST] **sts list**, [ST] **sts**
cumulative hazard ratio, *see* hazard ratio
cumulative incidence
    data, [ST] **epitab**
    estimator, [ST] **Glossary**, [ST] **stcrreg**
    function, [ST] **Glossary**, [ST] **stcrreg**, [ST] **stcurve**
cumulative subhazard, [ST] **Glossary**, [ST] **stcrreg**, [ST] **stcurve**
Cutler, S. J., [ST] **ltable**

D

data,
    case–cohort, *see* case–cohort data
    case–control, *see* case–control data
    count-time, *see* count-time data
    discrete survival, *see* discrete survival data
    multiple-record st data, *see* multiple-record st data
    nested case–control, *see* nested case–control data
    time-span, *see* time-span data
Day, N. E., [ST] **epitab**
De Stavola, B. L., [ST] **stcox**, [ST] **stset**
Desu, M. M., [ST] **stpower**, [ST] **stpower exponential**
deviance residual, [ST] **stcox postestimation**, [ST] **streg postestimation**
DFBETA, [ST] **Glossary**, [ST] **stcox postestimation**
Dickson, E. R., [ST] **stcrreg**
discrete survival data, [ST] **discrete**

Dobbin, K., [ST] **stpower**
Dohoo, I., [ST] **epitab**
Doll, R., [ST] **epitab**
Drukker, D. M., [ST] **stcox**, [ST] **streg**
Dupont, W. D., [ST] **epitab**, [ST] **stcox**, [ST] **stir**, [ST] **sts**

E

Ederer, F., [ST] **ltable**
effect size, [ST] **Glossary**, [ST] **stpower**, [ST] **stpower cox**, [ST] **stpower exponential**, [ST] **stpower logrank**
epidemiological tables, [ST] **epitab**
equality test, survivor functions, [ST] **sts test**
estat concordance command, [ST] **stcox postestimation**
estat phtest command, [ST] **stcox PH-assumption tests**
etiologic fraction, [ST] **epitab**
event, [ST] **Glossary**, [ST] **stpower logrank**
event of interest, [ST] **Glossary**
excess fraction, [ST] **epitab**
exponential, stpower subcommand, [ST] **stpower exponential**
exponential distribution, [ST] **streg**
exponential survival
    power, [ST] **stpower exponential**
    regression, [ST] **streg**
    sample size, [ST] **stpower exponential**
exponential test, [ST] **Glossary**
    power, [ST] **stpower exponential**
    sample size, [ST] **stpower exponential**
exposure variable, [ST] **Glossary**

F

failure event, [ST] **Glossary**
failure tables, [ST] **ltable**
failure-time model, [ST] **stcox**, [ST] **stcox PH-assumption tests**, [ST] **streg**
false-negative result, *see* type II error
false-positive result, *see* type I error
Feiveson, A. H., [ST] **stpower**
Feltbower, R., [ST] **epitab**
Fibrinogen Studies Collaboration, [ST] **stcox postestimation**
filling in values, [ST] **stfill**
Fine, J. P., [ST] **intro**, [ST] **stcrreg**
Fiocco, M., [ST] **stcrreg**, [ST] **stcrreg postestimation**
Fisher, L. D., [ST] **epitab**
Fisher, R. A., [ST] **streg**
Fisher's exact test, [ST] **epitab**
Fleiss, J. L., [ST] **epitab**
Fleming, T. R., [ST] **stcox**, [ST] **sts test**
follow-up period, [ST] **Glossary**
follow-up studies, *see* incidence studies

Foulkes, M. A., [ST] **stpower**, [ST] **stpower cox**, [ST] **stpower exponential**
fourfold tables, [ST] **epitab**
frailty, [ST] **Glossary**
frailty model, [ST] **streg**; [ST] **stcox**, [ST] **stcurve**
Freedman, L. S., [ST] **stpower**, [ST] **stpower cox**, [ST] **stpower exponential**, [ST] **stpower logrank**
Freeman, J. L., [ST] **epitab**
future history, [ST] **Glossary**, [ST] **stset**
Fyler, D. C., [ST] **epitab**
Fyles, A., [ST] **stcrreg**, [ST] **stcrreg postestimation**

G

Gail, M. H., [ST] **stcrreg**, [ST] **stpower**, [ST] **stpower exponential**, [ST] **strate**
gaps, [ST] **Glossary**
Garrett, J. M., [ST] **stcox PH-assumption tests**
Gart, J. J., [ST] **epitab**
Gauvreau, K., [ST] **ltable**, [ST] **sts**
Gehan, E. A., [ST] **sts test**
generalized gamma survival regression, [ST] **streg**
generate, sts subcommand, [ST] **sts generate**
generating variables, [ST] **stgen**, [ST] **sts generate**
George, S. L., [ST] **stpower**, [ST] **stpower exponential**
Geskus, R. B., [ST] **stcrreg**, [ST] **stcrreg postestimation**
Gichangi, A., [ST] **stcrreg**
Gill, R. D., [ST] **stcrreg**
Gini, R., [ST] **epitab**
Glass, R. I., [ST] **epitab**
Gleason, J. R., [ST] **epitab**
Glidden, D. V., [ST] **stcox**
Gloeckler, L., [ST] **discrete**
Goldblatt, A., [ST] **epitab**
Gompertz survival regression, [ST] **streg**
Gooley, T. A., [ST] **stcrreg**
Gould, W. W., [ST] **stcox**, [ST] **stcrreg**, [ST] **stdescribe**, [ST] **streg**, [ST] **stset**, [ST] **stsplit**, [ST] **stvary**, [ST] **survival analysis**
Grambsch, P. M., [ST] **stcox**, [ST] **stcox PH-assumption tests**, [ST] **stcox postestimation**, [ST] **stcrreg**
graph,
  adjusted Kaplan–Meier survivor curves, [ST] **sts**
  baseline hazard and survivor, [ST] **stcox**, [ST] **sts**
  cumulative hazard function, [ST] **stcurve**, [ST] **sts graph**
  hazard function, [ST] **ltable**, [ST] **stcurve**, [ST] **sts graph**
  Kaplan–Meier survivor curves, [ST] **stcox PH-assumption tests**, [ST] **sts**
  log-log curve, [ST] **stcox PH-assumption tests**
  survivor function, [ST] **stcurve**, [ST] **sts graph**
graph, sts subcommand, [ST] **sts graph**
Gray, R. J., [ST] **intro**, [ST] **stcrreg**
Greenhouse, J. B., [ST] **epitab**
Greenhouse, S. W., [ST] **epitab**
Greenland, S., [ST] **epitab**
Greenwood, M., [ST] **ltable**, [ST] **sts**
Greenwood confidence intervals, [ST] **sts**
Gross, A. J., [ST] **ltable**
Gutierrez, R. G., [ST] **stcox**, [ST] **stcrreg**, [ST] **stdescribe**, [ST] **streg**, [ST] **stset**, [ST] **stsplit**, [ST] **stvary**, [ST] **survival analysis**

H

Haenszel, W., [ST] **epitab**, [ST] **sts test**
Haldane, J. B. S., [ST] **epitab**
Halley, E., [ST] **ltable**
Hankey, B., [ST] **strate**
Harley, J. B., [ST] **stpower cox**
Harrell Jr., F. E., [ST] **stcox postestimation**
Harrell's C, [ST] **stcox postestimation**
Harrington, D. P., [ST] **stcox**, [ST] **sts test**
Harrison, J. M., [ST] **stcrreg**
Hassell, J., [ST] **sts**
Hastorf, A. H., [ST] **epitab**
hazard constrictions, [ST] **Glossary**
hazard function, [ST] **sts generate**, [ST] **sts list**, [ST] **sts**
  graph of, [ST] **ltable**, [ST] **stcurve**, [ST] **sts graph**
hazard ratio, [ST] **Glossary**
  minimal detectable difference, [ST] **stpower**
  minimal effect size, [ST] **stpower**
hazard tables, [ST] **ltable**
Heagerty, P. J., [ST] **epitab**
Hedley, D., [ST] **stcrreg**, [ST] **stcrreg postestimation**
Heinonen, O. P., [ST] **epitab**
Henry-Amar, M., [ST] **ltable**
Hertz, S., [ST] **stsplit**
Hess, K. R., [ST] **stcox PH-assumption tests**, [ST] **sts graph**
heterogeneity test, [ST] **epitab**
Hilgard, E. R., [ST] **epitab**
Hill, A. B., [ST] **epitab**
Hill, R. P., [ST] **stcrreg**, [ST] **stcrreg postestimation**
Hill, W. G., [ST] **epitab**
Hills, M., [ST] **epitab**, [ST] **stcox**, [ST] **stptime**, [ST] **strate**, [ST] **stset**, [ST] **stsplit**, [ST] **sttocc**
Hofman, A. F., [ST] **stcrreg**
Hogben, L., [ST] **sts**
Holmgren, J., [ST] **sts**
homogeneity test, [ST] **epitab**
Hooker, P. F., [ST] **streg**
Hosmer Jr., D. W., [ST] **stcox**, [ST] **stpower**, [ST] **stpower cox**, [ST] **streg**
Hossain, K. M., [ST] **epitab**
Hougaard, P., [ST] **streg**
Hsieh, F. Y., [ST] **stpower**, [ST] **stpower cox**, [ST] **stpower logrank**
Hu, M., [ST] **stcox**, [ST] **stset**
Huq, M. I., [ST] **epitab**

I

ID variable, [ST] **Glossary**
incidence, [ST] **Glossary**
incidence studies, [ST] **Glossary**
incidence-rate ratio, [ST] **epitab**, [ST] **stci**, [ST] **stir**, [ST] **stptime**, [ST] **stsum**
independence test, [ST] **epitab**
ir command, [ST] **epitab**
iri command, [ST] **epitab**

J

Jenkins, S. P., [ST] **discrete**, [ST] **stcox**
Jewell, N. P., [ST] **epitab**
Jick, H., [ST] **epitab**
joining time-span records, [ST] **stsplit**
Jones, B. S., [ST] **stcox**, [ST] **streg**
Jorgensen, R. A., [ST] **stcrreg**

K

Kahn, H. A., [ST] **epitab**, [ST] **ltable**, [ST] **stcox**
Kalbfleisch, J. D., [ST] **ltable**, [ST] **stcox**, [ST] **stcox PH-assumption tests**, [ST] **stcox postestimation**, [ST] **streg**, [ST] **sts test**, [ST] **sts**, [ST] **stset**
Kaplan, E. L., [ST] **stcrreg**, [ST] **stcrreg postestimation**, [ST] **sts**
Kaplan–Meier
    product-limit estimate, [ST] **Glossary**, [ST] **sts generate**, [ST] **sts graph**, [ST] **sts list**, [ST] **sts test**, [ST] **sts**
    survivor function, [ST] **sts**; [ST] **ltable**, [ST] **stcox PH-assumption tests**
Keiding, N., [ST] **stcrreg**, [ST] **stsplit**
Khan, M. R., [ST] **epitab**
Kimber, A. C., [ST] **streg**
Klein, J. P., [ST] **stci**, [ST] **stcox**, [ST] **stcox postestimation**, [ST] **stcrreg**, [ST] **stpower**, [ST] **stpower cox**, [ST] **streg**, [ST] **sts graph**, [ST] **sts test**, [ST] **sts**
Kleinbaum, D. G., [ST] **epitab**
Krakauer, H., [ST] **ltable**
Krall, J. M., [ST] **stpower cox**
Kreidberg, M. B., [ST] **epitab**
Kupper, L. L., [ST] **epitab**

L

Lachin, J. M., [ST] **stpower**, [ST] **stpower cox**, [ST] **stpower exponential**
Lakatos, E., [ST] **stpower**, [ST] **stpower exponential**, [ST] **stpower logrank**
Lambert, P. C., [ST] **stcrreg**
Lan, K. K. G., [ST] **stpower**, [ST] **stpower exponential**, [ST] **stpower logrank**
Lange, S. M., [ST] **stcrreg**
Langholz, B., [ST] **sttocc**

Lash, T. L., [ST] **epitab**
Latouche, A., [ST] **stcrreg**
Lavori, P. W., [ST] **stpower**, [ST] **stpower cox**
Lawless, J. F., [ST] **ltable**, [ST] **stpower**
Lee, E. T., [ST] **streg**
Lee, K. L., [ST] **stcox postestimation**
Lee, P., [ST] **streg**
Leisenring, W., [ST] **stcrreg**
Lemeshow, S., [ST] **stcox**, [ST] **stpower**, [ST] **stpower cox**, [ST] **streg**
LeSage, G., [ST] **stcrreg**
Levin, B., [ST] **epitab**
Levin, W., [ST] **stcrreg**, [ST] **stcrreg postestimation**
Lexis, W. H., [ST] **stsplit**
lexis command, [ST] **stsplit**
lexis diagram, [ST] **stsplit**
life tables, [ST] **Glossary**, [ST] **ltable**
likelihood displacement value, [ST] **Glossary**, [ST] **stcox postestimation**
Lilienfeld, D. E., [ST] **epitab**
Lin, D. Y., [ST] **stcox**, [ST] **stcrreg**
Lindor, K. D., [ST] **stcrreg**
Linhart, J. M., [ST] **sts**
list, sts subcommand, [ST] **sts list**
Liu, J.-P., [ST] **stpower**
LMAX value, [ST] **Glossary**, [ST] **stcox postestimation**
log-log plot, [ST] **stcox PH-assumption tests**
loglogistic survival regression, [ST] **streg**
lognormal survival regression, [ST] **streg**
log-rank,
    power, [ST] **stpower logrank**
    sample size, [ST] **stpower logrank**
logrank, stpower subcommand, [ST] **stpower logrank**
log-rank test, [ST] **stpower logrank**, [ST] **sts test**
longitudinal studies, see incidence studies
López-Vizcaíno, M. E., [ST] **epitab**
loss to follow-up, [ST] **Glossary**
ltable command, [ST] **ltable**
Ludwig, J., [ST] **stcrreg**
Lumley, T. S., [ST] **epitab**
Lunn, M., [ST] **stcrreg**

M

Machin, D., [ST] **stpower**, [ST] **stpower cox**, [ST] **stpower logrank**
MacMahon, B., [ST] **epitab**
Manchul, L., [ST] **stcrreg**, [ST] **stcrreg postestimation**
Mander, A., [ST] **stsplit**
Mantel, N., [ST] **epitab**, [ST] **sts test**
Mantel–Cox method, [ST] **strate**
Mantel–Haenszel
    method, [ST] **strate**
    test, [ST] **epitab**, [ST] **stir**

Marchenko, Y., [ST] **stcox**, [ST] **stcrreg**, [ST] **stdescribe**, [ST] **streg**, [ST] **stset**, [ST] **stsplit**, [ST] **stvary**, [ST] **survival analysis**
Mark, D. B., [ST] **stcox postestimation**
Marr, J. W., [ST] **stsplit**
Martin, W., [ST] **epitab**
martingale residual, [ST] **stcox postestimation**, [ST] **streg postestimation**
Marubini, E., [ST] **stcrreg**, [ST] **stpower**, [ST] **stpower logrank**, [ST] **sts test**
matched case–control data, [ST] **epitab**, [ST] **Glossary**, [ST] **sttocc**
May, S., [ST] **stcox**, [ST] **stpower**, [ST] **stpower cox**, [ST] **streg**
mcc command, [ST] **epitab**
mcci command, [ST] **epitab**
McCulloch, C. E., [ST] **stcox**
McGilchrist, C. A., [ST] **stcox**, [ST] **streg**
McNeil, D., [ST] **stcrreg**
McNemar, Q., [ST] **epitab**
McNemar's chi-squared test, [ST] **epitab**
Meier, P., [ST] **stcrreg**, [ST] **stcrreg postestimation**, [ST] **sts**
Meyer, B. D., [ST] **discrete**
mhodds command, [ST] **epitab**
Miettinen, O. S., [ST] **epitab**
Milosevic, M., [ST] **stcrreg**, [ST] **stcrreg postestimation**
minimal
 detectable difference, hazard ratio, [ST] **stpower**
 effect size, hazard ratio, [ST] **stpower**
Moeschberger, M. L., [ST] **stci**, [ST] **stcox**, [ST] **stcox postestimation**, [ST] **stcrreg**, [ST] **stpower**, [ST] **stpower cox**, [ST] **streg**, [ST] **sts graph**, [ST] **sts test**, [ST] **sts**
Monson, R. R., [ST] **epitab**
Morgenstern, H., [ST] **epitab**
Mori, M., [ST] **stcrreg**
Morris, J. N., [ST] **stsplit**
Morrow, A., [ST] **epitab**
mortality table, *see* life tables
Müller, H.-G., [ST] **sts graph**
multiarm trial, [ST] **Glossary**, [ST] **stpower**
multiple-record st data, [ST] **stfill**, [ST] **stvary**
Murtaugh, P. A., [ST] **stcrreg**

N

Nardi, G., [ST] **epitab**
Neff, R. K., [ST] **epitab**
Nelson, W., [ST] **stcrreg postestimation**, [ST] **sts**
Nelson–Aalen cumulative hazard, [ST] **sts generate**, [ST] **sts graph**, [ST] **sts list**, [ST] **sts**
nested case–control data, [ST] **sttocc**
Newman, S. C., [ST] **epitab**, [ST] **stcox**, [ST] **sts**
Newson, R., [ST] **stcox postestimation**
nonparametric test, equality of survivor functions, [ST] **sts test**

O

Oakes, D., [ST] **ltable**, [ST] **stcox**, [ST] **stcox PH-assumption tests**, [ST] **stpower**, [ST] **streg**, [ST] **sts**
odds ratio, [ST] **epitab**, [ST] **Glossary**
offset variable, [ST] **Glossary**
Olshansky, S. J., [ST] **streg**
Orsini, N., [ST] **epitab**

P

Pagano, M., [ST] **ltable**, [ST] **sts**
Paik, M. C., [ST] **epitab**
parametric survival models, [ST] **streg**
Parmar, M. K. B., [ST] **stpower**, [ST] **stpower cox**
partial DFBETA, [ST] **Glossary**
partial likelihood displacement value, [ST] **Glossary**
partial LMAX value, [ST] **Glossary**
Pasquini, J., [ST] **epitab**
past history, [ST] **Glossary**, [ST] **stset**
Pearce, M. S., [ST] **epitab**
penalized log-likelihood function, [ST] **Glossary**, [ST] **stcox**
Pepe, M. S., [ST] **stcrreg**
Pérez-Santiago, M. I., [ST] **epitab**
person time, [ST] **stptime**
Peto, J., [ST] **sts test**
Peto, R., [ST] **stcox**, [ST] **streg**, [ST] **sts test**
phtest, estat subcommand, [ST] **stcox PH-assumption tests**
Pike, M. C., [ST] **ltable**, [ST] **streg**
Pintilie, M., [ST] **stcrreg**, [ST] **stcrreg postestimation**
Pitblado, J. S., [ST] **sts**
Plummer, D., [ST] **epitab**
pooled estimates, [ST] **epitab**
population attributable risk, [ST] **epitab**
postestimation command, [ST] **stcurve**
power, [ST] **Glossary**
 Cox proportional hazards regression, [ST] **stpower**, [ST] **stpower cox**
 exponential survival, [ST] **stpower**, [ST] **stpower exponential**
 exponential test, [ST] **stpower**, [ST] **stpower exponential**
 log-rank, [ST] **stpower**, [ST] **stpower logrank**
Prentice, R. L., [ST] **discrete**, [ST] **ltable**, [ST] **stcox**, [ST] **stcox PH-assumption tests**, [ST] **stcox postestimation**, [ST] **streg**, [ST] **sts test**, [ST] **sts**, [ST] **stset**
prevalence studies, *see* cross-sectional studies
prevented fraction, [ST] **epitab**, [ST] **Glossary**
proportional hazards model, [ST] **Glossary**, *also see* Cox proportional hazards model
prospective study, [ST] **epitab**, [ST] **Glossary**
Pryor, D. B., [ST] **stcox postestimation**
Putter, H., [ST] **stcrreg**, [ST] **stcrreg postestimation**

R

Radmacher, R. D., [ST] **stpower**
Ramalheira, C., [ST] **ltable**
rate ratio, [ST] **epitab**, [ST] **stci**, [ST] **stir**, [ST] **stptime**, [ST] **stsum**
recruitment period, *see* accrual period
regression, competing risks, [ST] **stcrreg**
Reid, N., [ST] **stcox**
Reilly, M., [ST] **epitab**
relative risk, [ST] **epitab**
Richter, J. R., [ST] **stpower**
risk
    difference, [ST] **epitab**
    factor, [ST] **epitab**, [ST] **Glossary**
    pool, [ST] **Glossary**, [ST] **stcox**, [ST] **stcrreg**, [ST] **stset**
    ratio, [ST] **epitab**, [ST] **Glossary**
Robins, J. M., [ST] **epitab**
robust, Huber/White/sandwich estimator of variance,
    Cox proportional hazards model, [ST] **stcox**, [ST] **stcrreg**
    parametric survival models, [ST] **streg**
Rogers, W. H., [ST] **stcox PH-assumption tests**, [ST] **stcox postestimation**
Rosati, R. A., [ST] **stcox postestimation**
Rossi, S. S., [ST] **stcrreg**
Rothman, K. J., [ST] **epitab**
Royston, J. P., [ST] **epitab**, [ST] **stcox**, [ST] **stcox PH-assumption tests**, [ST] **stpower**, [ST] **stpower cox**, [ST] **streg**
Rubinstein, L. V., [ST] **stpower**, [ST] **stpower exponential**

S

Salim, A., [ST] **epitab**
sample size,
    Cox proportional hazards regression, [ST] **stpower**, [ST] **stpower cox**
    exponential survival, [ST] **stpower**, [ST] **stpower exponential**
    exponential test, [ST] **stpower**, [ST] **stpower exponential**
    log-rank, [ST] **stpower**, [ST] **stpower logrank**
Santner, T. J., [ST] **stpower**, [ST] **stpower exponential**
Sasieni, P., [ST] **stcox**
Savage, I. R., [ST] **sts test**
Schlesselman, J. J., [ST] **epitab**
Schoenfeld, D. A., [ST] **stcox**, [ST] **stcox postestimation**, [ST] **stpower**, [ST] **stpower cox**, [ST] **stpower exponential**, [ST] **stpower logrank**, [ST] **streg**
Schoenfeld residual, [ST] **stcox postestimation**, [ST] **streg postestimation**
Schumacher, M., [ST] **stcrreg**
Scotto, M. G., [ST] **streg**
Sears, R. R., [ST] **epitab**

Selvin, S., [ST] **ltable**, [ST] **stcox**
semiparametric model, [ST] **Glossary**, [ST] **stcox**, [ST] **stcrreg**, [ST] **stset**
Sempos, C. T., [ST] **epitab**, [ST] **ltable**, [ST] **stcox**
Serachitopol, D. M., [ST] **sts graph**
Shao, J., [ST] **stpower**, [ST] **stpower exponential**
Shapiro, S., [ST] **epitab**
Shiboski, S. C., [ST] **stcox**
Simon, R., [ST] **stpower**
singleton-group data, [ST] **Glossary**, [ST] **stcox**
Skovlund, E., [ST] **stpower**, [ST] **stpower cox**
Slone, D., [ST] **epitab**
Smith, R. L., [ST] **streg**
SMR, [ST] **epitab**, [ST] **Glossary**, [ST] **stptime**
Smythe, B., [ST] **sts**
snapshot data, [ST] **Glossary**, [ST] **snapspan**, [ST] **stset**
snapspan command, [ST] **snapspan**
Snell, E. J., [ST] **stcox**, [ST] **stcox PH-assumption tests**, [ST] **streg postestimation**
spell data, [ST] **Glossary**
splitting time-span records, [ST] **stsplit**
st data, [ST] **Glossary**
standardized
    mortality ratio, *see* SMR
    rates, [ST] **epitab**
stbase command, [ST] **stbase**
stci command, [ST] **stci**
stcox command, [ST] **stcox**
stcoxkm command, [ST] **stcox PH-assumption tests**
stcrreg command, [ST] **stcrreg**
st_ct, [ST] **st_is**
stcurve command, [ST] **stcurve**
stdescribe command, [ST] **stdescribe**
Sterne, J. A. C., [ST] **stcox**
Stewart, J., [ST] **ltable**
stfill command, [ST] **stfill**
stgen command, [ST] **stgen**
stir command, [ST] **stir**
st_is 2, [ST] **st_is**
stjoin command, [ST] **stsplit**
stmc command, [ST] **strate**
stmh command, [ST] **strate**
Stoll, B. J., [ST] **epitab**
Stolley, P. D., [ST] **epitab**
Storer, B. E., [ST] **stcrreg**
stphplot command, [ST] **stcox PH-assumption tests**
stpower cox command, [ST] **stpower cox**
stpower exponential command, [ST] **stpower exponential**
stpower logrank command, [ST] **stpower logrank**
stptime command, [ST] **stptime**
strate command, [ST] **strate**
stratification, [ST] **epitab**, [ST] **stcox**, [ST] **streg**
stratified tables, [ST] **epitab**
streg command, [ST] **streg**
Stryhn, H., [ST] **epitab**

sts command, [ST] **sts generate**, [ST] **sts graph**, [ST] **sts list**, [ST] **sts test**, [ST] **sts**, [ST] **stset**
sts generate command, [ST] **sts generate**, [ST] **sts**
sts graph command, [ST] **sts graph**, [ST] **sts**
sts list command, [ST] **sts list**, [ST] **sts**
sts test command, [ST] **sts test**, [ST] **sts**
stset command, [ST] **stset**
st_show, [ST] **st_is**
stsplit command, [ST] **stsplit**
stsum command, [ST] **stsum**
sttocc command, [ST] **sttocc**
stvary command, [ST] **stvary**
subhazard, [ST] **Glossary**
subhazard ratio, [ST] **Glossary**
survey sampling, [ST] **stcox**, [ST] **streg**, [ST] **sts**
survival analysis, [ST] **cttost**, [ST] **st**, [ST] **stdescribe**, [ST] **stset**, [ST] **sttoct**
survival clinical trial, [ST] **stpower**
survival-time data, see st data
survivor function, [ST] **Glossary**, [ST] **sts generate**, [ST] **sts list**, [ST] **sts test**, [ST] **sts**
  graph of, [ST] **stcurve**, [ST] **sts graph**
Svennerholm, A. M., [ST] **epitab**
Sweeting, T. J., [ST] **streg**

T

tables,
  actuarial, see life tables
  epidemiological, see epidemiological tables
  failure, see failure tables
  fourfold, see fourfold tables
  hazard, see hazard tables
  life, see life tables
tabodds command, [ST] **epitab**
Tan, S. B., [ST] **stpower**, [ST] **stpower logrank**
Tan, S. H., [ST] **stpower**, [ST] **stpower logrank**
Tarone, R. E., [ST] **epitab**, [ST] **sts test**
test,
  association, see association test
  Cox proportional hazards model, assumption, see Cox proportional hazards model, test of assumption
  equality of survivor functions, see equality test, survivor functions
  Fisher's exact, see Fisher's exact test
  heterogeneity, see heterogeneity test
  homogeneity, see homogeneity test
  independence, see independence test
  log-rank, see log-rank test
  Mantel–Haenszel, see Mantel–Haenszel test
  McNemar's chi-squared test, see McNemar's chi-squared test
  trend, see trend test
test, sts subcommand, [ST] **sts test**
test-based confidence intervals, [ST] **epitab**
Therneau, T. M., [ST] **stcox**, [ST] **stcox PH-assumption tests**, [ST] **stcox postestimation**, [ST] **stcrreg**

Thomas, D. C., [ST] **sttocc**
Thomas, D. G., [ST] **epitab**
Tilling, K., [ST] **stcox**
time-span data, [ST] **snapspan**
time-varying covariates, [ST] **Glossary**
Tippett, L. H. C., [ST] **streg**
Tobías, A., [ST] **streg**
trend test, [ST] **epitab**, [ST] **sts test**
Trichopoulos, D., [ST] **epitab**
truncation, [ST] **Glossary**
Tsiatis, A., [ST] **stcrreg**
type I error, [ST] **Glossary**
type II error, [ST] **Glossary**

U

under observation, [ST] **Glossary**
University Group Diabetes Program, [ST] **epitab**
Uthoff, V. A., [ST] **stpower cox**

V

Vach, W., [ST] **stcrreg**
Væth, M., [ST] **stpower**, [ST] **stpower cox**
Valsecchi, M. G., [ST] **stcrreg**, [ST] **stpower**, [ST] **stpower logrank**, [ST] **sts test**
van Belle, G., [ST] **epitab**
variables, generating, [ST] **stgen**
varying variables, [ST] **stvary**
Vidmar, S., [ST] **epitab**
Vittinghoff, E., [ST] **stcox**

W

Walker, A. M., [ST] **epitab**
Walker, S., [ST] **sts test**
Wang, H., [ST] **stpower**, [ST] **stpower exponential**
Wang, J.-L., [ST] **sts graph**
Wang, J. W., [ST] **streg**
Wang, Z., [ST] **epitab**
Ware, J. H., [ST] **sts test**
Warren, K., [ST] **epitab**
Weesie, J., [ST] **stsplit**
Wei, L. J., [ST] **stcox**, [ST] **stcrreg**
Weibull, W., [ST] **streg**
Weibull distribution, [ST] **streg**
Weibull survival regression, [ST] **streg**
West, S., [ST] **epitab**
White, I. R., [ST] **sts test**
Wiesner, R. H., [ST] **stcrreg**
Wilcoxon, F., [ST] **sts test**
Wilcoxon test (Wilcoxon–Breslow, Wilcoxon–Gehan, Wilcoxon–Mann–Whitney), [ST] **sts test**
Wilkinson, L., [ST] **sts**
withdrawal, [ST] **Glossary**
Wittes, J., [ST] **stpower**
Wolfram, S., [ST] **streg**

Wolk, A., [ST] **epitab**
Woolf, B., [ST] **epitab**
Woolf confidence intervals, [ST] **epitab**

Y

Yen, S., [ST] **epitab**